AF344489

Lecture Notes in Applied and Computational Mechanics

Volume 29

Series Editors

Prof. Dr.-Ing. Friedrich Pfeiffer
Prof. Dr.-Ing. Peter Wriggers

Lecture Notes in Applied and Computational Mechanics

Edited by F. Pfeiffer and P. Wriggers

Further volumes of this series found on our homepage: springer.com

Boundary Element Analysis

Mathematical Aspects and Applications

Martin Schanz
Olaf Steinbach

(Eds.)

 Springer

Prof. Dr.-Ing. Martin Schanz
Technische Universität Graz
Institut für Baumechanik
Technikerstr. 4
8010 Graz
Austria
m.schanz@tugraz.at

Prof. Dr. Olaf Steinbach
Technische Universität Graz
Institut für Numerische Mathematik
Steyrergasse 30
8010 Graz
Austria
o.steinbach@tugraz.at

With 118 Figures

Library of Congress Control Number: 2006936726

ISSN 1613-7736

ISBN-10   3-540-47465-X **Springer Berlin Heidelberg New York**
ISBN-13   978-3-540-47465-4 **Springer Berlin Heidelberg New York**

Springer is a part of Springer Science+Business Media
springer.com

© Springer-Verlag Berlin Heidelberg 2007
Printed in Germany

Typesetting: Data conversion by the editors.
Final processing by PTP-Berlin Protago-T_EX-Production GmbH, Germany (www.ptp-berlin.com)
Cover-Design: WMXDesign GmbH, Heidelberg
Printed on acid-free paper   89/3141/Yu - 5 4 3 2 1 0

Dedicated to Wolfgang L. Wendland

Preface

This volume on recent mathematical aspects and the state of the art applications of boundary element methods is dedicated to Wolfgang L. Wendland on the occasion of his 70th birthday in September 2006. Lectures related to the topics included in this book were given within a minisymposium held during the last Symposium of the International Association of Boundary Element Methods (IABEM) held in Graz, July 10–12, 2006.

Since the mid eighties there has been a remarkable development in both the mathematical analysis and engineering applications of boundary element methods. It turned out that most innovations grew up within strong cooperations between mathematicians and engineers. Both of us participated in the Priority Research Program of the German Research Foundation (DFG) on Boundary Element Methods (1989–1995), which was directed by W. L. Wendland. Within this program, a lot of new mathematical results were obtained and a lot of simulations of engineering applications has been realized. All results were finally published in the volume Boundary element topics (W. L. Wendland ed.), Springer, Berlin, 1997.

Whereas in these times the development of the method itself and the mathematical basis has been promoted, in the last decade there was another strong improvement in the analysis of boundary integral equation methods and in the numerical analysis and implementation of boundary element methods due to the overwhelming success of fast boundary element methods. Although the fast multipole method was already used for some time, their marriage with a profound numerical analysis of Galerkin boundary element methods was not considered before. Moreover, algebraic approximation methods such as the Adaptive Cross Approximation algorithm or the concept of Hierarchical Matrices contribute to the ongoing success of modern fast boundary element methods. It is worth to mention that almost all of those achievements are still obtained within strong cooperations between mathematicians and engineers, and with direct applications in industry.

Hence the spirit of the former DFG research program is still active and will hopefully initiate further collaborations leading to more impressive results. In particular, the aim is to solve more challenging real world applications. This strong cooperation between applied mathematics and engineering sciences was always one of the driving forces in the scientific work of Wolfgang Wendland. This spirit can also be observed in all presentations in this volume. We are happy to dedicate this book to him and to thank him for his long and ongoing support and encouragement of the boundary element community.

This volume contains eleven contributions showing the wide range of boundary integral equation and boundary element methods. Beside more analytical aspects in the formulation and analysis of boundary integral equations also the state of art of boundary element algorithms are described and analyzed from a mathematical point of view. In addition, engineering and industrial applications of those methods are presented showing the ability of modern boundary element methods to solve challenging problems.

We would like to thank all authors for their contributions to this volume. Moreover, we also thank all anonymous referees for their work, their criticism, and their proposals. These hints were very helpful to improve the contributions. Finally, we would like to thank Dr. T. Ditzinger of Springer Heidelberg for the continuing support and patience while preparing this volume.

Graz,   *Martin Schanz*
August 2006   *Olaf Steinbach*

Contents

Some Historical Remarks on the Positivity of Boundary Integral Operators

Martin Costabel

IRMAR, Université de Rennes 1, Campus de Beaulieu, 35042 Rennes, France
`costabel@univ-rennes1.fr`

Summary. Variational arguments go back a long time in the history of boundary integral equations. Energy methods have shown up very early, then virtually disappeared from the common knowledge and eventually resurfaced in the context of boundary element methods. We focus on some not so well known parts of classical works by well known classical authors and describe the relation of their ideas to modern variational principles in boundary element methods.

1 Introduction

The method of boundary integral equations has always had two important applications in the theory of boundary value problems for partial differential equations: As a theoretical tool for proving the existence of solutions and as a practical tool for the construction of solutions. This is one of the aspects that has remained constant since the times of Green and Gauss in the early 19th century until our times. Other ideas, in particular techniques of the analysis of integral equations, have of course greatly changed and evolved in the meantime, but it is curious to see how some of the very early questions and techniques are related to recent simple basic results about the structure of boundary integral equations.

This article has evolved from some observations made in the talk [6] about the scientific work of Wolfgang Wendland, connecting works by Carl Friedrich Gauss [11] and Carl Neumann [27] to the work by Wendland and his group on variational methods for boundary integral equations. In particular the curious case of "Gauss' missing theorem" on the positivity of the single layer potential operator – a proof of which could have been given by Gauss himself, but was in fact only given 135 years later by Nedelec and Planchard [26] – seemed to be sufficiently intriguing to merit a more detailed presentation. A secondary path concerning second kind boundary integral equations, leading from Neumann's observation of the contraction property of the double layer potential to the recent paper [31] by Steinbach and Wendland where energy methods were used to prove the contraction property in very general cases, seemed to be less straightforward on the level of analytical tools and mathematical ideas. Following the early twists of this path, however, one comes across the monumental paper [29] by Henri Poincaré which uses, indeed, energy methods for proving the contraction property of Neumann's operator. The historical trail of Poincaré's paper which, after having been instantly famous initially, seems to have disappeared from the common knowledge of the boundary element community, is a second curiosity on which we will try to shed some light here. By taking this look, we will even find some "new" mathematical results.

This paper does not present a serious research into the history of mathematics, which would require much more space, time and knowledge than available to the author. It rather stays within the narrow viewing angle of the history of the analytical foundations of boundary element methods, but it tries to illustrate how a fresh look, however biased, can reveal new details of old monuments. We will consider a domain spanned by the four papers by Gauss [11], Poincaré [29], Nedelec and Planchard [26], and Steinbach and Wendland [31]. If one prefers a hexagonal constellation, one can add Neumann's book [27] and the paper [7] in which the generality of energy methods was emphasized. For a justification of this combination of papers, suffice it to say that in the sky, for giving the perception of a well-balanced constellation, the more distant objects have to be much bigger stars than the objects closer to the observer... Within this constellation, there exists a myriad of other points of light, only some of which will make a short appearance in the following. Other very bright stars in the vicinity of our constellation, from Lebesgue and Fredholm over Hilbert to Calderón–Zygmund and Mazya, will not be considered here.

The papers we are trying to connect belong, in fact, to three quite different galaxies: There is ours, extending over roughly the last 40 years, characterized by the availability of many simple but very powerful tools like the basic theory of Hilbert and Banach spaces, distributions, Fourier transforms and Sobolev spaces. At the distant end there is the early 19th century with Gauss, where the first general tools in potential theory and partial differential equations were being forged. In between there is the end of the 19th century, roughly from 1870 to 1910 with a condensation around 1895–1900, in the center of which we see Poincaré, where in close relation with the emergence of modern physics the first steps were done in directions that led to the subsequent explosion of functional analysis beginning quite soon afterwards.

What is common to all three periods is the strong primary motivation of the mathematical research by applications, which then led to the discovery of beautiful structures that were investigated for their own sake, the result being the creation not only of fine new mathematics, but also of more powerful tools for the applications. Let us quote from Gauss' introduction to [10] where he talks about some of the ambivalence in the relation between mathematics and its applications:

> Der rastlose Eifer, womit man in neuerer Zeit in allen Theilen der
> Erdoberfläche die Richtung und Stärke der magnetischen Kraft der
> Erde zu erforschen strebt, ist eine um so erfreulichere Erscheinung,
> je sichtbarer dabei das rein wissenschaftliche Interesse hervortritt.
> Denn in der That, wie wichtig auch für die Schifffahrt die möglichst
> vollständige Kenntniss der Abweichungslinie ist, so erstreckt sich doch
> ihr Bedürfniss eben nicht weiter, und was darüber hinausliegt, bleibt
> für jene beinahe gleichgültig. Aber die Wissenschaft, wenn gleich
> auch dem materiellen Interesse förderlich, lässt sich nicht auf dieses
> beschränken, sondern fordert für Alle Elemente ihrer Forschung glei-
> che Anstrengung.[1]

An earlier quote is the following quite modern-sounding grumble from 1825
[13]:

> Ihr gütiger Brief hat mir um so mehr Vergnügen gemacht, je sel-
> tener jetzt in Deutschland warmes Interesse an Mathematik ist. So
> erfreulich die gegenwärtige hohe Blüthe der Astronomischen Wis-
> senschaften ist, so scheint doch die praktische Tendenz fast zu aus-
> schliesslich vorherrschend, und die meisten sehen die abstracte Ma-
> thematik höchstens als Magd der Astronomie an, die nur deswegen zu
> toleriren ist.[2]

On a more technical level, all three periods have in common that *variational
methods* play an important role. In Gauss' time, variational principles were
commonly used for existence proofs, such as in Gauss' existence proof for
the Dirichlet problem. In Poincaré's time, on the practical side their field
of applications had been enlarged to cover the construction of eigenfunction
systems via min-max principles, and on the theoretical side the problems
caused by the perceived inadequacies of too naïve applications of variational
principles (cf. Weierstrass' well-known criticism of Dirichlet's principle) were
beginning to find solutions. Hilbert [14] is often credited with having given
the first rigorous formulation and application of Dirichlet's principle. Here

---

[1]The restless eagerness with which in recent times one strives to investigate in
all parts of the surface of the earth the direction and strength of the magnetic force
of the earth, is a development which is all the more pleasing the clearer the purely
scientific interest is standing out. For, in fact, how important the most complete
knowledge of the deviation line may be for navigation, the need of the latter just does
not extend further, and it remains almost indifferent to anything that lies beyond.
But Science, albeit also beneficial to the material interest, cannot be restricted to
this, but requires for All elements of its research the same effort.

[2]Your kind letter has given me all the more pleasure the rarer there is now
warm interest in mathematics in Germany. As pleasant as the current high bloom
of the astronomical sciences may be, the practical tendency seems to be almost too
exclusively predominant, and most people consider abstract mathematics at most
as a servant of astronomy which is only therefore to be tolerated.

is, however, a quote from a recent paper [1] by one of the specialists in the calculus of variations:

> In 1900, D. Hilbert, in a celebrated address, followed by a (slightly) more detailed paper in 1904 [14, 15, 16], announced that he had solved the Dirichlet problem [...] via the Dirichlet principle which had been discovered by G. Green in 1833, with later contributions by C. F. Gauss (1837), W. Thomson (=Lord Kelvin) (1847) and G. Riemann (1853). [...]
> The announcement of Hilbert turned out to be a little premature. Instead, it became a program which stimulated many people during the period 1900-1940: B. Levi, H. Lebesgue, L. Tonelli, R. Courant, S. L. Sobolev and many others. In 1940, H. Weyl [40] completed Hilbert's program. By 1940 the Calculus of Variations had been placed on firm grounds [...]

Thus a closer look at history tends to blur the boundaries of what constitutes a formal and complete proof. In any case, nowadays we have clearcut basic tools like Hilbert spaces, the Riesz representation theorem, the Lax-Milgram lemma, and Sobolev spaces, which allow us to teach Dirichlet's principle in a first course on finite element methods.

A final bridge between the present and the past should be mentioned that allows us to approach those distant galaxies far more easily than had been possible for a long time: The Internet. Almost all references in this article are readily and freely available online, thanks to enterprises like actamathematica, Gallica, GDZ, JSTOR, NUMDAM, SpringerLink.

In the following we will first make some remarks about Gauss and the first kind integral equation of the single layer potential. Then we describe some of Poincaré's ideas about the double layer potential. In the final part we list a few known and unknown results related to these old ideas.

2 Gauss and the Single Layer Potential

In 1838-40, Carl Friedrich Gauss published three famous works which stand at the beginning of our curious history of boundary integral equation methods: In two of them, [10] and [12], he introduced boundary integral equations (of the first kind!) as a tool in numerical computations and published extensive tables and graphs of numerical results obtained in part by employing this tool. It is truly amazing to see how much could be achieved with numerical calculations by hand when powerful analytical tools were used. In [10, §32] Gauss gives a simple description of the principle of boundary reduction, an idea from which another track leads to later successful methods for proving existence for the Dirichlet problem, namely Schwarz' alternating method and Poincaré's sweeping or "balayage" method.

> [32.] Die Art der wirklichen Vertheilung der magnetischen Flüssigkei-
> ten in der Erde bleibt nothwendigerweise unbestimmt. In der That
> kann nach einem allgemeinen Theorem, welches bereits in der *Inten-*
> *sitas* Art. 2 erwähnt ist, und bei einer andern Gelegenheit ausführlich
> behandelt werden soll, anstatt jeder beliebigen Vertheilung der mag-
> netischen Flüssigkeiten innerhalb eines körperlichen Raumes allemal
> substituirt werden eine Vertheilung auf der Oberfläche dieses Raumes,
> so dass die Wirkung in jedem Punkte des äusseren Raumes genau
> dieselbe bleibt, woraus man leicht schliesst, dass *einerlei* Wirkung im
> ganzen äussern Raume aus unendlich vielen *verschiedenen* Vertheilun-
> gen der magnetischen Flüssigkeiten im Innern abzuleiten ist.[3]

After this, he gives, for the case of a ball, an expansion in spherical harmonics of the unknown density on the surface. The "other occasion" where the
mentioned Theorem was going to "be treated extensively", is the third paper
[11, §36].

In this paper, Gauss not only lays down the foundations of potential theory,
including the mean value property of harmonic functions (§20), the maximum
principle and the principle of unique continuation (§21), but he also studies in
detail the properties of single layer potentials. He presents the jump relations
(§15) and the basic integration by parts formula (§21; now known as Green's
first formula, because Green formulated this some years before Gauss, his
works not yet being widely known at the time of Gauss' paper). We will quote
these two results below in Gauss' own notation, as our pieces of evidence in
the curious case of "Gauss' missing theorem." Let us first see, however, how
Gauss considers the positivity of the single layer potential integral operator.
In his own words:

> [30.] Es ist von selbst klar, dass, wie auch immer eine Masse M über
> eine Fläche *gleichartig* vertheilt sein möge, das daraus entspringende
> überall positive Potential V in jedem Punkte der Fläche grösser sein
> wird, als $\frac{M}{r}$, wenn r die grösste Entfernung zweier Punkte der Fläche
> voneinander bedeutet: diesen Werth selbst könnte das Potential nur
> in einem Endpunkt der Linie r haben, wenn die ganze Masse in dem
> andern Endpunkte concentrirt wäre, ein Fall, der hier gar nicht in
> Frage kommt, indem nur von stetiger Vertheilung die Rede sein soll,
> wo jedem Element der Fläche ds nur eine unendlich kleine Masse m ds

---

[3]The specifics of the real distribution of the magnetic fluids in the earth remain
necessarily undetermined. Indeed, according to a general theorem which has already
been mentioned in the *Intensitas* Art. 2 and shall be treated extensively at another
occasion, one can always substitute instead of any arbitrary distribution of the
magnetic fluids inside a bodily space, a distribution on the surface of this space, so
that the effect in every point of the exterior space remains exactly the same, from
which one easily concludes that an *identical* effect in the entire exterior space is to
be derived from infinitely many *different* distributions of the magnetic fluids in the
interior.

> entspricht. Das Integral $\int Vm\,ds$ über die ganze Fläche ausgedehnt, ist also jedenfalls grösser als $\int \frac{M}{r}m\,ds$ oder $\frac{MM}{r}$, und so muss es nothwendig eine gleichartige Vertheilungsart geben, für welche jenes Integral einen Minimumwerth hat.[4]

The notion "gleichartig" (homogeneous) means not changing sign, in the case of a positive total mass M therefore non-negative.

In the paragraphs that follow, he considers a more general problem: Given a continuous function U on the surface, minimise the integral

$$\Omega = \int (V - 2U)m\,ds\,.$$

This is then seen to be equivalent to the integral equation problem: Find a non-negative mass density m of total mass M and a constant C such that the single layer potential V satisfies $V + C = U$ on the surface. He also considers the case where C is given and M is not fixed, thus the basic first kind integral equation with the $1/r$ kernel.

For this problem he gives a detailed proof of existence and uniqueness. From this result he then deduces an existence proof for the Dirichlet problem in potential theory.

What jumps out at us when we read this argument is, *of course*, that Gauss commits the freshman error of confusing infimum and minimum and that as a consequence he has, in reality, no existence proof. This whole piece of analysis was, indeed, far ahead of its time, and we all know that the crucial question of completeness was only seriously studied after Weierstrass had criticized this naïve use of variational arguments. Weierstrass' main victim was the Dirichlet principle, that is, the variational method involving minimization of the Dirichlet integral over the domain. It is worth while noting, however, that although Dirichlet's principle was apparently formulated by Green before Gauss' work, the first serious mathematical existence proof for the Dirichlet problem was the one discussed here, which used *a boundary integral equation of the first kind*.

The second weak point of the above argument is one noticed by Gauss himself: His positivity argument is of a simple *geometric* nature: Since r is bounded by the diameter of the surface, the positive kernel $1/r$ is bounded

---

[4]It is self-evident that, however a mass M may be distributed *homogeneously* over a surface, the resulting everywhere positive potential V will be, in every point of the surface, greater than $\frac{M}{r}$ if r designates the greatest distance between two points of the surface: this value itself could be attained by the potential only in an endpoint of the line r if the entire mass was concentrated in the other endpoint, a case which cannot appear here, because we will only consider a continuous distribution, where every surface element ds corresponds only to an infinitely small mass $m\,ds$. The integral $\int Vm\,ds$, extended over the whole surface, is therefore in any case greater than $\int \frac{M}{r}m\,ds$ or $\frac{MM}{r}$, and thus there must necessarily be a homogeneous kind of distribution for which that integral has a minimum value.

from below by a constant depending only on the domain. The quadratic form defined by the integral operator is therefore seen to be positive, but only for *non-negative* densities m. Having to respect this constraint makes the proof rather complicated: Only variations inside the positive cone are allowed, which means that in general, the solution of the minimisation problem solves only an integral *inequality*, turning into an *equation* only in those points where the solution is strictly positive. Gauss writes (Gauss' original emphasis as always):

> [33.] Der eigentliche Hauptnerv der im 32. Artikel entwickelten Beweisführung beruht auf der Evidenz, mit welcher die Existenz eines Minimumwerthes für Ω unmittelbar erkannt wird, solange man sich auf die gleichartigen Vertheilungen einer gegebenen Masse beschränkt. Fände die gleiche Evidenz auch ohne diese Beschränkung Statt, so würden die dortigen Schlüsse ohne weiteres zu dem Resultate führen, *dass es allemal, wenn nicht eine gleichartige, doch eine ungleichartige Vertheilung der gegebenen Masse gibt, für welche $W = V - U$ in allen Punkten der Fläche einen constanten Werth erhält*, indem dann die zweite Bedingung (Art. 31. II) wegfällt. Allein da jene Evidenz verloren geht, sobald wir die Beschränkung auf gleichartige Vertheilungen fallen lassen, so sind wir genöthigt, den strengen Beweis jenes wichtigsten Satzes unserer ganzen Untersuchung auf einem etwas künstlichern Wege zu suchen.[5]

Thus Gauss finds it desirable to prove the positivity of the quadratic form for not necessarily non-negative mass distributions. This would have given not only a much simpler proof, but even a much nicer theorem.

The truly odd observation is now that Gauss could easily have proved this general positivity himself by simply combining the jump relations and the integration by parts formula cited above. For completeness of this claim, here are Gauss' original formulations of these lemmas:

> [end of 15.] Man kann diesen wichtigen Satz auch so ausdrücken: der Grenzwerth von X, bei unendlich abnehmendem positiven x ist $X^0 - 2\pi k^0$, bei unendlich abnehmendem negativen x hingegen $X^0 + 2\pi k^0$, oder X ändert sich zweimal sprungsweise um $-2\pi k^0$, indem x aus einem negativen Werthe in einen positiven übergeht, das

---

[5] [33.] The actual main nerve of the line of proof developed in §32 rests on the self-evidence with which the existence of a minimum value for Ω is perceived immediately, as long as one restricts oneself to the homogeneous distributions of a given mass. If the same self-evidence held without this restriction, the above arguments would lead immediately to the result *that there is always, if not a homogeneous, then at least an inhomogeneous distribution of the given mass for which $W = V - U$ obtains in all points of the surface a constant value*, in that the second condition (§31. II) can then be omitted. However, since this self-evidence is lost as soon as we drop the restriction to homogeneous distributions, we are forced to search for the strict proof of this most important theorem of our whole investigation in a somewhat artificial way.

erstemal, indem x den Werth 0 erreicht, und das zweitemal, indem es ihn überschreitet.[6]

Here Gauss uses coordinates where the normal at a point on the surface coincides with the x axis and $X = \frac{\partial V}{\partial x}$ where V stands for the single layer potential with density k: $V = \int \frac{k\,ds}{r}$ with the surface element ds and the distance r between observation point and point of integration.

[24.] LEHRSATZ. Es ist

$$\int V \frac{dV}{dp}\,\mathrm{d}s = -\int qq\,\mathrm{d}T$$

wenn das erste Integral über die ganze Fläche, das zweite durch den ganzen Raum T ausgedehnt wird.[7]

Here Gauss denotes by q the gradient of the potential V, by T the interior domain, and $\frac{dV}{dp}$ is the interior normal derivative.

We see that he could have added the formulas from Lehrsatz 24 for the interior domain and the corresponding one for the exterior domain in order to get with Theorem 15 (in what would have been his formulation; he didn't write this, of course)

$$\int V m\,\mathrm{d}s = \frac{1}{4\pi}\int qq\,\mathrm{d}T > 0$$

where the second integral is extended over the whole space. This gives positivity for any m, positive or negative. It is also physically intuitive (in electro- or magnetostatic terms that were familiar to Gauss), stating equality between the potential energy stored in the surface and the total energy of the field.

We can only speculate why Gauss didn't write this. It is also strange that this result about the positivity of the quadratic form defined by the $1/r$ kernel, which was, as we have seen, formulated as a useful and non-trivial open problem in one of the most famous and widely studied papers of its time, did apparently not become the object of serious study for a long time. The reason cannot be that the simple argument of adding the interior and exterior Green formulas did not occur to anyone. As an example, here is a quote from a paper [32, p.216] by W. Steklov, written 1900 in the wake of Poincaré's paper [29].

Posons

$$V = \frac{1}{4\pi}\int \frac{W}{r}\,\mathrm{d}s\ ,$$

---

[6][end of 15.] One can express this important theorem also as follows: The limit of X for infinitely decreasing positive x is $X^0 - 2\pi k^0$, whereas for infinitely decreasing negative x it is $X^0 + 2\pi k^0$, or X jumps twice by $-2\pi k^0$ when x passes from a negative value to a positive one, the first time when x reaches the value 0, and the second time when it goes beyond it.

[7][24.] THEOREM. There holds [formula omitted] where the first integral is extended over the whole surface, the second one over the whole space T.

l'intégrale étant étendue à la surface (S) tout entière. Dans les suppositions faites par rapport à (S) nous pouvons employer le théorème connu de Green qui nous donne[8]

$$\int \sum (\frac{\partial V}{\partial x})^2 \mathrm{d}\tau + \int \sum (\frac{\partial V}{\partial x})^2 \mathrm{d}\tau' = \int V(\frac{\partial V_i}{\partial n} - \frac{\partial V_e}{\partial n}) \mathrm{d}s = \int VW \, \mathrm{d}s > 0 \,.$$

Steklov then uses this to prove that for a harmonic function the L^2 norm on the boundary is bounded by the diameter of the boundary times the H^1 seminorm on the domain. But he doesn't state this positivity as an interesting result in itself.

In 1935, Otto Frostman [8] finally formulated the positivity of this quadratic form as a theorem in order to complete Gauss' proof. But he considers the argument using Green's formula as easy to see, but too restrictive (p. 24: "Si le potentiel (newtonien) a des dérivées continues, cela résulte déjà des formules de Green et de Gauss; en effet on démontre facilement...".[9]). He then gives another proof using the composition property of Riesz potentials on the whole space which shows that the convolution with $1/r$ on $\mathbb{R}^3$ is a constant times the square of the convolution with $1/r^2$. This argument (which can easily be verified by taking Fourier transforms) is generalized by Frostman to other kernels of the form $1/r^\alpha$ with $\alpha > 0$. For these kernels, he then presents Gauss' complicated proof in the framework of positive measures using the maximum principle as a principal tool.

The, in our view, simpler and more general (because it applies to other equations of mathematical physics besides the Laplace equation) proof using the energy identity was not given before another 38 years, in 1973 by Nedelec and Planchard [26].

The difference between the two completions that have to be performed in order to complete Gauss' minimization argument is that on one hand, as Frostman showed, positive measures are complete in the energy norm. Thus in the well-understood framework of positive (Radon) measures, the infimum is indeed a minimum. One doesn't even have to know exactly what the finiteness of energy means for those measures (more about this point below); one can very well minimize a coercive lower semi-continuous functional that is not everywhere finite. On the other hand, as Nedelec and Planchard noticed, the space obtained by completion of a whole vector space (and not only the positive cone) in the energy norm is the Sobolev space $H^{-1/2}$ which is a space not of functions or measures, but of distributions.

Thus, whereas the efforts of Hilbert and others to complete the proof of Dirichlet's principle led to the introduction of the function spaces of Beppo

---

[8]Let [formula omitted], the integral being extended over the entire surface (S). With our assumptions on (S) we can apply the known theorem of Green which gives us [formula omitted].

[9]If the (Newtonian) potential has continuous derivatives, this follows already from the formulas of Green and Gauss; indeed one shows easily...

Levi and Sobolev already in the beginning of the 20th century (crudely stated: H^1 is a subspace of L^2, therefore a space of functions, once Lebesgue's notion of function is adopted), the energy space needed for Gauss' boundary integral form of the Dirichlet principle could only be constructed after the introduction of Schwartz' distributions and Sobolev spaces of fractional and negative index.

There is a glimpse of this difficulty in Henri Cartan's works in 1941 and 1945: In [2] he presents a proof of Frostman's theorem on the completeness of positive measures of finite energy (in fact a greatly generalized version thereof), but of the question of completeness of *all* signed measures of finite energy, he says (p.90) "C'est peu probable."[10]. In the paper [3] he gives a counterexample (p.87) showing that it is, indeed, not complete.

> On voit qu'en "complétant" l'espace $\mathcal{E}$ pour cette norme, on obtiendrait un espace de Hilbert. On vérifie sans peine que $\mathcal{E}$ lui-même *n'est pas complet ()*.[11]

But he does not want to quit the framework of measures (which he also calls "distributions") to investigate the nature of this Hilbert space.

Could it be that Gauss already had some intuition about the different nature of the minimizing objects that would appear when the condition of non-negativity was dropped? We can only speculate.

To finish this paragraph, here is another historic curiosity related to measures and their energy: As is well-known in the theory of the logarithmic single layer potential integral equation in two dimensions, the positivity is true there only under an additional condition on the boundary: Its capacity has to be less than one. It is also a classical result that the logarithmic capacity of a compact set in $\mathbb{R}^2$ is identical to its transfinite diameter and also to its exterior conformal radius (other names are Chebyshev constant or Robin constant). This was well known to Frostman in 1935, and the identity of transfinite diameter and exterior mapping radius for regular sets was already proved by Szegö in 1924 [39]. Now the standard reference (and the only available reference in book form, as far as I can tell) for a complete proof of this equivalence result is the book [17] by Einar Hille. Hille gives a detailed proof of all the equivalences, in particular (Theorem 16.4.4 p.284) a proof of the equality of transfinite diameter and logarithmic capacity by constructing a minimizing measure. He gets this measure as a limit of point measures supported by the Fekete points. This is also Corollary 1 (p. 285):

> Corollary 1. The equilibrium distribution $\nu(s)$ of E is the weak limit of the sequence of point distributions $\mu_n(S)$ associated with the zeros of the Fekete polynomials $F_n(z; E)$.

Unfortunately, in the proof it is used that the energy of μ_n is finite (and can even be given by a simple formula), which is not the case (Point measures are

---

[10] This is not very likely.

[11] One sees that by "completing" the space $\mathcal{E}$ in this norm, one would obtain a Hilbert space. One verifies with ease that $\mathcal{E}$ itself *is not complete.*

not in $H^{-1/2}$). Thus the standard reference for this basic (and well-known true) result has a hole that might still be open after more than 40 years!

3 Poincaré and the Double Layer Potential

After Gauss' work on the first kind integral equation of the single layer potential, the next major progress came with Carl Neumann's work on the double layer potential. Of his numerous publications on the subject of his "Methode des arithmetischen Mittels", we cite the book [27] from 1877 which is available online from the Gallica project of the BNF.

For convex domains, Neumann proves the convergence of the method of iterations which leads to the solution of the second kind integral equation by the Neumann series. The tool here is not positivity, but the contraction property of the integral operator in the maximum norm. Positivity comes in through the convexity of the domain which means that the measure defined by the double layer kernel

$$\mathrm{d}\theta_x(y) = \frac{1}{4\pi} \frac{n(y) \cdot (y - x)}{|x - y|^3} \mathrm{d}s(y) \tag{1}$$

is a positive measure of total mass 1. The idea that integration against such a measure should somehow level functions out and make iterations converge to a constant function seems to have been intuitive to physicists before Neumann. In a paper from 1856, quoted in its entirety by Neumann in his book (Chapter 6), the physicist Beer used an iterative method for the second kind integral equation of the normal derivative of the single layer potential (the adjoint equation to Neumann's). He formulates

> Dabei leuchtet ein, dass F' – welches innerhalb σ zwischen dem größten und kleinsten Werthe liegt, den die Funktion F auf der Fläche σ selbst annimmt – im Allgemeinen *gleichförmiger* als F verläuft.[12]

In a footnote, Neumann remarks that Beer does not offer any proof, and that the claim is not true, in general, unless the measure mentioned above is positive, that is, unless the domain is convex.

Neumann's proof of his result (and as a corollary also of Beer's result) uses highly non-trivial geometric and measure-theoretic arguments that constitute one of the early examples of "hard" analysis in potential theory. As a consequence, subsequent generalizations of his techniques were confined to hard harmonic analysis, too, see [24] and [25] for overviews.

Neumann's method of the arithmetic mean became famous, because it was at the time, besides Schwarz' alternating method and Poincaré's balayage method, the only rigorous way of proving existence for the Dirichlet

---

[12] Here it is clear that F' – which, inside σ, lies between the largest and smallest value that the function F takes on the surface σ itself – behaves in general *more uniformly* than F.

problem and for all the important theorems based on it like the Riemann mapping theorem. In addition, it looked like it was simpler to apply and more constructive than the other two methods. But the restrictive assumption of convexity of the domain was a mathematical challenge, and in 1895 Henri Poincaré published a paper [29] about Neumann's method which introduced a quite different argument for proving the contractivity that did not need convexity of the domain. The new method was based on positivity and energy identities.

In this paper, Poincaré presents an astonishing collection of techniques that were new at the time and that made the paper famous, at least for some years. Poincaré used this method only in one further work [30], a small paper on generalizations to elasticity theory which he himself characterizes as incomplete. But others continued and developed his ideas in various different directions, in particular Arthur Korn [19, 21, 22, 23], Vladimir Steklov [32, 33, 36, 37, 38] and Stanislaw Zaremba [41, 42, 43]. Korn and Steklov for some time engaged in a kind of race [34, 20, 35]. Here is a quote from [20] (our reference numbers):

> Dans une note [34] *sur la méthode de Neumann et le problème de Dirichlet*, M. W. Stekloff est arrivé à une démonstration de la méthode de la moyenne arithmétique de M. Neumann, qui est à peu près la même que celle que j'ai publiée il y a un an dans mon Cours sur la théorie du potentiel [19]. Ma démonstration, comme celle de M. Stekloff, a pour base le Mémoire ingénieux [29] de M. Poincaré, et nous avons éliminé tous les deux de la même manière la restriction de M. Poincaré, que l'existence d'une solution soit préalablement établie.[13]

The "fonctions fondamentales" mentioned in the titles of some of these papers, also called "universelle Funktionen" by Korn, are potentials generated by eigenfunctions of Neumann's integral operator or also by its adjoint, sometimes also the eigenfunctions of what is known as Steklov eigenvalue problem, or eigenfunctions of the Poincaré–Steklov operator.

These papers concentrated on eigenfunction expansions and eigenvalue estimates obtained by min-max principles as studied by Poincaré for the case of the eigenvalue problem for the Laplace operator with Dirichlet boundary conditions in his earlier important paper [28]. To prove existence of the eigenfunctions, regularity of the boundary had to be assumed, and after works by Hölder and Lyapunov, Hölder continuous functions on Lyapunov surfaces became the standard framework. During the same time, the new powerful Fredholm method for treating integral equations became widely accepted,

---

[13]In a note [34] *on Neumann's method and the Dirichlet problem*, Mr W. Stekloff arrived at a proof of Mr Neumann's method of the arithmetic mean which is more or less the same as the one that I have published a year ago in my Course on Potential Theory [19]. My proof, as the one of Mr Stekloff, is based on the ingenious paper [29] by Mr Poincaré, and we have both eliminated in the same manner the restriction of Mr Poincaré that the existence of a solution should be established beforehand.

and Hilbert published his book on integral equations which had the ambition to subsume all known results about integral equations. Hilbert and his group made big jumps forward by introducing the idea of function spaces and norms and developing the basics of modern functional analysis with the spectral theory of bounded and in particular compact selfadjoint operators in Hilbert spaces.

I mention all this well-known history as an explanation for the curious fact that the basic idea of Poincaré's paper on Neumann's method, namely to consider the convergence of the Neumann series in the energy norm, disappeared pretty much completely from the discussion. His estimates were only used for estimating the eigenvalues of the boundary integral operators considered as compact operators acting in spaces of continuous or Hölder continuous functions, and this remained the standard for a long time, see for example [25, Thm 12, p. 144]. One of the main advantages of Poincaré's method, namely its easy applicability to other elliptic problems having a positive energy, such as linear elasticity, remained present, but the other advantage, namely that it basically only uses Green's formula and is therefore valid for general Lipschitz domains, seems to have been forgotten.

Only very recently a similar point of view has been adopted in the paper by Steinbach and Wendland [31] where the contraction property of Neumann's operator in a norm related to the energy norm was proved for the first time for rather general positive second order elliptic systems on Lipschitz domains.

Poincaré's own estimates are being revisited and adapted to a modern standard in the very recent paper [18] which treats the same framework as Poincaré did, namely two- and three-dimensional potential theory on smooth domains. The full potential of Poincaré's main idea which easily generalizes to other positive elliptic operators and to domains with only Lipschitz continuous boundary, does not seem to have been exploited in a modern context yet. We will describe some of this in the next section.

Here is the basic estimate from [29, Chapter 2] in a notation similar to Poincaré's own notation: For a bounded domain Ω in $\mathbb{R}^3$ let W be a function harmonic in the domain and in the exterior domain $\Omega' = \mathbb{R}^3 \setminus \overline{\Omega}$, vanishing at infinity. Quantities related to the exterior domain are indicated by a prime. Let J and J' denote the interior and exterior Dirichlet integrals of W:

$$J = \int_\Omega |\nabla W|^2 \mathrm{d}x \ ; \quad J' = \int_{\Omega'} |\nabla W|^2 \mathrm{d}x \ .$$

Lemma 1. *There is a constant μ depending only on the domain such that*
(i) If W is a double layer potential, then

$$J \le \mu\, J' \quad and \quad J' \le \mu\, J \ . \tag{2}$$

(ii) If W is a single layer potential, then

$$J \le \mu\, J' \quad and\ if\ \int_{\partial\Omega} W\, \mathrm{d}s = 0,\ then \quad J' \le \mu\, J \ . \tag{3}$$

Here double and single layer potentials are defined by their jump properties: Single layer potentials are continuous across the surface $\partial\Omega$ and have a jump in their normal derivatives, whereas double layer potentials have a jump across the surface, but their normal derivatives from the interior and the exterior coincide. The difference between single and double layer potentials in the statement is caused by different behavior of potentials with vanishing Dirichlet integrals (constants): For double layer potentials, if W is constant in the interior domain, it is also constant (zero) in the exterior and vice versa, so that J and J' both vanish if one of them vanishes. For single layer potentials, W vanishing in the exterior implies W vanishing in the interior, so that $J' = 0$ implies $J = 0$, too, but there exists the non-trivial equilibrium density (Robin density) which has potential 1 in the interior and non-constant potential in the exterior, so that J' can be bounded by J only on a subspace of codimension one.

In 1900, Steklov [32, p.224], after stating the above estimate for single layer potentials, gets quite enthusiastic and writes (his emphasis):

> Nous appellerons ce théorème *théorème fondamental.*
>
> . . .
>
> Nous verrons dans ce qui va suivre, que *la solution de tous les problèmes fondamentaux de la Physique mathématique se ramène à la démonstration complète du théorème fondamental.*[14]

Writing this in a year when Planck introduced his quantum constant and Poincaré was already working on the theory of relativity seems, in hindsight, slightly exaggerated, but it underlines the importance of these estimates for potential theory and for related models of classical mathematical physics like elasticity, heat conduction, acoustics, electrostatics and electrodynamics, fluid dynamics and so on. Such applications were studied by Steklov, Korn, Zaremba and others, who also worked on removing some of the hypotheses Poincaré had to make in order to prove Lemma 1. They proved the lemma essentially for arbitrary connected Lyapunov (i.e. $C^{1,\alpha}$) surfaces.

Poincaré proved the lemma under the condition that the domain is diffeomorphic to a ball (actually for a simply connected smooth boundary; the question of the existence of a diffeomorphism to the ball is a first simple case of the famous Poincaré conjecture), and he used the diffeomorphism to reduce the estimates to the case of a ball where he could show them explicitly by expansion in spherical harmonics.

Nowadays, the lemma is easy to prove even for Lipschitz surfaces by noticing that the H^1 seminorm of a harmonic function on the interior or exterior domain is equivalent to both the $H^{1/2}$ seminorm of its trace and the $H^{-1/2}$ norm of its normal derivative on the boundary. This equivalence is seen immediately in one direction from the standard trace theorem (sometimes called

---

[14]We shall call this theorem the *fundamental theorem.. . . We* shall see in the following that *the solution of all the fundamental problems of Mathematical Physics can be reduced to a complete proof of the fundamental theorem.*

Gagliardo's trace theorem in the case of a merely Lipschitz continuous boundary) plus Poincaré's inequality (the one estimating the L^2 norm modulo constants by the H^1 seminorm) and the weak definition of the normal derivative, and in the other direction from the variational solution of the Dirichlet and the Neumann problems. But one should keep in mind that without the introduction of the fractional Sobolev space $H^{1/2}$ on the surface, which at first seems like overkill for proving a statement mentioning only Dirichlet integrals, one has no way of stating or proving that the trace spaces from the exterior and from the interior are the same, which is one of the crucial points in this argument. In fact, one can consider Poincaré's procedure of using a diffeomorphism to the sphere and estimating the coefficients of the expansion in spherical harmonics as an early definition of the space $H^{1/2}$, although the idea of function spaces and norms was not expressed in that paper.

Poincaré uses the estimates in Lemma 1 to prove the contraction property of Neumann's operator in the energy norm, and with this the convergence of Neumann's series solution for the Dirichlet problem in the same norm. He then shows trace estimates, first for the boundary L^2 norm modulo constants (Chapter 4), and then (Chapter 5) for the L^∞ norm of the double layer operator applied to the trace on the boundary. The latter estimate uses difficult geometric constructions, is not yet optimal, and is subsequently generalized by the above-mentioned authors and others like Lebesgue, Plemelj and Radon, one famous later observation being that whereas Neumann's operator is not a contraction in the L^∞ norm when the domain is not convex, the square of the operator is a contraction, at least when the domain is smooth. In any case, Poincaré completes the proof of the uniform convergence in the whole space of Neumann's series for general smooth domains.

Neumann's operator, as defined by Neumann himself and in the same way by Poincaré, is the mapping from the difference of the boundary traces of a double layer potential to the sum of the traces. If we denote the interior and exterior traces of the double layer potential W by V and V', respectively, then Neumann's operator N maps $V - V'$ to $V + V'$, which corresponds in our notation of the next section below to

$$N = -2K \, .$$

The problem studied by Poincaré (his equation (1)) is written not as an integral equation, but as a transmission problem with a parameter λ:

$$V - V' = \lambda(V + V') + 2\Phi \, . \tag{4}$$

The choice of $\lambda = 1$ corresponds to the exterior Dirichlet problem, and $\lambda = -1$ to the interior Dirichlet problem. Poincaré proves convergence (modulo constant functions) of the Neumann series solution of (4) for $|\lambda| < \frac{\mu+1}{\mu-1}$, where μ is the constant from Lemma 1.

4 Positive Boundary Integral Operators and the Convergence of Neumann's Series

In this section we will give a modern expression of Poincaré's idea that the estimate (2) of Lemma 1 implies that Neumann's operator is a contraction. We start by an abstract observation whose simple proof we leave to the reader. No tools more advanced than the Cauchy-Schwarz inequality are required for the proof.

Lemma 2. *Let A and B be bounded selfadjoint operators on a Hilbert space X satisfying $A + B = I$, where I is the identity operator.*
(i) *If $B - A$ is a contraction, then A and B are contractions with norms bounded by $(1 + \|B - A\|)/2$. The inverse A^{-1} can be represented in two different ways by convergent Neumann series*

$$A^{-1} = \sum_{\ell=0}^{\infty} B^{\ell} = 2 \sum_{\ell=0}^{\infty} (B - A)^{\ell} . \tag{5}$$

(ii) *If A is positive definite and B positive semidefinite:*

$$\exists \alpha > 0 \, , \exists \beta \geq 0 \, : \, \forall u \in X \, : \quad (Au, u) \geq \alpha \|u\|^2 \, ; \quad (Bu, u) \geq \beta \|u\|^2 \, ,$$

then B is a contraction with norm $\|B\| \leq 1 - \alpha$. If in addition $\beta > 0$, then $B - A$ is a contraction with norm $\|B - A\| \leq \max\{1 - 2\alpha, 1 - 2\beta\}$.

A situation where this lemma can easily be applied is the following:

Lemma 3. *Let a and b be symmetric bilinear forms on a vector space X_0. We assume that a and b are positive semidefinite and that a is non-degenerate:*

$$\forall u \in X_0 \, : \quad a(u, u) > 0 \text{ if } u \neq 0 \, ; \, b(u, u) \geq 0 \, .$$

Let X be the Hilbert space completion of X_0 with respect to the inner product

$$(u, v) = a(u, v) + b(u, v)$$

and let A and B be the operators on X defined by the bilinear forms a and b. If there exists $\mu > 0$ such that

$$\forall u \in X_0 \, : \quad b(u, u) \leq \mu \, a(u, u) \, ,$$

then A and B satisfy the hypothesis of Lemma 2 (ii) *with $\alpha = \frac{1}{\mu+1}$.*
In particular, B is a contraction with norm $\|B\| \leq \frac{\mu}{\mu+1}$. If, in addition,

$$\forall u \in X_0 \, : \quad a(u, u) \leq \mu \, b(u, u) \, ,$$

then $B - A$ is a contraction with norm $\|B - A\| \leq \frac{\mu-1}{\mu+1}$.

Note that the Riesz representation theorem implies that the existence of an estimate $b \leq \mu a$ is equivalent to the positive definiteness of a on the Hilbert space X.

Another remark which is easy to verify is that the non-degeneracy of a alone is sufficient to show that all *eigenvalues* of $B - A$ and of B are of absolute value strictly less than 1. One does not need the estimate $b \leq \mu a$ for this, but one also does not get the contractivity from it. If, however, $B - A$ has a pure point spectrum, for example if it is compact, then the contractivity follows. This may provide a partial explanation why Poincaré's mutual estimates of the interior and exterior energies were later forgotten: If the Fredholm-Riesz theory can be applied as is the case for Neumann's operator on a smooth surface, then they are not needed. They are then, in fact, a consequence of the Fredholm alternative: If a is positive semidefinite and non-degenerate and the corresponding operator A is Fredholm, then a is positive definite.

In the following, we present some applications of these simple estimates. In all cases, the quadratic forms a and b will correspond to the energy integrals in the exterior and interior domains, respectively, so that the Hilbert space X will be endowed with the norm of the total energy. Which concrete boundary integral operators correspond to the abstract operators A and B can vary, however, according to how the abstract vector space X_0 is represented by a concrete function space.

We choose the same general situation as considered in the paper [31] by Steinbach and Wendland. This covers some of the most important applications such as potential theory and elasticity theory (basically "every fundamental problem of mathematical physics" in the sense of Steklov quoted above).

The same ideas for proving the contraction property of second kind boundary integral operators could be applied to higher order strongly elliptic partial differential operators that have a positive energy form in the framework studied in [7], or to other situations where positivity of first kind integral operators has been shown by using Green's formulas like for parabolic problems in [5]. In this paper we will stay within the framework of positive second order systems as in [31]. This will allow an easy comparison in order to see similarities and differences with the arguments of [31]. Note, however, that while we consider the same objects as in [31], we will not always use the same letters to denote them.

Let then L be a second order selfadjoint elliptic partial differential operator on $\mathbb{R}^n$ with smooth, not necessarily constant coefficients about which we will make a certain number of further hypotheses. First we assume that L has a real-valued fundamental solution $U^*(x, y)$. Given a density ψ on the boundary Γ of the bounded Lipschitz domain Ω, the single layer potential $\mathcal{S}$ is defined in the interior domain Ω and in the exterior domain $\Omega^c = \mathbb{R}^n \setminus \overline{\Omega}$ by

$$\mathcal{S}\psi(x) = \int_{\Gamma} U^*(x, y)\psi(y)\,\mathrm{d}y \,. \tag{6}$$

Before defining the double layer potential, we need to assume that there exists a first Green formula

$$\int_{\Omega} (Lu(x))^{\top} v(x)\, \mathrm{d}x = \Phi(u,v) - \int_{\Gamma} (Tu(x))^{\top} v(x)\, \mathrm{d}s(x)\,. \qquad (7)$$

Here T is the conormal derivative, defined by this formula. The energy bilinear form Φ is a first order symmetric integro-differential form which we assume to be positive in the sense that it is non-negative and elliptic: There are constants α, c, k with $\alpha > 0$ such that $|\Phi(u,v)| \leq c\|u\|_{H^1(\Omega)}\|v\|_{H^1(\Omega)}$ and

$$\Phi(u,u) \geq 0 \quad \text{and} \quad \Phi(u,u) \geq \alpha\|u\|_{H^1(\Omega)}^2 - k\|u\|_{L^2(\Omega)}^2\,. \qquad (8)$$

As a consequence of the Gårding inequality (8) and the compact embedding of the Sobolev space $H^1(\Omega)$ into $L^2(\Omega)$, the space of functions of vanishing energy

$$\mathcal{R} = \{u \mid \Phi(u,u) = 0\} \qquad (9)$$

is finite-dimensional. For $u \in \mathcal{R}$ one has also $\Phi(u,v) = 0$ for all v, which according to (7) is the weak formulation of the homogeneous Neumann problem $Lu = 0$ in Ω, $Tu = 0$ on Γ, so that $\mathcal{R}$ can also be defined as solution space of the homogeneous Neumann problem.

For the exterior domain, we also assume the first Green formula

$$\int_{\Omega^c} (Lu(x))^{\top} v(x)\, \mathrm{d}x = \Phi^c(u,v) + \int_{\Gamma} (Tu(x))^{\top} v(x)\, \mathrm{d}s(x)\,. \qquad (10)$$

Whereas the previous equations (6)–(9) were assumed to be valid for all smooth functions – with the idea of extending the domain of validity by continuity to some larger Hilbert spaces of functions afterwards – in the Green formula (10) for the exterior domain we have to assume that u and v are, in addition, of compact support. For such functions, we assume then positivity of the exterior energy form:

$$\forall u \in C_0^{\infty}(\mathbb{R}^n)\ :\quad \Phi^c(u,u) > 0 \quad \text{unless } u \equiv 0\,. \qquad (11)$$

The final assumption we have to make is that *potentials have finite energy*. This is an assumption on the behavior of the fundamental solution at infinity which can be phrased as follows: If γ and δ are multi-indices and $\chi \in C^{\infty}(\mathbb{R}^n)$ is a cut-off function which is zero on a large enough ball and equal to one on a neighborhood of infinity, then the function u defined by

$$u(x) = \chi(x)\partial_x^{\gamma}\partial_y^{\delta}U^*(x,y)$$

satisfies $\Phi^c(u,u) < \infty$.

The assumptions made so far cover some important standard examples:
- The Laplace equation in dimension $n \geq 3$ with its standard fundamental solution. Here the conormal derivative T is the exterior normal derivative. The

space $\mathcal{R}$ consists of the constant functions on Ω. The condition that potentials have finite energy excludes the logarithmic potentials in the plane.

- The equations of linear elasticity in dimension $n \geq 3$. The conormal derivative T corresponds to the normal traction on the boundary, and the space $\mathcal{R}$ consists of the rigid motions.

- The mathematically simplest case is a strictly positive operator such as $-\Delta + \lambda I$ with $\lambda > 0$ in any dimension, or similarly any strongly elliptic constant coefficient operator plus λI with a sufficiently large λ. In this case, the energy form in the interior is positive definite, too, the space $\mathcal{R}$ is reduced to $\{0\}$, and the energy forms in both the interior and the exterior domain are equivalent to the square of the H^1 norm.

The double layer potential $\mathcal{D}$ with density φ is given for $x \notin \Gamma$ by

$$\mathcal{D}\varphi(x) = \int_\Gamma \left(T_y U^*(x,y)\right)^\top \varphi(y)\,\mathrm{d}s(y)\,. \tag{12}$$

It is well known [4] that the definitions (6) and (12) of the single and double layer potentials can be extended by continuity to densities $\psi \in H^{-1/2}(\Gamma)$ and $\varphi \in H^{1/2}(\Gamma)$, respectively, and that the potentials $v = \mathcal{S}\psi$ and $w = \mathcal{D}\varphi$ then satisfy

$$Lv = 0\,,\ Lw = 0 \text{ in } \Omega \cup \Omega^c\,;\ v \in H^1_{\mathrm{loc}}(\mathbb{R}^n)\,;\ w \in H^1(\Omega) \text{ and } w \in H^1_{\mathrm{loc}}(\overline{\Omega^c})\,.$$

If we denote the interior and exterior traces by γ and γ^c and the interior and exterior conormal derivatives (both taken with respect to the exterior normal) by γ_1 and γ_1^c, then these can also be extended by continuity to the potentials with this weak regularity, and there hold the jump relations

$$\begin{aligned}(\gamma^c - \gamma)\mathcal{S}\psi = 0\ ;\ (\gamma_1^c - \gamma_1)\mathcal{S}\psi = -\psi\ ;\\ (\gamma^c - \gamma)\mathcal{D}\varphi = \varphi\ ;\ (\gamma_1^c - \gamma_1)\mathcal{D}\varphi = 0\,.\end{aligned} \tag{13}$$

The four classical boundary integral operators are then defined as the operators of

- the single layer potential: $V = \gamma\mathcal{S} = \gamma^c\mathcal{S}$
- the normal derivative of the single layer potential: $K' = \frac{1}{2}(\gamma_1 + \gamma_1^c)\mathcal{S}$
- the double layer potential: $K = \frac{1}{2}(\gamma + \gamma^c)\mathcal{D}$
- the normal derivative of the double layer potential: $W = -\gamma_1\mathcal{D} = -\gamma_1^c\mathcal{D}$.

With these definitions, the traces of the single layer and double layer potentials take the form

$$\begin{aligned}\gamma\mathcal{S} &= \gamma^c\mathcal{S} = V\quad ;\ \gamma_1\mathcal{S} = \tfrac{1}{2}I + K'\ ;\ \gamma_1^c\mathcal{S} = -\tfrac{1}{2}I + K'\ ;\\ \gamma_1\mathcal{D} &= \gamma_1^c\mathcal{D} = -W\ ;\ \gamma\mathcal{D} = -\tfrac{1}{2}I + K\ ;\ \gamma^c\mathcal{D} = \ \tfrac{1}{2}I + K\,.\end{aligned} \tag{14}$$

As mentioned above, this way of defining the boundary integral operator K of the double layer potential corresponds to Neumann's and Poincaré's definitions for the case of potential theory. If one defines $K_0\varphi$ as the double layer potential of density φ *evaluated on the surface Γ in the sense of a Cauchy*

20   M. Costabel

principal value integral (which in potential theory is the same as integrating with respect to the solid angle measure (1)), then it is well known that $K\varphi(x) = K_0\varphi(x)$ for smooth boundary points x, but for corner points the two definitions differ. The operator whose contraction property is studied by Neumann is $N = -2K$. If N has a norm less than one in some function space, then the four operators $\frac{1}{2}I \pm K$ and $\frac{1}{2}I \pm K'$ will also have norms less than one.

We can now begin to apply Lemma 3 to various incarnations of vector space X_0 and bilinear forms a and b. We will always represent a by the energy integral Φ^c and b by Φ. According to the Green formulas (7) and (10), we have for a function u satisfying $Lu = 0$ in Ω and in Ω^c and any v:

$$\Phi(u,v) = \langle \gamma_1 u, \gamma v \rangle \; ; \qquad \Phi^c(u,v) = -\langle \gamma_1^c u, \gamma^c v \rangle \; . \tag{15}$$

Here we write $\langle \cdot, \cdot \rangle$ for the L^2 inner product (integral) on Γ, extended to the duality product between $H^{-1/2}(\Gamma)$ and $H^{1/2}(\Gamma)$.

4.1 Single Layer Potentials

The first possibility is to take for the space X_0 some space of integrable functions on Γ, for example the continuous functions, or $L^2(\Gamma)$. For $\varphi, \psi \in X_0$, we define the bilinear forms a and b as energy forms of the corresponding single layer potentials:

$$a(\varphi, \psi) = \Phi^c(\mathcal{S}\varphi, \mathcal{S}\psi) \; ; \qquad b(\varphi, \psi) = \Phi(\mathcal{S}\varphi, \mathcal{S}\psi) \; . \tag{16}$$

With the boundary reduction by Green's formula (15) and the expressions (14) for the traces of the single layer potential, we find the boundary integral forms

$$a(\varphi, \psi) = \langle (\frac{1}{2}I - K')\varphi, V\psi \rangle \; ; \qquad b(\varphi, \psi) = \langle (\frac{1}{2}I + K')\varphi, V\psi \rangle \; . \tag{17}$$

For the total energy $a + b$ we find the bilinear form defined by the single layer potential integral operator which is therefore positive definite (Gauss' missing theorem); and the Hilbert space X is the completion of our space X_0 in this energy norm which we know from Nedelec and Planchard [26] to be the Sobolev space $H^{-1/2}(\Gamma)$:

$$a(\varphi, \psi) + b(\varphi, \psi) = \langle \varphi, V\psi \rangle \; ; \; X = H^{-1/2}(\Gamma) \text{ with norm } \|\varphi\|_V^2 = \langle \varphi, V\varphi \rangle \; . \tag{18}$$

The operators A and B are defined by $(A\varphi, \psi)_V = a(\varphi, \psi)$ and $(B\varphi, \psi)_V = b(\varphi, \psi)$, hence

$$A = \frac{1}{2}I - K' \; ; \qquad B = \frac{1}{2}I + K' \; . \tag{19}$$

We conclude from our construction that the hypotheses of Lemma 2 are satisfied. In particular, $\frac{1}{2}I \pm K'$ are bounded operators in $H^{-1/2}(\Gamma)$, selfadjoint and positive semidefinite with respect to the inner product $(\varphi, \psi)_V = \langle \varphi, V\psi \rangle$. As we explained after Lemma 1, the positive definiteness of A or, equivalently, the Poincaré estimate $b \leq \mu a$ is a simple consequence of the identity between X and $H^{-1/2}(\Gamma)$: $b(\varphi, \varphi)$ is the energy integral $\Phi(\mathcal{S}\varphi, \mathcal{S}\varphi)$, and $u = \mathcal{S}\varphi$ is the solution of the Dirichlet problem $Lu = 0$ in Ω, $u = V\varphi$ on Γ, hence $\Phi(\mathcal{S}\varphi, \mathcal{S}\varphi)$ is bounded by $\|V\varphi\|^2_{H^{1/2}(\Gamma)}$. Now V is continuous from $H^{-1/2}(\Gamma)$ to $H^{1/2}(\Gamma)$, so we get an estimate by $\|\varphi\|^2_{H^{-1/2}(\Gamma)}$. That this in turn can be estimated by $a(\varphi, \varphi)$ is an a-priori estimate for the solution of the exterior Neumann problem which follows from its variational formulation.

In this way we obtain that B is a contraction. If we want to show that A is a contraction, too, or even stronger that $B - A$ is a contraction, we need the positive definiteness of B, and this is not satisfied, in general, if the space $\mathcal{R}$ of functions of vanishing energy in Ω is non-trivial. The nullspace of the form b consists of densities whose single layer potential has vanishing energy on Ω:

$$b(\psi, \psi) = 0 \iff \mathcal{S}\psi \in \mathcal{R} \iff V\psi \in \gamma\mathcal{R} .$$

To make B positive definite, we have to factor this kernel out, which is done by the definition [31]

$$\begin{aligned}
H_0^{-1/2}(\Gamma) &= \{\varphi \in H^{-1/2}(\Gamma) \mid \forall \psi \in \ker B : (\varphi, \psi)_V = 0\} \\
&= \{\varphi \in H^{-1/2}(\Gamma) \mid \forall u \in \mathcal{R} : \langle \varphi, \gamma u \rangle = 0\}
\end{aligned} \tag{20}$$

Equivalently, we could have passed to the quotient space $H^{-1/2}(\Gamma)/\gamma\mathcal{R}$. In any case, we then find that B is positive definite, which by Lemma 2 implies that both $B - A$ and A are contractions. We also note that since A and B commute, $\ker B$ and its orthogonal complement are invariant subspaces of A. We summarize these results:

Theorem 1. *The operators* $A = \frac{1}{2}I - K'$ *and* $B = \frac{1}{2}I + K'$ *are positive semidefinite bounded selfadjoint operators on the Hilbert space* $H^{-1/2}(\Gamma)$ *equipped with the inner product* $(\cdot, \cdot)_V$. *The operator* $\frac{1}{2}I - K'$ *is positive definite, and the operator* $\frac{1}{2}I + K'$ *is a contraction. The Neumann series*

$$(\frac{1}{2}I - K')^{-1} = \sum_{\ell=0}^{\infty} (\frac{1}{2}I + K')^{\ell}$$

converges in $H^{-1/2}(\Gamma)$ *in the operator norm associated with the norm* $\|\cdot\|_V$. *On the subspace* $H_0^{-1/2}(\Gamma)$, *the operator* $\frac{1}{2}I + K'$ *is positive definite, and the operators* $\frac{1}{2}I - K'$ *and* $B - A = 2K'$ *are contractions. On this subspace, there are the convergent Neumann series:*

$$\left(\frac{1}{2}I - K'\right)^{-1} = 2\sum_{\ell=0}^{\infty}(2K')^{\ell}$$

$$\left(\frac{1}{2}I + K'\right)^{-1} = \sum_{\ell=0}^{\infty}\left(\frac{1}{2}I - K'\right)^{\ell}$$

$$\left(\frac{1}{2}I + K'\right)^{-1} = 2\sum_{\ell=0}^{\infty}(-2K')^{\ell}$$

4.2 Double Layer Potentials

As a second possibility, we now look at double layer potentials. In order to have finite energy, we have to take a space of more regular functions for our classical departure space X_0, Hölder continuous functions for example. For $\varphi, \psi \in X_0$, we now define the bilinear forms a and b as energy forms of the corresponding double layer potentials:

$$a(\varphi, \psi) = \Phi^c(\mathcal{D}\varphi, \mathcal{D}\psi) \; ; \qquad b(\varphi, \psi) = \Phi(\mathcal{D}\varphi, \mathcal{D}\psi) \; . \tag{21}$$

With the boundary reduction by Green's formula (15) and the expressions (14) for the traces of the double layer potential, we find the boundary integral forms

$$a(\varphi, \psi) = \langle W\varphi, (\frac{1}{2}I + K)\psi\rangle \; ; \qquad b(\varphi, \psi) = \langle W\varphi, (\frac{1}{2}I - K)\psi\rangle \; . \tag{22}$$

The total energy $a+b$ is now given by the bilinear form defined by the operator W of the conormal derivative of the double layer potential. It is easy to see that the nullspace of W is given by the traces of the zero-energy fields $\mathcal{R}$. Densities in $\gamma\mathcal{R}$ generate double layer potentials that are identically zero in the exterior domain Ω^c and belong to $\mathcal{R}$ in Ω. In order to be able to apply our program, to get a positive definite bilinear form a and hence Hilbert space X, we have to factor these densities out from the beginning. Our Hilbert space is therefore a quotient space

$$X = H^{1/2}(\Gamma)/\gamma\mathcal{R} \text{ with norm } \|\varphi\|_W^2 = \langle W\varphi, \varphi\rangle \; . \tag{23}$$

This is the natural dual space of $H_0^{-1/2}(\Gamma)$ with respect to $L^2(\Gamma)$ duality. We know from the variational solution of the Dirichlet problem $Lu = 0$ in Ω or Ω^c, $\gamma u = \varphi$ or $\gamma^c u = \varphi$, that on this space the square of the (quotient) norm is equivalent to each one of the energy forms $\Phi(u, u)$ and $\Phi^c(u, u)$. Thus both quadratic forms a and b can be mutually estimated, and we get the full result of Lemmas 3 and 2.

It remains to identify the operators A and B. We have for all $\varphi, \psi \in X$:

$$\langle W\varphi, (\frac{1}{2}I + K)\psi\rangle = (\varphi, A\psi)_W = \langle W\varphi, A\psi\rangle$$

and similarly for B. This shows that if $\pi_{\mathcal{R}} : H^{1/2}(\Gamma) \to X$ is the canonical projection on the quotient space, we have

$$A = \pi_{\mathcal{R}}\left(\frac{1}{2}I + K\right) ; \qquad B = \pi_{\mathcal{R}}\left(\frac{1}{2}I - K\right) .$$

In the case of the operator A, we can omit the extra factor $\pi_{\mathcal{R}}$, because $\ker(\frac{1}{2}I + K) = \gamma\mathcal{R}$, and therefore $\frac{1}{2}I + K$ is well-defined on the quotient space and commutes with the projector. This remark does not apply in the same way to the operator B, but since $\frac{1}{2}I - K$ commutes with $\frac{1}{2}I + K$, the kernel $\gamma\mathcal{R}$ of the latter is an invariant subspace of the former, so that $\frac{1}{2}I - K$ is also defined in a natural way on the quotient space. The operator $\frac{1}{2}I - K$ actually acts as the identity on the subspace $\gamma\mathcal{R}$, so that its inverse on the whole space $H^{1/2}(\Gamma)$ can be obtained from the inverse on the quotient space. Altogether, we can simply write without ambiguity

$$A = \frac{1}{2}I + K ; \qquad B = \frac{1}{2}I - K . \tag{24}$$

We can now summarize the conclusion of Lemma 2 in this case:

Theorem 2. *The operators $A = \frac{1}{2}I + K$ and $B = \frac{1}{2}I - K$ are positive definite bounded selfadjoint operators on the quotient space $H^{1/2}(\Gamma)/\gamma\mathcal{R}$ equipped with the inner product $(\cdot, \cdot)_W$. Both operators, as well as the operator $B - A = -2K$ (Neumann's operator) are contractions in the corresponding operator norm. The Neumann series*

$$\left(\frac{1}{2}I - K\right)^{-1} = \sum_{\ell=0}^{\infty}\left(\frac{1}{2}I + K\right)^{\ell}$$

$$\left(\frac{1}{2}I - K\right)^{-1} = 2\sum_{\ell=0}^{\infty}(2K)^{\ell}$$

$$\left(\frac{1}{2}I + K\right)^{-1} = \sum_{\ell=0}^{\infty}\left(\frac{1}{2}I - K\right)^{\ell}$$

$$\left(\frac{1}{2}I + K\right)^{-1} = 2\sum_{\ell=0}^{\infty}(-2K)^{\ell}$$

all converge in the operator norm in the quotient space, which corresponds to convergence in $H^{1/2}(\Gamma)$ modulo the traces $\gamma\mathcal{R}$ of the zero-energy fields in Ω. The first Neumann series for the operator $(\frac{1}{2}I - K)^{-1}$ converges in the whole Sobolev space $H^{1/2}(\Gamma)$.

4.3 Single Layer Potentials via Dirichlet Data

The bijectivity of the single layer integral operator V offers another possible interpretation of the results of Section 4.1: Instead of representing a single

24   M. Costabel

layer potential $v = \mathcal{S}\psi$ by its density ψ, one can represent it by its Dirichlet trace $\gamma v = V\psi$. Since $V : H^{-1/2}(\Gamma) \to H^{1/2}(\Gamma)$ is bijective, it can be used to transport the Hilbert space structure on $H^{-1/2}(\Gamma)$ which we considered before to $H^{1/2}(\Gamma)$. From the relation

$$(\varphi, \psi)_V = \langle \varphi, V\psi \rangle = \langle V^{-1}V\varphi, V\psi \rangle$$

we see that if we define the inner product on $H^{1/2}(\Gamma)$ by

$$(u, v)_{V^{-1}} = \langle V^{-1}u, v \rangle \,,$$

then $V : H^{-1/2}(\Gamma) \to H^{1/2}(\Gamma)$ becomes an isometry. Instead of writing our whole program once again with a new space X, we can simply transport all the results of Section 4.1 via this Hilbert space isomorphism. Positivity, operator norms and convergence of Neumann series are conserved, the only question that has to be settled is the form of the operators A and B in this new representation.

The answer to this question is provided by the well-known relation

$$KV = VK'$$

which is one of the relations that give the projection property of the Calderón projector, obtained from the representation of a single layer potential as a sum of a single layer potential and a double layer potential of its own Cauchy data.

The operator $A = \frac{1}{2}I - K'$ on $H^{-1/2}(\Gamma)$ is therefore transported to the operator $VAV^{-1} = V(\frac{1}{2}I - K')V^{-1} = \frac{1}{2}I - K$, and $B = \frac{1}{2}I + K'$ is transported to the operator $\frac{1}{2}I + K$. In this way, we can transport all of Theorem 1. In particular, $\frac{1}{2}I + K$ is a contraction on $H^{1/2}(\Gamma)$ equipped with the norm $\|\cdot\|_{V^{-1}}$. For the other results we have to transport the subspace $H_0^{-1/2}(\Gamma)$. We find

$$VH_0^{-1/2}(\Gamma) = \{\varphi \in H^{1/2}(\Gamma) \mid \forall u \in \mathcal{R} : (\varphi, \gamma u)_{V^{-1}} = 0\}$$

On this space, the operator $\frac{1}{2}I - K$ and Neumann's operator $-2K$ are contractions.

Thus we get similar results as in Section 4.2, with a different norm on $H^{1/2}(\Gamma)$. The results in this form (except for the operator $-2K$) were first proved by Steinbach and Wendland in [31].

4.4 Final Remarks

Although our results obtained here from Poincaré's estimates are largely similar to the results of Steinbach and Wendland in [31], their method for proving the contraction property of $\frac{1}{2}I \pm K$ and $\frac{1}{2}I \pm K'$ is different:

The simple idea here was that if two positive numbers add up to 1, then both of them must be smaller than 1; with Lemma 2 as a transposition of this idea to the class of selfadjoint operators on a Hilbert space.

The corresponding simple idea in [31] is that if a number is bigger than its square, then it must lie between 0 and 1. For operators, this idea can be stated as follows:

Let A and B be bounded selfadjoint operators on a Hilbert space. If

$$B = B^2 + A \text{ and } A \text{ is positive definite, } A \geq \alpha I,$$

then B is a contraction with norm $\|B\| \leq \frac{1}{2} + \sqrt{\frac{1}{4} - \alpha}$.

This lemma can be applied to the well-known relations

$$(\frac{1}{2}I + K)(\frac{1}{2}I - K) = VW \ ; \qquad (\frac{1}{2}I + K')(\frac{1}{2}I - K') = WV$$

which are a consequence of the symmetry of the energy form $\Phi(u, v)$ between a double layer potential u and a single layer potential v, or also of the projection property of the Calderón projector. Since WV is positive semi-definite in the inner product $(\cdot, \cdot)_V$ and VW is positive definite in the inner product $(\cdot, \cdot)_W$ and positive semi-definite in the inner product $(\cdot, \cdot)_{V^{-1}}$, the respective contraction properties for $\frac{1}{2}I \pm K$ and $\frac{1}{2}I \pm K'$ follow.

References

1. H. Brezis: The interplay between analysis and topology in some nonlinear PDE problems. Bull. AMS 40(2) (2003) 179–201.
2. H. Cartan: Sur les fondements de la théorie du potentiel. Bull. S.M.F. 69 (1941) 71–96.
3. H. Cartan: Théorie du potentiel newtonien: énergie, capacité, suite de potentiels. Bull. S.M.F. 73 (1945) 74–106.
4. M. Costabel: Boundary integral operators on Lipschitz domains: Elementary results. SIAM J. Math. Anal. 19 (1988) 613–626.
5. M. Costabel: Boundary integral operators for the heat equation. Integral Equations Oper. Theory 13 (1990) 498–552.
6. M. Costabel: Zum wissenschaftlichen Werk Wolfgang L. Wendlands. Vortrag, Universität Stuttgart, 22. Oktober 2004.
7. M. Costabel, W. L. Wendland: Strong ellipticity of boundary integral operators. J. Reine Angew. Math. 372 (1986) 39–63.
8. O. Frostman: Potentiel d'équilibre et capacité des ensembles. Thesis, University of Lund, 1935.
9. C. F. Gauss: Werke. Dieterich, Göttingen, 1863.
   URL: http://www-gdz.sub.uni-goettingen.de/cgi-bin/digbib.cgi?PPN235957348
10. C. F. Gauss: Allgemeine Theorie des Erdmagnetismus (1838). Werke 5 (1867) 127–193.
11. C. F. Gauss: Allgemeine Lehrsätze in Beziehung auf die im verkehrten Verhältnisse des Quadrats der Entfernung wirkenden Anziehungs- und Abstossungs-Kräfte (1839). Werke 5 (1867) 197–244.
12. C. F. Gauss: Atlas des Erdmagnetismus 1840. Werke 12 (1929) 326–408.
13. C. F. Gauss: Letter to Hansen 11. December 1825. Werke 12 (1929) 6–9.

14. D. Hilbert: Über das Dirichletsche Prinzip. Jahresbericht Deut. Math.-Ver. 8 (1900) 184–188.
15. D. Hilbert: Über das Dirichletsche Prinzip. Math. Ann. 59 (1904) 161–184.
16. D. Hilbert: Über das Dirichletsche Prinzip. J. Reine Angew. Math. 129 (1905) 63–67.
17. E. Hille: Analytic function theory. Vol. II. Introductions to Higher Mathematics. Ginn and Co., Boston, Mass.-New York-Toronto, Ont., 1962.
18. D. Khavinson, M. Putinar, H. Shapiro: On Poincaré's variational problem in potential theory. Preprint, UCSB 2006.
19. A. Korn: Lehrbuch der Potentialtheorie. Allgemeine Theorie des Potentials und der Potentialfunctionen im Raume. Ferd. Dümmler, Berlin, 1899.
20. A. Korn: Sur la méthode de Neumann et le problème de Dirichlet. C. R. 130 (1900) 557.
21. A. Korn: Lehrbuch der Potentialtheorie. II. Allgemeine Theorie des logarithmischen Potentials und der Potentialfunctionen in der Ebene. Ferd. Dümmler, Berlin, 1901.
22. A. Korn: Sur les équations de l'élasticité. Ann. Sci. Éc. Norm. Sup. (3) 24 (1907) 9–75.
23. A. Korn: Sur certaines questions qui se rattachent au problème des efforts dans la thórie de l'élasticité. Ann. Fac. Sci.Toulouse (3) 2 (1910) 7–18.
24. J. Král: Integral operators in potential theory. Lecture Notes in Mathematics, Vol. 823. Springer, Berlin, 1980.
25. V. G. Maz'ya: Boundary integral equations. In: Analysis, IV. Encyclopaedia Math. Sci., Vol. 27, pp. 127–222. Springer, Berlin, 1991.
26. J.-C. Nedelec, J. Planchard: Une méthode variationnelle d'éléments finis pour la résolution numérique d'un problème extérieur dans $\mathbb{R}^3$. RAIRO 7 (1973) 105–129.
27. C. Neumann: Untersuchungen über das Logarithmische und Newtonsche Potential. B. G. Teubner, Leipzig, 1877.
28. H. Poincaré: Sur les équations aux dérivées partielles de la physique mathématique. Amer. J. Math. 12 (1890) 211–294.
29. H. Poincaré: La méthode de Neumann et le problème de Dirichlet. Acta Math. 20 (1896) 59–142.
30. H. Poincaré: Sur l'équilibre d'un corps élastique. C. R. A. S. 122 (1896) 154–159.
31. O. Steinbach, W. L. Wendland: On C. Neumann's method for second-order elliptic systems in domains with non-smooth boundaries. J. Math. Anal. Appl. 262 (2001) 733–748.
32. W. Stekloff: Les méthodes générales pour résoudre les problèmes fondamentaux de la physique mathématique. Ann Fac. Sci. Toulouse (2) 2 (1900) 207–272.
33. W. Stekloff: Mémoire sur les fonctions harmoniques de M. H. Poincaré. Ann Fac. Sci. Toulouse (2) 2 (1900) 273–303.
34. W. Stekloff: Sur la méthode de Neumann et le problème de Dirichlet. C. R. 130 (1900) 396–399.
35. W. Stekloff: Remarque à une note de M. A. Korn: "Sur la méthode de Neumann et le problème de Dirichlet.". C. R. 130 (1900) 826–827.
36. W. Stekloff: Sur les problèmes fondamentaux de la physique mathématique. Ann de l'Éc. Norm. (3) 19 (1902) 191–259.
37. W. Stekloff: Sur les problèmes fondamentaux de la physique mathématique (suite et fin). Ann de l'Éc. Norm. (3) 19 (1902) 455–490.

38. W. Stekloff: Théorie générale des fonctions fondamentales. Ann Fac. Sci. Toulouse (2) 6 (1904) 351–475.
39. G. Szegö: Bemerkungen zu einer Arbeit von Herrn M. Fekete: Über die Verteilung der Wurzeln bei gewissen algebraischen Gleichungen mit ganzzahligen Koeffizienten. Math. Z. 21(1) (1924) 203–208.
40. H. Weyl: The method of orthogonal projection in potential theory. Duke Math. J. 7 (1940) 411–440.
41. S. Zaremba: Sur le problème de Dirichlet. Ann. Sci. Éc. Norm. Sup. (3) 14 (1897) 251–258.
42. S. Zaremba: Sur l'équation aux dérivées partielles $\Delta u + \xi u + f = 0$ et sur les fonctions harmoniques. Ann. Sci. Éc. Norm. Sup. (3) 16 (1899) 427–464.
43. S. Zaremba: Contribution à la théorie des fonctions fondamentales. Ann. Sci. Éc. Norm. Sup. (3) 20 (1903) 9–26.

Averaging Techniques for a Posteriori Error Control in Finite Element and Boundary Element Analysis

Carsten Carstensen[1] and Dirk Praetorius[2]

[1] Institut für Mathematik, Humboldt–Universität zu Berlin,
Unter den Linden 6, 10099 Berlin, Germany
cc@math.hu-berlin.de[*]

[2] Institut für Analysis und Wissenschaftliches Rechnen, Technische Universität
Wien, Wiedner Hauptstraße 8–10, 1040 Wien, Austria
Dirk.Praetorius@tuwien.ac.at

Summary. Averaging techniques for a posteriori error control are established for differential and integral equations within a unifying setting. The reliability and efficiency of the introduced estimator results from two grids $\mathcal{T}_h$ and $\mathcal{T}_H$ with different polynomial degrees for a smooth exact solution. The proofs are based on first order approximation operators and inverse estimates. For a finer and finer fine mesh $\mathcal{T}_h$, the estimator becomes asymptotically exact. The abstract framework is applicable to a finite element method for the Laplace equation, boundary element methods for Symm's and the hypersingular integral equation or transmission problems.

1 Introduction

The striking simplicity of averaging techniques in a posteriori error control as well as their amazing accuracy in many numerical examples have made them an extremely popular tool in scientific computing over the last decade. Given a discrete stress or flux p_h and a post-processed (smoothened) approximation $\mathcal{A}p_h$, the a posteriori error estimator reads

$$\eta_A := \|p_h - \mathcal{A}p_h\|.$$

There is not even a need for an equation to compute the estimator η_A and hence averaging techniques are easily employed everywhere. The most prominent example is occasionally named after Zienkiewicz and Zhu [36], and also called *gradient recovery* but preferably called *averaging technique* in the literature. The most frequently quoted paper is [36] for the $P1$ finite element

[*]Supported by the DFG Research Center MATHEON "Mathematics for key technologies" in Berlin.

method for some Laplace equation on some domain ω and some local averaging operator $\mathcal{A}p_h$ on the piecewise constant gradients $p_h = Du_h$ followed by linear interpolation. The estimator $\eta_A = \|p_h - \mathcal{A}p_h\|$ is then computed with respect to the norm $\|\cdot\|$ on $L^2(\Omega)$.

In the work of Zienkiewicz and Zhu [36], there was no rigorous justification to interpret η_A as some computable approximation of the (rigorous) exact error $\|p - p_h\|$ with $p = Du$, but there arose quite some numerical evidence for that.

The first mathematical justification of the error estimator η_A as a computable approximation of the (unknown) error $\|p - p_h\|$ involved the concept of superconvergence points. For highly structured meshes and a very smooth exact solution p, the error $\|p - \mathcal{A}p_h\|$ of the post-processed approximation $\mathcal{A}p_h$ may be (much) smaller than $\|p - p_h\|$ of the given p_h. Under the assumption that $\|p - \mathcal{A}p_h\|$ is sufficiently small in relative terms, written $\|p - \mathcal{A}p_h\| = $ h.o.t $=$ higher-order terms, the triangle inequality immediately verifies reliability, i.e.,

$$\|p - p_h\| \leq C_{\mathrm{rel}}\,\eta_A + \text{ h.o.t.,}$$

and efficiency, i.e.,

$$\eta_A \leq C_{\mathrm{eff}}\,\|p - p_h\| + \text{ h.o.t.,}$$

of the averaging error estimator η_A (even with $C_{\mathrm{rel}} = C_{\mathrm{eff}} = 1$). However, the required assumptions on the symmetry of the mesh and the smoothness of the solution essentially contradict the use of adaptive grid refinement when p is singular. Moreover, the proper treatment of boundary conditions remains unclear.

The first mathematical verification by Rodriguez on reliability of η_A on unstructured grids has been indicated in the literature [6, 25, 26, 27] but was not mentioned in the (otherwise comprehensive) works [1, 2, 20, 33]. The first author was unaware of Rodriguez's result [27] when he started to work on the mathematical justification [17] that ended in the surprizing and new conclusion that, in fact, all averaging techniques are reliable [4, 5, 7, 8, 9, 10, 11].

A corresponding technique for the boundary element method was initiated with extraction and recovery techniques in [29, 30, 31, 32, 34] and was proposed thereafter in a small series of works of the two authors [12, 13] and in [21]. In the latter works, an approximation $\mathcal{A}p_h$ is computed as some best approximation of p_h based on a higher-order spline space on some coarser mesh. For some smooth exact solution, the resulting approximation error is of higher order. The corresponding error estimator is therefore efficient. Reliability follows provided the quotient of the mesh-sizes is sufficiently small. These two arguments, called approximation assumption (AA) and discrete property (DP), allow a unified analysis of reliability and efficiency of η_A.

This paper links the two discretization methods, namely the finite element method and the boundary element method, in that there is one abstract setting provided in which an averaging scheme is seen to be reliable and efficient without any reference to some saturation assumption or superconvergence. The paper is roughly organized in two mayor parts: In Section 2–4, we provide and analyze the analytical setting for our averaging method, while the remaining Sections 5–8 of the paper discuss concrete applications. Namely, in Section 2 we state and prove our abstract main result in Theorem 1, which is commented in Section 3. The essential condition for Theorem 1 is a discrete property (DP). We stress the difference of (DP) and a saturation assumption and remark on further generalizations of Theorem 1. In Section 4, the essential condition is studied in detail and characterized as some strengthened Cauchy inequality of related spaces. Section 5 considers the introduced averaging technique for the finite element method for a model example. Section 6 is an overview of a recent work [12] on averaging for Symm's integral equation. In Section 7, we treat the hypersingular integral equation following [13, 21]. Finally, the last application of our abstract analysis concerns the boundary integral formulation of a transmission problem in Section 8.

2 Abstract Setting

We consider the abstract framework of the Lax-Milgram lemma with a finite dimensional subspace $\mathcal{S}_h$ of a real Hilbert space $\mathcal{H}$ with corresponding norm $\|\cdot\|_{\mathcal{H}}$. Let $\langle\cdot,\cdot\rangle$ be an elliptic and bounded (but possibly non-symmetric) bilinear form on $\mathcal{H}$, i.e., there are constants $0 < C_{\text{ell}} \le C_{\text{bd}}$ such that

$$C_{\text{ell}}\|u\|_{\mathcal{H}}^2 \le \langle u,u\rangle \quad \text{and} \quad \langle u,v\rangle \le C_{\text{bd}}\|u\|_{\mathcal{H}}\|v\|_{\mathcal{H}} \quad \text{for all } u,v \in \mathcal{H}. \tag{1}$$

The (linear) Galerkin projection $\mathbb{G}_h : \mathcal{H} \to \mathcal{S}_h$ is characterized by the Galerkin orthogonality

$$\langle v - \mathbb{G}_h v, v_h\rangle = 0 \quad \text{for all } v_h \in \mathcal{S}_h \text{ and } v \in \mathcal{H}. \tag{2}$$

An immediate consequence is the quasi-optimal convergence, also known as Céa's lemma:

$$\|v - \mathbb{G}_h v\|_{\mathcal{H}} \le (C_{\text{bd}}/C_{\text{ell}}) \min_{v_h \in \mathcal{S}_h} \|v - v_h\|_{\mathcal{H}} \quad \text{for all } v \in \mathcal{H}. \tag{3}$$

Given an unknown solution $u \in \mathcal{H}$ for a prescribed right-hand side $f = \langle u,\cdot\rangle \in \mathcal{H}^*$, the discrete solution $u_h := \mathbb{G}_h u$ is computed. In order to approximate the energy norm of the (unknown) error

$$e := u - u_h, \tag{4}$$

we are given a second finite-dimensional subspace $\mathcal{S}_H$ of $\mathcal{H}$. Then, the a posteriori error estimator for $\|u - u_h\|_{\mathcal{H}}$ reads

$$\eta_M := \min_{v_H \in \mathcal{S}_H} \|u_h - v_H\|_{\mathcal{H}}. \tag{5}$$

The justification below is based on one approximation assumption (AA) and some discrete property (DP) of $\mathcal{S}_h$ and $\mathcal{S}_H$ where, in applications below, $\mathcal{S}_h$ corresponds to a lower polynomial degree ansatz but a finer mesh when compared to $\mathcal{S}_H$, and u is smooth. Moreover, as the triangulation $\mathcal{T}_h$ corresponding to $\mathcal{S}_h$ will be a uniform refinement of the triangulation $\mathcal{T}_H$ corresponding to $\mathcal{S}_H$, we assume that $\mathcal{S}_h$ and $\mathcal{S}_H$ are linked through the mesh-sizes h and H:

$$\delta_{hH} := \min_{v_H \in \mathcal{S}_H} \|u - v_H\|_{\mathcal{H}} / \min_{v_h \in \mathcal{S}_h} \|u - v_h\|_{\mathcal{H}} = o(1), \tag{AA}$$

$$q := \max_{v_H \in \mathcal{S}_H \setminus \{0\}} \min_{v_h \in \mathcal{S}_h} \frac{\|v_H - v_h\|_{\mathcal{H}}}{\|v_H\|_{\mathcal{H}}} < C_{\mathrm{ell}}/C_{\mathrm{bd}}. \tag{DP}$$

Theorem 1. *With the notation from* (AA) *and under assumption* (DP) *there holds*

$$\eta_M/(1 + \delta_{hH}) \le \|e\|_{\mathcal{H}} \le C_{\mathrm{rel}}(\eta_M + \min_{v_H \in \mathcal{S}_H} \|u - v_H\|_{\mathcal{H}}) \tag{6}$$

with

$$C_{\mathrm{rel}} := C_{\mathrm{bd}}/(C_{\mathrm{ell}} - qC_{\mathrm{bd}}). \tag{7}$$

Proof. The lower estimate (efficiency of η_M) is an immediate consequence of the triangle inequality: For any $v_H \in \mathcal{S}_H$, there holds

$$\eta_M \le \|e\|_{\mathcal{H}} + \|u - v_H\|_{\mathcal{H}}.$$

A passage of v_H to the minimum in (AA) yields

$$\eta_M \le \|e\|_{\mathcal{H}} + \delta_{hH} \min_{v_h \in \mathcal{S}_h} \|u - v_h\|_{\mathcal{H}} \le \|e\|_{\mathcal{H}}(1 + \delta_{hH}).$$

This establishes efficiency of η_M. To prove the reliability of η_M, let $e_H \in \mathcal{S}_H$ be the best approximation of e, i.e.

$$\|e - e_H\|_{\mathcal{H}} = \min_{v_H \in \mathcal{S}_H} \|e - v_H\|. \tag{8}$$

By the definition of q in the discrete property (DP), there holds

$$\min_{v_h \in \mathcal{S}_h} \|e_H - v_h\|_{\mathcal{H}} \le q\|e_H\|_{\mathcal{H}}.$$

The Galerkin orthogonality of $\mathbb{G}_h$ and the boundedness of the bilinear form $\langle \cdot, \cdot \rangle$ followed by the aforementioned estimate lead to

$$\langle e, e_H \rangle = \min_{v_h \in \mathcal{S}_h} \langle e, e_H - v_h \rangle \le q\, C_{\mathrm{bd}} \|e\|_{\mathcal{H}} \|e_H\|_{\mathcal{H}}.$$

Combining this with the ellipticity and boundedness of $\langle \cdot, \cdot \rangle$, we obtain

$$C_{\mathrm{ell}}\|e\|_{\mathcal{H}}^2 \le \langle e, e \rangle = \langle e, e - e_H \rangle + \langle e, e_H \rangle \le C_{\mathrm{bd}}\|e\|_{\mathcal{H}}(\|e - e_H\|_{\mathcal{H}} + q\|e_H\|_{\mathcal{H}}).$$

Now, the stability estimate $\|e_H\|_\mathcal{H} \le \|e\|_\mathcal{H}$ proves

$$\|e\|_\mathcal{H} \le \frac{C_{\mathrm{ell}}^{-1} C_{\mathrm{bd}}}{1 - qC_{\mathrm{ell}}^{-1} C_{\mathrm{bd}}} \|e - e_H\|_\mathcal{H} = C_{\mathrm{rel}} \min_{v_H \in \mathcal{S}_H} \|e - v_H\|_\mathcal{H}.$$

If u_H and u_{hH} denote the best approximations of u resp. u_h in $\mathcal{S}_H$, the special choice of $v_H = u_H - u_{hH}$ and a triangle inequality yield

$$\|e\|_\mathcal{H} \le C_{\mathrm{rel}}(\|u - u_H\|_\mathcal{H} + \|u_{hH} - u_h\|_\mathcal{H}) = C_{\mathrm{rel}}\Big(\min_{v_H \in \mathcal{S}_H} \|u - v_H\|_\mathcal{H} + \eta_M\Big).$$

This concludes the proof of the reliability. $\square$

3 Comments

Some remarks are in order before a list of applications enlightens the abstract results of the preceeding chapter.

3.1 Efficiency and Reliability

The discrete property (DP) is *not* necessary for efficiency of η_M. The reliability depends essentially on the discrete property (DP) in that, up to some approximation error

$$\mathrm{h.o.t.} := \min_{v_H \in \mathcal{S}_H} \|u - v_H\|_\mathcal{H},$$

there holds reliability in the sense of

$$\|e\|_\mathcal{H} \le C_{\mathrm{rel}}(\eta_M + \mathrm{h.o.t.}).$$

However, this is reasonable only if $\mathrm{h.o.t.} \sim \delta_{hH}\|e\|_\mathcal{H}$ is indeed of higher order. In fact, there holds

$$\|e\|_\mathcal{H} \le C_{\mathrm{rel}}(\eta_M + \delta_{hH}\|e\|_\mathcal{H}).$$

Then, for $\delta_{hH} < C_{\mathrm{rel}}^{-1}$, there holds

$$\|e\|_\mathcal{H} \le C_{\mathrm{rel}}/(1 - \delta_{hH}C_{\mathrm{rel}})\,\eta_M.$$

3.2 Constants in the Symmetric Case

In the important case that the bilinear form $\langle \cdot\,,\cdot \rangle$ is symmetric, it is a scalar product. The induced norm $\|v\| := \langle v\,, v \rangle^{1/2}$ is an equivalent Hilbert norm on $\mathcal{H}$. Moreover, $\mathbb{G}_h$ is the orthogonal projection onto $\mathcal{S}_h$ with respect to $\langle \cdot\,,\cdot \rangle$. Then, (3) holds with $(C_{\mathrm{bd}}/C_{\mathrm{ell}})^{1/2}$ replacing $C_{\mathrm{bd}}/C_{\mathrm{ell}}$, and $\mathbb{G}_h$ is characterized by the best approximation property $\|v - \mathbb{G}_h v\| = \min_{v_h \in \mathcal{S}_h} \|v - v_h\|$ for all $v \in \mathcal{H}$.

In the symmetric case, one usually states (6) with respect to the energy norm $\|\cdot\|_{\mathcal{H}} = \|\cdot\|$, i.e. $C_{\mathrm{bd}} = 1 = C_{\mathrm{ell}}$. Asymptotic exactness of η_M then follows for $q \to 0$ in the sense of $C_{\mathrm{rel}} \to 1$. Moreover, the reliability constant $C_{\mathrm{rel}} = 1/(1-q)$ from (7) can be improved to $C_{\mathrm{rel}} = 1/(1-q^2)^{1/2}$ by the following refined stability estimate: Using the symmetry of orthogonal projections and the same arguments as in the proof of Theorem 1, we obtain

$$\|e_H\|^2 = \langle e_H, e_H \rangle = \langle e, e_H \rangle = \min_{v_h \in \mathcal{S}_h} \langle e, e_H - v_h \rangle \leq q\|e\|\|e_H\|.$$

This implies the refined stability estimate $\|e_H\| \leq q\|e\|$. Together with the Pythagoras theorem, there holds

$$\|e\|^2 = \|e - e_H\|^2 + \|e_H\|^2 \leq \|e - e_H\|^2 + q^2\|e\|^2.$$

This yields $\|e\| \leq \|e_H\|/(1-q^2)^{1/2}$, and we obtain the reliability of η_M with the improved constant $C_{\mathrm{rel}} = 1/(1-q^2)^{1/2}$.

3.3 Remarks on the Saturation Assumption

Assumption (DP) is just a definition of δ_{hH} with the possible interpretation discussed in Section 3.1. A much stronger statement is the *saturation assumption* of the form

$$\delta_{hH} = \|u - \mathbb{G}_H u\|/\|e\| \leq C_{\mathrm{sat}} < 1 \tag{SA}$$

in the symmetric case $\|\cdot\|_{\mathcal{H}} = \|\cdot\|$ etc. of the preceding subsection. Recall that $\mathbb{G}_H$ denote the Galerkin projection onto $\mathcal{S}_H$. With $u_H := \mathbb{G}_H u$, a triangle inequality for $e = u - u_H + u_H - u_h$ plus (SA) leads to the reliable a posteriori error estimate

$$\|e\| \leq \|u_h - u_H\|/(1 - C_{\mathrm{sat}})$$

for the different hierarchical estimator $\|u_h - u_H\|$. It has been the starting point of our analysis to avoid a strong assumption on the actual size of δ_{hH} like (SA) because it is hard to check in practise.

3.4 Verification of Assumption (DP)

This subsection outlines the arguments sufficient for (DP) in an abstract (and non-local) framework. Examples follow in the remaining applications of this paper. For an appropriate seminorm $|\cdot|$ and the mesh-size parameter $H > 0$ associated with $\mathcal{S}_H$, an inverse estimate is of the form

$$|v_H| \leq c_{\mathrm{inv}} H^{-\alpha}\|v_H\|_{\mathcal{H}} \quad \text{for all } v_H \in \mathcal{S}_H.$$

The exponent $\alpha > 0$ depends only on the energy (Sobolev) space, e.g., $\mathcal{H} = H^{\alpha}$ or $\mathcal{H} = H^{-\alpha}$. Moreover, $|\cdot|$ may allow an approximation estimate of the form

$$\min_{v_h \in \mathcal{S}_h} \|v_H - v_h\|_{\mathcal{H}} \leq c_{\mathrm{apx}} h^{\alpha} |v_H| \quad \text{for all } v_H \in \mathcal{S}_H.$$

The combination of the two estimates yields

$$q := \max_{v_H \in \mathcal{S}_H \setminus \{0\}} \min_{v_h \in \mathcal{S}_h} \frac{\|v_H - v_h\|_{\mathcal{H}}}{\|v_H\|_{\mathcal{H}}} \leq c_{\mathrm{apx}} c_{\mathrm{inv}} (h/H)^{\alpha}.$$

Hence, for any mesh-size h sufficiently small relative to H, (DP) follows.

3.5 Other Averaging Techniques

Under assumptions (AA)–(DP), we obtain reliable error estimators η_A whenever we replace the minimum of the best approximation by an *arbitrary* operator $\mathcal{A}_H : \mathcal{H} \to \mathcal{S}_H$,

$$\eta_A := \|u_h - \mathcal{A}_H u_h\|_{\mathcal{H}} \geq \min_{v_H \in \mathcal{S}_H} \|u_h - v_H\|_{\mathcal{H}} =: \eta_M. \tag{9}$$

Thus, *each averaging technique yields a reliable error estimator* [4]. Clearly, the efficiency of η_A is some further property of the chosen operator $\mathcal{A}_H$. According to Céa's lemma (3), the Galerkin projection $\mathcal{A}_H = \mathbb{G}_H$ always leads to an efficient and reliable error estimator since

$$(C_{\mathrm{ell}}/C_{\mathrm{bd}}) \|v - \mathbb{G}_H v\|_{\mathcal{H}} \leq \min_{v_H \in \mathcal{S}_H} \|v - v_H\|_{\mathcal{H}} \leq \|v - \mathbb{G}_H v\|_{\mathcal{H}}.$$

3.6 Generalizations

Theorem 1 can be generalized in several ways. In the following, we give some simple examples, for which the analysis from Section 2 also works: (i) For the Hilbert space $\mathcal{H}$, there holds $e_H = u_H - u_{hH}$ for the best approximations in the proof of Theorem 1. However, the linearity of the best approximation is not needed, and the argument remains valid in the case that $\mathcal{H}$ only is a reflexive Banach space: There still holds the Lax-Milgram lemma, and the best approximation problem (8) still allows for a (in general non-unique) solution e_H. Finally, a triangle inequality proves stability $\|e_H\|_{\mathcal{H}} \leq 2\|e\|_{\mathcal{H}}$. We must therefore assume $2q C_{\mathrm{ell}}^{-1} C_{\mathrm{bd}} < 1$ in (DP) and are led to reliability with $C_{\mathrm{rel}} = 2C_{\mathrm{bd}}/(C_{\mathrm{ell}} - 2q C_{\mathrm{bd}})$.

(ii) Theorem 1 also holds when we consider weakly non-linear problems. More precisely, let $A : \mathcal{H} \to \mathcal{H}^*$ be a uniformly monotone and Lipschitz continuous operator on the Hilbert space $\mathcal{H}$, i.e. there holds, for all $u, v \in \mathcal{H}$,

$$C_{\mathrm{ell}} \|u - v\|_{\mathcal{H}}^2 \leq \langle Au - Av, u - v \rangle_{\mathcal{H}^* \times \mathcal{H}} \quad \text{and} \quad \|Au - Av\|_{\mathcal{H}^*} \leq C_{\mathrm{bd}} \|u - v\|_{\mathcal{H}},$$

where $\langle \cdot, \cdot \rangle_{\mathcal{H}^* \times \mathcal{H}}$ denote the duality brackets. Also in this context, there holds the Lax-Milgram lemma. The (nonlinear) Galerkin projection $\mathbb{G}_h : \mathcal{H} \to \mathcal{S}_h$ is characterized by the Galerkin orthogonality

$$\langle Av - A(\mathbb{G}_h v), v_h \rangle_{\mathcal{H}^* \times \mathcal{H}} = 0 \quad \text{for all } v_h \in \mathcal{S}_h \text{ and } v \in \mathcal{H}.$$

There still holds Céa's lemma (3), and we prove Theorem 1 with the same techniques.

(iii) A generalization of our averaging method in the context of the FEM-BEM coupling and saddle point problems which allow an LBB condition is slightly more involved and shall therefore appear elsewhere [14].

4 Characterizations of Discrete Property (DP) in Hilbert Spaces

In this section, let V and W be closed subspaces of the real Hilbert space $\mathcal{H}$ and let $V^\perp$ denote the orthogonal complement of V,

$$V^\perp := \{ x \in \mathcal{H} : \forall v \in V \quad \langle x, v \rangle_{\mathcal{H}} = 0 \}.$$

The main focus is on the uniform estimate

$$\min_{v \in V} \|v - w\|_{\mathcal{H}} \leq c \|w\|_{\mathcal{H}} \quad \text{for all } w \in W. \tag{10}$$

Obviously, there holds $c \leq 1$, and we discuss the case of $c < 1$ in the following. This plus the optimal constant is characterized in Theorem 2 in terms of

$$\gamma_{V^\perp, W} := \sup_{v^\perp \in V^\perp \setminus \{0\}} \sup_{w \in W \setminus \{0\}} \frac{\langle v^\perp, w \rangle_{\mathcal{H}}}{\|v^\perp\|_{\mathcal{H}} \|w\|_{\mathcal{H}}}$$

and

$$q_{V,W} := \sup_{w \in W \setminus \{0\}} \min_{v \in V} \frac{\|v - w\|_{\mathcal{H}}}{\|w\|_{\mathcal{H}}}.$$

Notice that $q_{\mathcal{S}_H, \mathcal{S}_h}$ is called q in the discrete property (DP) of Section 2. The estimate $\gamma_{V^\perp, W} < 1$ is known as *strengthened Cauchy inequality* between $V^\perp$ and W. (In fact $0 \leq \cos(\sphericalangle(V^\perp, W)) := \gamma_{V^\perp, W} \leq 1$ defines the angle $\sphericalangle(V^\perp, W)$ between the spaces $V^\perp$ and W.)

The following result, which is essentially taken from [3], states that the optimal constant in (10) equals $c = q_{V,W} = \gamma_{V^\perp, W}$ and the estimates (ii)-(iv) are in fact equivalent characterizations of $c < 1$.

Theorem 2. *There holds $q_{V,W} = \gamma_{V^\perp, W} \leq 1$, and for any constant $c \geq 0$ with $c < 1$ the assertions* (i), (ii), (iii), (iv) *are pairwise equivalent.*

(i) $\gamma_{V^\perp, W} = q_{V,W} \leq c$,

(ii) *there holds* $\sqrt{1 - c^2}\, \|v^\perp\|_{\mathcal{H}} \leq \min_{w \in W} \|v^\perp - w\|_{\mathcal{H}}$ *for all* $v^\perp \in V^\perp$,

(iii) *there holds* $\sqrt{(1 - c^2)/2}\, (\|v^\perp\|_{\mathcal{H}} + \|w\|_{\mathcal{H}}) \leq \|v^\perp + w\|_{\mathcal{H}}$ *for all* $(v^\perp, w) \in V^\perp \times W$,

(iv) *there holds* $\min_{v \in V} \|v - w\|_{\mathcal{H}} \leq c \|w\|_{\mathcal{H}}$ *for all* $w \in W$.

Proof. The equivalence of $\gamma_{V^\perp,W} \leq c < 1$ with (ii) and (iii), respectively, can be found in [3, Lemma 3.1], where V is substituted by $V^\perp$. The equivalence of $q_{V,W} \leq c$ and (iv) is obvious since $q_{V,W}$ is, by definition, the optimal constant in (iv). Thus, it only remains to prove the equality $\gamma_{V^\perp,W} = q_{V,W}$:
Given $v^\perp \in V^\perp$, $v \in V$, and $w \in W$, there holds

$$\langle v^\perp, w \rangle_{\mathcal{H}} = \langle v^\perp, w - v \rangle_{\mathcal{H}} \leq \|v^\perp\|_{\mathcal{H}} \|w - v\|_{\mathcal{H}}.$$

Since $v \in V$ is arbitrary, we obtain

$$\langle v^\perp, w \rangle_{\mathcal{H}} \leq \|v^\perp\|_{\mathcal{H}} \min_{v \in V} \|v - w\|_{\mathcal{H}} \leq q_{V,W} \|v^\perp\|_{\mathcal{H}} \|w\|_{\mathcal{H}} \quad \text{for all } w \in W,$$

whence $\gamma_{V^\perp,W} \leq q_{V,W}$. To prove the converse inequality, we construct sequences $v_j^\perp \in V^\perp \backslash \{0\}$ and $w_j \in W$ such that $\|w_j\|_{\mathcal{H}} = 1$ and

$$\lim_{j \to \infty} \langle v_j^\perp, w_j \rangle_{\mathcal{H}} / \|v_j^\perp\|_{\mathcal{H}} = q_{V,W}.$$

Without loss of generality we assume $q_{V,W} \neq 0$ since $q_{V,W} = 0$ implies $V = W$ and thus $\gamma_{V^\perp,W} = 0$ as well. For $q_{V,W} > 0$, let $w_j \in W$ be a sequence with

$$\|w_j\|_{\mathcal{H}} = 1, \quad \lim_{j \to \infty} \min_{v \in V} \|v - w_j\|_{\mathcal{H}} = q_{V,W} > 0, \quad \text{and} \quad \min_{v \in V} \|v - w_j\|_{\mathcal{H}} > 0.$$

Let $\Pi : \mathcal{H} \to V$ denote the orthogonal projection onto V and choose $v_j := \Pi w_j$. Then, there holds

$$\|v_j - w_j\|_{\mathcal{H}} = \min_{v \in V} \|v - w_j\|_{\mathcal{H}},$$

and $v_j^\perp := w_j - v_j$ satisfies $v_j^\perp \in V^\perp \backslash \{0\}$ and

$$\langle v_j^\perp, w_j \rangle_{\mathcal{H}} = \langle v_j^\perp, w_j - v_j \rangle_{\mathcal{H}} = \|w_j - v_j\|_{\mathcal{H}}^2 = \|w_j - v_j\|_{\mathcal{H}} \|v_j^\perp\|_{\mathcal{H}}.$$

Finally, we obtain

$$\gamma_{V^\perp,W} \geq \lim_{j \to \infty} \frac{\langle v_j^\perp, w_j \rangle_{\mathcal{H}}}{\|v_j^\perp\|_{\mathcal{H}}} = \lim_{j \to \infty} \|w_j - v_j\|_{\mathcal{H}} = q_{V,W}.$$

This concludes the proof. $\qquad\square$

5 Finite Element Method for the Laplace Problem

We consider the following model example on a bounded Lipschitz domain $\Omega \subset \mathbb{R}^d$, $d = 2, 3$,

$$\begin{aligned} -\Delta u &= f && \text{in } \Omega, \\ u &= 0 && \text{on } \Gamma_D \subseteq \partial\Omega, \\ \partial u / \partial \nu &= g && \text{on } \Gamma_N = \partial\Omega \backslash \Gamma_D. \end{aligned} \tag{11}$$

We assume that Γ_D is closed and that the right-hand side f and the given normal flux g allow for a weak solution

$$u \in \mathcal{H} = H_D^1(\Omega) := \{u \in H^1(\Omega) \,:\, u|_{\Gamma_D} = 0, \tag{12}$$

of (11). Provided Γ_D has positive surface measure, the Friedrichs' inequality shows that

$$\langle u, v \rangle = \int_\Omega \nabla u \cdot \nabla v \, dx \tag{13}$$

defines the energy scalar product with equivalent norm $\|\cdot\|_{\mathcal{H}} := \|\cdot\| \sim \|\cdot\|_{H^1(\Omega)}$ on $\mathcal{H}$. The weak form of (11) allows for a unique solution $u \in \mathcal{H}$ in the usual sense

$$\langle u, v \rangle = \int_\Omega f v \, dx + \int_{\Gamma_N} g v \, ds_x \quad \text{for all } v \in \mathcal{H}. \tag{14}$$

The lowest order conforming FE discretization of (14) uses $\mathcal{T}_h$-piecewise affine and globally continuous functions: Let $\mathcal{T}_h$ be a regular triangulation (in the sense of Ciarlet) which consists of triangles, for $d = 2$, and tetrahedra, for $d = 3$, respectively. For $p \in \mathbb{N}$, let $\mathcal{P}^p(\mathcal{T}_h)$ denote the vector space of functions $w_h \in \mathcal{P}^p(\mathcal{T}_h)$ which are polynomials of total degree $\leq p$ on each element $T \in \mathcal{T}_h$. Let $h \in L^\infty(\Omega)$ denote the local mesh-size of $\mathcal{T}_h$ defined by $h|_T = \mathrm{diam}(T)$ for $T \in \mathcal{T}_h$.

To apply the averaging technique, let $\mathcal{T}_H$ be a regular triangulation of Ω and let $\mathcal{T}_h$ be obtained from $\ell \in \mathbb{N}$ red-refinements of $\mathcal{T}_H$, i.e., we recursively refine each element $T \in \mathcal{T}_H$ ℓ-times into 4 (resp. 8 in case of $d = 3$) congruent elements. In particular, $H/h = 2^\ell$. With

$$\mathcal{S}_D^p(\mathcal{T}_h) := \{u_h \in \mathcal{P}^p(\mathcal{T}_h) \cap \mathcal{C}(\Omega) \,:\, u_h|_{\Gamma_D} = 0\} \subset \mathcal{H},$$

set

$$\mathcal{S}_h = \mathcal{S}_D^1(\mathcal{T}_h) \quad \text{and} \quad \mathcal{S}_H = \mathcal{S}_D^2(\mathcal{T}_H). \tag{15}$$

Finally, we denote by $H^s(\mathcal{T})$ the space of all $\mathcal{T}$-piecewise H^s functions for $s \geq 0$.

Theorem 3. *Provided $u \in \mathcal{H} \cap H^{2+\varepsilon}(\mathcal{T}_H)$ for some $\varepsilon > 0$ and ℓ large enough, Assumptions (AA)–(DP) hold and therefore Theorem 1 applies with $\eta_M = \|u_h - \mathbb{G}_H u_h\|$.*

Proof. Recall the local inverse estimate

$$\|H \, w_H\|_{L^2(\Omega)} \leq c_{\mathrm{inv}} \|\nabla w_H\|_{L^2(\Omega)} \quad \text{for all } w_H \in \mathcal{P}^1(\mathcal{T}_H),$$

where $c_{\mathrm{inv}} > 0$ depends only on the shape of the elements in $\mathcal{T}_H$ and the gradient ∇ is evaluated elementwise. In particular, this holds with $w_H = \nabla v_H$ for all $v_H \in \mathcal{P}^2(\mathcal{T}_H)$. Moreover, the Bramble-Hilbert lemma implies

$$\|\nabla v - \nabla(\mathbb{P}_h v)\|_{L^2(\Omega)} \leq c_{\mathrm{apx}} \|h\, D^2 v\|_{L^2(\Omega)}$$

for all continuous $v \in H^1(\Omega) \cap H^2(\mathcal{T}_h)$ and $\mathbb{P}_h$ the nodal interpolation operator. Together with $H/h = 2^\ell$, the combination of both estimates proves

$$q := \max_{v_H \in \mathcal{S}_H \setminus \{0\}} \min_{v_h \in \mathcal{S}_h} \frac{\|v_H - v_h\|}{\|v_H\|} \leq c_{\mathrm{apx}} c_{\mathrm{inv}}/2^\ell$$

Therefore, (DP) is satisfied for ℓ sufficiently large. Note the best approximation result $\|u - \mathbb{G}_h u\| = \mathcal{O}(h)$ and $\|u - \mathbb{G}_H u\| = \mathcal{O}(H^{1+\varepsilon})$. Given a fixed parameter ℓ, (AA) follows. $\square$

Remark 1. Since the energy norm is based on the local L^2-norm, we can write η_M as a sum of local contributions

$$\eta_M = \left(\sum_{T_j \in \mathcal{T}_H} \eta_{M,j}^2 \right)^{1/2} \quad \text{with} \quad \eta_{M,j} := \|\nabla u_h - \nabla(\mathbb{G}_H u_h)\|_{L^2(T_j)}. \tag{16}$$

The refinement indicators $\eta_{M,j}$ can be used for an adaptive mesh-refining strategy.

Remark 2. With Π_H the L^2 projection onto $\mathcal{P}^1(\mathcal{T}_H)^d$, we define

$$\mu_\Pi := \min_{q_H \in \mathcal{P}^1(\mathcal{T}_H)^d} \|\nabla u_h - q_H\|_{L^2(\Omega)} = \|\nabla u_h - \Pi_H(\nabla u_h)\|_{L^2(\Omega)}. \tag{17}$$

Since $\nabla(\mathbb{G}_H u_h) \in \mathcal{P}^1(\mathcal{T}_H)^d$, there holds $\mu_\Pi \leq \eta_M$. Therefore, μ_Π is efficient up to terms of higher order under the assumptions of Theorem 3. The mathematical analysis of the reliability of μ_Π — although supported by numerical evidence — remains open.

6 Symm's Integral Equation

In this section, we consider Symm's integral equation

$$V u = f \quad \text{on } \Gamma \tag{18}$$

with a relatively open subset $\Gamma \subseteq \partial\Omega$ of the boundary $\partial\Omega$ of a bounded Lipschitz domain Ω in $\mathbb{R}^d$, $d = 2, 3$. The operator V is the single-layer potential

$$V u(x) = \int_\Gamma \kappa(x, y) u(y)\, ds_y, \tag{19}$$

where ds denotes the integration on the manifold Γ, and $\kappa(x, y)$ denotes (up to a multiplicative constant) the fundamental solution of the Laplace operator,

$$\kappa(x, y) = \begin{cases} -\dfrac{1}{\pi} \log |x - y| & \text{for } d = 2, \\[2ex] +\dfrac{1}{2\pi} |x - y|^{-1} & \text{for } d = 3. \end{cases} \tag{20}$$

The variational formulation of (19) needs Sobolev spaces on the boundary. First, the space

$$H^{1/2}(\partial\Omega) = \{u|_{\partial\Omega} : u \in H^1(\mathbb{R}^d)\}$$

of traces of H^1 functions associated with the trace norm

$$\|u\|_{H^{1/2}(\partial\Omega)} = \inf\{\|\widehat{u}\|_{H^1(\mathbb{R}^d)} : \widehat{u} \in H^1(\mathbb{R}^d) \text{ with } \widehat{u}|_\Gamma = u\}.$$

Moreover, we consider the subspace

$$H^{1/2}(\Gamma) = \{u|_\Gamma : u \in H^{1/2}(\partial\Omega)\},$$

where the norm of $u \in H^{1/2}(\Gamma)$ is defined as the minimal norm of any extension, i.e.

$$\|u\|_{H^{1/2}(\Gamma)} = \inf\{\|\widehat{u}\|_{H^{1/2}(\partial\Omega)} : \widehat{u} \in H^{1/2}(\partial\Omega) \text{ with } \widehat{u}|_\Gamma = u\}.$$

Furthermore, there are Sobolev spaces

$$\widetilde{H}^{1/2}(\Gamma) = \{u \in H^{1/2}(\partial\Omega) : \operatorname{supp}(u) \subseteq \overline{\Gamma}\}$$

associated with the usual $H^{1/2}(\Gamma)$ norm. Finally, the corresponding spaces of negative order are defined by duality with respect to the extended L^2 scalar product,

$$H^{-1/2}(\Gamma) = \widetilde{H}^{1/2}(\Gamma)^* \quad \text{and} \quad \widetilde{H}^{-1/2}(\Gamma) = H^{1/2}(\Gamma)^*.$$

Remark 3. There are other equivalent definitions of the involved Sobolev spaces, e.g., by real oder complex interpolation, a Fourier norm, or Sobolev-Slobodeckij norms [35, 24].

For a particular right-hand side f in (18) and $\Gamma = \partial\Omega$, Symm's integral equation is an equivalent formulation of the Laplace problem (11) with $\Gamma_D = \partial\Omega$, cf. [24]. For $d = 3$ and provided additionally $\operatorname{diam}(\Omega) < 1$ for $d = 2$, the operator

$$V : \widetilde{H}^{-1/2}(\Gamma) \to H^{1/2}(\Gamma) \tag{21}$$

is an isomorphism between the two Hilbert spaces $\widetilde{H}^{-1/2}(\Gamma)$ and $H^{1/2}(\Gamma)$ which build a dual pairing with respect to the extended L^2 scalar product $\langle \cdot, \cdot \rangle$. The energy scalar product

$$\langle\!\langle u, v \rangle\!\rangle := \langle Vu, v \rangle \quad \text{for } u, v \in \widetilde{H}^{-1/2}(\Gamma) \tag{22}$$

induces an equivalent norm $\|\cdot\|_{\mathcal{H}} := \|\!|\cdot|\!\|$ on $\mathcal{H} = \widetilde{H}^{-1/2}(\Gamma)$.

Let $\mathcal{T}_h = \{\Gamma_1, \ldots, \Gamma_n\}$ be a regular triangulation of Γ with local mesh-size $h \in L^\infty(\Gamma)$, $h|_{\Gamma_j} = \mathrm{diam}(\Gamma_j)$. Each element Γ_j of $\mathcal{T}_h$ is supposed to be a connected (affine) boundary piece for $d = 2$ and a (flat) triangle for $d = 3$, respectively.

For an integer $p \geq 0$, $\mathcal{P}^p(\mathcal{T}_h)$ denotes the space of all piecewise polynomials of degree $\leq p$ (defined on reference elements $\Gamma_{\mathrm{ref}}^{\mathrm{2D}} = [0,1]$ and $\Gamma_{\mathrm{ref},3}^{\mathrm{3D}} = \mathrm{conv}\{(0,0),(0,1),(1,0)\}$ and $\Gamma_{\mathrm{ref},4}^{\mathrm{3D}} = \mathrm{conv}\{(0,0),(0,1),(1,0),(1,1)\}$ for $d = 2, 3$, respectively).

For the averaging error estimation, we consider again the lowest order case: Let $\mathcal{T}_H$ be a regular triangulation of Γ and obtain $\mathcal{T}_h$ by $\ell \in \mathbb{N}$ red-refinements of $\mathcal{T}_H$. Adopt the aforegoing notations for $\mathcal{T}_H$ and $\mathcal{T}_h$ accordingly and define the discrete spaces

$$\mathcal{S}_h = \mathcal{P}^0(\mathcal{T}_h) \quad \text{and} \quad \mathcal{S}_H = \mathcal{P}^1(\mathcal{T}_H). \tag{23}$$

Theorem 4. *Provided $u \in \mathcal{H} \cap H^{1+\varepsilon}(\mathcal{T}_H)$ for some $\varepsilon > 0$ and ℓ large enough, Assumptions (AA)–(DP) hold and therefore Theorem 1 applies with $\eta_M = \|u_h - \mathbb{G}_H u_h\|$.*

Proof. Local inverse estimates for fractional order Sobolev spaces [19, 22] read

$$\|H^{\alpha+\kappa} v_H\|_{L^2(\Gamma)} \leq c_{\mathrm{inv}}^{H,p} \|H^\kappa v_H\|_{H^{-\alpha}(\Gamma)} \quad \text{for all } v_H \in \mathcal{P}^p(\mathcal{T}_H) \text{ and } \kappa \in \mathbb{R}. \tag{24}$$

The constant $c_{\mathrm{inv}}^{H,p} > 0$ depends only on the shape of the elements in $\mathcal{T}_H$, the polynomial degree $p \in \mathbb{N}_0$, and the parameter $\alpha \geq 0$. Since $\widetilde{H}^\alpha(\Gamma)$ is a closed subspace of $H^\alpha(\Gamma)$, the corresponding dual spaces $H^{-\alpha}(\Gamma) = \widetilde{H}^\alpha(\Gamma)^*$ and $\widetilde{H}^{-\alpha}(\Gamma) = H^\alpha(\Gamma)^*$ satisfy $\widetilde{H}^{-\alpha}(\Gamma) \subseteq H^\alpha(\Gamma)^*$ with $\|v\|_{H^{-\alpha}(\Gamma)} \leq \|v\|_{\widetilde{H}^{-\alpha}(\Gamma)}$. Therefore, we may apply (24) for the energy norm $\|\cdot\| \sim \|\cdot\|_{\widetilde{H}^\alpha(\Gamma)}$. This leads to

$$\|H^{1/2} v_H\|_{L^2(\Gamma)} \leq c_{\mathrm{inv}}^{H,p} \|v_H\| \quad \text{for all } v_H \in \mathcal{P}^p(\mathcal{T}_H). \tag{25}$$

(Note that, for a closed boundary $\Gamma = \partial\Omega$, there holds $H^\alpha(\Gamma) = \widetilde{H}^\alpha(\Gamma)$ with equal norms.) Moreover, with the L^2-projection Π_h^p onto $\mathcal{P}^p(\mathcal{T}_h)$, there holds [12]

$$\|v - \Pi_h^p v\|_{\widetilde{H}^{-\alpha}(\Gamma)} \leq c_{\mathrm{apx}}^{h,p} \|h^\alpha v\|_{L^2(\Gamma)} \quad \text{for all } v \in L^2(\Gamma). \tag{26}$$

Here, $c_{\mathrm{apx}}^{h,p} > 0$ depends only on the shape of the elements in $\mathcal{T}_h$, the polynomial degree $p \in \mathbb{N}_0$, and $\alpha \geq 0$. Together with $H/h = 2^\ell$, the combination of (25) and (26), for $\alpha = 1/2$ and $\|\cdot\| \sim \|\cdot\|_{\widetilde{H}^{-1/2}(\Gamma)}$, proves

$$q := \max_{v_H \in \mathcal{S}_H \setminus \{0\}} \min_{v_h \in \mathcal{S}_h} \frac{\|v_H - v_h\|}{\|v_H\|} \leq c_{\mathrm{apx}}^{h,0} c_{\mathrm{inv}}^{H,1} / 2^{-\ell/2}.$$

This proves (DP) for ℓ sufficiently large. Assumption (AA) follows from best approximation results $\|u - \mathbb{G}_h u\| = \mathcal{O}(h^{3/2})$, $\|u - \mathbb{G}_H u\| = \mathcal{O}(H^{3/2+\varepsilon})$, cf. [28]. $\qquad\square$

In contrast to the FE method from the previous section with H^m norms, the energy norm $\|\cdot\| \sim \|\cdot\|_{\widetilde{H}^{-1/2}(\Gamma)}$ is non-local, i.e., it cannot be written as a sum over non-interacting local contributions. The following theorem asserts the equivalence of the energy norm based error estimator η_M and the weighted L^2 norm based error estimator

$$\mu_M := \|H^{1/2}(u_h - \mathbb{G}_H u_h)\|_{L^2(\Gamma)}. \tag{27}$$

This leads to the equivalent error estimators

$$\eta_\Pi := \|u_h - \Pi_H^1 u_h\| \quad \text{and} \quad \mu_\Pi := \|H^{1/2}(u_h - \Pi_H^1 u_h)\|_{L^2(\Gamma)}, \tag{28}$$

where Π_H^1 denotes the L^2 projection onto $\mathcal{P}^1(\mathcal{T}_H)$. Under the assumptions of Theorem 4, μ_M, μ_Π, and η_Π are reliable and efficient in the following sense.

Theorem 5. *There are constants $C_1, C_2 > 0$ which only depend on the shape of the elements in $\mathcal{T}_H$ and the quotient $H/h = 2^\ell$ such that*

$$\eta_M \leq \eta_\Pi \leq C_1\,\mu_\Pi \quad and \quad \mu_\Pi \leq \mu_M \leq C_2\,\eta_M. \tag{29}$$

Proof. The estimate $\eta_M \leq \eta_\Pi$ follows from the best approximation property of $\mathbb{G}_H$ and was already mentioned in the introduction. Since we consider globally discontinuous polynomials, Π_H^1 is also $\mathcal{T}_H$-elementwise orthogonal. Hence,

$$\|u_h - \Pi_H^1 u_h\|_{L^2(\Gamma_j)} \leq \|u_h - \mathbb{G}_H u_h\|_{L^2(\Gamma_j)}.$$

This proves $\mu_\Pi \leq \mu_M$. According to the mesh generation of $\mathcal{T}_h$ from $\mathcal{T}_H$, there holds $u_h - \mathbb{G}_H u_h \in \mathcal{P}^1(\mathcal{T}_h)$. An inverse estimate (25) yields

$$\|h^{1/2}(u_h - \mathbb{G}_H u_h)\|_{L^2(\Gamma)} \leq c_{\mathrm{inv}}^{h,1} \|u_h - \mathbb{G}_H u_h\|$$

and, therefore, with $H/h = 2^\ell$, that

$$\mu_M = 2^{\ell/2}\,\|h^{1/2}(u_h - \mathbb{G}_H u_h)\|_{L^2(\Gamma)} \leq 2^{\ell/2}\,c_{\mathrm{inv}}^{h,1}\eta_M.$$

To prove $\eta_\Pi \leq c_{\mathrm{apx}}^{H,1}\mu_\Pi$, define $v = u_h - \Pi_H^1 u_h \in L^2(\Gamma)$. With $\mathbb{1}$ the identity on $L^2(\Gamma)$, the operator $(\mathbb{1} - \Pi_H^1)$ is a projection, whence $v = (\mathbb{1} - \Pi_H^1)v$. An application of (26) proves

$$\eta_\Pi = \|v\| = \|(\mathbb{1} - \Pi_H^1)v\| \leq c_{\mathrm{apx}}^{H,1}\|H^{1/2}v\|_{L^2(\Gamma)} = \mu_\Pi. \qquad \square$$

Remark 4. For an adaptive mesh-refining algorithm, one may localize the error estimators μ_M and μ_Π, respectively, to obtain refinement indicators, e.g.

$$\mu_\Pi = \left(\sum_{\Gamma_j \in \mathcal{T}_H} \mu_{\Pi,j}^2\right)^{1/2} \quad \text{with} \quad \mu_{\Pi,j} = \|H^{1/2}(u_h - \Pi_H^1 u_h)\|_{L^2(\Gamma_j)}. \tag{30}$$

The computation of the error estimators η_M, μ_M, and η_Π needs the computation of dense matrices which stem from the Galerkin projection $\mathbb{G}_H$ (explicitly or implicitly for the computation of the energy norm). Matrix compression techniques, e.g., hierarchical matrices or panel clustering provide an effective implementation. The error estimator μ_Π avoids the computation of $\mathbb{G}_H$ and can be computed in linear complexity with respect to the number N of elements.

7 Hypersingular Integral Equation

With the notation from Section 6, we consider the hypersingular integral equation

$$Wu = f \quad \text{on } \Gamma \tag{31}$$

and the hypersingular integral operator

$$Wu(x) = -\frac{\partial}{\partial \nu_x} \int_\Gamma \frac{\partial}{\partial \nu_y} \kappa(x, y) u(y) \, ds_y, \tag{32}$$

where ν_x and ν_y denote the outer normal vectors on Γ at x and y, respectively. For particular right-hand sides and $\Gamma = \partial\Omega$, the hypersingular integral equation (31) is equivalent to the Laplace problem (11) with pure Neumann boundary condition $\Gamma_N = \partial\Omega$.

For an open boundary piece $\Gamma \subsetneqq \partial\Omega$, the operator

$$W : \widetilde{H}^{1/2}(\Gamma) \to H^{-1/2}(\Gamma)$$

is an isomorphism. For a closed boundary $\Gamma = \partial\Omega$, one has to consider the factor spaces $H_0^\alpha(\Gamma) = H^\alpha/\mathbb{R}(\Gamma) = \{u \in H^\alpha(\Gamma) : \int_\Gamma u \, ds = 0\}$ to neglect constant functions. Then,

$$W : H_0^{1/2}(\Gamma) \to H_0^{-1/2}(\Gamma)$$

is isomorphic. In both cases, W maps the energy space $\mathcal{H} = \widetilde{H}^{1/2}(\Gamma)$ resp. $\mathcal{H} = H_0^{1/2}(\Gamma)$ onto its dual, and

$$\langle u, v \rangle := \langle Wu, v \rangle \quad \text{for } u, v \in \mathcal{H} \tag{33}$$

defines a scalar product with equivalent norm $\|\cdot\|_{\mathcal{H}} := \|\|\cdot\|\|$ on $\mathcal{H}$. The discretization is based on subspaces of $\mathcal{S}^p(\mathcal{T}_h) := \mathcal{P}^p(\mathcal{T}_h) \cap \mathcal{C}(\Gamma)$ for a regular triangulation $\mathcal{T}_h$ of Γ and

$$S_0^p(\mathcal{T}_h) = \begin{cases} \{v_h \in \mathcal{S}^p(\mathcal{T}_h) : v_h|_{\partial\Gamma} = 0\} & \text{if } \Gamma \subset \partial\Omega; \\ \{v_h \in \mathcal{S}^p(\mathcal{T}_h) : \int_\Gamma v_h \, ds = 0\} & \text{if } \Gamma = \partial\Omega. \end{cases}$$

With respect to the abstract setting in Section 2, let $\mathcal{T}_H$ be a shape-regular triangulation of Γ and $\mathcal{T}_h$ obtained from $\mathcal{T}_H$ by $\ell \in \mathbb{N}$ red-refinements and set

$$\mathcal{S}_h = \mathcal{S}_0^1(\mathcal{T}_h) \quad \text{and} \quad \mathcal{S}_H = \mathcal{S}_0^2(\mathcal{T}_H). \tag{34}$$

Theorem 6. *Provided $u \in \mathcal{H} \cap H^{2+\varepsilon}(\mathcal{T}_H)$ for some $\varepsilon > 0$ and ℓ large enough, Assumptions (AA)–(DP) hold and therefore Theorem 1 applies with $\eta_M = \|u_h - \mathbb{G}_H u_h\|$.*

Proof. Note that there holds the local inverse estimate [13]

$$\|H^{1-\alpha} \nabla v_H\|_{L^2(\Gamma)} \leq c_{\mathrm{inv}}^{H,p} \|v_H\|_{H^\alpha(\Gamma)} \quad \text{for all } v_H \in \mathcal{S}^p(\mathcal{T}_H), \tag{35}$$

where ∇ denotes the arc-length derivative ∇ for $d = 2$ and the surface gradient for $d = 3$, respectively. The constant $c_{\mathrm{inv}}^{H,p} > 0$ depends only on the shape of the elements in $\mathcal{T}_h$, the polynomial degree $p \in \mathbb{N}$, and the parameter $\alpha \geq 0$. In [21] it is proven that the Galerkin projection $\mathbb{G}_h^p$ onto $\mathcal{S}_0^p(\mathcal{T}_h)$ satisfies, for all $v \in \mathcal{H} \cap H^1(\Gamma)$,

$$\|v - \mathbb{G}_h^p v\| \leq c_{\mathrm{apx}}^{h,p} \min \left\{ \|h^{1/2} \nabla v\|_{L^2(\Gamma)}, \|h^{1/2} \nabla (v - \mathbb{G}_h^p v)\|_{L^2(\Gamma)} \right\}. \tag{36}$$

The constant $c_{\mathrm{apx}}^{h,p} > 0$ depends only on the shape of the elements in $\mathcal{T}_h$. As before, Assumption (DP) is satisfied, provided ℓ is large enough,

$$q := \max_{v_H \in \mathcal{S}_H \setminus \{0\}} \min_{v_h \in \mathcal{S}_h} \frac{\|v_H - v_h\|}{\|v_H\|} \leq c_{\mathrm{apx}}^{h,1} c_{\mathrm{inv}}^{H,2} / 2^{\ell/2}.$$

Assumption (AA) follows from best approximation results $\|u - \mathbb{G}_h u\| = \mathcal{O}(h^{3/2})$ and $\|u - \mathbb{G}_H u\| = \mathcal{O}(H^{3/2+\varepsilon})$ [28]. $\qquad\square$

As for Symm's integral equation, the energy norm $\|\!\|\cdot\|\!\|$ for the hypersingular equation is non-local and has to be localized. This can be done by $H^{1/2}$-weighted H^1-seminorms. The following theorem states the efficiency and reliability of the error estimator

$$\mu_M := \|H^{1/2} \nabla (u_h - \mathbb{G}_H u_h)\|_{L^2(\Gamma)} \tag{37}$$

under the assumptions of Theorem 6.

Theorem 7. *There are constants $C_3, C_4 > 0$ which only depend on the shape of the elements in $\mathcal{T}_H$ and the quotient $H/h = 2^\ell$ such that*

$$C_3^{-1} \mu_M \leq \eta_M \leq C_4 \, \mu_M. \tag{38}$$

Proof. The follows from an inverse estimate with constant $C_3 = c_{\mathrm{inv}}^{h,2} \ell^{1/2}$ and the approximation result (36) with $C_4 = c_{\mathrm{apx}}^{H,2}$. $\qquad\square$

The computation of μ_M involves the dense stiffness matrix corresponding to the Galerkin projection $\mathbb{G}_H$. To avoid this numerical effort, one can consider the estimator

$$\mu_\Pi := \|H^{1/2} (\nabla u_h - \Pi_H^1 (\nabla u_h))\|_{L^2(\Gamma)} \tag{39}$$

with the L^2 projection Π_H^1 onto $\mathcal{P}^1(\mathcal{T}_H)$, which is efficient under the assumptions of Theorem 6.

Corollary 1. *There holds $\mu_\Pi \le \mu_M$.*

Remark 5. The reliability of μ_Π, which is observed numerically [13, 21], remains open — as for the finite element method in Section 5.

Another computationally challenging variant might be to consider the H_0^1 projection $\mathbb{P}_H : \mathcal{H} \cap H^1(\Gamma) \to \mathcal{S}_H$, i.e. the gradient L^2 projection defined by

$$\int_\Gamma \nabla(u - \mathbb{P}_H u) \cdot \nabla v_H = 0 \quad \text{for all } v_H \in \mathcal{S}_H. \tag{40}$$

The numerical realization only involves the sparse stiffness matrix from the P^1 finite element method.

$$\eta_\mathbb{P} := \|u_h - \mathbb{P}_H u_h\| \quad \text{and} \quad \mu_\mathbb{P} := \|H^{1/2}\nabla(u_h - \mathbb{P}_H u_h)\|_{L^2(\Gamma)} \tag{41}$$

Clearly, $\eta_M \le \eta_\mathbb{P}$, and therefore $\eta_\mathbb{P}$ is reliable under the assumptions of Theorem 6. The analysis for fractional order Sobolev spaces $H^\alpha(\Gamma)$ and $\alpha > 0$ is more involved than for $\alpha < 0$, i.e. for Symm's integral equation: For quasi-uniform meshes, there holds $\mu_\mathbb{P} \le C\,\mu_M$ since

$$\|\nabla(u_h - \mathbb{P}_H u_h)\|_{L^2(\Gamma)} \le \|\nabla(u_h - \mathbb{G}_H u_h)\|_{L^2(\Gamma)}.$$

An estimate of the type $\mu_\mathbb{P} \le C\,\mu_M$ remains open for adaptively generated meshes. For $d = 2$, it is proven that $\eta_\mathbb{P}$ and $\mu_\mathbb{P}$ are equivalent [13].

Theorem 8. *For $d = 2$, there are constants $C_5, C_6 > 0$ such that*

$$C_5^{-1}\mu_\Pi \le \eta_\mathbb{P} \le C_6\,\mu_\mathbb{P}. \tag{42}$$

Proof. The lower estimate follows as in Theorem 7. We recall from [13] that the H_0^1 projection $\mathbb{P}_h^p$ onto $\mathcal{S}_0^p(\mathcal{T}_h)$ satisfies, for all $v \in \mathcal{H} \cap H^1(\Gamma)$,

$$\|v - \mathbb{P}_h^p v\| \le c_{\text{apx}}^{h,p} \min\left\{\|h^{1/2}\nabla v\|_{L^2(\Gamma)}, \|h^{1/2}\nabla(v - \mathbb{P}_h^p v)\|_{L^2(\Gamma)}\right\}. \tag{43}$$

The constant $c_{\text{apx}}^{h,p}$ only depends on p and the local mesh-ratio

$$\varrho(\mathcal{T}_h) := \max\{h_j/h_k : \Gamma_j, \Gamma_k \in \mathcal{T}_h \text{ s.t. } \Gamma_j \text{ is a neighbour of } \Gamma_k\}. \tag{44}$$

From (43), we obtain the upper estimate with $C_6 = c_{\text{apx}}^{H,2}$. $\square$

Remark 6. If $\mathcal{A}_H$ denotes the L^2 projection onto $\mathcal{S}_0^2(\mathcal{T}_H)$, define

$$\eta_A := \|u_h - \mathcal{A}_H u_h\| \quad \text{and} \quad \mu_A := \|H^{1/2}\nabla(u_h - \mathcal{A}_H u_h)\|_{L^2(\Gamma)}.$$

Then, η_A is reliable, and one can prove that η_A and μ_A are equivalent. Unfortunately, the L^2 projection $\mathcal{A}_H$ onto $\mathcal{S}_0^2(\mathcal{T}_H)$ is, in general, not H^1 stable. Thus, one does neither analytically obtain nor numerically observe efficiency of η_A and μ_A, cf. [13].

8 Integral Equation for a Transmission Problem

This section is devoted to a transmission problem which involves the integral operators of Section 6 and 7, from where notation is adopted. Given $(f, g) \in H^{1/2}(\Gamma) \times H^{-1/2}(\Gamma)$ along the boundary $\Gamma = \partial\Omega$ of a bounded Lipschitz domain $\Omega \subset \mathbb{R}^d$, the strong form of the transmission problem reads: Find $u^- \in H^1(\Omega)$ and $u^+ \in H^1_{\ell oc}(\Omega)$ with

$$\Delta u^- = 0 \text{ in } \Omega, \quad \Delta u^+ = 0 \text{ in } \mathbb{R}^d \backslash \Omega \tag{45}$$

with some radiation condition on u^+ at infinity and

$$u^- = u^+ + f, \quad \frac{\partial u^-}{\partial \nu} = \frac{\partial u^+}{\partial \nu} + g \quad \text{on } \Gamma. \tag{46}$$

This is equivalently formulated by the boundary integral equation [18]

$$A \begin{pmatrix} u \\ \phi \end{pmatrix} = \left(\frac{1}{2} + A \right) \begin{pmatrix} f \\ g \end{pmatrix} \quad \text{in } \mathcal{H} \subset H^{1/2}(\Gamma) \times H^{-1/2}(\Gamma) \tag{47}$$

with the Calderón projector (in symbolic form)

$$A = \begin{pmatrix} -K & V \\ W & K' \end{pmatrix}. \tag{48}$$

The operator V is defined in (19), and W is defined in (32) with kernel $\kappa(x, y)$ from (20). Moreover, K denotes the double layer potential operator and K' its adjoint defined by

$$K : H^{1/2}(\Gamma) \to H^{1/2}(\Gamma), \qquad Kv(x) = \int_\Gamma v(y) \frac{\partial}{\partial \nu_y} \kappa(x, y) \, ds_y, \tag{49}$$

$$K' : H^{-1/2}(\Gamma) \to H^{-1/2}(\Gamma), \qquad K'\phi(x) = \int_\Gamma \phi(y) \frac{\partial}{\partial \nu_x} \kappa(x, y) \, ds_y. \tag{50}$$

Duality is understood with respect to the extended L^2 scalar product,

$$\left\langle \begin{pmatrix} u \\ \phi \end{pmatrix}, \begin{pmatrix} v \\ \psi \end{pmatrix} \right\rangle_{\mathcal{H}} = \langle u, \psi \rangle + \langle v, \phi \rangle \tag{51}$$

for $(u, \phi), (v, \psi) \in \mathcal{H} := H^{1/2}_0(\Gamma) \times H^{-1/2}_0(\Gamma)$.

The transmission problem (45)–(46) and the boundary integral formulation (47) are equivalent in the following sense [18, 16]: If $(u^-, u^+) \in H^1(\Omega) \times H^1_{\ell oc}(\mathbb{R}^d \backslash \Omega)$ solves the transmission problem, then $(u, \phi) \in \mathcal{H}$ solves (47), where $u := u^-|_\Gamma - \int_\Gamma u^- \, ds \in H^{1/2}_0(\Gamma)$ and $\phi := \partial u^- / \partial \nu|_\Gamma \in H^{-1/2}_0(\Gamma)$. Conversely, if $(u, \phi) \in \mathcal{H}$ solves (47), then the Cauchy data of u^- are given by $(u^-, \partial u^- / \partial \nu)|_\Gamma = (u + u_0, \phi)$ with

$$u_0 = \frac{\int_\Gamma \left(\frac{1}{2} (K - 1)f - \frac{1}{2} Vg + V\phi - Ku \right) ds}{\int_\Gamma 1 \, ds} \in \mathbb{R}.$$

The solution (u^-, u^+) is then obtained from the representation formulae in Ω and $\mathbb{R}^d \backslash \Omega$.

The mapping properties of the involved boundary operators [24] shows that $A : \mathcal{H} \to \mathcal{H}$ is continuous and $\mathcal{H}$-elliptic with respect to the canonical norm $\|(v, \psi)\|_{\mathcal{H}}^2 := \|v\|_{H^{1/2}(\Gamma)}^2 + \|\psi\|_{H^{-1/2}(\Gamma)}^2$. In fact, elementary calculations show that the (non-symmetric) bilinear form

$$\langle\!\langle (u, \phi), (v, \psi) \rangle\!\rangle = \langle A \begin{pmatrix} u \\ \phi \end{pmatrix}, \begin{pmatrix} v \\ \psi \end{pmatrix} \rangle_{\mathcal{H}} \tag{52}$$

induces an equivalent norm $\|\!|\cdot|\!\|$ which satisfies

$$\|\!|(u, \phi)|\!\|^2 = \|\phi\|_V^2 + \|u\|_W^2 \geq C_{\mathrm{ell}} \|(u, \phi)\|_{\mathcal{H}}^2 \quad \text{for all } (u, \phi) \in \mathcal{H} \tag{53}$$

with the energy norms $\|\cdot\|_V$ and $\|\cdot\|_W$ from Section 6 and 7, respectively. Note that $\|\!|\cdot|\!\|$ is indeed a Hilbert norm, but $\langle\!\langle \cdot, \cdot \rangle\!\rangle$ is *not* the corresponding scalar product! Let $\mathcal{T}_H$ be a shape-regular triangulation of Γ and let $\mathcal{T}_h$ be obtained from $\mathcal{T}_H$ by $\ell \in \mathbb{N}$ red-refinements. Set

$$\mathcal{P}_0^p(\mathcal{T}) := \{ v_h \in \mathcal{P}^p(\mathcal{T}) : \int_\Gamma v_h \, ds = 0 \},$$

set

$$\mathcal{S}_h = \mathcal{S}_0^1(\mathcal{T}_h) \times \mathcal{P}_0^0(\mathcal{T}_h) \quad \text{and} \quad \mathcal{S}_H = \mathcal{S}_0^2(\mathcal{T}_H) \times \mathcal{P}_0^1(\mathcal{T}_H).$$

Theorem 9. *Provided $(u, \phi) \in \mathcal{H} \cap \left(H^{2+\varepsilon}(\mathcal{T}_H) \times H^{1+\varepsilon}(\mathcal{T}_H) \right)$ for some $\varepsilon > 0$ and ℓ large enough, Assumptions (AA) and (DP) hold and therefore Theorem 1 applies with $\eta_M = \min\limits_{(v_H, \psi_H) \in \mathcal{S}_H} \|\!|(u_h, \phi_h) - (v_H, \phi_H)|\!\|$.*

Proof. Assumption (AA) follows from the regularity of (u, ϕ). The inverse estimates (25) and (35) lead to

$$\|H^{1/2}(\nabla v_H, \psi_H)\|_{L^2(\Gamma)} \leq c_{\mathrm{inv}}^{H,2,1} \|\!|(v_H, \psi_H)|\!\| \quad \text{for all } (v_H, \phi_H) \in \mathcal{S}_H.$$

Since the L^2-projection $\Pi_h^0 : L^2(\Gamma) \to \mathcal{P}^0(\mathcal{T}_h)$ preserves the vanishing integral mean (i.e., $\Pi_h^0 \psi_H \in \mathcal{P}_0^0(\mathcal{T}_h)$ provided $\int_\Gamma \psi_H \, ds = 0$), (26) and (36) yield

$$\|\!|(v_H, \psi_H) - (\mathbb{G}_h^W v_H, \Pi_h^0 \psi_H)|\!\| \leq c_{\mathrm{apx}}^{h,1,0} \|h^{1/2}(\nabla v_H, \psi_H)\|_{L^2(\Gamma)},$$

where $\mathbb{G}_h^W : H_0^{1/2}(\Gamma) \to \mathcal{S}_0^1(\mathcal{T}_h)$ denotes the Galerkin projection with respect to W from Section 7. The combination of the previous two inequalities results in

$$q := \max_{(v_H, \psi_H) \in \mathcal{S}_H \backslash \{0\}} \min_{(v_h, \psi_h) \in \mathcal{S}_h} \frac{\|\!|(v_H, \psi_H) - (v_h, \psi_h)|\!\|}{\|\!|(v_H, \psi_H)|\!\|} \leq c_{\mathrm{apx}}^{h,1,0} c_{\mathrm{inv}}^{H,2,1} / 2^{\ell/2}.$$

This implies (DP) for sufficiently large ℓ. $\qquad\square$

Remark 7. For an adaptive mesh-refinement, the non-local energy norm is localized via the localization arguments from the previous sections; further details are straightforward and hence omitted.

9 Numerical Experiments

This section provides some numerical experiments for the proposed error estimation. We only consider the symmetric case, where $\langle \cdot, \cdot \rangle$ defines a scalar product and give the numerical results with respect to the energy norm, cf. Section 3.1–3.2. Throughout, we compare uniform mesh-refinement with an adaptive mesh-refinement, which is based on the local contributions of our averaging error estimators as refinement indicators.

9.1 Adaptive Mesh-Refinement

The mesh-refinement strategy is formulated in the following adaptive algorithm from [12], which is stated for the finite element method from Section 5.

Algorithm 1 *Choose a regular initial coarse mesh $\mathcal{T}_H^{(0)}$, $k = 0$, $\ell \in \mathbb{N}$ and $0 \le \theta \le 1$.*

(i) *Obtain $\mathcal{T}_h^{(k)} = \{T_1, \ldots, T_n\}$ from $\mathcal{T}_H^{(k)} = \{\tau_1, \ldots, \tau_N\}$ by ℓ uniform refinements.*

(ii) *Compute the approximation $u_h^{(k)}$ for the current mesh $\mathcal{T}_h^{(k)}$.*

(iii) *Compute the error estimator η_M and the corresponding refinement indicators $\eta_{M,j}$ from (16).*

(iv) *Mark element τ_j for red-refinement provided the corresponding refinement indicator satisfies $\eta_{M,j} \ge \theta \max\{\eta_{M,1}, \ldots, \eta_{M,N}\}$.*

(v) *Use a red-green-blue mesh-refinement strategy to obtain a regular coarse mesh $\mathcal{T}_H^{(k+1)}$, update k, and go to (i).*

Note that we do the adaptive mesh-refinement on the coarse grid level to obtain a sequence of meshes $\mathcal{T}_H^{(k)}$. Surprisingly, our numerical experiments give empirical evidence that one may choose $\ell = 1$ in Algorithm 1. That is, the corresponding fine mesh $\mathcal{T}_h^{(k)}$, on which we compute our discrete solution u_h, is obtained by *one* uniform refinement of $\mathcal{T}_H^{(k)}$. We remark that the choice of $\theta = 0$ leads to uniform mesh-refinement. To obtain an adaptive mesh-refinement, we choose $\theta = 0.5$ in the subsequent experiments.

In the formulation of Algorithm 1, we consider the local contributions $\eta_{M,j}$ of η_M as refinement indicators. Alternatively, one may choose the local contributions of the (efficient) error estimator μ_Π from (17),

$$\mu_{\Pi,j} := \min_{q \in \mathcal{P}^1(\tau_j)} \|\nabla u_h - q\|_{L^2(\tau_j)} = \|\nabla u_h - \Pi_H(\nabla u_h)\|_{L^2(\tau_j)}. \qquad (54)$$

9.2 Visualization of Numerical Results

In all experiments we plot the Galerkin error $\|u - u_h\|$ and the error estimators η_M and μ_Π against the number $n = \#\mathcal{T}_h$ of fine grid elements for uniform ($\theta = 0$) and adaptive ($\theta = 0.5$) mesh-refinement, respectively. Throughout, we choose the parameter $\ell = 1$ in Algorithm 1. The error is computed by use of the Galerkin orthogonality

$$\|u - u_h\|^2 = \|u\|^2 - \|u_h\|^2. \tag{55}$$

The squared energy norm of the discrete solution u_h reads $\|u_h\|^2 = \mathbf{x} \cdot \mathbf{A}\mathbf{x}$ with the stiffness matrix $\mathbf{A}$ and the coefficient vector $\mathbf{x}$ corresponding to u_h. The norm $\|u\|^2$ can, in principle, be computed exactly. However, we use the value $\|u\|^2$ which is obtained by Aitkin's Δ^2-extrapolation as follows: For a sequence $\mathcal{T}_h^{(k)}$ of uniformly refined meshes, we compute the sequence of energies $E_k := \|u_h^{(k)}\|^2$, where $u_h^{(k)}$ is the discrete solution corresponding to the triangulation $\mathcal{T}_h^{(k)}$. Extrapolation of the sequence E_k then yields a good approximation of $\|u\|^2$.

From our analysis in Section 2 and Section 5, respectively, we know that η_M and μ_Π are efficient, i.e. there holds

$$\mu_\Pi \leq \eta_M \leq C_{\text{eff}} \|u - u_h\|$$

with efficiency constant $C_{\text{eff}} \leq 1 + \delta_{hH}$ and the approximation constant $\delta_{hH} = \|u - \mathbb{G}_H u\| / \|u - u_h\|$ from Assumption (AA). Provided δ_{hH} stays bounded, we therefore expect that the curves corresponding to η_M and μ_Π have at least the same slope as the curve corresponding to $\|u - u_h\|$. For smooth u, δ_{hH} tends to zero with h. Therefore, the experimental efficiency constant $C_{\text{eff}} := \eta_M / \|u - u_h\| \leq 1 - \delta_{hH}$ is expected to satisfy $C_{\text{eff}} \leq 1$ at least for the limit case for a finer and finer mesh-size h. Therefore, the absolute values and hence the curves of the error estimators should be below the curve of the error. Provided η_M is also reliable, i.e. $\|u - u_h\| \leq C_{\text{rel}}\eta_M$, the quotient $\|u - u_h\|/\eta_M$ is bounded. In this case, the slopes of the curves corresponding to $\|u - u_h\|$ and η_M are the same, i.e. the curves are parallel.

To study the efficiency and reliability of η_M even in the case that the solution u is non-smooth, we plot the experimental reliability constant $C_{\text{rel}} := \|u - u_h\|/\eta_M$ and the approximation constant δ_{hH} in dependence on the number $n = \#\mathcal{T}_h$ of fine grid elements. The Galerkin error $\|u - \mathbb{G}_H u\|$ for the higher-order method is computed as in (55).

9.3 Finite Element Method with Smooth Solution

For our first numerical experiment, we adopt the notation from Section 5. We consider the Dirichlet problem (11) on the unit square $\Omega = (0,1)^2 \subset \mathbb{R}^2$ with $\Gamma_D = \partial\Omega$ and

$$f(x) = (k^2\pi^2/2)\sin(x_1 k\pi/2)\sin(x_2 k\pi/2).$$

The exact solution is then given by

$$u(x) = \sin(x_1 k\pi/2)\sin(x_2 k\pi/2),$$

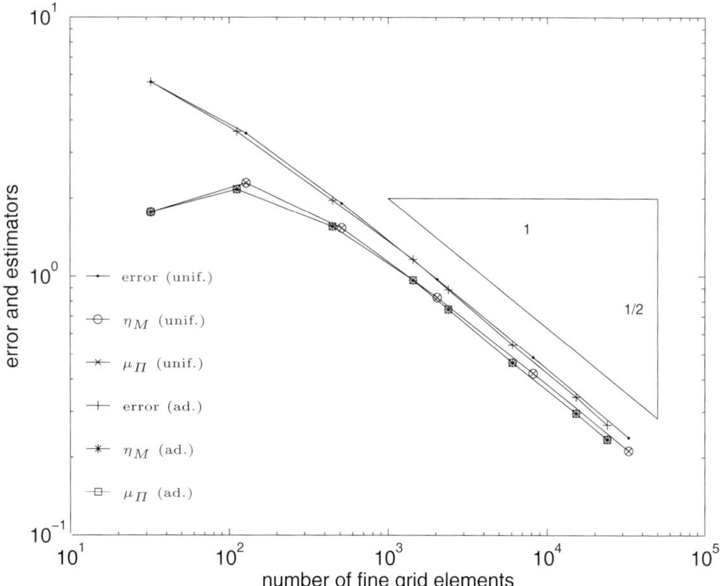

Fig. 1. Error $\|u - u_h\|$ and error estimators η_M and μ_Π in Example 9.3 in dependence on the number of fine grid elements $n = \#\mathcal{T}_h$. We observe optimal order of convergence $\mathcal{O}(n^{-1/2})$ for error and error estimators and independent of uniform (indicated by *unif.*) and adaptive mesh-refinement (indicated by *ad.*). The values of the error estimators η_M and μ_Π coincide up to rounding errors. The error estimation is reliable and efficient.

and therefore u satisfies the smoothness assumptions of Theorem 3. According to the Bramble-Hilbert lemma, we expect that uniform mesh-refinement leads to the optimal order of convergence $\mathcal{O}(h)$ for the error $\|u - u_h\|$, which is computed by (55). Aitkin's Δ^2-extrapolation yields $\|u\|^2 = 44.4132$.

In Fig. 1 we plot the error $\|u - u_h\|$ as well as the estimators η_M and μ_Π. Note that the optimal order of convergence $\mathcal{O}(h)$ for P^1-elements corresponds to $\mathcal{O}(n^{-1/2})$ in terms of elements $n = \#\mathcal{T}_h$. Both, uniform and adaptive mesh-refinement, lead to the optimal order of convergence for the error. Moreover, we observe that η_M and μ_Π coincide and that both are efficient and reliable. We stress the reliability of η_M which is analytically only predicted for sufficiently large $\ell \in \mathbb{N}$, whereas we use the minimal possible choice $\ell = 1$. Moreover, note that we have only proven $\mu_\Pi \leq \eta_M$. In our experiment, there holds even $\mu_\Pi = \eta_M$ up to rounding errors.

In Fig. 2 we plot the approximation quotient δ_{hH}. From standard approximation results and $h \sim H$ for the local mesh-sizes, we know that the nominator converges like $\mathcal{O}(h^2)$, whereas the denominator is $\mathcal{O}(h)$, i.e. we expect $\delta_{hH} = \mathcal{O}(h)$. This is what is observed experimentally in Fig. 2. Moreover, we plot the experimental reliability constant $C_{\mathrm{rel}} := \|u - u_h\|/\eta_M$. We observe

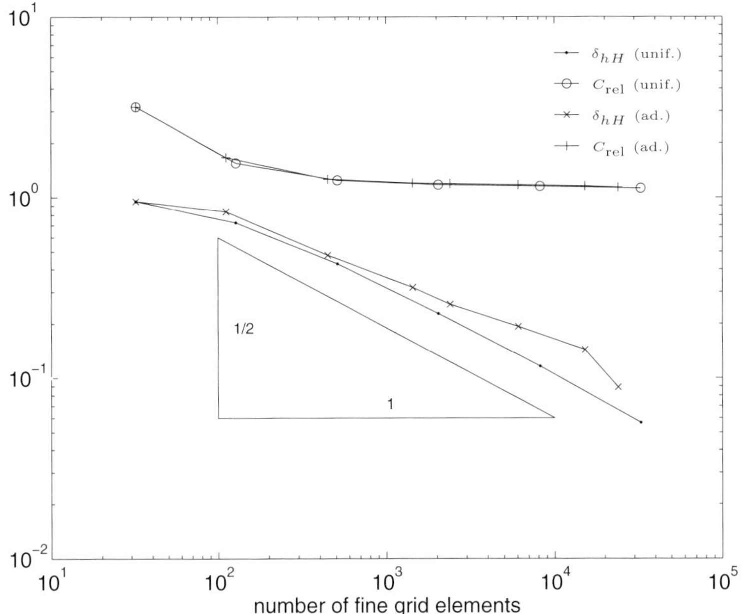

Fig. 2. Quotient $\delta_{hH} = \|u - \mathbb{G}_H u\|/\|u - u_h\|$ in approximation assumption (AA) and experimental reliability constant $C_{\mathrm{rel}} := \|u - u_h\|/\eta_M$ for Example 9.3. For both, uniform (indicated by *unif.*) and adaptive mesh-refinement (indicated by *ad.*), δ_{hH} tends to zero with the theoretically expected order $\mathcal{O}(n^{-1/2})$ with $n = \#\mathcal{T}_h$. The experimental reliability constant C_{rel} is slowly decreasing with absolute values ≈ 1.13 at the end of the computations ($n = 32768$ resp. $n = 24016$)

that it is slowly decreasing with absolute values about 1.13 at the end of our computations.

9.4 Finite Element Method with Weakly Singular Solution

For our second example, we again adopt the notation from Section 5 and consider the Dirichlet problem (11) on the L-shaped domain $\overline{\Omega} = [-1,0]^2 \cup [-1,0] \times [0,1] \cup [0,1]^2$ with $\Gamma_D = \partial\Omega$, cf. Fig. 3 which also shows the initial coarse mesh $\mathcal{T}_H^{(0)}$. The right-hand side is constant $f(x) = 1$. The solution $u(x)$ is known to be a bubble $u \in H^{1+2/3-\varepsilon}(\Omega)$, for all $\varepsilon > 0$, with singularity at the reentrant corner $(0,0)$. Therefore, uniform mesh-refinement is expected to lead to a suboptimal (experimental) convergence rate for the error $\|u - u_h\| = \mathcal{O}(h^{2/3})$ which can usually be cured by adaptive mesh-refinement.

In Fig. 4 we plot the error $\|u - u_h\|$ and the error estimators η_M and μ_Π, where the error is computed by (55) with the extrapolated value $\|u\|^2 = 0.214076$. As in Example 9.3, we observe that for both, uniform and

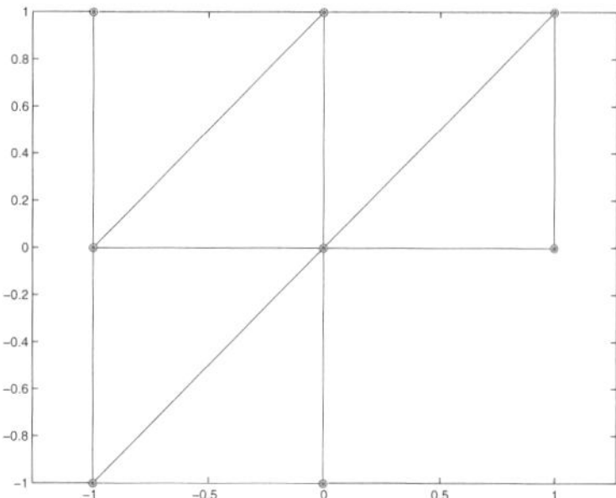

Fig. 3. In Example 9.4, we consider the L-shaped domain $\overline{\Omega} = [-1,0]^2 \cup [-1,0] \times [0,1] \cup [0,1]^2$. The initial coarse mesh $\mathcal{T}_H^{(0)}$ consists of $N = 6$ rectangular triangles.

adaptive mesh-refinement, the error estimators η_M and μ_Π coincide up to rounding errors. Independent of the mesh-refining strategy, the error estimators are reliable and efficient. For uniform mesh-refinement, we observe a suboptimal order of convergence $\mathcal{O}(n^{-2/5})$ which corresponds to $\mathcal{O}(h^{4/5})$. This is slightly better than the expected order of $\mathcal{O}(h^{2/3})$. For adaptive mesh-refinement, we retain the optimal order of convergence $\mathcal{O}(n^{-1/2})$ after a pre-asymptotic phase (up to about $n = 900$ elements), where we observe the same order of convergence as for the uniform refinement.

In Fig. 5 we plot the approximation quotient δ_{hH} and the experimental reliability constant $C_{\mathrm{rel}} := \|u - u_h\|/\eta_M$. For uniform mesh-refinement, the corner singularity of u dominates the convergence behavior so that we observe $\delta_{hH} = \mathcal{O}(1)$. For adaptive mesh-refinement, however, we obtain the optimal order $\delta_{hH} = \mathcal{O}(n^{-1/2})$. The experimental reliability constant C_{rel} is slowly decreasing in case of adaptive mesh-refinement with absolute value about 1.15 at the end of our computation ($n = 43040$). In contrast, for uniform mesh-refinement, C_{rel} is slowly increasing and is about 1.39 at the end of our computation ($n = 24565$).

9.5 Symm's Integral Equation

Finally, we consider the integral formulation of the Poisson problem

$$\Delta U = 0 \text{ in } \Omega \quad \text{and} \quad U = g \text{ on } \Gamma = \partial\Omega, \tag{56}$$

which is formulated as Symm's integral equation [24]

$$Vu = (K + 1)g, \tag{57}$$

where V is the single-layer and K is the double-layer potential from (19) and (49), respectively. Then, the exact solution of (57) is just the normal derivative $u = \partial U/\partial n$ of the solution U from (56) on the boundary Γ.

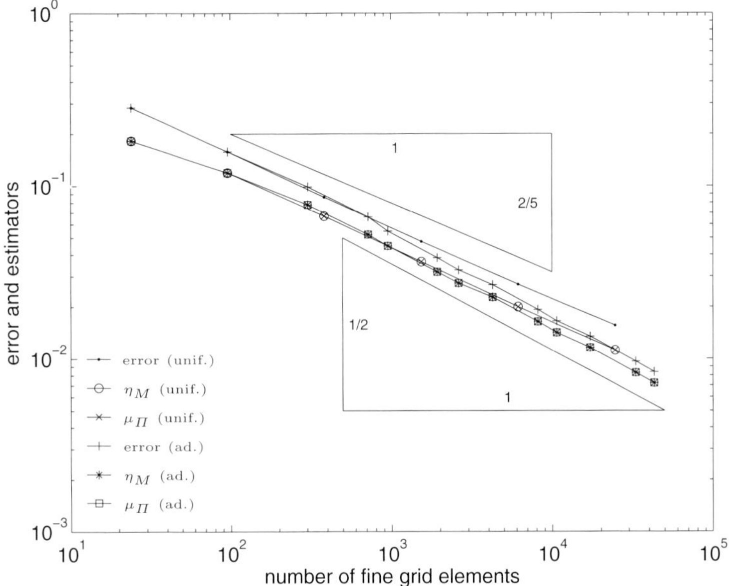

Fig. 4. Error $\|u - u_h\|$ and error estimators η_M and μ_Π in Example 9.4 in dependence on the number of fine grid elements $n = \#\mathcal{T}_h$. For uniform mesh-refinement [indicated by *unif.*], we observe a suboptimal order of convergence $\mathcal{O}(n^{-2/5})$ for error and error estimators. This is cured by our adaptive mesh-refining strategy [indicated by *ad.*], which leads to optimal order of convergence $\mathcal{O}(n^{-1/2})$. The values of the error estimators η_M and μ_Π coincide up to rounding errors. Independent of the mesh-refinement, the error estimation is reliable and efficient.

We adopt the notation from Section 6. The presented numerical results are taken from [12]: We consider a rotated L-shaped domain shown in Fig. 6. The Dirichlet data are chosen such that the exact solution $U \in H^1(\Omega)$ of (56) reads

$$U(x) = r^{2/3}\cos(2\varphi/3) \quad \text{in polar coordinates} \quad x = r\,(\cos\varphi, \sin\varphi).$$

Then, the exact solution $u \in H^{-1/2}(\Gamma)$ of Symm's integral equation (57) is given by

$$u(x) = \frac{2}{3}\,(w(\varphi)\cdot n(x))\,r^{-1/3} \tag{58}$$

with

$$w(\varphi) := \begin{pmatrix} \cos(\varphi)\cos(2\varphi/3) + \sin(\varphi)\sin(2\varphi/3) \\ \sin(\varphi)\cos(2\varphi/3) - \cos(\varphi)\sin(2\varphi/3) \end{pmatrix}. \tag{59}$$

Fig. 6 shows the initial coarse mesh $\mathcal{T}_H^{(0)}$ as well as the exact solution u from (58) plotted against the arclength of Γ. The singularity of u at $(0,0)$

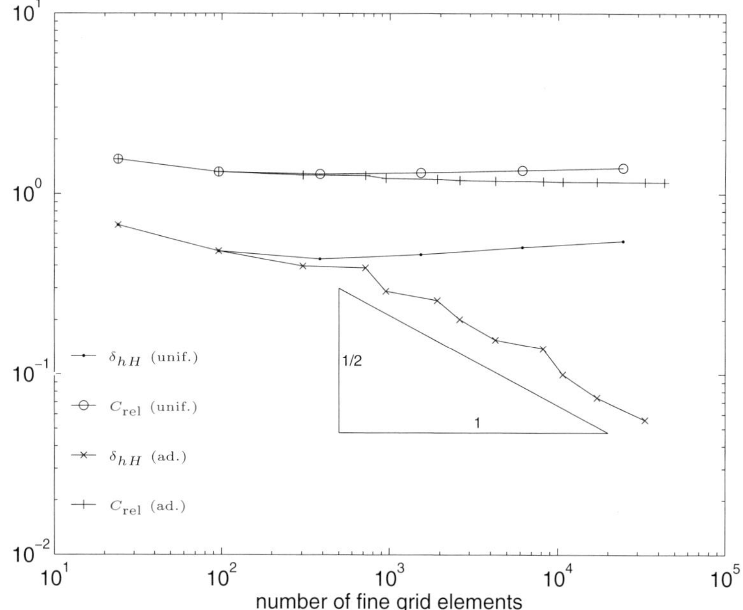

Fig. 5. Quotient $\delta_{hH} = \|u - \mathbb{G}_H u\| / \|u - u_h\|$ in approximation assumption (AA) and experimental reliability constant $C_{\mathrm{rel}} := \|u - u_h\| / \eta_M$ for Example 9.4. For uniform mesh-refinement [indicated by *unif.*], the corner singularity of u dominates the convergence behavior so that we observe $\delta_{hH} = \mathcal{O}(1)$. For adaptive mesh-refinement [indicated by *ad.*], we observe optimal convergence of $\delta_{hH} = \mathcal{O}(n^{-1/2})$. The experimental reliability constant C_{rel} is slowly decreasing in case of adaptive mesh-refinement with absolute value ≈ 1.15 at the end of the computation ($n = 43040$). However, for uniform mesh-refinement, C_{rel} is slowly increasing with absolute value ≈ 1.39 at the end of the computation ($n = 24576$).

is visible at arc-length parameter $s = 0$ and $s = 2$ by periodicity. Aitkin's Δ^2-method gives $\|u\|^2 = 0.404116$.

We consider uniform ($\theta = 0$) and adaptive mesh-refinement ($\theta = 1/2$), where we use the local contributions of the error estimator μ_Π from (30) as refinement indicators in Algorithm 1. Again, we restrict to the minimal choice $\ell = 1$ to obtain $\mathcal{T}_h$ from $\mathcal{T}_H$.

Fig. 7 shows the numerical results on the convergence of the error $\|u - u_h\|$ and of the error estimators $\eta_M = \|u_h - \mathbb{G}_H u_h\|$ and μ_M, η_Π and μ_Π from (27)–(28), respectively. We plot the error and the error estimators in dependence on the number of fine grid elements $n = \#\mathcal{T}_h$. Note that an experimental convergence rate $\mathcal{O}(h^\kappa)$ now corresponds to $\mathcal{O}(n^{-\kappa})$ in terms of fine grid elements, since we are dealing with a 1D discretization.

Uniform mesh-refinement leads to a suboptimal order of convergence $\mathcal{O}(h^{2/3})$ which is due to the singularity of the exact solution at the reentrant

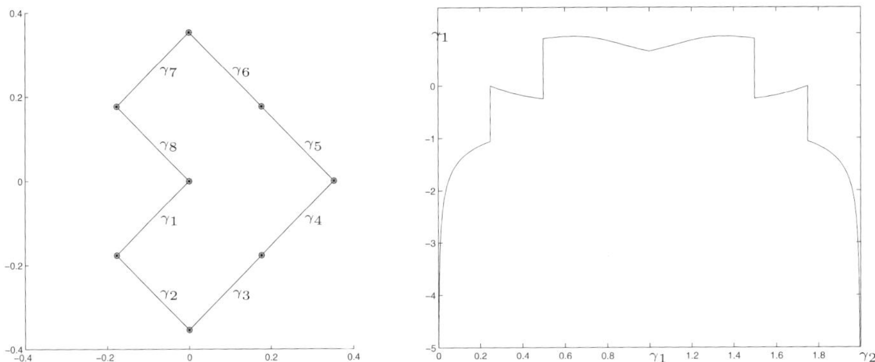

Fig. 6. In Example 9.5, we consider a rotated L-shaped domain Ω (left). Furthermore, the plot shows the initial coarse mesh $\mathcal{T}_H^{(0)}$ with $N = 8$ elements and uniform mesh size $H = 1/4$. The exact solution u from (58) is plotted over the arc-length $s = 0, \ldots, 2$ (right), where $s = 0$ and $s = 2$ correspond to the reentrant corner $(0, 0)$, where u is singular.

corner and which can be predicted theoretically. The fact that the slope of the corresponding error estimators even is $2/3$ gives empirical evidence that the estimators are reliable and efficient although the solution lacks the regularity assumed in Section 6. The proposed adaptive algorithm cures that shortcoming in the sense that it leads to the optimal order of convergence $\mathcal{O}(n^{-3/2})$ for the error, where we used the local contributions of μ_Π as refinement indicators. Due to numerical instabilities in the computation of the matrices corresponding to $\mathbb{G}_H$, we can only present the results for μ_M, η_M and η_Π up to about $n = 300$ elements in the case of adaptive mesh-refinement. This corresponds to an error about $10^{-7/2}$ for the higher order method. The explicit values of η_M and η_Π as well as the explicit values of μ_M and μ_Π coincide up to 2% so that there is no difference visible in the corresponding curves. Moreover, all four estimators show numerical evidence for efficiency and reliability. The computation of μ_Π is stable as it only involves the computation of some L^2-mass matrices, and the condition numbers of which are $\mathcal{O}(1)$ under some mild restrictions on the triangulation. The μ_Π steered mesh-refinement retains the optimal order of convergence $\mathcal{O}(n^{-3/2})$.

10 Conclusions

In this paper we provided an abstract analytical setting for the study of the reliability and efficiency of a posteriori averaging error estimators. The abstract setting applies to the Galerkin method for both, differential and integral equations, under weak assumptions on the finite elements or boundary elements used. The strongest assumption is a (piecewise) high regularity of

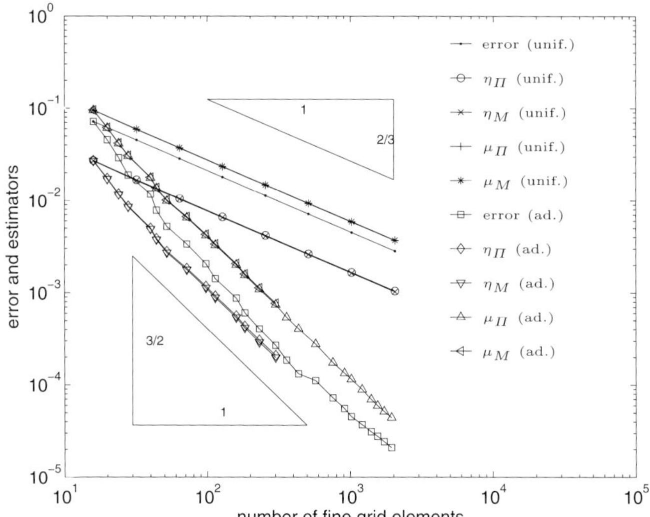

Fig. 7. Error $\|u - u_h\|$ and error estimators η_M, η_Π, μ_M, and μ_Π for uniform (indicated by *unif.*) and μ_Π-adaptive (indicated by *ad.*) mesh-refinement in Example 9.5. Uniform mesh-refinement leads to a suboptimal order of convergence. This is improved by the proposed adaptive strategy, which retains the optimal order of convergence. In both cases, the error estimation is reliable and efficient. The error estimators η_M and η_Π as well as μ_M and μ_Π coincide up to 2%.

the exact solution u. We recalled an adaptive algorithm from [12] which steers the mesh-refinement with respect to some localized error estimators. In the numerical experiments we considered examples with different regularity. In our experiments and in the experiments of [12, 13, 21] the adaptive strategy retains the optimal order of convergence and is therefore superior to uniform mesh-refinement.

However, there are still some gaps in the analysis: First, the introduced error estimators are only proven to be reliable if the parameter $\ell \in \mathbb{N}$ in Algorithm 1 is large enough. In the experiments we used the minimal choice $\ell = 1$ throughout. Nevertheless, we always observed the reliability. Second, the analytical verification of the introduced error estimators needs a high regularity assumption on u. However, this regularity assumption might be nonsatisfied in practice. Since our numerical experiments indicate that this assumption can be weakened, it would be desirable to have a refined analysis that covers these cases as well, i.e. which either avoids a regularity assumption on u or explains the good performance of the indicator-based adaptive strategy analytically.

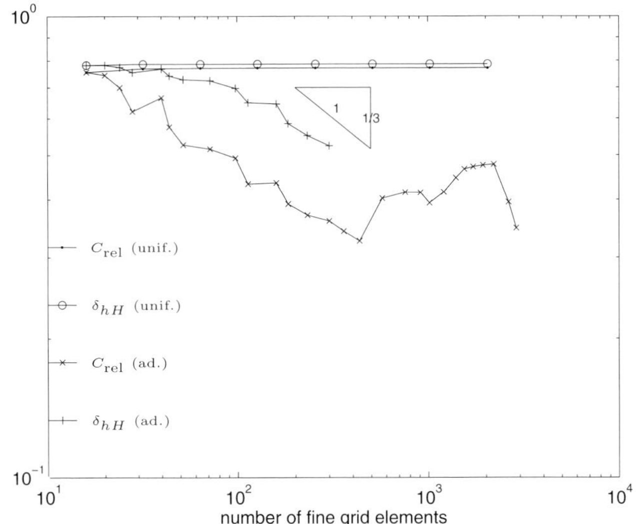

Fig. 8. Quotient $\delta_{hH} = \|u - \mathbb{G}_H u\|/\|u - u_h\|$ in approximation assumption (AA) and experimental reliability constant $C_{\rm rel} := \|u - u_h\|/\eta_M$ for uniform (indicated by *unif.*) and μ_Π-adaptive (indicated by *ad.*) mesh-refinement in Example 9.5. Note that according to the scaling of the y-axis, $C_{\rm rel}$ is almost constant.

References

1. M. Ainsworth, J. T. Oden: A posteriori error estimation in finite element analysis. John Wiley & Sons, New York, 2000.
2. I. Babuška, T. Strouboulis: The finite element method and its reliability. The Clarendon Press, Oxford University Press, New York, 2001.
3. D. Braess: Enhanced assumed strain elements and locking in membrane problems. Comput. Methods Appl. Mech. Engrg. 165 (1998) 155–174.
4. S. Bartels, C. Carstensen: Each averaging technique yields reliable a posteriori error control in FEM on unstructured grids. I. Low order conforming, nonconforming, and mixed FEM. Math. Comp. 71 (2002) 945–969. II. Higher order FEM. Math. Comp. 71 (2002) 971–994.
5. S. Bartels, C. Carstensen: Averaging techniques yield reliable a posteriori finite element error control for obstacle problems. Numer. Math. 99 (2004) 225–249.
6. R. Becker, R. Rannacher: A feed-back approach to error control in finite element methods: basic analysis and examples. East-West J. Numer. Math. 4 (1996) 237–264.
7. C. Carstensen: Some remarks on the history and future of averaging techniques in a posteriori finite element error analysis. Z. Angew. Math. Mech. 84 (2004) 3–21.
8. C. Carstensen: Reliable and efficient averaging techniques as universal tool for a posteriori finite element error control on unstructured grids. Int. J. Numer. Anal. Model. 3 (2006) 333–347.

9. C. Carstensen, J. Alberty: Averaging techniques for reliable a posteriori FE-error control in elastoplasticity with hardening. Comput. Methods Appl. Mech. Engrg. 192 (2003) 1435–1450.

10. C. Carstensen, S. A. Funken: Averaging technique for FE a posteriori error control in elasticity. I. Conforming FEM. Comput. Methods Appl. Mech. Engrg. 190 (2001) 2483–2498. II. λ-independend estimates. Comput. Methods Appl. Mech. Engrg. 190 (2001) 4663–4675. III. Locking-free nonconforming FEM. Comput. Methods Appl. Mech. Engrg. 191 (2001) 861–877.

11. C. Carstensen, S. A. Funken: A posteriori error control in low-order finite element discretisations of incompressible stationary flow problems. Math. Comp. 70 (2001) 1353–1381.

12. C. Carstensen, D. Praetorius: Averaging Techniques for the Effective Numerical Solution of Symm's Integral Equation of the First Kind. SIAM J. Sci. Comp. 27 (2006) 1226–1260.

13. C. Carstensen, D. Praetorius: Averaging Techniques for the A Posteriori BEM Error Control for a Hypersingular Integral Equation in 2D. SIAM J. Sci. Comp., accepted for publication, 2005.

14. C. Carstensen, D. Praetorius: A Unified Theory on Averaging Techniques for the Effective Numerical Solution of Differential and Integral Equations. Work in progress (2006).

15. C. Carstensen, E. P. Stephan: Adaptive coupling of boundary elements and finite elements. RAIRO Modél. Math. Anal. Numér. 29 (1995) 779–817.

16. C. Carstensen, E. P. Stephan: Adaptive boundary element methods for transmission problems. J. Austr. Math. Soc. Ser. B 38 (1997) 336–367.

17. C. Carstensen, R. Verfürth: Edge residuals dominate a posteriori error estimates for low order finite element methods. SIAM J. Numer. Anal. 36 (1999) 1571–1587.

18. M. Costabel, E. P. Stephan: A direct boundary integral equation method for transmission problems. J. Math. Anal. Appl. 106 (1985) 367–413.

19. W. Dahmen, B. Faermann, I. G. Graham, W. Hackbusch, S. A. Sauter: Inverse inequalities on non-quasi-uniform meshes and application to the mortar element method. Math. Comp. 73 (2004) 1107–1138.

20. K. Eriksson, D. Estep, P. Hansbo, C. Johnson: Computational differential equations. Cambridge University Press, 1996.

21. S. A. Funken, D. Praetorius: Averaging on Large Patches for First Kind Integral Equations in 3D. Work in progress (2006).

22. I. G. Graham, W. Hackbusch, S. A. Sauter: Finite elements on degenerate meshes: inverse-type inequalities and applications. IMA J. Numer. Anal. 25 (2005) 379–407.

23. W. Hoffmann, A. H. Schatz, L. B. Wahlbin, G. Wittum: Asymptotically exact a posteriori estimators for the pointwise gradient error on each element in irregular meshes. I. A smooth problem and globally quasi-uniform meshes. Math. Comp. 70 (2001) 897–909.

24. W. McLean: Strongly elliptic systems and boundary integral equations. Cambridge University Press, Cambridge, 2000.

25. R. H. Nochetto: Removing the saturation assumption in a posteriori error analysis. Istit. Lombardo Accad. Sci. Lett. Rend. A 127 (1994) 67–82.

26. R. Rodríguez: Some remarks on Zienkiewicz-Zhu estimator. Numer. Methods Partial Differential Equations 10 (1994) 625–635.

27. R. Rodríguez: A posteriori error analysis in the finite element method. Lecture Notes in Pure and Appl. Math., Vol 164, Dekker, New York, pp. 389–397, 1994.
28. S. A. Sauter, C. Schwab: Randelementmethoden. Analyse, Numerik und Implementierung schneller Algorithmen. B. G. Teubner, Stuttgart, Leipzig, Wiesbaden, 2004.
29. M. Schulz, W. L. Wendland: A general approach to a posteriori error estimates for strictly monotone and Lipschitz continuous nonlinear operators illustrated in elasto-plasticity. ENUMATH 97, World Sci. Publishing, pp. 572–579, 1998.
30. H. Schulz, W. L. Wendland: Local a posteriori error estimates for boundary element methods. ENUMATH 97, World Sci. Publishing, pp. 564–571, 1998.
31. H. Schulz, C. Schwab, W. L. Wendland: An extraction technique for boundary element methods. In: Boundary elements: implementation and analysis of advanced algorithms. Notes Numer. Fluid Mech., Vol. 54, Vieweg, Braunschweig, pp. 219–231, 1996.
32. C. Schwab, W. L. Wendland: On the extraction technique in boundary integral equations. Math. Comp. 68 (1999) 91–122.
33. R. Verfürth: A review of a posteriori error estimation and adaptive mesh-refinement techniques. Wiley, Teubner, 1996.
34. W. L. Wendland, H. Schulz, C. Schwab: On the computation of derivatives up to the boundary and recovery techniques in BEM. In: IUTAM Symposium on Discretization Methods in Structural Mechanics (Vienna, 1997), Solid Mech. Appl., Vol. 68, Kluwer, pp. 155–164, 1997.
35. W. L. Wendland: Elliptic systems in the plane. Monographs and Studies in Mathematics, Vol. 3, Boston, London, 1979.
36. O. C. Zienkiewicz, J. Z. Zhu: A simple error estimator and adaptive procedure for practical engineering analysis. Internat. J. Numer. Methods Engrg. 24 (1987) 337–357.

Coupled Finite and Boundary Element Domain Decomposition Methods

Ulrich Langer[1,2] and Olaf Steinbach[3]

[1] Institut für Numerische Mathematik, Johannes Kepler Universität Linz,
Altenberger Strasse 69, 4040 Linz, Austria
`ulanger@numa.uni-linz.ac.at`
[2] Johann Radon Institute for Computational and Applied Mathematics (RICAM),
Austrian Academy of Sciences, Altenberger Strasse 69, 4040 Linz, Austria
`ulrich.langer@assoc.oeaw.uni-linz.ac.at`
[3] Institut für Numerische Mathematik, Technische Universität Graz,
Steyrergasse 30, 8010 Graz, Austria
`o.steinbach@tugraz.at`

Summary. The finite element method and the boundary element method often
have complementary properties in different situations. The domain decomposition
technique allows to use the discretization method which is most appropriate for the
subdomain under consideration. The coupling is based on the transmission conditions. The Dirichlet to Neumann (D2N) and Neumann to Dirichlet (N2D) maps are
playing a crucial role in representing the transmission conditions. In this paper we
study the D2N and N2D maps and their finite and boundary element approximations. Different formulations of the transmission conditions lead to different domain
decomposition schemes with different properties. In any case we have to solve large
scale systems of coupled finite and boundary element equations. The efficiency of
iterative methods heavily depends on the availability of efficient preconditioners. We
consider various solution strategies and provide appropriate preconditioners resulting in asymptotically almost optimal solvers.

1 Introduction

Domain Decomposition (DD) Methods provide not only the basic technology
for parallelizing numerical algorithms for solving partial differential equations
(PDEs) but also for coupling different physical fields and different discretization techniques. Beside the Finite Volume Method (FVM) and the Finite
Element Method (FEM), the Boundary Element Method (BEM) is certainly
one of the most popular discretization techniques for PDEs. If we compare
the FEM with the BEM, then we observe that both methods have advantages
and disadvantages in different situations. It is commonly known that the BEM
can easily treat unbounded regions whereas the FEM requires special modi-
fications for this case. On the other hand, the FEM is very flexible and can

be applied to very general problems including PDEs with varying coefficients and non–linear problems. However, beside unbounded computational regions, there are a lot of other problems where we can benefit from using boundary element discretization. Large air subdomains or rotating subdomains which are typical for electromagnetical problems, e.g., electrical machines, belong to this class of problems. Also the approximation of singularities can be handled much easier by a boundary mesh than by a volume mesh. Sometime only the complete Cauchy date are needed on the boundary of the computational domain or on the skeleton of a domain decomposition. In this situation, we would like to avoid the meshing of the domain or of the subdomains. A similar situation arises if we are only interested in the solution or in derivatives of the solution in some points or in some small subdomains. Therefore, it is certainly very attractive to develop coupling algorithms and software that can handle both the finite element and the boundary element technologies. There are many early contributions to the FEM–BEM coupling in the engineering literature, see, e.g., [6, 58, 59]. Most of them are using the collocation technique on the boundary element side that does not really fit to the finite element Galerkin technique. Moreover, there was some other drawback of the classical boundary element methods. They produce dense matrices. The breakthrough through this complexity barrier was achieved by developing data–sparse approximation techniques like the fast multipole method [9, 42], panel clustering [21], $\mathcal{H}$–matrix approaches [20], Adaptive Cross Approximation (ACA) methods [2, 3], and wavelet approximations [14, 43].

In the mathematical literature, there are also some early works on unsymmetric Galerkin BEM–FEM couplings by F. Brezzi and C. Johnson [8], C. Johnson and J. Nédélec [27] and others at the end of the 70ies and at the beginning of the 80ies. These results are based on the use of the first boundary integral equation using the single and double layer potentials only. In fact, the analysis requires the compactness of the double layer potential and therefore smooth boundaries have to be assumed. Since we are interested in domain decomposition techniques with Lipschitz subdomains, we heavily rely on the symmetric coupling that was first proposed by M. Costabel in [11]. This approach makes also use of the second boundary integral equation with the hypersingular boundary integral operator. The symmetric formulation can also be generalized to non–linear problems such as elastoplastic problems [13, 40]. G. C. Hsiao and W. L. Wendland first used the symmetric coupling technique for constructing symmetric boundary element domain decomposition equations [26]. The first fast solvers for coupled finite and boundary element domain decomposition equations were proposed and analyzed by U. Langer [32]. The classical Finite Element Tearing and Interconnecting (FETI) methods, which were introduced by C. Farhat and F.–X. Roux [17] in 1991 as a dual version of the classical iterative substructuring methods, and, in particular, the more recently developed dual–primal FETI (FETI–DP) and BDDC (Balanced Domain Decomposition by Constraints) methods are now well–established as efficient and robust parallel solvers for large–scale finite element equations. We refer the reader to the recently published monograph [55] by A. Toselli and O. Widlund for more informations about the relevant references and for the analysis of FETI methods. U. Langer and O. Steinbach have recently introduced the Boundary Element Tearing and Interconnecting (BETI) methods [36] and the coupled BETI/FETI methods [37]. Inexact data–sparse BETI methods were discussed in [33]. The hybrid coupling of finite element methods and boundary element methods as a macro element was considered by G. C. Hsiao, E. Schnack and W. L. Wendland in [24] for general second order elliptic systems, and in [23] for applications in elasticity. Hybrid domain decomposition methods based on the approximation of the local Dirichlet to Neumann mappings by finite and boundary element methods and a related stability and error analysis were given by O. Steinbach in [48].

This paper provides a unified approach to the construction, analysis and solution of coupled finite and boundary domain decomposition equations. The potential equation with piecewise constant coefficients serves as a simple model problem. On an appropriate domain decomposition, such a special potential problem and similar elliptic boundary value problems in general can be reformulated as variational problems defined on the skeleton of the domain decomposition. These skeleton variational formulations reflect the transmission conditions which can be incorporated in different ways. The local Steklov–Poincaré (D2N) and Poincaré–Steklov (N2D) operators play an important role in these formulations. These operators can locally be approximated by finite and boundary element methods. We discuss and analyse these approximations. Finally we have to solve large scale coupled finite and boundary domain decomposition equations which are in general symmetric, but indefinite. Reductions to symmetric and positive definite Schur complement problems are always possible, but not always recommendable for efficiency reasons. Primal, primal–dual and dual iterative substructuring solvers require asymptotically almost optimal and robust preconditioners. Such preconditioners can be constructed by the use of boundary element technologies for both the boundary element and the finite element blocks.

The rest of the paper is organized as follows: In Section 2, we consider the Dirichlet boundary value problem for the potential equation with piecewise constant coefficients as a typical model problem and study the local Steklov–Poincaré and Poincaré–Steklov operators as well as their finite and boundary element approximations. Section 3 is devoted to different domain decomposition methods. We consider two types of symmetric coupling techniques. The Dirichlet domain decomposition methods presented in Subsection 3.1 require the strong continuity of the primal variable (the potentials) whereas the Neumann domain decomposition methods studied in Subsection 3.2 require the strong continuity of the dual variables (the fluxes). The tearing and interconnecting technology allows us to develop a unique approach to both domain decomposition techniques. In contrast to the primal–dual tearing and interconnecting methods, we prefer the all–floating technique that was introduced by G. Of [38]. In Section 4, we discuss the iterative solution of the linear systems arising in Section 3 and provide preconditioners leading to asymptotically almost optimal and robust solvers. Finally, we draw some conclusions in Section 5.

2 Boundary Value Problems

Let $\Omega \subset \mathbb{R}^3$ be a bounded domain with Lipschitz boundary $\Gamma = \partial\Omega$. As a model problem, we consider the Dirichlet boundary value problem for the potential equation,

$$-\mathrm{div}[\alpha(x)\nabla u(x)] = f(x) \quad \text{for } x \in \Omega, \quad u(x) = g(x) \quad \text{for } x \in \Gamma \qquad (1)$$

where $g \in H^{1/2}(\Gamma) \cap C(\Gamma)$ is a given continuous function. We assume that $\alpha(\cdot)$ is piecewise constant with $\alpha(x) = \alpha_i > 0$ for $x \in \Omega_i$ and for $i = 1, \ldots, p$, where we have given a non–overlapping domain decomposition

$$\overline{\Omega} = \bigcup_{i=1}^{p} \overline{\Omega}_i, \quad \Omega_i \cap \Omega_j = \emptyset \quad \text{for } i \neq j, \quad \Gamma_i = \partial\Omega_i, \quad \overline{\Gamma}_{ij} = \overline{\Gamma}_i \cap \overline{\Gamma}_j$$

of the computational domain Ω into p Lipschitz subdomains Ω_i. Moreover,

$$\Gamma_S = \bigcup_{i=1}^{p} \Gamma_i = \Gamma \cup \Gamma_I \quad \text{and} \quad \overline{\Gamma}_I = \bigcup_{i<j} \overline{\Gamma}_{ij}$$

denote the skeleton and the interface of the domain decomposition, respectively. Instead of the global boundary value problem (1), we now consider the local boundary value problems

$$-\alpha_i \Delta u_i(x) = f_i(x) \quad \text{for } x \in \Omega_i, \quad u_i(x) = g(x) \quad \text{for } x \in \Gamma_i \cap \Gamma \qquad (2)$$

together with the transmission conditions

$$u_i(x) = u_j(x), \quad \alpha_i \frac{\partial}{\partial n_i} u_i(x) + \alpha_j \frac{\partial}{\partial n_j} u_j(x) = 0 \quad \text{for } x \in \Gamma_{ij}, \qquad (3)$$

where $f_i(x) = f(x)$ for $x \in \Omega_i$. In what follows, we will describe some variational formulations for domain decomposition methods which are based on the local solution of either Dirichlet or Neumann boundary value problems. The idea behind is that all solutions u_i of the local boundary value problems (2) are known as soon as the Cauchy data along the coupling boundaries Γ_{ij} satisfying the transmission conditions (3) are determined.

2.1 Dirichlet Boundary Value Problems

We start with the local Dirichlet boundary value problem for a given continuous function $g_i \in H^{1/2}(\Gamma_i) \cap C(\Gamma_i)$

$$-\alpha_i \Delta u_i(x) = f_i(x) \quad \text{for } x \in \Omega_i, \quad u_i(x) = g_i(x) \quad \text{for } x \in \Gamma_i \qquad (4)$$

where the weak solution $u_i \in H^1(\Omega_i)$ is well defined. Moreover, the normal derivative $t_i = n_i \cdot \nabla u_i$ defines the associated Neumann datum. The solution of the local Dirichlet boundary value problem (4) therefore defines the local Dirichlet to Neumann map $g_i \longmapsto t_i$. Hence, we have to find the correct Dirichlet datum g_i such that the transmission boundary conditions (3), i.e.,

$$u_i(x) = u_j(x), \quad \alpha_i t_i(x) + \alpha_j t_j(x) = 0 \quad \text{for } x \in \Gamma_{ij},$$

are satisfied along the coupling interfaces Γ_{ij}. To describe the local Dirichlet to Neumann map we may consider either a domain variational formulation or boundary integral equations to obtain explicit representations of the local Steklov–Poincaré (Dirichlet to Neumann) operators involved.

Domain Variational Formulation

The associated variational formulation of the local Dirichlet boundary value problem (4) is to find $u_i \in H^1(\Omega_i)$, $u_i(x) = g_i(x)$ for $x \in \Gamma_i$, such that

$$\int_{\Omega_i} \alpha_i \nabla u_i(x) \cdot \nabla v_i(x) dx = \int_{\Omega_i} f_i(x) v_i(x) dx \qquad (5)$$

is satisfied for all test functions $v_i \in H_0^1(\Omega_i)$. As usual, $H^1(\Omega_i)$ is the closure of $C^\infty(\Omega_i)$ with respect to the norm

$$\|v_i\|_{H^1(\Omega_i)} = \left[\|v_i\|^2_{L_2(\Omega_i)} + \|\nabla v_i\|^2_{L_2(\Omega)} \right]^{1/2}.$$

However, in what follows we will use an equivalent norm in $H^1(\Omega_i)$ which is given by

$$\|v_i\|_{H^1(\Omega_i),\Gamma_i} = \left[\left(\int_{\Gamma_i} v_i(x) ds_x \right)^2 + \|\nabla v_i\|^2_{L_2(\Omega_i)} \right]^{1/2}.$$

Moreover, $H_0^1(\Omega_i) = \{v_i \in H^1(\Omega_i), v_i(x) = 0 \text{ for } x \in \Gamma_i\}$. The bilinear form

$$a_{\Omega_i}(v_i, v_i) = \int_{\Omega_i} |\nabla v_i(x)|^2 dx = \|\nabla v_i\|^2_{L_2(\Omega_i)} = \|v_i\|^2_{H^1(\Omega_i),\Gamma_i} \quad \text{for } v_i \in H_0^1(\Omega_i)$$

defines an equivalent norm in $H_0^1(\Omega_i)$, i.e., $(a_{\Omega_i}(v_i, v_i))^{1/2}$ is the energy norm in $H_0^1(\Omega_i)$. By taking the trace of $H^1(\Omega_i)$ we may define the trace space $H^{1/2}(\Gamma_i)$ which is equipped with the norm

$$\|w_i\|_{H^{1/2}(\Gamma_i)} = \min_{v_i \in H^1(\Omega_i), v_i|_{\Gamma_i} = w_i} \|v_i\|_{H^1(\Omega_i),\Gamma_i}.$$

For $g_i \in H^{1/2}(\Gamma_i)$, there exists a bounded extension $u_{g_i} = \mathcal{E}_i g_i \in H^1(\Omega_i)$ satisfying

$$\|u_{g_i}\|_{H^1(\Omega_i),\Gamma_i} = \|\mathcal{E}_i g_i\|_{H^1(\Omega_i),\Gamma_i} \leq c_{\mathcal{E}_i}\|g_i\|_{H^{1/2}(\Gamma_i)}.$$

A particular choice would be to consider the harmonic extension $u_{g_i} \in H^1(\Omega_i)$ as the unique solution of the variational problem

$$\int_{\Omega_i} \nabla u_{g_i}(x) \cdot \nabla v_i(x)dx = 0 \quad \text{for all } v_i \in H_0^1(\Omega_i).$$

It remains to find $u_{i,0} \in H_0^1(\Omega_i)$ such that the homogenized variational problem

$$\int_{\Omega_i} \alpha_i \nabla u_{i,0}(x) \cdot \nabla v_i(x)dx = \int_{\Omega_i} f_i(x)v_i(x)dx - \int_{\Omega_i} \alpha_i \nabla u_{g_i}(x) \cdot \nabla v_i(x)dx \quad (6)$$

is satisfied for all $v_i \in H_0^1(\Omega_i)$. For $u_{i,0}, v_i \in H_0^1(\Omega_i)$, the bilinear form

$$a_{\Omega_i}(u_{i,0}, v_i) = \int_{\Omega_i} \nabla u_{i,0}(x) \cdot \nabla v_i(x)dx = \langle A_{\Omega_i,0} u_{i,0}, v_i \rangle_{\Omega_i},$$

induces, by the Riesz representation theorem, a bounded linear operator

$$A_{\Omega_i,0} : H_0^1(\Omega_i) \to H^{-1}(\Omega_i) = [H_0^1(\Omega_i)]'.$$

In addition, for $u_{g_i} \in H^1(\Omega_i)$ and $v_i \in H_0^1(\Omega_i)$ we define the bounded operator $A_{\Gamma_i} : H^1(\Omega_i) \to H^{-1}(\Omega_i)$ satisfying

$$\langle A_{\Gamma_i} u_{g_i}, v_i \rangle_{\Omega_i} = \int_{\Omega_i} \nabla u_{g_i}(x) \cdot \nabla v_i(x)dx.$$

Hence, we can write the variational problem (6) as an operator equation to find $u_{i,0} \in H_0^1(\Omega_i)$ such that

$$\alpha_i A_{\Omega_i,0} u_{i,0} = f_i - \alpha_i A_{\Gamma_i} \mathcal{E}_i g_i \in H^{-1}(\Omega_i). \quad (7)$$

The operator $A_{\Omega_i,0} : H_0^1(\Omega_i) \to H^{-1}(\Omega_i)$ is $H_0^1(\Omega_i)$–elliptic, i.e., for all $v_i \in H_0^1(\Omega_i)$, we have

$$\langle A_{\Omega_i,0} v_i, v_i \rangle_{\Omega_i} = \int_{\Omega_i} |\nabla v_i(x)|^2 dx = \|\nabla v_i\|^2_{L_2(\Omega_i)} = \|v_i\|^2_{H^1(\Omega_i),\Gamma_i}.$$

Hence, there exists the unique solution of the operator equation (7),

$$u_{i,0} = \frac{1}{\alpha_i} A_{\Omega_i,0}^{-1} f_i - A_{\Omega_i,0}^{-1} A_{\Gamma_i} \mathcal{E}_i g_i \in H_0^1(\Omega_i),$$

and, therefore, $u_i = u_{0,i} + u_{g_i} \in H^1(\Omega_i)$ is the weak solution of the Dirichlet boundary value problem (4). In particular, from

$$\|f_i\|_{H^{-1}(\Omega_i)} = \sup_{0 \neq v_i \in H_0^1(\Omega_i)} \frac{\langle f_i, v_i \rangle_{\Omega_i}}{\|v_i\|_{H^1(\Omega_i),\Gamma_i}} \geq \frac{\langle f_i, u_{i,0} \rangle_{\Omega_i}}{\|u_{i,0}\|_{H^1(\Omega_i),\Gamma_i}}$$

$$= \frac{\alpha_i}{\|u_{i,0}\|_{H^1(\Omega_i),\Gamma_i}} \int_{\Omega_i} \nabla[u_{i,0}(x) + u_{g_i}(x)] \nabla u_{i,0}(x)\,dx$$

$$\geq \frac{\alpha_i}{\|u_{i,0}\|_{H^1(\Omega_i),\Gamma_i}} \left[\|\nabla u_{i,0}\|_{L_2(\Omega_i)}^2 - \|\nabla u_{i,0}\|_{L_2(\Omega_i)} \|\nabla u_{g_i}\|_{L_2(\Omega_i)} \right]$$

$$= \alpha_i \left[\|\nabla u_{i,0}\|_{L_2(\Omega_i)} - \|\nabla u_{g_i}\|_{L_2(\Omega_i)} \right],$$

we find

$$\|\nabla u_{i,0}\|_{L_2(\Omega_i)} \leq \frac{1}{\alpha_i} \|f_i\|_{H^{-1}(\Omega_i)} + \|\nabla u_{g_i}\|_{L_2(\Omega_i)}.$$

In particular, for $f_i = 0$, we therefore have

$$\|u_i\|_{H^1(\Omega_i),\Gamma_i}^2 = \left(\int_{\Gamma_i} g_i(x)ds_x \right)^2 + \|\nabla u_{i,0}\|_{L_2(\Omega_i)}^2$$

$$\leq \left(\int_{\Gamma_i} g_i(x)ds_x \right)^2 + \|\nabla u_{g_i}\|_{L_2(\Omega_i)}^2$$

$$= \|u_{g_i}\|_{H^1(\Omega_i),\Gamma_i}^2 \leq c_{\mathcal{E}_i} \|g_i\|_{H^{1/2}(\Gamma_i)}^2,$$

i.e.,

$$\|u_i\|_{H^1(\Omega_i),\Gamma_i} \leq c_{\mathcal{E}_i} \|g_i\|_{H^{1/2}(\Gamma_i)}. \tag{8}$$

It remains to find the associated Neumann datum $t_i = n_i \cdot \nabla u_i \in H^{-1/2}(\Gamma_i)$, where $H^{-1/2}(\Gamma_i) = [H^{1/2}(\Gamma_i)]'$ is the dual space which is equiped with the norm

$$\|\tau_i\|_{H^{-1/2}(\Gamma_i)} = \sup_{0 \neq w_i \in H^{1/2}(\Gamma_i)} \frac{\langle \tau, w_i \rangle_{\Gamma_i}}{\|w_i\|_{H^{1/2}(\Gamma_i)}}.$$

Using Green's first formula, $t_i \in H^{-1/2}(\Gamma_i)$ solves the variational problem

$$\int_{\Gamma_i} \alpha_i t_i(x) w_i(x) ds_x = \int_{\Omega_i} \alpha_i \nabla u_i(x) \cdot \nabla \mathcal{E}_i w_i(x) dx - \int_{\Omega_i} f_i(x) \mathcal{E}_i w_i(x) dx \tag{9}$$

for all test functions $w_i \in H^{1/2}(\Gamma_i)$. Using duality arguments, we then obtain

$$\|t_i\|_{H^{-1/2}(\Gamma_i)} = \sup_{0 \neq w_i \in H^{1/2}(\Gamma_i)} \frac{\langle t_i, w_i \rangle_{\Gamma_i}}{\|w_i\|_{H^{1/2}(\Gamma_i)}}$$

$$= \sup_{0 \neq w_i \in H^{1/2}(\Gamma_i)} \frac{1}{\|w_i\|_{H^{1/2}(\Gamma_i)}} \left[\langle \nabla u_i, \nabla \mathcal{E}_i w_i \rangle_{\Omega_i} - \frac{1}{\alpha_i} \langle f_i, \mathcal{E}_i w_i \rangle_{\Omega_i} \right]$$

$$\leq \sup_{0 \neq w_i \in H^{1/2}(\Gamma_i)} \frac{\|\mathcal{E}_i w_i\|_{H^1(\Omega_i)}}{\|w_i\|_{H^{1/2}(\Gamma_i)}} \left[\|u_i\|_{H^1(\Omega_i)} + \frac{1}{\alpha_i} \|f_i\|_{\widetilde{H}^{-1}(\Omega_i)} \right]$$

$$\leq c_{\mathcal{E}_i} \left[\|u_i\|_{H^1(\Omega_i)} + \frac{1}{\alpha_i} \|f_i\|_{\widetilde{H}^{-1}(\Omega_i)} \right],$$

where $\widetilde{H}^{-1}(\Omega_i) = [H^1(\Omega_i)]'$. The local Neumann datum t_i therefore depends only on the given right hand side f_i and on the prescribed Dirichlet datum g_i. Hence we have given a Dirichlet to Neumann map as

$$\alpha_i t_i(x) = \alpha_i (S_i g_i)(x) - (N_i f_i)(x) \quad \text{for } x \in \Gamma_i,$$

where we have used the local Steklov–Poincaré operator $S_i : H^{1/2}(\Gamma_i) \to H^{-1/2}(\Gamma_i)$, and the Newton potential $N_i : \widetilde{H}^{-1}(\Omega_i) \to H^{-1/2}(\Gamma_i)$ as given below. In particular, for $f_i = 0$, we therefore have $t_i = S_i g_i$ satisfying

$$\|S_i g_i\|_{H^{-1/2}(\Gamma_i)} = \|t_i\|_{H^{-1/2}(\Gamma_i)} \leq c_{\mathcal{E}_i}^2 \|g_i\|_{H^{1/2}(\Gamma_i)}. \tag{10}$$

If we define the linear operator $A_{\Omega_i} : H^1(\Omega_i) \to \widetilde{H}^{-1}(\Omega_i)$ via the Riesz representation theorem as

$$\langle A_{\Omega_i} u_i, v_i \rangle_{\Omega_i} = \int_{\Omega_i} \nabla u_i(x) \cdot \nabla v_i(x) dx \quad \text{for } u_i, v_i \in H^1(\Omega_i),$$

we can rewrite the variational formulation (9) as

$$\alpha_i \langle t_i, w_i \rangle_{\Gamma_i} = \alpha_i \int_{\Omega_i} \nabla u_i(x) \cdot \nabla \mathcal{E}_i w_i(x) dx - \int_{\Omega_i} f_i(x) \mathcal{E}_i w_i(x) dx$$

$$= \alpha_i \langle A_{\Omega_i} u_i, \mathcal{E}_i w_i \rangle_{\Omega_i} - \langle f_i, \mathcal{E}_i w_i \rangle_{\Omega_i}$$

$$= \alpha_i \langle A_{\Omega_i} [u_{i,0} + u_{g_i}], \mathcal{E}_i w_i \rangle_{\Omega_i} - \langle f_i, \mathcal{E}_i w_i \rangle_{\Omega_i}$$

$$= \alpha_i \langle A'_{\Gamma_i} u_{i,0}, \mathcal{E}_i w_i \rangle_{\Omega_i} + \alpha_i \langle A_{\Omega_i} u_{g_i}, \mathcal{E}_i w_i \rangle_{\Omega_i} - \langle f_i, \mathcal{E}_i w_i \rangle_{\Omega_i}$$

$$= \langle \alpha_i A'_{\Gamma_i} u_{i,0} + \alpha_i A_{\Omega_i} u_{g_i} - f_i, \mathcal{E}_i w_i \rangle_{\Omega_i},$$

and, therefore, as the following operator equation,

$$\alpha_i t_i = \mathcal{E}'_i \left[\alpha_i A_{\Omega_i} u_{g_i} + \alpha_i A'_{\Gamma_i} u_{i,0} - f_i \right]$$

$$= \mathcal{E}'_i \left[\alpha_i A_{\Omega_i} u_{g_i} + \alpha_i A'_{\Gamma_i} \left(\frac{1}{\alpha_i} A_{\Omega_i,0}^{-1} f_i - A_{\Omega_i,0}^{-1} A_{\Gamma_i} u_{g_i} \right) - f_i \right]$$

$$= \alpha_i \mathcal{E}'_i \left[A_{\Omega_i} - A'_{\Gamma_i} A_{\Omega_i,0}^{-1} A_{\Gamma_i} \right] \mathcal{E}_i g_i + \mathcal{E}'_i \left[A'_{\Gamma_i} A_{\Omega_i,0}^{-1} - I \right] f_i,$$

where A'_{Γ_i} and $\mathcal{E}'_i$ are the corresponding adjoint operators. Hence we can represent the Steklov–Poincaré operator as

$$S_i = \mathcal{E}'_i \left[A_{\Omega_i} - A'_{\Gamma_i} A^{-1}_{\Omega_i,0} A_{\Gamma_i} \right] \mathcal{E}_i \; : \; H^{1/2}(\Gamma_i) \to H^{-1/2}(\Gamma_i) \tag{11}$$

and the Newton potential as

$$N_i = \mathcal{E}'_i \left[A'_{\Gamma_i} A^{-1}_{\Omega_i,0} - I \right] \; : \; \widetilde{H}^{-1}(\Omega_i) \to H^{-1/2}(\Gamma_i). \tag{12}$$

Theorem 1. *The Steklov–Poincaré operator* $S_i \; : \; H^{1/2}(\Gamma_i) \to H^{-1/2}(\Gamma_i)$ *as defined in* (11) *is bounded,*

$$\|S_i g_i\|_{H^{-1/2}(\Gamma_i)} \leq c_{\mathcal{E}_i} \|g_i\|_{H^{1/2}(\Gamma)} \quad \textit{for all } g_i \in H^{1/2}(\Gamma_i),$$

and $H^{1/2}(\Gamma_i)$*–semi–elliptic, i.e.,*

$$\langle S_i g_i, g_i \rangle_{\Gamma_i} \geq \|g_i\|^2_{H^{1/2}(\Gamma_i)} \quad \textit{for all } g_i \in H_0^{1/2}(\Gamma_i),$$

where

$$H_0^{1/2}(\Gamma_i) = \left\{ w_i \in H^{1/2}(\Gamma_i) \; : \; \int\limits_{\Gamma_i} w_i(x)ds_x = 0 \right\}.$$

In particular, for $g_i \equiv 1$, *we have* $S_i g_i = 0$.

Proof. The boundedness of the Steklov–Poincaré operator S_i is just the estimate (10). Using (9) with $f_i = 0$, for $g_i \in H_0^{1/2}(\Gamma_i)$, we get

$$\langle S_i g_i, g_i \rangle_{\Gamma_i} = \langle t_i, g_i \rangle_{\Gamma_i} = \int\limits_{\Omega_i} \nabla u_i(x) \cdot \nabla u_{g_i}(x)dx$$

$$= \int\limits_{\Omega_i} \nabla u_i(x) \cdot \nabla[u_{g_i}(x) + u_{i,0}(x)]dx \; = \; |u_i|^2_{H^1(\Omega_i)}$$

$$= \left(\int\limits_{\Gamma_i} g_i(x)ds_x \right)^2 + |u_i|^2_{H^1(\Omega_i)} \; = \; \|u_i\|^2_{H^1(\Omega_i),\Gamma_i}.$$

Now the $H^{1/2}(\Gamma_i)$–semi–ellipticity follows from the trace theorem. $\square$

Finite Element Approximation

To define an approximate Dirichlet to Neumann map we first introduce the local finite element trial spaces

$$S_h^1(\Omega_i) \; = \; \mathrm{span}\{\phi^1_{i,k}\}_{k=1}^{\widetilde{M}_i} \subset H^1(\Omega_i)$$

and

$$S^1_{h,0}(\Omega_i) \;=\; S^1_h(\Omega_i) \cap H^1_0(\Omega_i) \;=\; \operatorname{span}\{\phi^1_{i,k}\}^{\widetilde{M}_i}_{k=M_i+1}$$

of piecewise linear basis functions $\phi^1_{i,k}$ with respect to some regular finite element mesh $\Omega_{i,h}$ characterized by the local mesh–size parameter h_i. Note that the basis functions $\phi^1_{i,k}$, $k = M_i + 1, \ldots, \widetilde{M}_i$ correspond to the interior degrees of freedom, while the remaining basis functions $\phi^1_{i,k}$, $k = 1, \ldots, M_i$ are associated to degrees of freedom on the boundary.

Let

$$u_{g_i,h}(x) = \sum_{k=1}^{\widetilde{M}_i} u_{g_i}(x_{i,k})\phi^1_{i,k}(x)$$

be the piecewise linear interpolation of the continuous extension u_{g_i}. The finite element approximation of the local variational problem (6) is to find $u_{i,0,h} \in S^1_{h,0}(\Omega_i)$ such that

$$\int_{\Omega_i} \alpha_i \nabla u_{i,0,h} \cdot \nabla v_{i,h}(x)dx = \int_{\Omega_i} f_i(x)v_{i,h}(x)dx - \int_{\Omega_i} \alpha_i \nabla u_{g_i,h}(x) \cdot \nabla v_{i,h}(x)dx$$

$$\tag{13}$$

is satisfied for all $v_{i,h} \in S^1_{h,0}(\Omega_i)$. This is equivalent to the Galerkin equations

$$\sum_{k=M_i+1}^{\widetilde{M}_i} [u_{i,0,k} + u_{g_i}(x_{i,k})] \int_{\Omega_i} \alpha_i \nabla \phi^1_{i,k}(x) \cdot \nabla \phi^1_{i,\ell}(x)dx$$

$$= \int_{\Omega_i} f_i(x)\phi^1_{i,\ell}(x)dx - \sum_{k=1}^{M_i} g_i(x_{i,k}) \int_{\Omega_i} \alpha_i \nabla \phi^1_{i,k}(x) \cdot \nabla \phi^1_{i,\ell}(x)dx$$

for all $\ell = M_i + 1, \ldots, \widetilde{M}_i$. Introducing the nodal values

$$u_{I,i,k} = u_{i,0,k} + u_{g_i}(x_{i,k}) \quad \text{for } k = M_i + 1, \ldots, \widetilde{M}_i,$$
$$u_{C,i,k} = g_i(x_{i,k}) \quad \text{for } k = 1, \ldots, M_i$$

as new unknowns, this is equivalent to a linear system

$$\alpha_i K_{II,i}\underline{u}_{I,i} \;=\; \underline{f}_{I,i} - \alpha_i K_{CI,i}\underline{u}_{C,i},$$

where the local stiffness matrix is given by

$$K_{II,i}[\ell, k] = \int_{\Omega_i} \nabla \phi^1_{i,k}(x) \cdot \nabla \phi^1_{i,\ell}(x)dx$$

for $k, \ell = M_i + 1, \ldots, \widetilde{M}_i$, while the vector of the right hand side is determined by

$$f_{I,i,\ell} = \int_{\Omega_i} f_i(x)\phi^1_{i,\ell}(x)dx.$$

In addition,

$$K_{CI,i}[\ell, k] = \int_{\Omega_i} \nabla\phi^1_{i,k}(x) \cdot \nabla\phi^1_{i,\ell}(x)dx$$

for $k = 1, \ldots, M_i$, $\ell = M_i + 1, \ldots, \widetilde{M}_i$. The solution vector

$$\underline{u}_{I,i} = \frac{1}{\alpha_i} K^{-1}_{II,i}\underline{f}_{I,i} - K^{-1}_{II,i}K_{CI,i}\underline{u}_{C,i}$$

defines an approximate solution $u_{i,h} = u_{i,0,h} + u_{g_i,h}$ for which the error estimate

$$\|u_i - u_{i,h}\|_{H^1(\Omega_i),\Gamma_i} \leq c\, h_i\, |u_i|_{H^2(\Omega_i)}$$

provided that the regularity assumption $u_i \in H^2(\Omega_i)$ holds.

Now, instead of the variational problem (9), we have to consider a perturbed formulation to find $\widetilde{t}_i \in H^{-1/2}(\Gamma_i)$ such that

$$\int_{\Gamma_i} \alpha_i\widetilde{t}_i(x)w_i(x)ds_x = \int_{\Omega_i} \alpha_i\nabla u_{i,h}(x) \cdot \nabla\mathcal{E}_i w_i(x)dx - \int_{\Omega_i} f_i(x)\mathcal{E}_i w_i(x)dx \quad (14)$$

is satisfied for all test functions $w_i \in H^{1/2}(\Gamma_i)$. This implies an approximated Dirichlet to Neumann map

$$\alpha_i\widetilde{t}_i(x) = \alpha_i(\widetilde{S}_i g_i)(x) - (\widetilde{N}_i f_i)(x) \quad \text{for } x \in \Gamma_i,$$

where $\widetilde{S}_i$ is an approximate Steklov–Poincaré operator which is defined via the solution of the Galerkin variational formulation (13).

Theorem 2. [48] *The approximate Steklov–Poincaré operator $\widetilde{S}_i : H^{1/2}(\Gamma_i) \to H^{-1/2}(\Gamma_i)$ as defined above is bounded and $H^{1/2}_0(\Gamma_i)$–elliptic. Moreover, there holds the a priori error estimate*

$$\|(S_i - \widetilde{S}_i)g_i\|_{H^{-1/2}(\Gamma_i)} \leq c\, h_i\, |u_i|_{H^2(\Omega_i)}$$

when assuming $u_i \in H^2(\Omega_i)$.

When choosing in (14) $\phi^1_{i,\ell}$, $\ell = 1, \ldots, M_i$ as a test function, this gives

$$\alpha_i \int_{\Gamma_i} \widetilde{t}_i(x)\phi^1_{i,\ell}(x)ds_x = \alpha_i \int_{\Omega_i} \nabla u_{i,h}(x) \cdot \nabla\phi^1_{i,\ell}(x)dx - \int_{\Omega_i} f_i(x)\phi^1_{i,\ell}(x)dx$$

$$= \sum_{k=1}^{\widetilde{M}_i} \alpha_i u_{I,i,k} \int_{\Omega_i} \nabla\phi^1_{i,k}(x) \cdot \nabla\phi^1_{i,\ell}(x)dx$$

$$+ \sum_{k=\widetilde{M}_i+1}^{M_i} u_{C,i,k} \int_{\Omega_i} \nabla\phi^1_{i,k}(x) \cdot \nabla\phi^1_{i,\ell}(x)dx - \int_{\Omega_i} f_i(x)\phi^1_{i,\ell}(x)dx$$

$$= \sum_{k=1}^{\widetilde{M}_i} \alpha_i K_{IC,i}[\ell,k] u_{I,i,k} + \sum_{k=\widetilde{M}_i+1}^{M_i} \alpha_i K_{CC,i}[\ell,k] u_{C,i,k} - f_{C,i,\ell}$$

$$= \alpha_i \left(K_{CC,i}\underline{u}_{C,i} + K_{IC,i}\underline{u}_{I,i} \right)_\ell - f_{C,i,\ell}$$

$$= \alpha_i \left(\left[K_{CC,i} - K_{IC,i} K_{II,i}^{-1} K_{CI,i} \right] \underline{u}_{C,i} \right)_\ell + \left(K_{CI,i} K_{II,i}^{-1} \underline{f}_{I,i} \right)_\ell - f_{C,i,\ell}.$$

Hence we obtain the discrete Dirichlet to Neumann map

$$\alpha_i \underline{\widetilde{t}}_i = \alpha_i \widetilde{S}_{i,h}^{\text{FEM}} \underline{u}_C + K_{CI,i} K_{II,i}^{-1} \underline{f}_{I,i} - \underline{f}_{C,i} \tag{15}$$

with the finite element approximation of the Steklov–Poincaré operator

$$\widetilde{S}_{i,h}^{\text{FEM}} = K_{CC,i} - K_{IC,i} K_{II,i}^{-1} K_{CI,i}. \tag{16}$$

Boundary Integral Equations

Instead of using finite element discretizations of domain variational formulations for the numerical solution of the local Dirichlet boundary value problem (4), we now consider boundary integral formulations and their boundary element discretization. The starting point is the representation formula

$$u_i(x) = \int_{\Gamma_i} U^*(x,y) t_i(y) ds_y - \int_{\Gamma_i} \frac{\partial}{\partial n_y} U^*(x,y) g_i(y) ds_y + \frac{1}{\alpha_i} \int_{\Omega_i} U^*(x,y) f_i(y) dy,$$

that holds for $x \in \Omega_i$, where

$$U^*(x,y) = \frac{1}{4\pi} \frac{1}{|x-y|}$$

is the fundamental solution of the Laplace operator. To find the yet unknown Neumann datum $t_i \in H^{-1/2}(\Gamma_i)$, we first consider the boundary integral equation which results from the representation formula for $x \to \Gamma_i$,

$$\int_{\Gamma_i} U^*(x,y) t_i(y) ds_y = \frac{1}{2} g_i(x) + \int_{\Gamma_i} \frac{\partial}{\partial n_y} U^*(x,y) g_i(y) ds_y - \frac{1}{\alpha_i} \int_{\Omega_i} U^*(x,y) f(y) dy,$$

or,

$$(V_i t_i)(x) = (\frac{1}{2} I + K_i) g_i(x) - \frac{1}{\alpha_i} (\widetilde{N}_{i,0} f_i)(x) \quad \text{for } x \in \Gamma_i. \tag{17}$$

Here, $x \in \Gamma_i$ is assumed to be on a smooth part of the boundary Γ_i. Since we are using a Galerkin approach, such an assumption is sufficient. Moreover, $V_i : H^{-1/2}(\Gamma_i) \to H^{1/2}(\Gamma_i)$ is the single layer potential, $K_i : H^{1/2}(\Gamma_i) \to H^{1/2}(\Gamma_i)$ is the double layer potential, and $\widetilde{N}_{i,0} : \widetilde{H}^{-1}(\Omega_i) \to H^{1/2}(\Gamma_i)$ is the Newton potential. Since the single layer potential operator is $H^{-1/2}(\Gamma_i)$–elliptic and therefore invertible, we find the Dirichlet to Neumann map

$$\alpha_i t_i(x) = \alpha_i V_i^{-1}(\frac{1}{2}I + K_i)g_i(x) - V_i^{-1}\widetilde{N}_{0,i}f(x)$$
$$= \alpha_i(S_i g_i)(x) - (N_i f)(x), \quad x \in \Gamma_i$$

with the boundary integral operator representation of the Steklov–Poincaré operator

$$S_i = V_i^{-1}(\frac{1}{2}I + K_i) : H^{1/2}(\Gamma_i) \to H^{-1/2}(\Gamma_i), \tag{18}$$

and with the operator

$$N_i = V_i^{-1}\widetilde{N}_{0,i} : \widetilde{H}^{-1}(\Omega_i) \to H^{-1/2}(\Gamma_i).$$

Although the Steklov–Poincaré operator (18) is self–adjoint in the continuous case, an approximation of this composed operator results in a non–symmetric stiffness matrix in general. Hence we are interested in alternative representations which result in symmetric boundary element approximations.

Since the solution of the local Dirichlet boundary value problem (4) is given by the representation formula, the application of the normal derivative to the representation formula gives a second, the so–called hypersingular boundary integral equation,

$$t_i(x) = \frac{1}{2}t_i(x) + \int_{\Gamma_i} \frac{\partial}{\partial n_x}U^*(x,y)t_i(y)ds_y - \frac{\partial}{\partial n_x}\int_{\Gamma_i} \frac{\partial}{\partial n_y}U^*(x,y)g_i(y)ds_y$$
$$- \frac{1}{\alpha_i}\frac{\partial}{\partial n_x}\int_{\Omega_i} U^*(x,y)f_i(y)dy,$$

or,

$$t_i(x) = \frac{1}{2}t_i(x) + (K_i't_i)(x) + (D_i g_i)(x) - \frac{1}{\alpha_i}(\widetilde{N}_{i,1}f_i)(x) \quad \text{for } x \in \Gamma_i. \tag{19}$$

Here, $K_i' : H^{-1/2}(\Gamma_i) \to H^{-1/2}(\Gamma_i)$ is the adjoint double layer potential, $D_i : H^{1/2}(\Gamma_i) \to H^{-1/2}(\Gamma_i)$ is the hypersingular boundary integral operator, and $\widetilde{N}_{i,1} : \widetilde{H}^{-1}(\Omega_i) \to H^{-1/2}(\Gamma_i)$ is the normal derivative of the Newton potential. Inserting the first boundary integral representation of the Dirichlet to Neumann map into (19) gives the relations

$$\alpha_i t_i(x) = \alpha_i(D_i g_i)(x) + (\frac{1}{2}I + K_i')(\alpha_i t_i)(x) - (\widetilde{N}_{i,1}f_i)(x)$$

$$= \alpha_i(D_i g_i)(x) + (\frac{1}{2}I + K_i')\left[\alpha_i V_i^{-1}(\frac{1}{2}I + K_i)g_i(x) - V_i^{-1}\widetilde{N}_{i,0}f_i(x)\right]$$

$$- (\widetilde{N}_{1,i}f)(x)$$

$$= \alpha_i \left[D_i + (\frac{1}{2}I + K_i')V_i^{-1}(\frac{1}{2}I + K_i) \right] g_i(x)$$

$$-(\widetilde{N}_{i,1}f_i)(x) - (\frac{1}{2}I + K_i')V_i^{-1}\widetilde{N}_{i,0}f_i(x)$$

$$= \alpha_i(S_i g_i)(x) - (N_i f)(x) \quad \text{for } x \in \Gamma_i$$

with the so–called symmetric boundary integral operator representation of the Steklov–Poincaré operator,

$$S_i = D_i + (\frac{1}{2}I + K_i')V_i^{-1}(\frac{1}{2}I + K_i) \; : \; H^{1/2}(\Gamma_i) \to H^{-1/2}(\Gamma_i), \qquad (20)$$

and with an alternative representation of

$$N_i = \widetilde{N}_{i,1} + (\frac{1}{2}I + K_i')V_i^{-1}\widetilde{N}_{i,0} \; : \; \widetilde{H}^{-1}(\Omega_i) \to H^{-1/2}(\Gamma_i).$$

While the Steklov–Poincaré operator $S_i : H^{1/2}(\Gamma_i) \to H^{-1/2}(\Gamma_i)$ is bounded, it is not obvious which Sobolev norms in $H^{1/2}(\Gamma_i)$ and $H^{-1/2}(\Gamma_i)$ have to be used, respectively. When using appropriate norms, explicit estimates can be derived as in Theorem 1, where we used a trace norm to characterize $H^{1/2}(\Gamma_i)$.

In the case of boundary integral operators a natural choice is to use norms which are induced by the single layer potential and its inverse. In particular,

$$\|w_i\|_{V_i} = \sqrt{\langle V_i w_i, w_i \rangle_{\Gamma_i}} \quad \text{and} \quad \|v_i\|_{V_i^{-1}} = \sqrt{\langle V_i^{-1} v_i, v_i \rangle_{\Gamma_i}}$$

define equivalent norms of the Sobolev spaces $H^{-1/2}(\Gamma_i)$ and $H^{1/2}(\Gamma_i)$, respectively. Using both boundary integral representations (18) and (20) of the Steklov–Poincaré operator S_i, we obtain the estimate [52]

$$\|(\frac{1}{2}I + K_i)v_i\|_{V_i^{-1}} \leq c_K(\Gamma_i) \|v_i\|_{V_i^{-1}} \quad \text{for all } v_i \in H^{1/2}(\Gamma_i), \qquad (21)$$

where

$$c_K(\Gamma_i) = \frac{1}{2} + \sqrt{\frac{1}{4} - c_1^{V_i} c_1^{D_i}} < 1$$

is the contraction constant of the double layer potential $\frac{1}{2}I + K_i$ defined by the ellipticity constants $c_1^{V_i}$ and $c_1^{D_i}$ of the single layer potential V_i and of the hypersingular boundary integral operator D_i, respectively.

Using (21) we find the boundedness estimate [52]

$$\|S_i g_i\|_{V_i} \leq c_K(\Gamma_i) \|g_i\|_{V_i^{-1}} \quad \text{for all } g_i \in H^{1/2}(\Gamma_i) \qquad (22)$$

as well as the ellipticity estimate

$$\langle S_i g_i, g_i \rangle_{\Gamma_i} \geq [1 - c_K(\Gamma_i)] \|g_i\|_{V_i^{-1}}^2 \quad \text{for all } g_i \in H_0^{1/2}(\Gamma_i). \qquad (23)$$

Since all representations of the local Steklov–Poincaré operators S_i coincide, the boundedness estimate (22) and the ellipticity estimate (23) are also true for the definition (11) based on a domain variational formulation. Note that the contraction rate $c_K(\Gamma_i)$ only reflects the shape of the subdomain, but does not reflect the size or the diameter of the subdomain Ω_i.

Boundary Element Methods

For $g_i \in H^{1/2}(\Gamma_i)$, the application of the Steklov–Poincaré operator S_i in its symmetric representation (20) can be rewritten as

$$S_i g_i = D_i g_i + (\frac{1}{2}I + K'_i)V_i^{-1}(\frac{1}{2}I + K_i)g_i = D_i g_i + (\frac{1}{2}I + K'_i)w_i,$$

where $w_i = V_i^{-1}(\frac{1}{2}I + K_i)g_i \in H^{-1/2}(\Gamma_i)$ is the unique solution of the local variational problem

$$\langle V_i w_i, \tau_i \rangle_{\Gamma_i} = \langle (\frac{1}{2}I + K_i)g_i, \tau_i \rangle_{\Gamma_i} \quad \text{for all } \tau_i \in H^{-1/2}(\Gamma_i).$$

Let

$$S_h^0(\Gamma_i) = \text{span}\{\psi_{i,n}^0\}_{n=1}^{N_i} \subset H^{-1/2}(\Gamma_i)$$

be the boundary element space of piecewise constant basis functions $\psi_{i,n}^0$. Using the Galerkin solution $w_{i,h} \in S_h^0(\Gamma_i)$ satisfying

$$\langle V_i w_{i,h}, \tau_{i,h} \rangle_{\Gamma_i} = \langle (\frac{1}{2}I + K_i)g_i, \tau_{i,h} \rangle_{\Gamma_i} \quad \text{for all } \tau_{i,h} \in S_h^0(\Gamma_i), \qquad (24)$$

we may define an approximate Steklov–Poincaré operator by the relation

$$\widetilde{S}_i g_i = D_i g_i + (\frac{1}{2}I + K'_i)w_{i,h}. \qquad (25)$$

Theorem 3. [48] *The approximate Steklov–Poincaré operator $\widetilde{S}_i : H^{1/2}(\Gamma_i) \to H^{-1/2}(\Gamma_i)$ as defined in (25) is bounded and $H_0^{1/2}(\Gamma_i)$–elliptic, i.e.,*

$$\langle \widetilde{S}_i g_i, g_i \rangle_{\Gamma_i} \geq \langle D_i g_i, g_i \rangle_{\Gamma_i} \geq c_1^{D_i} \|g_i\|_{H^{1/2}(\Gamma_i)}^2 \quad \textit{for all } g_i \in H_0^{1/2}(\Gamma_i).$$

Moreover, there holds the a priori error estimate

$$\|(S_i - \widetilde{S}_i)g_i\|_{H^{-1/2}(\Gamma_i)} \leq c\, h_i^{3/2} \|S_i g_i\|_{H_{pw}^1(\Gamma_i)}$$

when assuming $S_i g_i \in H_{\mathrm{pw}}^1(\Gamma_i)$, i.e., $u_i \in H^{5/2}(\Omega_i)$. Note that $H_{\mathrm{pw}}^1(\Gamma_i)$ is the Sobolev space which is defined piecewise.

In the same way as above we may also introduce some approximation of the volume potential $N_i f = V_i^{-1}\widetilde{N}_{0,i}f$. In particular, $N_{i,h}f \in S_h^0(\Gamma_i)$ is defined as the unique solution of the Galerkin variational problem

$$\langle V_i N_{i,h} f, \tau_{i,h} \rangle_{\Gamma_i} = \langle \widetilde{N}_{0,i} f, \tau_{i,h} \rangle_{\Gamma_i} \quad \text{for all } \tau_{i,h} \in S_h^0(\Gamma_i).$$

Let

$$g_{i,h} = \sum_{k=1}^{M_i} u_{C,i} \varphi_{i,k}^1 \in S_h^1(\Gamma_i)$$

be some piecewise linear approximation of the given Dirichlet datum g_i. For the approximate Dirichlet to Neumann map we then find

$$\int_{\Gamma_i} \alpha_i \widetilde{t}_i(x) \varphi_\ell^1(x)(x) ds_x$$

$$= \int_{\Gamma_i} \left[\alpha_i (D_i g_{i,h})(x) + \alpha_i (\frac{1}{2} I + K_i') w_{i,h}(x) - N_{i,h} f \right] \varphi_{i,\ell}^1(x) ds_x$$

$$= \sum_{k=1}^{M_i} u_{C,i,k} \alpha_i \langle D_i \varphi_{i,k}^1, \varphi_{i,\ell}^1 \rangle_{\Gamma_i} + \sum_{n=1}^{N_i} w_{i,n} \alpha_i \langle (\frac{1}{2} I + K_i') \psi_{i,n}^0, \varphi_{i,\ell}^1 \rangle_{\Gamma_i}$$

$$- \sum_{n=1}^{N_i} N_{i,h,n} \langle \psi_{i,n}^0, \varphi_{i,\ell}^1 \rangle_{\Gamma_i}$$

for $\ell = 1, \ldots, M_i$ where

$$\sum_{n=1}^{N_i} w_{i,n} \langle V_i \psi_{i,n}^0, \psi_{i,m}^0 \rangle_{\Gamma_i} = \sum_{k=1}^{M_i} u_{C,i,k} \langle (\frac{1}{2} I + K_i) \varphi_{i,k}^1, \psi_{i,m}^0 \rangle_{\Gamma_i}$$

and

$$\sum_{n=1}^{N_i} N_{i,h,n} \langle V_i \psi_{i,n}^0, \psi_{i,m}^0 \rangle_{\Gamma_i} = \langle \widetilde{N}_{0,i} f, \psi_{i,m}^0 \rangle_{\Gamma_i} = f_{N,i,m}$$

for $m = 1, \ldots, N_i$. Hence we obtain the discrete Dirichlet to Neumann map

$$\alpha_i \underline{\widetilde{t}}_i = \alpha_i \widetilde{S}_{i,h}^{\text{BEM}} \underline{u}_{C,i} - M_{i,h}^\top V_{i,h}^{-1} \underline{f}_{N,i} \tag{26}$$

with the boundary element approximation of the Steklov–Poincaré operator

$$\widetilde{S}_{i,h}^{\text{BEM}} = D_{i,h} + \widetilde{K}_{i,h}^\top V_{i,h}^{-1} \widetilde{K}_{i,h} \tag{27}$$

and

$$V_{i,h}[m, n] = \langle V_i \psi_{i,n}^0, \psi_{i,m}^0 \rangle_{\Gamma_i},$$

$$D_{i,h}[\ell, k] = \langle D_i \varphi_{i,k}^1, \varphi_{i,\ell}^1 \rangle_{\Gamma_i},$$

$$\widetilde{K}_{i,h}[m, k] = \langle (\frac{1}{2} I + K_i) \varphi_{i,k}^1, \psi_{i,m}^0 \rangle_{\Gamma_i},$$

$$M_{i,h}[m, k] = \langle \varphi_{i,k}^1, \psi_{i,m}^0 \rangle_{\Gamma_i}$$

for $m, n = 1, \ldots, N_i$, $k, \ell = 1, \ldots, M_i$.

Instead of using the symmetric representation (20) we may use also the first representation (18) to define an approximate Steklov–Poincaré operator as

$$\widetilde{S}_i g_i = w_{i,h} \qquad (28)$$

where $w_{i,h} \in S_h^0(\Gamma_i)$ is the unique solution of the variational problem (24). Although this approximated Steklov–Poincaré operator $\widetilde{S}_i : H^{1/2}(\Gamma_i) \to H^{-1/2}(\Gamma_i)$ is bounded and satisfies an approximation property as in Theorem 3, it is in general not stable. Let $S_h^1(\Gamma_i) \subset H^{1/2}(\Gamma_i)$ be some boundary element space of piecewise linear basis functions. To ensure the $S_h^1(\Gamma_i)$–semi–ellipticity of the approximate Steklov–Poincaré operator $\widetilde{S}_i$ as defined in (28), we need to assume the discrete stability condition

$$c_S \, \|g_{i,h}\|_{H^{1/2}(\Gamma_i)} \leq \sup_{0 \neq \tau_{i,h} \in S_h^0(\Gamma_i)} \frac{\langle g_{i,h}, \tau_{i,h} \rangle_{\Gamma_i}}{\|\tau_{i,h}\|_{H^{-1/2}(\Gamma_i)}} \quad \text{for all } g_{i,h} \in S_h^1(\Gamma_i). \quad (29)$$

Note that the discrete stability condition (29) is satisfied, for example, when using a sufficiently small mesh size to define the trial space $S_h^0(\Gamma_i)$ compared to the mesh size or $S_h^1(\Gamma_i)$ [56], or when using piecewise linear basis functions to define both trial spaces [47].

2.2 Neumann Boundary Value Problems

We now consider the local Neumann boundary value problem

$$-\alpha_i \Delta u_i(x) = f_i(x) \quad \text{for } x \in \Omega_i, \quad \alpha_i \frac{\partial}{\partial n_i} u_i(x) = \lambda_i(x) \quad \text{for } x \in \Gamma_i, \quad (30)$$

where we have to assume the solvability condition

$$\int_{\Omega_i} f_i(x) dx + \int_{\Gamma_i} \lambda_i(x) ds_x = 0. \qquad (31)$$

Moreover, the solution of the local Neumann boundary value problem (30) is only unique up to an additive constant, i.e., if u_i is a solution of (30), then $u_i + \gamma_i$ is also a solution of (30) for any constant $\gamma_i \in \mathbb{R}$.

Domain Variational Formulation

The associated variational formulation of the Neumann boundary value problem (30) is to find $u_i \in H_*^1(\Omega_i)$ such that

$$\alpha_i \int_{\Omega_i} \nabla u_i(x) \nabla v_i(x) dx = \int_{\Omega_i} f_i(x) v_i(x) dx + \int_{\Gamma_i} \lambda_i(x) v_i(x) ds_x \qquad (32)$$

is satisfied for all $v_i \in H_*^1(\Omega_i)$, where

$$H^1_*(\Omega_i) = \left\{ v_i \in H^1(\Omega_i) \ : \ \int_{\Omega_i} v_i(x)dx = 0 \right\}$$

is a suitable choosen subspace of $H^1(\Omega_i)$. Since

$$\|v_i\|^2_{H^1(\Omega_i),\Omega_i} = \left[\int_{\Omega_i} v_i(x)dx \right]^2 + \int_{\Omega_i} |\nabla v_i(x)|^2 dx$$

defines an equivalent norm in $H^1(\Omega)$, the operator $A_{\Omega_i} : H^1(\Omega_i) \to \widetilde{H}^{-1}(\Omega_i)$ defined via the Riesz representation theorem, i.e.,

$$\langle A_{\Omega_i} u_i, v_i \rangle_{\Omega_i} = \int_{\Omega_i} \nabla u_i(x)\nabla v_i(x)dx \quad \text{for all } u_i, v_i \in H^1(\Omega),$$

is $H^1_*(\Omega_i)$–elliptic. Hence there exists a unique solution $u_i \in H^1_*(\Omega_i)$ of the variational problem (32). Using the trace operator $B_i : H^1(\Omega_i) \to H^{1/2}(\Gamma_i)$ and its adjoint, $B'_i : H^{-1/2}(\Gamma_i) \to \widetilde{H}^{-1}(\Omega)$, we can write the general solution of the Neumann boundary value problem (30) as

$$u_i = A^+_{\Omega_i}[f_i + B'_i\lambda_i] + \gamma_i, \quad \gamma_i \in \mathbb{R}$$

where $A^+_{\Omega_i}$ is the associated pseudoinverse. From this we obtain the Neumann to Dirichlet map as

$$g_i = B_i A^+_{\Omega_i}[B'_i\lambda_i + f_i] + \gamma_i, \quad \gamma_i \in \mathbb{R}, \quad \langle f_i, 1 \rangle_{\Omega_i} + \langle \lambda_i, 1 \rangle_{\Gamma_i} = 0. \tag{33}$$

Instead of the variational formulation (32), where the side condition $\langle v_i, 1 \rangle_{\Omega_i}=0$ was included in the definition of the function space $H^1_*(\Omega)$, we now consider an extended variational problem to find $u_i \in H^1(\Omega_i)$ satisfying

$$\alpha_i \int_{\Omega_i} u_i(x)dx \int_{\Omega_i} v_i(x)dx + \alpha_i \int_{\Omega_i} \nabla u_i(x)\nabla v_i(x)dx \tag{34}$$

$$= \int_{\Omega_i} f_i(x)v_i(x)dx + \int_{\Gamma_i} \lambda_i(x)v_i(x)ds_x$$

for all $v_i \in H^1(\Omega_i)$. Since the operator $\bar{A}_{\Omega_i} : H^1(\Omega_i) \to \widetilde{H}^{-1}(\Omega)$ defined by

$$\langle \bar{A}_{\Omega_i} u_i, v_i \rangle_{\Omega_i} = \int_{\Omega_i} u_i(x)dx \int_{\Omega_i} v_i(x)dx + \int_{\Omega_i} \nabla u_i(x)\nabla v_i(x)dx$$

for all $u_i, v_i \in H^1(\Omega_i)$ is $H^1(\Omega_i)$–elliptic, we find

$$u_i = \frac{1}{\alpha_i} \bar{A}^{-1}_{\Omega_i}[f_i + B'_i\lambda_i]$$

as the unique solution of the variational problem (34) for any given data $f_i \in \widetilde{H}^{-1}(\Omega_i)$ and $\lambda_i \in H^{-1/2}(\Gamma_i)$. Moreover, when assuming the solvability condition (31), we obtain $u_i \in H^1_*(\Omega)$, and therefore, the general Neumann to Dirichlet map

$$g_i = \frac{1}{\alpha_i} B_i \bar{A}_{\Omega_i}^{-1}[B_i'\lambda_i + f_i] + \gamma_i, \quad \gamma_i \in \mathbb{R}, \quad \langle f_i, 1 \rangle_{\Omega_i} + \langle \lambda_i, 1 \rangle_{\Gamma_i} = 0. \quad (35)$$

The involved Poincaré–Steklov operator

$$T_i = B_i \bar{A}_{\Omega_i}^{-1} B_i' : H^{-1/2}(\Gamma_i) \to H^{1/2}(\Gamma_i)$$

is bounded and $H^{-1/2}(\Gamma_i)$–elliptic.

Finite Element Approximation

Let
$$S_h^1(\Omega_i) = \mathrm{span}\{\phi_{i,k}^1\}_{k=1}^{\widetilde{M}_i} \subset H^1(\Omega_i)$$

be the local finite element space of piecewise linear basis functions $\phi_{i,k}^1$ which are again defined with respect to some regular finite element mesh $\Omega_{i,h}$ with the mesh–size parameter h_i. In addition, let

$$\lambda_{i,h} \in S_h^0(\Gamma_i) = \mathrm{span}\{\psi_{i,n}^0\}_{n=1}^{N_i}$$

be some approximation of the given Neumann data by using piecewise constant basis functions $\psi_{i,n}^0$. The Galerkin formulation of the extended variational problem (34) is to find $u_{i,h} \in S_h^1(\Omega_i)$ such that

$$\alpha_i \int_{\Omega_i} u_{i,h}(x)dx \int_{\Omega_i} v_{i,h}(x)dx + \alpha_i \int_{\Omega_i} \nabla u_{i,h}(x)\nabla v_{i,h}(x)dx \qquad (36)$$

$$= \int_{\Omega_i} f_i(x)v_{i,h}(x)dx + \int_{\Gamma_i} \lambda_{i,h}(x)v_{i,h}(x)ds_x$$

is satisfied for all $v_{i,h} \in S_h^1(\Omega_i)$. This is equivalent to a linear system of algebraic equations,

$$\alpha_i \bar{A}_{\Omega_i,h}\underline{u}_i = \underline{f}_i + B_{i,h}^\top \underline{\lambda}_i,$$

with

$$\bar{A}_{\Omega_i,h}[\ell, k] = \int_{\Omega_i} \phi_{i,k}^1(x)dx \int_{\Omega_i} \phi_{i,\ell}^1(x)dx + \int_{\Omega_i} \nabla\phi_{i,k}^1(x)\nabla\phi_{i,\ell}^1(x)dx,$$

$$f_{i,\ell} = \int_{\Omega_i} f_i(x)\phi_{i,\ell}^1(x)dx,$$

$$B_{i,h}[n, k] = \int_{\Gamma_i} \phi_{i,k}^1(x)\psi_{i,n}^0(x)ds_x$$

for $k, \ell = 1, \ldots, \widetilde{M}_i$, $n = 1, \ldots, N_i$. Hence we find

$$\underline{u}_i = \bar{A}_{\Omega_i,h}^{-1}[\underline{f}_i + B_{i,h}^\top \underline{\lambda}_i]$$

yielding the approximate solution $u_{i,h} \in S_h^1(\Omega_i)$. Taking the trace $u_{i,h|\Gamma_i}$, this defines an approximation of the Neumann to Dirichlet map (35), i.e., an approximate Poincaré–Steklov operator $\widetilde{T}_i$.

Theorem 4. *The approximate Poincaré–Steklov operator $\widetilde{T}_i : H^{-1/2}(\Gamma_i) \to H^{1/2}(\Gamma_i)$ as introduced above is bounded and $H^{-1/2}(\Gamma_i)$–elliptic. Moreover, there holds the a priori error estimate*

$$\|(T_i - \widetilde{T}_i)\lambda_i\|_{H^{1/2}(\Gamma_i)} \leq c\,h_i\,\|u_i\|_{H^2(\Omega_i)}$$

when assuming $u_i \in H^2(\Omega_i)$.

Boundary Integral Equations

Using the hypersingular boundary integral equation (19) the unknown Dirichlet datum $g_i \in H^{1/2}(\Gamma_i)$ is a solution of

$$\alpha_i(D_i g_i)(x) = \frac{1}{2}\lambda_i(x) - (K_i'\lambda_i)(x) + (\widetilde{N}_{i,1}f_i)(x) \quad \text{for } x \in \Gamma_i.$$

The local hypersingular boundary integral operator $D_i : H^{1/2}(\Gamma_i) \to H^{-1/2}(\Gamma_i)$ is only $H^{1/2}(\Gamma_i)$–semi–elliptic, i.e.

$$\langle D_i v_i, v_i \rangle_{\Gamma_i} \geq c_1^{D_i} \|v_i\|_{H^{1/2}(\Gamma_i)}^2 \quad \text{for all } v_i \in H_0^{1/2}(\Gamma_i).$$

As for the extended variational formulation (34) we may consider an extended variational problem [39] to find $g_i \in H^{1/2}(\Gamma_i)$ such that

$$\alpha_i\left[\langle D_i u_i, v_i \rangle_{\Gamma_i} + \langle u_i, 1 \rangle_{\Gamma_i}\langle v_i, 1 \rangle_{\Gamma_i}\right] = \langle(\frac{1}{2}I - K_i')\lambda_i, v_i \rangle_{\Gamma_i} + \langle \widetilde{N}_{i,1}f_i, v_i \rangle_{\Gamma_i} \quad (37)$$

is satisfied for all $v_i \in H^{1/2}(\Gamma_i)$. Since the modified hypersingular boundary integral operator $\widetilde{D}_i : H^{1/2}(\Gamma_i) \to H^{-1/2}(\Gamma_i)$ which is defined via the bilinear form

$$\langle \widetilde{D}_i u_i, v_i \rangle_{\Gamma_i} = \langle D_i u_i, v_i \rangle_{\Gamma_i} + \langle u_i, 1 \rangle_{\Gamma_i}\langle v_i, 1 \rangle_{\Gamma_i},$$

is $H^{1/2}(\Gamma_i)$–elliptic, the extended variational problem (37) has a unique solution $u_i \in H^{1/2}(\Gamma_i)$ for any given data $f_i \in \widetilde{H}^{-1}(\Omega_i)$ and $\lambda_i \in H^{-1/2}(\Gamma_i)$. If the solvability condition (31) is satisfied, then we have $u_i \in H_0^{1/2}(\Gamma_i)$ and the general solution of the local Neumann boundary value problem is given by

$$u_i = \frac{1}{\alpha_i}\widetilde{D}_i^{-1}(\frac{1}{2}I - K_i')\lambda_i + \frac{1}{\alpha_i}\widetilde{D}_i^{-1}\widetilde{N}_{i,1}f_i + \gamma_i, \quad \gamma_i \in \mathbb{R}.$$

Inserting this into the first boundary integral equation (17), we obtain

$$\alpha_i u_i = V_i \lambda_i + \alpha_i(\frac{1}{2}I - K_i)u_i + \widetilde{N}_{i,0}f_i$$

$$= V_i \lambda_i + \alpha_i(\frac{1}{2}I - K_i)\left[\frac{1}{\alpha_i}\widetilde{D}_i^{-1}(\frac{1}{2}I - K_i')\lambda_i + \frac{1}{\alpha_i}\widetilde{D}_i^{-1}\widetilde{N}_{i,1}f_i\right] + \widetilde{N}_{i,0}f_i$$

$$= \left[V_i + (\frac{1}{2}I - K_i)\widetilde{D}_i^{-1}(\frac{1}{2}I - K_i')\right]\lambda_i + (\frac{1}{2}I - K_i)\widetilde{D}_i^{-1}\widetilde{N}_{i,1}f_i + \widetilde{N}_{i,0}f_i$$

and therefore the Neumann to Dirichlet map

$$u_i(x) = \frac{1}{\alpha_i}(T_i\lambda_i)(x) + \frac{1}{\alpha_i}(\overline{N}_i f_i)(x) + \gamma_i \quad \text{for } x \in \Gamma_i, \gamma_i \in \mathbb{R},$$

where

$$T_i = V_i + (\frac{1}{2}I - K_i)\widetilde{D}_i^{-1}(\frac{1}{2}I - K_i') : H^{-1/2}(\Gamma_i) \to H^{1/2}(\Gamma_i)$$

is again the Poincaré–Steklov operator.

Boundary Element Methods

For $\lambda_i \in H^{-1/2}(\Gamma_i)$, the application of the Poincaré–Steklov operator T_i in its symmetric representation reads as

$$T_i\lambda_i = V_i\lambda_i + (\frac{1}{2}I - K_i)\widetilde{D}_i^{-1}(\frac{1}{2}I - K_i')\lambda_i = V_i\lambda_i + (\frac{1}{2}I - K_i)z_i,$$

where $z_i = \widetilde{D}_i^{-1}(\frac{1}{2}I - K_i')\lambda_i \in H^{1/2}(\Gamma_i)$ is the unique solution of the local variational problem

$$\langle \widetilde{D}_i z_i, v_i \rangle_{\Gamma_i} = \langle (\frac{1}{2}I - K_i')\lambda_i, v_i \rangle_{\Gamma_i} \quad \text{for all } v_i \in H^{1/2}(\Gamma_i).$$

Let

$$S_h^1(\Gamma_i) = \text{span}\{\varphi_{i,k}^1\}_{k=1}^{M_i} \subset H^{1/2}(\Gamma_i)$$

be some boundary element space of piecewise linear basis functions $\varphi_{i,k}^1$. Using the Galerkin solution $z_{i,h} \in S_h^1(\Gamma_i)$ satisfying

$$\langle \widetilde{D}_i z_{i,h}, v_{i,h} \rangle_{\Gamma_i} = \langle (\frac{1}{2}I - K_i')\lambda_i, v_{i,h} \rangle_{\Gamma_i} \quad \text{for all } v_{i,h} \in S_h^1(\Gamma_i),$$

we may define an approximate Poincaré–Steklov operator as

$$\widetilde{T}_i\lambda_i = V_i\lambda_i + (\frac{1}{2}I - K_i)z_{i,h}. \tag{38}$$

Theorem 5. *The approximate Poincaré–Steklov operator $\widetilde{T}_i : H^{-1/2}(\Gamma_i) \to H^{1/2}(\Gamma_i)$ as defined in (38) is bounded and $H^{-1/2}(\Gamma_i)$–elliptic. Moreover, there holds the a priori error estimate*

$$\|(T_i - \widetilde{T}_i)\lambda_i\|_{H^{1/2}(\Gamma_i)} \leq c\, h_i^{3/2}\, \|T_i\lambda_i\|_{H^2(\Gamma_i)}$$

when assuming $T_i\lambda_i \in H^2(\Gamma_i)$, i.e. $u_i \in H^{5/2}(\Omega_i)$.

3 Domain Decomposition Methods

Using the local Dirichlet to Neuman map

$$\alpha_i t_i(x) = \alpha_i(S_i u_i)(x) - (N_i f_i)(x) \quad \text{for } x \in \Gamma_i$$

with the Steklov–Poincaré operator S_i as defined in (11), (18) or in (20), we can reformulate the coupled domain decomposition formulation (2) and (3) as

$$\begin{aligned}
\alpha_i t_i(x) &= \alpha_i(S_i u_i)(x) - (N_i f)(x) &&\text{for } x \in \Gamma_i, \\
u_i(x) &= g(x) &&\text{for } x \in \Gamma_i \cap \Gamma, \\
u_i(x) &= u_j(x) &&\text{for } x \in \Gamma_{ij}, \\
\alpha_i t_i(x) + \alpha_j t_j(x) &= 0 &&\text{for } x \in \Gamma_{ij}.
\end{aligned} \tag{39}$$

3.1 Dirichlet Domain Decomposition Methods

Eliminating the local Neumann data t_i in (39) gives the transmission conditions

$$u_i(x) = u_j(x), \quad \alpha_i(S_i u_i)(x) + \alpha_j(S_j u_j)(x) = (N_i f_i)(x) + (N_j f_j)(x)$$

for $x \in \Gamma_{ij}$. Let $H^{1/2}(\Gamma_S)$ be the skeleton trace space of $H^1(\Omega)$. To ensure the Dirichlet transmission condition $u_i(x) = u_j(x)$ for $x \in \Gamma_{ij}$ we may define $u_i(x) = u(x)$, $x \in \Gamma_i$, as the restriction of a globally defined funtion $u \in H^{1/2}(\Gamma_S)$ with $u(x) = g(x)$ for $x \in \Gamma$. Hence we have to find $u \in H^{1/2}(\Gamma_S)$, $u(x) = g(x)$ for $x \in \Gamma$, such that

$$\alpha_i(S_i u_{|\Gamma_i})(x) + \alpha_j(S_j u_{|\Gamma_j})(x) = (N_i f_i)(x) + (N_j f_j)(x) \quad \text{for } x \in \Gamma_{ij}.$$

The associated variational problem is to find $u \in H^{1/2}(\Gamma_S)$ such that $u = g$ on Γ and

$$\sum_{i=1}^{p} \int_{\Gamma_i} \alpha_i(S_i u_{|\Gamma_i})(x) v_{|\Gamma_i}(x) ds_x = \sum_{i=1}^{p} \int_{\Gamma_i} (N_i f_i)(x) v_{|\Gamma_i}(x) ds_x \tag{40}$$

is satisfied for all $v \in H^{1/2}(\Gamma_S)$ vanishing on Γ.

Let

$$S_h^1(\Gamma_S) = \text{span}\{\varphi_k^1\}_{k=1}^{M} \subset H^{1/2}(\Gamma_S)$$

be some global finite element space of piecewise linear basis functions φ_k^1 which are defined with respect to some regular finite element mesh $\Gamma_{S,h}$ of the skeleton Γ_S. By $S_h^1(\Gamma_i)$ we denote the restriction of $S_h^1(\Gamma_S)$ onto the local subdomain boundary Γ_i. In particular, for any $v_h \in S_h^1(\Gamma_S)$ we find the local restriction $v_{i,h} \in S_h^1(\Gamma_i)$ via a transformation of the associated coefficients, $\underline{v}_i = A_i \underline{v}$, where $A_i : \mathbb{R}^M \to \mathbb{R}^{M_i}$ is the connectivity matrix. Moreover, let

$S_h^1(\Gamma)$ be the restriction of $S_h^1(\Gamma_S)$ onto the Dirichlet boundary $\Gamma = \partial\Omega$, where the associated connectivity matrix is $A_0 \in \mathbb{R}^{M_0 \times M}$. Let $\underline{g} \in \mathbb{R}^{M_0}$ result from some piecewise linear approximation $g_h \in S_h^1(\Gamma)$ of the given Dirichlet datum g.

Using one of the previous introduced approximate Dirichlet to Neumann maps, the Galerkin variational formulation of (40) is to find $u_h \in S_h^1(\Gamma_S)$ satisfying the Dirichlet boundary condition $u_h(x_k) = g(x_k)$ for $x_k \in \Gamma$ such that

$$\sum_{i=1}^{p} \int_{\Gamma_i} \alpha_i (\widetilde{S}_i u_{h|\Gamma_i})(x) v_{h|\Gamma_i}(x) ds_x = \sum_{i=1}^{p} \int_{\Gamma_i} (\widetilde{N}_i f_i)(x) v_{h|\Gamma_i}(x) ds_x \qquad (41)$$

is satisfied for all $v_h \in S_h^1(\Gamma_S)$ with $v_h(x) = 0$, $x \in \Gamma$. This is equivalent to a linear system of algebraic equations to find $\underline{u} \in \mathbb{R}^M$ such that

$$\sum_{i=1}^{p} \alpha_i A_i^\top \widetilde{S}_{i,h} A_i \underline{u} = \sum_{i=1}^{p} A_i^\top \underline{f}_i, \qquad A_0 \underline{u} = \underline{g}. \qquad (42)$$

In (42), the approximate stiffness matrices $\widetilde{S}_{i,h}$ and the local vectors $\underline{f}_i$ of the right hand side correspond to the discretization of the locally defined approximate Steklov–Poincaré operators $\widetilde{S}_i$. In particular, when using the finite element approximation (15) this gives

$$\widetilde{S}_{i,h}^{\text{FEM}} = K_{CC,i} - K_{IC,i} K_{II,i}^{-1} K_{CI,i}, \qquad \underline{f}_i^{\text{FEM}} = \underline{f}_{C,i} - K_{CI,i} K_{II,i}^{-1} \underline{f}_{I,i}.$$

When using the symmetric boundary element approximation (27) of the Steklov–Poincaré operator this gives

$$\widetilde{S}_{i,h}^{\text{BEM}} = D_{i,h} + \widetilde{K}_{i,h}^\top V_{i,h}^{-1} \widetilde{K}_{i,h} \qquad \underline{f}_i^{\text{BEM}} = M_{i,h}^\top V_{i,h}^{-1} \underline{f}_{N,i}.$$

When using a boundary element approximation in the first $q \le p$ subdomains Ω_i, and a finite element approximation in the remaining subdomains, the linear system (42) can be written as

$$\sum_{i=1}^{q} \alpha_i A_i^\top \widetilde{S}_{i,h}^{\text{BEM}} A_i \underline{u} + \sum_{i=q+1}^{p} \alpha_i A_i^\top \widetilde{S}_{i,h}^{\text{FEM}} A_i \underline{u} = \sum_{i=1}^{q} A_i^\top \underline{f}_i^{\text{BEM}} + \sum_{i=q+1}^{p} A_i^\top \underline{f}_i^{\text{FEM}}$$

$$(43)$$

together with the side condition $A_0 \underline{u} = \underline{g}$.

The solution $\underline{u} \in \mathbb{R}^M$ of the assembled linear system (43) is also characterized as the unique solution of the constrained minimization problem

$$F(\underline{u}) = \min_{\underline{v} \in \mathbb{R}^M, A_0 \underline{v} = \underline{g}} F(\underline{v}), \qquad (44)$$

where

$$F(v \qquad \_) = \sum_{i=1}^{p} \left\{ \frac{\alpha_i}{2} (\widetilde{S}_{i,h}^{\mathrm{BEM/FEM}} A_i \underline{v}, A_i \underline{v}) - (\underline{f}_i^{\mathrm{BEM/FEM}}, A_i \underline{v}) \right\}.$$

By introducing the local vectors $\underline{v}_i = A_i \underline{v} \in \mathbb{R}^{M_i}$ we have to minimize

$$\widetilde{F}(\underline{v}_1, \ldots, \underline{v}_p) = \sum_{i=1}^{p} \left\{ \frac{\alpha_i}{2} (\widetilde{S}_{i,h}^{\mathrm{BEM/FEM}} \underline{v}_i, \underline{v}_i) - (\underline{f}_i^{\mathrm{BEM/FEM}}, \underline{v}_i) \right\}$$

where we have to add the constraints $A_0 \underline{v} = g$ due to the Dirichlet boundary condition and $\underline{v}_i = A_i \underline{v}$ which ensures the global continuity $v_{i,i_k} = v_{j,j_k}$ along Γ_{ij}. Here, v_{i,i_k} is the local degree of freedom which belongs to a global node $x_k \in \Gamma_S$, i.e. $A_i[i_k, k] = 1$. Now we can formulate all above constraints as

$$\sum_{i=1}^{p} B_i \underline{v}_i = A_0^{\top} g = \widetilde{g} \in \mathbb{R}^{\overline{M}}$$

where the nonzero elements of the matrices $B_i \in \mathbb{R}^{\overline{M} \times M_i}$ are defined as follows:

- $x_k \in \Gamma_i \cap \Gamma$ is on the Dirichlet boundary:

$$B_i[k, i_k] = 1;$$

- $x_k \in \Gamma_{ij} = \Gamma_i \cap \Gamma_j$ is on the interface:

$$B_i[\ell_k, i_k] = 1, \quad B_j[\ell_k, j_k] = -1, \quad i < j.$$

Note that the above constraints are defined in a redundant manner, i.e. ℓ_k corresponds to the multiplicity of constraints which are associated to the node x_k. Now, instead of the minimization problem (44), we have to solve a modified constrained minimization problem, i.e.,

$$\widetilde{F}(\underline{u}_1, \ldots, \underline{u}_p) = \inf_{\sum_{i=1}^{p} B_i \underline{v}_i = \widetilde{g}} \widetilde{F}(\underline{v}_1, \ldots, \underline{v}_p). \tag{45}$$

By introducing the Lagrange multiplier $\underline{\lambda} \in \mathbb{R}^{\overline{M}}$, we have to minimize the extended functional

$$\widetilde{F}_\lambda(\underline{v}_1, \ldots, \underline{v}_p) = \widetilde{F}(\underline{v}_1, \ldots, \underline{v}_p) - (\underline{\lambda}, \sum_{i=1}^{p} B_i \underline{v}_i - \widetilde{g}).$$

The necessary conditions give the equations

$$\alpha_i \widetilde{S}_{i,h}^{\mathrm{BEM/FEM}} \underline{u}_i - \underline{f}_i^{\mathrm{BEM/FEM}} - B_i^{\top} \underline{\lambda} = \underline{0} \tag{46}$$

by taking the derivative with respect to $\underline{v}_i$ for $i = 1, \ldots, p$, and,

$$\sum_{i=1}^{p} B_i \underline{v}_i = \widetilde{\underline{g}}$$

by taking the derivative with respect to $\underline{\lambda}$. Hence, we have to solve the linear system

$$\begin{pmatrix} S_{\mathrm{BEM}} & & -B_{\mathrm{BEM}}^{\top} \\ & S_{\mathrm{FEM}} & -B_{\mathrm{FEM}}^{\top} \\ B_{\mathrm{BEM}} & B_{\mathrm{FEM}} & \end{pmatrix} \begin{pmatrix} \underline{u}_{\mathrm{BEM}} \\ \underline{u}_{\mathrm{FEM}} \\ \underline{\lambda} \end{pmatrix} = \begin{pmatrix} \underline{f}_{\mathrm{BEM}} \\ \underline{f}_{\mathrm{FEM}} \\ \widetilde{\underline{g}} \end{pmatrix}, \tag{47}$$

where

$$S_{\mathrm{BEM}} = \mathrm{diag}\left(\alpha_i [D_{i,h} + \widetilde{K}_{i,h}^{\top} V_{i,h}^{-1} \widetilde{K}_{i,h}] \right)_{i=1,\dots,q},$$

$$S_{\mathrm{FEM}} = \mathrm{diag}\left(\alpha_i [K_{CC,i} - K_{IC,i} K_{II,i}^{-1} K_{CI,i}] \right)_{i=q+1,\dots,p}.$$

In what follows we proceed as for the solution of a local Neumann boundary value problem. Due to

$$S_{i,h}^{\mathrm{BEM}} \underline{1}_i = S_{i,h}^{\mathrm{FEM}} \underline{1}_i = \underline{0},$$

we can write the local variables $\underline{u}_i \in \mathbb{R}^{M_i}$ as

$$\underline{u}_i = \underline{u}_{i,0} + \gamma_i \underline{1}_i, \quad (\underline{u}_{i,0}, \underline{1}_i) = 0 \tag{48}$$

and therefore we have to solve

$$\alpha_i S_{i,h}^{\mathrm{BEM/FEM}} \underline{u}_{i,0} - B_i^{\top} \underline{\lambda} = \underline{f}_i^{\mathrm{BEM/FEM}} \quad \text{for } i = 1, \dots, p$$

as well as

$$\sum_{i=1}^{p} B_i \underline{u}_{i,0} + \sum_{i=1}^{p} \gamma_i B_i \underline{1}_i = \widetilde{\underline{g}}_i.$$

On the other hand, for $i = 1, \dots, p$, we find

$$(B_{i,h}^{\top} \underline{\lambda} + \underline{f}_i^{\mathrm{BEM/FEM}}, \underline{1}_i) = \alpha_i (S_{i,h}^{\mathrm{BEM/FEM}} \underline{u}_i, \underline{1}_i) = \alpha_i (\underline{u}_i, S_{i,h}^{\mathrm{BEM/FEM}} \underline{1}_i) = 0$$

and, therefore, the additional constraints

$$(\underline{\lambda}, B_{i,h} \underline{1}_i) = -(\underline{f}_i^{\mathrm{BEM/FEM}}, \underline{1}_i) \quad \text{for } i = 1, \dots, p.$$

Hence, we obtain $\underline{u}_{i,0} \in \mathbb{R}^{M_i}$ as the unique solution of

$$\alpha_i [S_{i,h}^{\mathrm{BEM/FEM}} + \underline{1}_i \underline{1}_i^{\top}] \underline{u}_{i,0} - B_i^{\top} \underline{\lambda} = \underline{f}_i^{\mathrm{BEM/FEM}}$$

for $i = 1, \dots, p$. Now, instead of (47), we may solve the extended system

$$\begin{pmatrix} \overline{S}_{\mathrm{BEM}} & & -B_{\mathrm{BEM}}^{\top} & \\ & \overline{S}_{\mathrm{FEM}} & -B_{\mathrm{FEM}}^{\top} & \\ B_{\mathrm{BEM}} & B_{\mathrm{FEM}} & & G \\ & & G^{\top} & \end{pmatrix} \begin{pmatrix} \underline{u}_{\mathrm{BEM},0} \\ \underline{u}_{\mathrm{FEM},0} \\ \underline{\lambda} \\ \underline{\gamma} \end{pmatrix} = \begin{pmatrix} \underline{f}_{\mathrm{BEM}} \\ \underline{f}_{\mathrm{FEM}} \\ \widetilde{\underline{g}} \\ \underline{e} \end{pmatrix}, \tag{49}$$

where

$$\overline{S}_{\text{BEM}} = \text{diag}\left(\alpha_i[D_{i,h} + \widetilde{K}_{i,h}^\top V_{i,h}^{-1}\widetilde{K}_{i,h} + \underline{1}_i\underline{1}_i^\top]\right)_{i=1,\dots,q},$$

$$\overline{S}_{\text{FEM}} = \text{diag}\left(\alpha_i[K_{CC,i} - K_{IC,i}K_{II,i}^{-1}K_{CI,i} + \underline{1}_i\underline{1}_i^\top]\right)_{i=q+1,\dots,p}$$

and

$$G = (B_1\underline{1}_1,\dots,B_p\underline{1}_p), \quad e_i = -(\underline{f}_i^{\text{BEM/FEM}},\underline{1}_i) \quad \text{for } i=1,\dots,p.$$

The local boundary element equations in (49) can be written as

$$\alpha_i[D_{i,h} + \underline{1}_i\underline{1}_i^\top]\underline{u}_{i,0} + \alpha_i\widetilde{K}_{i,h}^\top\underline{w}_i - B_i^\top\underline{\lambda} = \underline{f}_i^{\text{BEM}},$$

$$\alpha_i V_{i,h}\underline{w}_i - \alpha_i\widetilde{K}_{i,h}\underline{u}_{i,0} = \underline{0},$$

while the finite element equations are equivalent to

$$\alpha_i[K_{CC,i} + \underline{1}_i\underline{1}_i^\top]\underline{u}_{i,0} + \alpha_i K_{IC,i}\underline{u}_{I,i} - B_i^\top\underline{\lambda} = \underline{f}_{C,i},$$

$$\alpha_i K_{II,i}\underline{u}_{I,i} + \alpha_i K_{CI,i}\underline{u}_{i,0} = \underline{f}_{I,i}.$$

Hence, we have to solve the linear system

$$\begin{pmatrix} V_h & -\widetilde{K}_h & & & \\ \widetilde{K}_h^\top & \widetilde{D}_h & & & -B_{\text{BEM}}^\top \\ & & K_{II} & K_{CI} & \\ & & K_{CI} & \widetilde{K}_{CC} & -B_{\text{FEM}}^\top \\ & B_{\text{BEM}} & & B_{\text{FEM}} & G \\ & & & & G^\top \end{pmatrix}\begin{pmatrix} \underline{w} \\ \underline{u}_{\text{BEM},0} \\ \underline{u}_I \\ \underline{u}_{\text{FEM},0} \\ \underline{\lambda} \\ \underline{\gamma} \end{pmatrix} = \begin{pmatrix} \underline{0} \\ \underline{f}_{\text{BEM}} \\ \underline{f}_I \\ \underline{f}_C \\ \underline{\widetilde{g}} \\ \underline{e} \end{pmatrix}, \quad (50)$$

where

$$V_h = \text{diag}\left(\alpha_i V_{i,h}\right)_{i=1}^q, \qquad \widetilde{D}_h = \text{diag}\left(\alpha_i[D_{i,h} + \underline{1}_i\underline{1}_i^\top]\right)_{i=1}^q,$$

$$K_{II} = \text{diag}\left(\alpha_i K_{II,i}\right)_{i=q+1}^p, \qquad \widetilde{K}_{CC} = \text{diag}\left(\alpha_i[K_{CC,i} + \underline{1}_i\underline{1}_i^\top]\right)_{i=q+1}^p,$$

$$\widetilde{K}_h = \begin{pmatrix} \alpha_1\widetilde{K}_{1,h} \\ \vdots \\ \alpha_q\widetilde{K}_{q,h} \end{pmatrix}, \qquad K_{CI} = \begin{pmatrix} \alpha_{q+1}\widetilde{K}_{CI,1} \\ \vdots \\ \alpha_p\widetilde{K}_{CI,p} \end{pmatrix}.$$

Next we use a subspace projection in order to separate the determination of $\underline{\gamma}$ from the determination of the rest of the unknowns in (50). Thus, we introduce the orthogonal projection

$$P = I - QG(G^\top QG)^{-1}G^\top \quad (51)$$

where Q is some suitable diagonal scaling matrix [7, 30]. Since $P^\top G = 0$, the application of $P^\top$ to the fifth equation of (50) gives $P^\top G\underline{\gamma} = 0$ that excludes $\underline{\gamma}$ from the first five equations of (50). Let us represent $\underline{\lambda}$ in the form

$$\underline{\lambda} = T_0 \underline{\lambda}_0 + \underline{\lambda}_e, \quad \underline{\lambda}_e = QG(G^\top QG)^{-1}\underline{e}. \tag{52}$$

Hence we have to find $T_0\underline{\lambda}_0 \in \ker G^\top$, i.e. $G^\top T_0\underline{\lambda}_0 = 0$. In particular, the columns of T_0 span a basis of $\ker G^\top = (\operatorname{range} G)^\perp$. Hence we also conclude $T_0\underline{\lambda}_0 = PT_0\underline{\lambda}_0$. Therefore we have to solve the reduced linear system

$$\begin{pmatrix} V_h & -\widetilde{K}_h & & & \\ \widetilde{K}_h^\top & \widetilde{D}_h & & & -B_{\mathrm{BEM}}^\top PT_0 \\ & & K_{II} & K_{CI} & \\ & & K_{CI} & \widetilde{K}_{CC} & -B_{\mathrm{FEM}}^\top PT_0 \\ & T_0^\top P^\top B_{\mathrm{BEM}} & & T_0^\top P^\top B_{\mathrm{FEM}} & \end{pmatrix} \begin{pmatrix} \underline{w} \\ \underline{u}_{\mathrm{BEM},0} \\ \underline{u}_I \\ \underline{u}_{\mathrm{FEM},0} \\ \underline{\lambda}_0 \end{pmatrix} \tag{53}$$

$$= \begin{pmatrix} \underline{0} \\ \underline{f}_{\mathrm{BEM}} + B_{\mathrm{BEM}}^\top \underline{\lambda}_e \\ \underline{f}_I \\ \underline{f}_C + B_{\mathrm{FEM}}^\top \underline{\lambda}_e \\ T_0^\top P^\top \underline{\widetilde{g}} \end{pmatrix}.$$

Once the vectors $\underline{w}$, $\underline{u}_{\mathrm{BEM},0}$, $\underline{u}_I$, $\underline{u}_{\mathrm{FEM},0}$ and $\underline{\lambda}_0$ are defined from (53), we get $\underline{\lambda}$ from (52), $\underline{\gamma}$ from the fifth equation in (50), i.e.,

$$\underline{\gamma} = (G^\top QG)^{-1}G^\top Q\left[\underline{\widetilde{g}} - B_{\mathrm{BEM}}\underline{u}_{\mathrm{BEM},0} - B_{\mathrm{FEM}}\underline{u}_{\mathrm{FEM},0}\right],$$

and, finally, $\underline{u}$ from (48).

3.2 Neumann Domain Decomposition Methods

Instead of eliminating the Neumann data in (39) we are now going to eliminate the Dirichlet data. For this, we introduce global Neumann data as follows: For any interface $\Gamma_{ij} = \Gamma_i \cap \Gamma_i$, we introduce $t_{ij} \in \widetilde{H}^{-1/2}(\Gamma_{ij})$ and set

$$t_i(x) = \frac{1}{\alpha_i}t_{ij}(x), t_j(x) = -\frac{1}{\alpha_j}t_{ij}(x) \quad \text{for } x \in \Gamma_{ij}, i < j,$$

and for the Dirichlet boundary we introduce $t_0 \in H^{-1/2}(\Gamma)$ and set $t_i = t_{0|\Gamma_i}/\alpha_i$. Hence, we have satisfied the Neumann transmission condition in (39) in a strong sense. Therefore, we have to impose the Dirichlet transmission conditions and the Dirichlet boundary conditions in some weak sense, i.e.,

$$\int_{\Gamma_{ij}} [u_i(x) - u_j(x)]\tau_{ij}(x)ds_x = 0 \quad \text{for all } \tau_{ij} \in \widetilde{H}^{-1/2}(\Gamma_{ij}),$$

and

$$\int\limits_{\Gamma_i \cap \Gamma} [u_i(x) - g(x)]\tau_{0|\Gamma_i} ds_x = 0 \quad \text{for all } \tau_0 \in H^{-1/2}(\Gamma),$$

where $\widetilde{H}^{-1/2}(\Gamma_{ij}) = (\widetilde{H}^{1/2}(\Gamma_{ij}))'$. For the interfaces Γ_{ij}, we find from the weak formulations of the Dirichlet to Neumann map

$$\langle t_{ij}, v_{i|\Gamma_{ij}} \rangle_{\Gamma_{ij}} = \langle \alpha_i t_i, v_{i|\Gamma_{ij}} \rangle_{\Gamma_{ij}} = \langle \alpha_i S_i u_i - N_i f, v_{i|\Gamma_{ij}} \rangle_{\Gamma_{ij}}, \quad v_i \in H^{1/2}(\Gamma_i),$$

and

$$-\langle t_{ij}, v_{j|\Gamma_{ij}} \rangle_{\Gamma_{ij}} = \langle \alpha_j t_j, v_{j|\Gamma_{ij}} \rangle_{\Gamma_{ij}} = \langle \alpha_j S_j u_j - N_j f, v_{j|\Gamma_{ij}} \rangle_{\Gamma_{ij}}, \quad v_j \in H^{1/2}(\Gamma_j).$$

Hence,

$$\langle \alpha_i S_i u_i, v_{i|\Gamma_{ij}} \rangle_{\Gamma_{ij}} + \langle \alpha_j S_j u_j, v_{j|\Gamma_{ij}} \rangle_{\Gamma_{ij}} - \langle t_{ij}, v_{i|\Gamma_{ij}} - v_{j|\Gamma_{ij}} \rangle_{\Gamma_{ij}}$$
$$= \langle N_i f, v_{i|\Gamma_{ij}} \rangle_{\Gamma_{ij}} + \langle N_j f, v_{j|\Gamma_{ij}} \rangle_{\Gamma_{ij}}.$$

Moreover, on the local Dirichlet boundaries $\Gamma_i \cap \Gamma$, we have

$$\langle \alpha_i S_i u_i, v_{i|\Gamma_i \cap \Gamma} \rangle_{\Gamma_i \cap \Gamma} - \langle t_0, v_{i|\Gamma_i \cap \Gamma} \rangle_{\Gamma_i \cap \Gamma} = \langle N_i f, v_{i|\Gamma_i \cap \Gamma} \rangle_{\Gamma_i \cap \Gamma}.$$

The associated variational formulation is to find $u_i \in H^{1/2}(\Gamma_i)$ for $i = 1, \ldots, p$, $t_{ij} \in \widetilde{H}^{-1/2}(\Gamma_{ij})$ for all $i < j$ and $t_0 \in H^{-1/2}(\Gamma)$ such that

$$\sum_{i=1}^{p} \langle \alpha_i S_i u_i, v_i \rangle_{\Gamma_i} - \sum_{i<j} \langle t_{ij}, v_{i|\Gamma_{ij}} - v_{j|\Gamma_{ij}} \rangle_{\Gamma_{ij}} - \sum_{i=1}^{p} \langle t_0, v_i \rangle_{\Gamma_i \cap \Gamma} = \sum_{i=1}^{p} \langle N_i f, v_i \rangle_{\Gamma_i}$$

$$\langle u_{i|\Gamma_{ij}} - u_{j|\Gamma_{ij}}, \tau_{ij} \rangle_{\Gamma_{ij}} = 0 \tag{54}$$

$$\langle u_i, \tau_0 \rangle_{\Gamma_i \cap \Gamma} = \langle g, \tau_0 \rangle_{\Gamma_i \cap \Gamma}$$

is satisfied for all $v_i \in H^{1/2}(\Gamma_i)$, $\tau_{ij} \in \widetilde{H}^{-1/2}(\Gamma_{ij})$, and $\tau_0 \in H^{-1/2}(\Gamma)$.

The saddle point formulation (54) describes a hybrid domain decomposition method [1] which is also known as a mortar domain decomposition method to couple locally different trial spaces [4].

For a Galerkin discretization of (54), we introduce local boundary element spaces

$$S_h^1(\Gamma_i) = \text{span}\{\varphi_{i,k}^1\}_{k=1}^{M_i} \subset H^{1/2}(\Gamma_i)$$

of, e.g., piecewise linear basis functions $\varphi_{i,k}^1$. Moreover, for each coupling boundary Γ_{ij}, we consider a trial space to discretize the local Neumann datum t_{ij},

$$S_h(\Gamma_{ij}) = \text{span}\{\psi_{ij,n}\}_{n=1}^{N_{ij}} \subset \widetilde{H}^{-1/2}(\Gamma_{ij})$$

where $\psi_{ij,n}$ are some basis functions to be defined in an appropriate way. In the same manner we introduce

$$S_h(\Gamma) = \mathrm{span}\{\psi_{0,n}\}_{n=1}^{N_0} \subset H^{-1/2}(\Gamma)$$

to discretize the unknown Neumann datum on Γ. The choice of the basis functions $\psi_{ij,n}$ and $\psi_{0,n}$ is very sensitive, since we have to ensure local inf–sup conditions which are related to the saddle point formulation (54), i.e.,

$$c_S \|\tau_{ij,h}\|_{\widetilde{H}^{-1/2}(\Gamma_{ij})}$$

$$\leq \sup_{(v_{i,h},v_{j,h})\in S_h^1(\Gamma_i)\times S_h^1(\Gamma_j)} \frac{\langle \tau_{ij}, v_{i|\Gamma_{ij}} - v_{j|\Gamma_{ij}}\rangle_{\Gamma_{ij}}}{\sqrt{\|v_{i,h|\Gamma_{ij}}\|^2_{H^{1/2}(\Gamma_{ij})} + \|v_{j,h|\Gamma_{ij}}\|^2_{H^{1/2}(\Gamma_{ij})}}}.$$

For appropriate choices of the trial spaces $S_h(\Gamma_{ij})$ and $S_h(\Gamma)$ see, for example, [57] and the references given therein.

The Galerkin discretization of the variational problem (54) is equivalent to a set of linear equations which can be written as

$$\sum_{i=1}^{p} \widetilde{S}_{i,h}^{\mathrm{FEM/BEM}} \underline{u}_i - \sum_{i<j}(M_{ij,h}^\top - M_{ji,h}^\top)\underline{t}_{ij} - \sum_{i=1}^{p} M_{0i,h}^\top \underline{t}_0 = \underline{f}_0,$$

$$M_{ij,h}\underline{u}_i - M_{ji,h}\underline{u}_j = \underline{0},$$

$$M_{0i,h}\underline{u}_i = \underline{g}$$

with the discrete Steklov–Poincaré operator as defined in (16) for a finite element approximation, and as given in (27) for a boundary element discretization. Moreover,

$$M_{ij,h}[m,k] = \langle \varphi_{i,k}^1, \psi_{ij,m}\rangle_{\Gamma_{ij}},$$

$$M_{ji,h}[m,k] = \langle \varphi_{j,k}^1, \psi_{ij,m}\rangle_{\Gamma_{ij}},$$

$$M_{0i,h}[m,k] = \langle \varphi_{i,k}^1, \psi_{0,m}\rangle_{\Gamma_{ij}}.$$

By reordering all degrees of freedom we then obtain the coupled linear system

$$\begin{pmatrix} S_{\mathrm{BEM}} & & -M_{\mathrm{BEM}}^\top \\ & S_{\mathrm{FEM}} & -M_{\mathrm{FEM}}^\top \\ M_{\mathrm{BEM}} & M_{\mathrm{FEM}} & \end{pmatrix} \begin{pmatrix} \underline{u}_{\mathrm{BEM}} \\ \underline{u}_{\mathrm{FEM}} \\ \underline{t} \end{pmatrix} = \begin{pmatrix} \underline{f}_{\mathrm{BEM}} \\ \underline{f}_{\mathrm{FEM}} \\ \underline{g} \end{pmatrix} \tag{55}$$

which is of the same structure as the linear system (47). In fact, when considering conforming local trial spaces $S_h^1(\Gamma_i)$ and choosing $S_h(\Gamma_{ij})$ and $S_h(\Gamma)$ to be spanned by biorthogonal basis functions $\psi_{ij,n}$ and $\psi_{0,n}$, respectively, both linear systems (47) and (55) will coincide. In general, we may apply all the transformations which are used to reformulate the linear system (47) to solve the linear system (55) in a similar way, we skip the details.

4 Preconditioned Iterative Solution Techniques

In this section we describe some preconditioned CG–like iterative methods for solving the linear system (53),

$$\begin{pmatrix} V_h & -\widetilde{K}_h & & & \\ \widetilde{K}_h^\top & \widetilde{D}_h & & & -B_{\mathrm{BEM}}^\top PT_0 \\ & & K_{II} & K_{CI} & \\ & & K_{CI} & \widetilde{K}_{CC} & -B_{\mathrm{FEM}}^\top PT_0 \\ & T_0^\top P^\top B_{\mathrm{BEM}} & & T_0^\top P^\top B_{\mathrm{FEM}} & \end{pmatrix} \begin{pmatrix} \underline{w} \\ \underline{u}_{\mathrm{BEM},0} \\ \underline{u}_I \\ \underline{u}_{\mathrm{FEM},0} \\ \underline{\lambda}_0 \end{pmatrix}$$

$$= \begin{pmatrix} \underline{0} \\ \underline{f}_{\mathrm{BEM}} + B_{\mathrm{BEM}}^\top \underline{\lambda}_e \\ \underline{f}_I \\ \underline{f}_C + B_{\mathrm{FEM}}^\top \underline{\lambda}_e \\ T_0^\top P^\top \underline{\widetilde{g}} \end{pmatrix}.$$

Since $V_h = \mathrm{diag}(\alpha_i V_{i,h})$ and $K_{II} = \mathrm{diag}(\alpha_i K_{II,i})$ are block diagonal and therefore easily invertible we may first eliminate the vectors $\underline{w}$ and $\underline{u}_I$ to obtain

$$\begin{pmatrix} \overline{S}_{\mathrm{BEM}} & & -B_{\mathrm{BEM}}^\top PT_0 \\ & \overline{S}_{\mathrm{FEM}} & -B_{\mathrm{FEM}}^\top PT_0 \\ T_0^\top P^\top B_{\mathrm{BEM}} & T_0^\top P^\top B_{\mathrm{FEM}} & \end{pmatrix} \begin{pmatrix} \underline{u}_{\mathrm{BEM},0} \\ \underline{u}_{\mathrm{FEM},0} \\ \underline{\lambda}_0 \end{pmatrix} \qquad (56)$$

$$= \begin{pmatrix} \underline{f}_{\mathrm{BEM}} + B_{\mathrm{BEM}}^\top \underline{\lambda}_e \\ \underline{f}_C - K_{CI} K_{II}^{-1} \underline{f}_I + B_{\mathrm{FEM}}^\top \underline{\lambda}_e \\ T_0^\top P^\top \underline{\widetilde{g}} \end{pmatrix}.$$

Eliminating $\underline{u}_{\mathrm{BEM},0}$ and $\underline{u}_{\mathrm{FEM},0}$ we have to solve the Schur complement system of (56),

$$F\underline{\lambda}_0 = T_0^\top P^\top B\overline{S}B^\top PT_0\underline{\lambda}_0 = \underline{\widetilde{f}}. \qquad (57)$$

Since the system matrix in (57) is symmetric and positive definite one may use a preconditioned conjugate gradient scheme to solve (57). For this, an appropriate preconditioner C_F is needed, which is spectrally equivalent to the Schur complement matrix F. Another choice is the application of a Bramble–Pasciak conjugate gradient scheme [5] to the one–fold saddle point problem (56). For this, besides C_F also preconditioners $C_S = \mathrm{diag}(C_{S_i})$ for the local discrete Steklov–Poincaré operators $\overline{S}_{i,h}^{\mathrm{BEM/FEM}}$ are needed. A third possibility is to use a CG–like iterative method to solve the two–fold saddle point problem (53), see [33, 60]. Then, also preconditioners C_{V_i} and C_{K_i} for the local matrices $V_{i,h}$ and $K_{II,i}$ are needed, respectively.

Following [36] we can define the scaled hypersingular BETI preconditioner

$$C_F^{-1} = (BC_\alpha^{-1}B^\top)^{-1} BC_\alpha^{-1} \overline{D}_h C_\alpha^{-1} B^\top (BC_\alpha^{-1}B^\top)^{-1} \qquad (58)$$

where C_α is some diagonal scaling. Note that there hold the spectral equivalence inequalities [36, Theorem 3.2]

$$c_1^F \, (C_F \underline{\mu}, \underline{\mu}) \ \leq \ (F \underline{\mu}, \underline{\mu}) \ \leq \ c_2^F \, (1 + \log(H/h))^2 (C_F \underline{\mu}, \underline{\mu})$$

for all $\underline{\mu} \in \ker G^\top$ where the positive constants c_1^F and c_2^F are independent of the local mesh size h, the subdomain diameter H, the number p of subdomains, and of the coefficient jumps. The preconditioner (58) is based on local realizations of the discrete stabilized hypersingular boundary integral operator with respect to all subdomains, independent of whether a finite or boundary element discretization is used locally.

To construct preconditioning matrices C_{S_i} for the local discrete Schur complement matrices $\widetilde{S}_{i,h}^{\mathrm{BEM/FEM}}$ we will apply the concept of boundary integral operators of the opposite order [51]. Based on the local trial space $S_h^1(\Gamma_i)$ of piecewise linear basis functions $\varphi_{i,k}^1$ as used for the Galerkin discretization of the local hypersingular boundary integral operators D_i we define the Galerkin matrices

$$\bar{V}_{i,h}[\ell, k] \ = \ \langle V_i \varphi_{i,k}^1, \varphi_{i,\ell}^1 \rangle_{\Gamma_i}, \quad \bar{M}_{i,h}[\ell, k] \ = \ \langle \varphi_{i,k}^1, \varphi_{i,\ell}^1 \rangle_{\Gamma_i}$$

for $k, \ell = 1, \ldots, M_i$ and the application of the resulting preconditioning matrix is given by

$$C_{S_i}^{-1} \ = \ \bar{M}_{i,h}^{-1} \bar{V}_{i,h} \bar{M}_{i,h}^{-1} \quad \text{for } i = 1, \ldots, p. \tag{59}$$

Moreover, there hold the spectral equivalence inequalities

$$c_1^{S_i} \, (C_{S_i} \underline{v}_i, \underline{v}_i) \ \leq \ (\overline{S}_i^{\mathrm{BEM/FEM}} \underline{v}_i, \underline{v}_i) \ \leq \ c_2^{S_i} \, (C_{S_i} \underline{v}_i, \underline{v}_i)$$

for all $\underline{v}_i \in \mathbb{R}^{M_i}$.

For the definition of preconditioners C_{V_i} for the local discrete single layer potentials $V_{i,h}$, there exists a wide variety of different possible choices. Here, we only mention multilevel methods [18, 53] which are based on a given mesh hierarchy or algebraic multilevel techniques [35, 38, 50].

For finite element subdomains one may also use geometric or algebraic multigrid methods to construct preconditioners C_{K_i} for the local finite element stiffness matrices $K_{II,i}$, see, for example, [15] and the references given therein.

5 Conclusions

In this paper we have provided a unique approach to both the Dirichlet and the Neumann domain decomposition techniques. The all–floating tearing and interconnecting technology is a very general and powerful technique. Eliminating more or less variables results in symmetric and positive definite Schur complement problems, one–fold or two–fold saddle point problems which can be solved by preconditioned conjugate gradient methods. We have used boundary element technologies for constructing the required block preconditioners for both the boundary element and the finite element blocks. There are many papers showing the efficiency of FETI methods including the efficiency in large–scale parallel computations, see, e.g., [16, 29, 44]. Numerical results for BETI and coupled BETI–FETI methods can be found in [33, 34, 38].

The methods and techniques discussed in this paper are not restricted to the potential problem. They can be extended to linear elasticity problems as well [38]. The generalization to three–dimensional electromagnetic problems usually considered in $H(\mathrm{curl})$ is certainly more challenging, see [22] for the symmetric coupling and [54] for FETI–DP methods. Coupled finite and boundary element tearing and interconnecting solvers for nonlinear potential problems were discussed in [34].

References

1. A. Agouzal, J.–M. Thomas: Une méthode d'éléments finis hybrides en décomposition de domaines. Math. Modell. Numer. Anal. 29 (1995) 749–764.
2. M. Bebendorf: Approximation of boundary element matrices. Numer. Math. 86 (2000) 565–589.
3. M. Bebendorf, S. Rjasanow: Adaptive low–rank approximation of collocation matrices. Computing 70 (2003) 1–24.
4. C. Bernardi, Y. Maday, A. T. Patera: A new nonconforming approach to domain decompostion: the mortar element method. In: Nonlinear partial differential equations and their applications. Collége de France Seminar, Vol. XI (H. Brezis, J. L. Lions eds.), Pitmans Research Notes Mathematics Series, Vol. 299, Longman, pp. 13–51, 1994.
5. J. H. Bramble, J. E. Pasciak: A preconditioning technique for indefinite systems resulting from mixed approximations of elliptic problems. Math. Comp. 50 (1988) 1–17.
6. C. A. Brebbia, J. C. F. Telles, L. C. Wrobel: Boundary Element Techniques. Springer, Berlin, Heidelberg, New York, Tokyo, 1984.
7. S. C. Brenner: An additive Schwarz preconditioner for the FETI method. Numer. Math. 94 (2003) 1–31.
8. F. Brezzi, C. Johnson: On the coupling of boundary integral and finite element methods. Calcolo 16 (1979) 189–201.
9. J. Carrier, L. Greengard, V. Rokhlin: A fast adaptive multipole algorithm for particle simulations. SIAM J. Sci. Stat. Comput. 9 (1988) 669–686.
10. C. Carstensen, M. Kuhn, U. Langer: Fast parallel solvers for symmetric boundary element domain decomposition equations. Numer. Math. 79 (1998) 321–347.
11. M. Costabel: Symmetric methods for the coupling of finite elements and boundary elements. In: Boundary Elements IX (C. A. Brebbia, G. Kuhn, W. L. Wendland eds.), Springer, Berlin, pp. 411–420, 1987.
12. M. Costabel: Boundary integral operators on Lipschitz domains: Elementary results. SIAM J. Math. Anal. 19 (1988) 613–626.
13. M. Costabel, E. P. Stephan: Coupling of finite and boundary element methods for an elastoplastic interface problem. SIAM J. Numer. Anal. 27 (1990) 1212–1226.
14. W. Dahmen, S. Prössdorf, R. Schneider: Wavelet approximation methods for pseudodifferential equations I: Stability and convergence. Math. Z. 215 (1994) 583–620.

15. C. C. Douglas, G. Haase, U. Langer: A tutorial on elliptic PDE solvers and their parallelization. Software, Environments, and Tools, 16. SIAM, Philadelphia, 2003.
16. C. Farhat, M. Lesoinne, K. Pierson: A scalable dual-primal domain decomposition method. Numer. Linear Algebra Appl. 7 (2000) 687–714.
17. C. Farhat, F.-X. Roux: A method of finite element tearing and interconnecting and its parallel solution algorithm. Internat. J. Numer. Methods Engrg. 32 (1991) 1205–1227.
18. S. A. Funken, E. P. Stephan: The BPX preconditioner for the single layer potential operator. Appl. Anal. 67 (1997) 327–340.
19. G. Haase, B. Heise, M. Kuhn, U. Langer: Adaptive domain decomposition methods for finite and boundary element equations. In: Boundary Element Topics (W. L. Wendland ed.), Springer, Berlin, Heidelberg, New York, 1997.
20. W. Hackbusch: A sparse matrix arithmetic based on $\mathcal{H}$–matrices. Computing 62 (1999) 89–108.
21. W. Hackbusch, Z. P. Nowak: On the fast matrix multiplication in the boundary element method by panel clustering. Numer. Math. 54 (1989) 463–491.
22. R. Hiptmair: Symmetric coupling for eddy current problems. SIAM J. Numer. Anal. 40 (2002) 41–65.
23. G. C. Hsiao, E. Schnack, W. L. Wendland: A hybrid coupled finite–boundary element method in elasticity. Comput. Methods Appl. Mech. Engrg. 173 (1999) 287–316.
24. G. C. Hsiao, E. Schnack, W. L. Wendland: Hybrid coupled finite–boundary element methods for elliptic systems of second order. Cpmput. Methods Appl. Mech. Engrg. 190 (2000) 431–485.
25. G. C. Hsiao, O. Steinbach, W. L. Wendland: Domain decomposition methods via boundary integral equations. J. Comput. Appl. Math. 125 (2000) 523–539.
26. G. C. Hsiao, W. L. Wendland: Domain decomposition methods in boundary element methods. In: Domain Decomposition Methods for Partial Differential Equations (R. Glowinski et. al. eds.), Proceedings of the Fourth International Conference on Domain Decomposition Methods, SIAM, Baltimore, pp. 41–49, 1990.
27. C. Johnson, J. C. Nedelec: On coupling of boundary integral and finite element methods. Math. Comp. 35 (1980) 1063–1079.
28. B. N. Khoromskij, G. Wittum: Numerical Solution of Elliptic Differential Equations by Reduction to the Interface. Lecture Notes in Computational Science and Engineering 36, Springer, Berlin, 2004.
29. A. Klawonn, O. Rheinbach: Some computational results for robust FETI-DP methods applied to heterogeneous elasticity problems in 3D. In: Proceedings of the 16th International Conference on Domain Decomposition Methods in Science and Engineering (D. Keyes and O. Widlund, eds), New York, USA, January 12-15, 2005. Lecture Notes in Computational Science and Engineering, Springer, Berlin, 2006, to appear.
30. A. Klawonn, O. B. Widlund: FETI and Neumann–Neumann iterative substructuring methods: Connections and new results. Comm. Pure Appl. Math. 54 (2001) 57–90.
31. M. Kuhn, O. Steinbach: Symmetric coupling of finite and boundary elements for exterior magnetic field problems. Math. Methods Appl. Sci. 25 (2002) 357–371.
32. U. Langer: Parallel iterative solution of symmetric coupled fe/be equations via domain decomposition. Contemp. Math. 157 (1994) 335–344.

33. U. Langer, G. Of, O. Steinbach, W. Zulehner: Inexact data–sparse boundary element tearing and interconnecting methods. SIAM J. Sci. Comput., to appear.

34. U. Langer, C. Pechstein: Coupled finite and boundary element tearing and interconnecting solvers for nonlinear potential problems. ZAMM, to appear.

35. U. Langer, D. Pusch: Data–sparse algebraic multigrid methods for large scale boundary element equations. Appl. Numer. Math. 54 (2005) 406–424.

36. U. Langer, O. Steinbach: Boundary element tearing and interconnecting methods. Computing 71 (2003) 205–228.

37. U. Langer, O. Steinbach: Coupled boundary and finite element tearing and interconnecting methods. In: Domain decomposition methods in science and engineering. Selected papers of the 15th international conference on domain decomposition (R. Kornhuber et. al. eds.), Berlin, Germany, July 21-25, 2003. Lecture Notes in Computational Science and Engineering 40, pp. 83–97. Springer, Berlin, 2005.

38. G. Of: BETI Gebietszerlegungsmethoden mit schnellen Randelementverfahren und Anwendungen. Doctoral Thesis, Universität Stuttgart, 2006.

39. G. Of, O. Steinbach: A fast multipole boundary element method for a modified hypersingular boundary integral equation. In: Proceedings of the International Conference on Multifield Problems (M. Efendiev, W. L. Wendland eds.), Springer Lecture Notes in Applied Mechanics vol. 12, Springer, Berlin, pp. 163–169, 2003.

40. C. Polizzotto: A symmetric-definite BEM formulation for the elastoplastic rate problem. Boundary elements IX, Vol. 2 (Stuttgart, 1987), pp. 315–334, Comput. Mech., Southampton, 1987.

41. A. Quarteroni, A. Valli: Domain Decomposition Methods for Partial Differential Equations. Oxford Science Publications, 1999.

42. V. Rokhlin: Rapid solution of integral equations of classical potential theory. J. Comput. Phys. 60 (1985) 187–207.

43. R. Schneider: Multiskalen- und Wavelet-Matrixkompression: Analysisbasierte Methoden zur effizienten Lösung grosser vollbesetzter Gleichungssysteme. Advances in Numerical Mathematics, B. G. Teubner, Stuttgart, 1998.

44. D. Stefanica: A numerical study of FETI algorithms for mortar finite element methods. SIAM J. Sci. Comput. 23 (2001) 1135–1160.

45. O. Steinbach: Boundary elements in domain decomposition. Contemp. Math. 180 (1994) 343–348.

46. O. Steinbach: Gebietszerlegungsmethoden mit Randintegralgleichungen und effiziente numerische Lösungsverfahren für gemischte Randwertprobleme. Doctoral Thesis, Universität Stuttgart, 1996.

47. O. Steinbach: Mixed approximations for boundary elements. SIAM J. Numer. Anal. 38 (2000) 401–413.

48. O. Steinbach: Stability estimates for hybrid coupled domain decomposition methods. Lecture Notes in Mathematics vol. 1809, Springer, Berlin, Heidelberg, New York, 2003.

49. O. Steinbach: Numerische Näherungsverfahren für elliptische Randwertprobleme. Finite Elemente und Randelemente. B. G. Teubner, Stuttgart, Leipzig, Wiesbaden, 2003.

50. O. Steinbach: Artificial multilevel boundary element preconditioners. Proc. Appl. Math. Mech. 3 (2003) 539–542.

51. O. Steinbach, W. L. Wendland: The construction of some efficient preconditioners in the boundary element method. Adv. Comput. Math. 9 (1998) 191–216.

52. O. Steinbach, W. L. Wendland: On C. Neumann's method for second order elliptic systems in domains with non–smooth boundaries. J. Math. Anal. Appl. 262 (2001) 733–748.
53. E. P. Stephan: Multilevel methods for $h-$, $p-$, and $hp-$versions of the boundary element method. J. Comput. Appl. Math. 125 (2000) 503–519.
54. A. Toselli: Dual-primal FETI algorithms for edge finite-element approximations in 3D. IMA J. Numer. Anal. 26 (2006) 96–130.
55. A. Toselli, O. Widlund: Domain Decomposition Methods – Algorithms and Theory. Springer Series in Computational Mathematics, vol. 34, Springer, New York, 2004.
56. W. L. Wendland: On asymptotic error estimates for comined BEM and FEM. In: Finite Element and Boundary Element Techniques from Mathematical and Engineering Point of View (E. Stein, W. L. Wendland eds.), CISM Courses and Lectures 301, Springer, Wien, New York, pp. 273–333, 1988.
57. B. I. Wohlmuth: Discretization Methods and Iterative Solvers Based on Domain Decomposition. Lecture Notes in Computational Science and Engineering 17, Springer, Berlin, 2001.
58. O. C. Zienkiewicz, D. M. Kelly, P. Bettes: The coupling of the finite element method and boundary solution procedures. Int. J. Numer. Meth. Eng. 11 (1977) 355–376.
59. O. C. Zienkiewicz, D. M. Kelly, P. Bettes: Marriage a la mode? the best of both worlds (Finite elements and boundary integrals). In: Energy Methods in Finite Element Analysis (R. Glowinski, E. Y. Rodin, O. C. Zienkiewicz eds.), Chapter 5, J. Wiley and Son, London, pp. 81-106, 1979.
60. W. Zulehner: Uzawa–type methods for block–structured indefinite linear systems. SFB–Report 2005–05, SFB F013, Johannes Kepler University Linz, Austria, 2005.

The hp–Version of the Boundary Element Method for the Lamé Equation in 3D

Matthias Maischak and Ernst P. Stephan

Institut für Angewandte Mathematik, Universität Hannover,
Welfengarten 1, 30167 Hannover, Germany
{maischak,stephan}@ifam.uni-hannover.de

Summary. We analyze the h-p version of the BEM for Dirichlet and Neumann problems of the Lamé equation on open surface pieces. With given regularity of the solution in countably normed spaces we show that the boundary element Galerkin solution of the h-p version converges exponentially fast on geometrically graded meshes. We describe in detail how to use an analytic integration for the computation of the entries of the Galerkin matrix. Numerical benchmarks correspond to our theoretical results.

1 Introduction

It is well-known that an appropriate combination of mesh refinement and polynomial degree distribution (the hp-version with geometrically refined graded meshes) may lead to an exponential rate of convergence, even in the presence of singularities (for the FEM see [6, 7], and for the BEM see [8, 10, 11, 17]). The approximation strategy for such hp-methods is to use polynomial degrees of lowest order where solutions behave singularly and to use high order polynomials where solutions are smooth. This strategy has the advantage that it completely avoids the approximation analysis of singular functions by high order polynomials. This differs from the situation for a pure p-version, see [3, 2].

In this paper we consider the hp-version of the boundary element method (BEM) for Dirichlet and Neumann problems of the Lamé equation in $\Omega_\Gamma := \mathbb{R}^3 \backslash \bar{\Gamma}$, where Γ is a smooth open surface piece with a piecewise smooth boundary curve. That is:

For given $\mathbf{u}_1, \mathbf{u}_2 \in (H^{1/2}(\Gamma))^3$ with $\mathbf{u}_1 - \mathbf{u}_2 \in (\tilde{H}^{1/2}(\Gamma))^3$ (Dirichlet) or for given $\mathbf{t}_1, \mathbf{t}_2 \in (H^{-1/2}(\Gamma))^3$ with $\mathbf{t}_1 - \mathbf{t}_2 \in (\tilde{H}^{-1/2}(\Gamma))^3$ (Neumann) find $\mathbf{u}$ satisfying

$$\Delta^{*}\mathbf{u} := \mu\boldsymbol{\Delta}\mathbf{u} + (\lambda + \mu)\operatorname{grad}\operatorname{div}\mathbf{u} = 0 \text{ in } \Omega_{\Gamma}, \tag{1}$$

$$\mathbf{u}|_{\Gamma_1} = \mathbf{u}_1, \mathbf{u}|_{\Gamma_2} = \mathbf{u}_2 \text{ (Dirichlet)} \tag{2}$$

$$\mathbf{T}(\mathbf{u})|_{\Gamma_1} = \mathbf{t}_1, \mathbf{T}(\mathbf{u})|_{\Gamma_2} = \mathbf{t}_2 \text{ (Neumann)} \tag{3}$$

$$\mathbf{u}(x) = o(1), \frac{\partial}{\partial x_j}\mathbf{u}(x) = o(|x|^{-1}), j = 1, 2, 3, \ |x| \to \infty. \tag{4}$$

Here, Γ_i, $i = 1, 2$, are the two sides of Γ and $\mu > 0$, $\lambda > -2/3\mu$ are the given Lamé constants.

The corresponding Neumann data of the linear elasticity problem are the tractions

$$\mathbf{T}(\mathbf{u}) = \lambda(\operatorname{div}\mathbf{u})\mathbf{n} + 2\mu\frac{\partial\mathbf{u}}{\partial\mathbf{n}} + \mu\mathbf{n} \times \operatorname{curl}\mathbf{u} \text{ on } \Gamma_i, i = 1, 2, \tag{5}$$

where $\mathbf{n}$ is the normal vector exterior to a bounded domain Ω, such that $\Gamma \subset \partial\Omega$.

Let $G(x, y) \in \mathbb{R}^{3\times 3}$ denote the fundamental solution of the differential operator Δ^{*}, i.e.

$$G(x, y) = \frac{\lambda + 3\mu}{4\pi\mu(\lambda + 2\mu)} \left\{ \frac{1}{|x - y|} I_{3\times 3} + \frac{\lambda + \mu}{\lambda + 3\mu} \frac{(x - y)(x - y)^{\top}}{|x - y|^3} \right\}. \tag{6}$$

The problem (1)–(4) can be formulated as an integral equation of the first kind, see, e.g. [4, 5, 20, 21]:

Dirichlet:

$\mathbf{u} \in (H_{\mathrm{loc}}^1(\mathbb{R}^3\backslash\bar{\Gamma}))^3$ is the solution of the Dirichlet problem (1), (2) and (4) if and only if the jump of the traction $\mathbf{t} := \mathbf{T}(u)|_{\Gamma_1} - \mathbf{T}(u)|_{\Gamma_2} \in (\tilde{H}^{-1/2}(\Gamma))^3$ solves the weakly singular integral equation

$$\mathbf{V}\mathbf{t}(x) := \int_{\Gamma} G(x, y)\mathbf{t}(y)\, ds_y = \mathbf{g}(x), \quad x \in \Gamma \tag{7}$$

where

$$\mathbf{g}(x) = \frac{1}{2}(\mathbf{u}_1 + \mathbf{u}_2)(x) + \int_{\Gamma} \mathbf{T}_y G(x, y)(\mathbf{u}_1 - \mathbf{u}_2)(y)\, ds_y. \tag{8}$$

The solution $\mathbf{t}$ of (7) yields the solution of the Dirichlet problem (1), (2) and (4) via the representation or Betti's formula

$$\mathbf{u}(x) = \int_{\Gamma} \left(G(x, y)\mathbf{t}(y) - (\mathbf{T}_y G(x, y))^t (\mathbf{u}_1(y) - \mathbf{u}_2(y)) \right) ds_y, \quad x \notin \Gamma.$$

The Galerkin scheme for (7) is given by: Find $\mathbf{t}_N \in S^{p,0}(\Gamma_{\sigma}^n) \subset (\tilde{H}^{-1/2}(\Gamma))^3$ such that for all $v \in S^{p,0}(\Gamma_{\sigma}^n)$

$$\langle \mathbf{V}\mathbf{t}, v \rangle = \langle \mathbf{g}, v \rangle \tag{9}$$

where $\langle \cdot, \cdot \rangle$ denotes the duality pairing of $(H^{1/2}(\Gamma))^3$ and $(\tilde{H}^{-1/2}(\Gamma))^3$. The symmetric bilinear form $\langle \mathbf{V}\cdot, \cdot \rangle$ is positive definite on $(\tilde{H}^{-1/2}(\Gamma))^3 \times (\tilde{H}^{-1/2}(\Gamma))^3$ giving the energy norm $\|t\|_V = \langle \mathbf{V}t, t \rangle^{1/2}$.

Neumann:

$\mathbf{u} \in (H^1_{\mathrm{loc}}(\mathbb{R}^3 \setminus \bar{\Gamma}))^3$ is the solution of the Neumann problem (1), (3) and (4) if and only if the jump of the displacement $\phi := \mathbf{u}|_{\Gamma_1} - \mathbf{u}|_{\Gamma_2} \in (\tilde{H}^{1/2}(\Gamma))^3$ solves the hyper-singular integral equation

$$\mathbf{W}\phi(x) := -\mathbf{T}_x \int_\Gamma (\mathbf{T}_y G(x,y))^t \, \phi(y) \, ds_y = \mathbf{f}(x), \quad x \in \Gamma \qquad (10)$$

where

$$\mathbf{f}(x) = \frac{1}{2}(\mathbf{t}_1 + \mathbf{t}_2)(x) - \mathbf{T}_x \int_\Gamma G(x,y)(\mathbf{t}_1 - \mathbf{t}_2)(y) \, ds_y. \qquad (11)$$

The solution ϕ of (10) yields the solution of the Neumann problem (1), (3) and (4) via the representation or Betti's formula

$$\mathbf{u}(x) = \int_\Gamma \left(G(x,y)(\mathbf{t}_1(y) - \mathbf{t}_2(y)) - (\mathbf{T}_y G(x,y))^t \phi(y) \right) ds_y, \quad x \notin \Gamma.$$

The Galerkin scheme for (10) is given by: Find $\phi_N \in S^{p,1}(\Gamma^n_\sigma) \subset (\tilde{H}^{1/2}(\Gamma))^3$ such that for all $\psi \in S^{p,1}(\Gamma^n_\sigma)$

$$\langle \mathbf{W}\phi, \psi \rangle = \langle \mathbf{f}, \psi \rangle \qquad (12)$$

where $\langle \cdot, \cdot \rangle$ denotes the duality pairing of $(H^{-1/2}(\Gamma))^3$ and $(\tilde{H}^{1/2}(\Gamma))^3$. The symmetric bilinear form $\langle \mathbf{W} \cdot, \cdot \rangle$ is positive definite on $(\tilde{H}^{1/2}(\Gamma))^3 \times (\tilde{H}^{1/2}(\Gamma))^3$ giving the energy norm $\|\phi\|_W = \langle \mathbf{W}\phi, \phi \rangle^{1/2}$.

Both Galerkin schemes (9) and (12) converge quasi-optimally in the energy norm with algebraic orders of convergence for the h- and p-versions, namely of order $\mathcal{O}(h^{1/2}p^{-1})$. This follows by extending corresponding results for the Laplacian [1, 3, 19, 20, 22, 26]. These low convergence rates result from the singular behavior of the solutions $\mathbf{t}$ of (7) and ϕ of (10) near the boundary of Γ; this describes the well-known behavior of the displacement and traction near the edges of the crack [24, 26], cf. [25]. On the other hand, if we use an hp-version with a geometrically refined mesh towards the edges of the surface Γ we obtain even exponentionally fast convergence (cf. Fig. 3 and Fig. 4). Especially, as shown below, there hold the following error estimates for the exact solutions $\mathbf{t}$ of (7) and ϕ of (10) and the Galerkin solutions $\mathbf{t}_N \in S^{p,0}(\Gamma^n_\sigma)$ of (9) and $\phi_N \in S^{p,1}(\Gamma^n_\sigma)$ of (12), i.e.

$$\|\mathbf{t} - \mathbf{t}_N\|_V \leq Ce^{-bN^{1/4}}, \qquad \|\phi - \phi_N\|_W \leq Ce^{-bN^{1/4}} \qquad (13)$$

with constants $C, b > 0$ independent of N (see Theorems 4 and 5 below, c.f. [10, 13, 18, 23]).

Another important issue is the implementation of the hp-version for the Galerkin equations itself. In this paper we explicitly describe how analytic integration can be used in the computation of the entries of the Galerkin matrices. The trick is to reduce the integrals for Lamé-case to simpler ones which already have been used for the computations of the integral operators belonging to the Laplacian [16]. Numerical benchmarks underline our theoretical results.

2 The hp-Version with Geometric Mesh

In this section we introduce the boundary element spaces for the hp-version together with countably normed spaces.

Now we define the geometric mesh on a triangle F. This is no loss of generality because every polygonal domain can be decomposed into triangles. We divide this triangle into three parallelograms and three triangles where each parallelogram lies in a corner of F and each triangle lies at an edge of F but does not touch the corners (see Fig. 1). By linear transformations φ_i we can map the parallelograms onto the reference square $Q = [0,1]^2$ such that the vertices of F are mapped to $(0,0)$. The triangles can be mapped by linear transformations $\tilde{\varphi}_i$ onto the reference triangle $\tilde{Q} = \{(x,y) \in Q \,|\, y \le x\}$ such that the corner point of the triangle in the interior of F is mapped to $(1,1)$ of the reference triangle. By Definition 1 the geometric mesh and appropriate spline spaces are defined on the reference element Q. Analogously the geometric mesh can be defined on the reference triangle $\tilde{Q}$ (see Fig. 1).

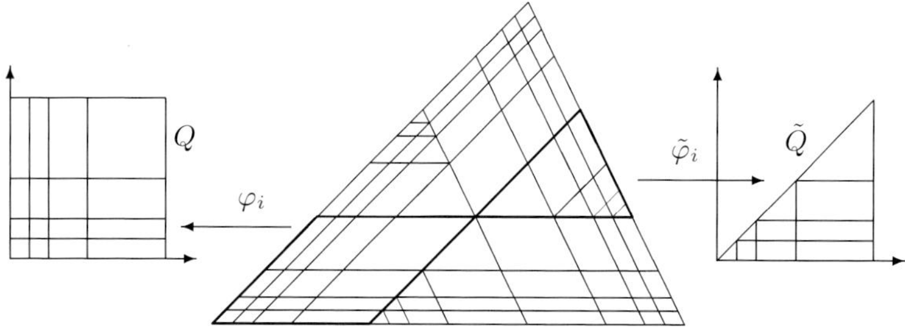

Fig. 1. Geometric mesh with $\sigma = 0.5$ on the triangle F

Via the transformations $\varphi_i^{-1}, \tilde{\varphi}_i^{-1}$ the geometric mesh Γ_σ^n can also be defined on the faces of a polyhedron. The approximation on the reference square is the more interesting case because it handles the corner-edge singularities. Therefore we deal in the following only with the approximation on the reference square.

Definition 1 (geometric mesh). *Let $I = [0,1]$. For $0 < \sigma < 1$ we use the partition I_σ^n of I into n subintervals $[x_{k-1}, x_k]$, $k = 1, \ldots, n$, where*

$$x_0 = 0, \qquad x_k = \sigma^{n-k}, \quad k = 1, \ldots, n. \tag{14}$$

With I_σ^n we associate a degree-vector $p = (p_1, \ldots, p_n)$ and define $S^{p,r}(I_\sigma^n) \subset H^r(I)$ as the vector space of all piecewise polynomials w on I having degree p_j on (x_{j-1}, x_j), $j = 1, \ldots, n$, i.e. $w|_{(x_{j-1}, x_j)} \in P_{p_j}((x_{j-1}, x_j))$.

Let $Q = [0,1] \times [0,1]$. For $0 < \sigma < 1$ we use the partition Q_σ^n of Q into n^2 subsquares R_{kl}

$$R_{kl} = [x_{k-1}, x_k] \times [x_{l-1}, x_l], \quad (k,l = 1, \ldots, n), \qquad Q = \bigcup_{k,l=1}^{n} R_{kl}. \quad (15)$$

With Q_σ^n we associate a degree vector $p = (p_1, \ldots, p_n)$ and define $S^{p,r}(Q_\sigma^n) \subset H^r(Q)$ as the vector space of all piecewise polynomials $v(x,y)$ on Q having degree p_k in x and p_l in y on R_{kl}, $k,l = 1, \ldots, n$, i.e. $v|_{R_{kl}} \in P_{p_k, p_l}(R_{kl})$. The index $r \in \{0,1\}$ in $S^{p,r}(I_\sigma^n)$ and $S^{p,r}(Q_\sigma^n)$ determines the regularity of the piecewise polynomials, i.e. discontinuity in case of $r = 0$ and continuity in case of $r = 1$. For the differences $h_k = x_k - x_{k-1}$ we have with $\lambda = (1 - \sigma)/\sigma$

$$h_k = x_k - x_{k-1} = x_{k-1}\left(\frac{1}{\sigma} - 1\right) \leq x\left(\frac{1}{\sigma} - 1\right) = x\lambda, \quad \forall x \in [x_{k-1}, x_k] \ (2 \leq k \leq n) \quad (16)$$

Then we have by construction:

$$S^{p,r}(I_\sigma^n) \times S^{p,r}(I_\sigma^n) \subset S^{p,r}(Q_\sigma^n) \quad (17)$$

Fig. 2 shows the geometric meshes for $\sigma = 1/2$ and $n = 4$.

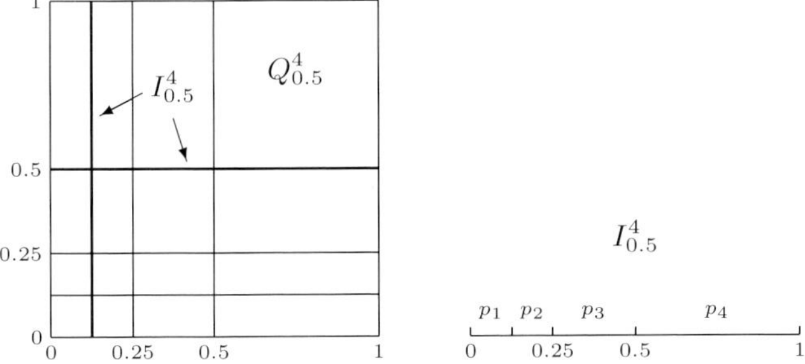

Fig. 2. Geometric mesh on the square plate ($\sigma = 0.5$, $n = 4$).

Now we define countably normed spaces on the reference element Q using Cartesian coordinates.

Definition 2 (countably normed spaces $B_\beta^l(Q)$). *Let β be a real number with $0 < \beta < 1$. The weight function $\Phi_{\beta,\alpha,l} = \Phi_{\beta,\alpha,l}(x,y)$ is for $\alpha = (\alpha_1, \alpha_2)$ and an integer $l \geq 1$ defined by*

$$\Phi_{\beta,\alpha,l} = x^\beta \sum_{\gamma_1 = \max(\alpha_1 - l, 0)}^{\min(\alpha_1 - 1, \alpha_1 + \alpha_2 - l)} x^{\gamma_1} y^{\alpha_1 + \alpha_2 - l - \gamma_1} + y^\beta \sum_{\gamma_2 = \max(\alpha_2 - l, 0)}^{\min(\alpha_2 - 1, \alpha_1 + \alpha_2 - l)} x^{\alpha_1 + \alpha_2 - l - \gamma_2} y^{\gamma_2}. \quad (18)$$

Let

$$D^\alpha = \frac{\partial^{|\alpha|}}{\partial x^{\alpha_1} \partial y^{\alpha_2}} = \partial_x^{\alpha_1} \partial_y^{\alpha_2}.$$

The weighted Sobolev spaces for integers m, l with $m \geq l \geq 1$ are defined by

$$H_\beta^{m,l}(Q) = \Big\{ u : u \in H^{l-1}(Q) \text{ for } l > 0, \tag{19}$$

$$\|\Phi_{\beta,\alpha,l} D^\alpha u\|_{L^2(Q)} < \infty \text{ for } l \leq |\alpha| \leq m \Big\},$$

with the norm

$$\|u\|_{H_\beta^{m,l}(Q)}^2 = \|u\|_{H^{l-1}(Q)}^2 + \sum_{k=l}^{m} \sum_{|\alpha|=k} \int_Q |D^\alpha u(x,y)|^2 \Phi_{\beta,\alpha,l}^2(x,y) \, dy \, dx \tag{20}$$

and the semi norm

$$|u|_{H_\beta^{m,l}(Q)}^2 = \sum_{k=l}^{m} \sum_{|\alpha|=k} \int_Q |D^\alpha u(x,y)|^2 \Phi_{\beta,\alpha,l}^2(x,y) \, dy \, dx. \tag{21}$$

The countably normed spaces for $l \geq 1$ are defined by

$$B_\beta^l(Q) = \Big\{ u : u \in H^{l-1}(Q), \|\Phi_{\beta,\alpha,l} D^\alpha u\|_{L^2(Q)} \leq C \, d^{k-l}(k-l)!$$

$$\text{for } |\alpha| = k = l, l+1, \ldots, ; \quad C \geq 1, d \geq 1 \text{ independent of } k \Big\}. \tag{22}$$

If we would like to emphasize the dependence on the constants C, d we will write $B_\beta^l(Q) = B_{\beta,C,d}^l(Q)$, etc.

Theorem 1. [12] *Let Q be the reference element and let φ be the linear transformation from a parallelogram, lying in a corner of the triangle F, to the reference element Q. Then, for $l = 1, 2$, $u \in B_{\beta,\tilde{C},\tilde{d}}^l(\varphi(Q))$ implies $u \circ \varphi^{-1} \in B_{\beta,C,d}^l(Q)$ where C, d (resp. $\tilde{C}, \tilde{d}$) are the constants in the definition of $B_\beta^l(Q)$ (resp. $B_\beta^l(\varphi(Q))$). For the case $l = 1$ the reverse implication holds as well.*

The exponentially good approximation properties of splines on our geometric meshes for general functions $u \in B_\beta^l(Q)$ ($l = 1, 2$) are given by the following theorem (see also [12, 15, 17, 18]).

Theorem 2.

(i) Let $u \in B_\beta^1(Q)$ with $0 < \beta < 1$. Let Q_σ^n be a geometric mesh and assume $p = (p_1, \ldots, p_n)$, $p_k = [\mu(k-1)]$ for some $\mu > 0$. Set $N = \dim S^{p,0}(Q_\sigma^n)$. Then there exist constants $C_1, b_1 > 0$ independent of N, but depending on σ, μ, β, such that the L^2-projection $u_N \in S^{p,0}(Q_\sigma^n)$ of u satisfies

$$\|u - u_N\|_{L^2(Q)} \leq C_1 \, e^{-b_1 \sqrt[4]{N}}. \tag{23}$$

(ii) Let $v \in B_\beta^2(Q)$ with $0 < \beta < 1$. Let Q_σ^n be a geometric mesh and assume $p = (p_1, \ldots, p_n)$, $p_1 = 1$, $p_k = \max(2, [\mu(k-1)] + 1)$ $(k > 1)$ for some $\mu > 0$. Set $N = \dim S^{p,1}(Q_\sigma^n)$. Then there is a spline function $v_N \in S^{p,1}(Q_\sigma^n)$ and constants $C_2, b_2 > 0$ independent of N, but dependent on σ, μ, β, such that

$$\|v - v_N\|_{H^1(Q)} \le C_2\, e^{-b_2 \sqrt[4]{N}}. \tag{24}$$

(iii) Let $v \in B_\beta^1(Q) \cap C^0(\bar{Q})$, $v|_{\partial Q} = 0$ with $0 < \beta < 1/2$. Let Q_σ^n be a geometric mesh and assume $p = (p_1, \ldots, p_n)$, $p_1 = 1$, $p_k = \max(2, [\mu(k-1)] + 1)$ $(k > 1)$ for some $\mu > 0$. Set $N = \dim S^{p,1}(Q_\sigma^n)$. Then there is a spline function $v_N \in S^{p,1}(Q_\sigma^n)$ and constants $C_3, b_3 > 0$ independent of N, but dependent on σ, μ, β, such that

$$\|v - v_N\|_{\tilde{H}^{1/2}(Q)} \le C_3\, e^{-b_3 \sqrt[4]{N}}. \tag{25}$$

Now, we want to recall the typical structure of the solutions of our problems for sufficiently smooth right-hand side functions g and f.

Theorem 3. *[24, Theorem 2.3, 2.4 and 2.5] Let V and E denote the sets of vertices and edges of Γ, respectively. For $v \in V$, let $E(v)$ denote the set of edges with v as an end point. Then, the solution $\mathbf{t}$ of (7) has the form*

$$\mathbf{t} = \mathbf{t}_{\mathrm{reg}} + \sum_{e \in E} \mathbf{t}^e + \sum_{v \in V} \mathbf{t}^v + \sum_{v \in V} \sum_{e \in E(v)} \mathbf{t}^{ev}, \tag{26}$$

with a regular part $\mathbf{t}_{\mathrm{reg}}$, edge singularities $\mathbf{t}^e$, vertex singularities $\mathbf{t}^v$ and edge-vertex singularities $\mathbf{t}^{ev}$. These terms result from applying boundary traction to the corresponding decomposition of the solution.

Accordingly, the solution ϕ of (10) has the form

$$\phi = \phi_{\mathrm{reg}} + \sum_{e \in E} \phi^e + \sum_{v \in V} \phi^v + \sum_{v \in V} \sum_{e \in E(v)} \phi^{ev}. \tag{27}$$

Checking the specific terms (26) and (27) , which are given in [24], one realizes that these terms $\mathbf{t}^e, \mathbf{t}^v, \mathbf{t}^{ev}$ and $\phi^e, \phi^v, \phi^{ev}$ belong to countably normed spaces. Therefore we can argue as done in [10] and obtain the following convergence results.

Theorem 4. *Let the right hand side g in equation (7) be piecewise analytic, let $\mathbf{t}$ be the solution of (7) and let $\mathbf{t}_N \in S^{p,0}(\Gamma_\sigma^n)$ be its Galerkin approximation defined by (9). Then, with $N = \dim S^{p,0}(\Gamma_\sigma^n)$, there holds for any $\alpha > 0$*

$$\|\mathbf{t} - \mathbf{t}_N\|_{(\tilde{H}^{-1/2}(\Gamma))^3} \le C\, e^{-b \sqrt[4]{N}} + \mathcal{O}(N^{-\alpha}) \tag{28}$$

for constants $C, b > 0$, depending on σ, μ and α, but independent of N.

Theorem 5. *Let the right hand side f in equation (10) be piecewise analytic and let ϕ be the solution of (10) and let $\phi_N \in S^{p,1}(\Gamma_\sigma^n)$ be its Galerkin approximation defined by (12). Then there holds for all $\alpha > 0$*

$$\|\phi - \phi_N\|_{(\tilde{H}^{1/2}(\Gamma))^3} \leq C\, e^{-b\sqrt[4]{N}} + \mathcal{O}(N^{-\alpha}) \tag{29}$$

for constants $C, b > 0$ depending on σ, μ and α, but not depending on $N = \dim S^{p,1}(\Gamma_\sigma^n)$.

Remark 1. Due to the splittings (26) and (27) into finitely many singularity terms the regular remainder terms $\mathbf{t}_{\mathrm{reg}}$ and ϕ_{reg} have only restricted regularity, even for given smooth right hand sides. On the other hand, even taking infinitely many singularity terms, would not automatically guarantee that the solutions $\mathbf{t}$ and ϕ themselves belong to countably normed spaces. To our knowledge this is an open problem. Therefore we get the additional $\mathcal{O}(N^{-\alpha})$-terms in the estimates (28) and (29).

3 Implementation of Galerkin Scheme

Assume that the surface piece $\Gamma \subset \mathbb{R}^3$ can be decomposed into triangles and parallelograms, i.e. $\Gamma = \bigcup_{i=1}^N \Gamma_i$, with Γ_i pairwise disjoint and Γ_i is the affine image of the reference square $\square = [-1, 1]^2$ or the reference triangle $\Delta = \{(t_1, t_2) : 0 \leq t_1 \leq 1 - t_2 \leq 1\}$. That means

$$\Gamma_i = \{a_i t_1 + b_i t_2 + x_i : (t_1, t_2) \in Q\}, \quad Q \in \{\Delta, \square\} \tag{30}$$

depending on whether Γ_i is a triangle or a parallelogram, with $a_i, b_i, x_i \in \mathbb{R}^3$, $i = 1, \ldots, N$. Here we investigate only basis functions whose restriction to Γ_i are polynomials. Effectively, we compute the integrals only for monomials as test- and trial-functions, from which all other basis functions can be constructed.

For $Q \in \{\Delta, \square\}$ let

$$F_i : \begin{cases} Q \Rightarrow \Gamma_i \\ t = (t_1, t_2) \to x = a_i t_1 + b_i t_2 + x_i \end{cases} \tag{31}$$

be the affine transformation from the reference element Δ or $\square$ to Γ_i with $|\frac{\partial F_i}{\partial t}| = |a_i \times b_i|$. We will write Q for Δ or $\square$, respectively, if the expressions hold for both cases. Then the basis functions on Γ_i are defined by

$$\varphi_{kl}^i(x) = \tilde{\varphi}_{kl}(F_i^{-1}(x)) = \tilde{\varphi}_{kl} \circ F_i^{-1}(x) \tag{32}$$

with $\tilde{\varphi}_{kl}(t_1, t_2) = t_1^k t_2^l$ for $x \in \Gamma_i$ and $\varphi_{kl}^i(x) = 0$ otherwise. The vector valued test and trial functions ϕ restricted to an element Γ_i can be represented as linear combination of this monomial basis functions $\varphi_{kl}^i(x)$, i.e. we have

$$\phi(x)|_{\Gamma_i} = \sum_{r=1}^3 \mathbf{e}_r \phi_r(x)|_{\Gamma_i}, \quad \phi_r(x)|_{\Gamma_i} = \sum_{kl} c_{kl}^{i,r} \varphi_{kl}^i(x)$$

with $\mathbf{e}_1 = (1, 0, 0), \mathbf{e}_2 = (0, 1, 0), \mathbf{e}_3 = (0, 0, 1)$.

Single layer potential

Using (6) the single layer potential is then given by

$$\mathbf{V}\phi(x) := \int_\Gamma G(x,y)\phi(y)\,ds_y = \sum_{i=1}^N \sum_{r=1}^3 \mathbf{e}_r \sum_{s=1}^3 \sum_{kl} c_{kl}^s \int_{\Gamma_i} G_{rs}(x,y)\varphi_{kl}^i(y)\,ds_y$$

$$(33)$$

and the corresponding bilinear form reads

$$\langle \mathbf{V}\phi, \psi \rangle = \int_\Gamma \int_\Gamma \psi_r(x) G_{rs}(x,y)\phi_s(y)\,ds_y\,ds_x. \tag{34}$$

In the following we are interested in the computation of the term

$$V_{kl}^{i,rs}(x) := \int_{\Gamma_i} G_{rs}(x,y)\varphi_{kl}^i(y)\,ds_y. \tag{35}$$

We will use the following form of the fundamental solution (6)

$$G_{rs}(x,y) = \frac{\lambda + 3\mu}{4\pi\mu(\lambda + 2\mu)} \left\{ \frac{1}{|x-y|}\delta_{rs} + \frac{\lambda + \mu}{\lambda + 3\mu}\frac{(x_r - y_r)(x_s - y_s)}{|x-y|^3} \right\}$$

$$= \frac{1}{4\pi\mu}\frac{1}{|x-y|}\delta_{rs} - \frac{\lambda + \mu}{8\pi\mu(\lambda + 2\mu)}\frac{\partial}{\partial y_r}\frac{y_s - x_s}{|y-x|}. \tag{36}$$

By extending the affine transformation F_i to

$$F_i(t_1, t_2, t_3) = a_i t_1 + b_i t_2 + n_i t_3 + x_i,$$

where n_i is the normal direction on the patch Γ_i, we obtain the following integral

$$V_{kl}^{i,rs}(x) = \frac{1}{4\pi\mu}\int_{\Gamma_i}\frac{1}{|x-y|}\delta_{rs}\varphi_{kl}^i(y)\,ds_y$$

$$\quad - \frac{\lambda + \mu}{8\pi\mu(\lambda + 2\mu)}\int_{\Gamma_i}\frac{\partial}{\partial y_r}\frac{y_s - x_s}{|y-x|}\varphi_{kl}^i(y)\,ds_y$$

$$= \frac{1}{4\pi\mu}\left|\frac{\partial F_i}{\partial t}\right|\delta_{rs}\int_Q \frac{1}{|F_i(t)-x|}\tilde\varphi_{kl}(t)\,dt$$

$$\quad - \frac{\lambda + \mu}{8\pi\mu(\lambda + 2\mu)}\left|\frac{\partial F_i}{\partial t}\right|\int_Q \sum_{p=1}^3 \frac{\partial t_p}{\partial y_r}\frac{\partial}{\partial t_p}\frac{(F_i(t)-x)_s}{|F_i(t)-x|}\tilde\varphi_{kl}(t)\,dt$$

$$=: \frac{1}{4\pi\mu}\left|\frac{\partial F_i}{\partial t}\right|\delta_{rs}A_{kl}^i(x) - \frac{\lambda + \mu}{8\pi\mu(\lambda + 2\mu)}\left|\frac{\partial F_i}{\partial t}\right|B_{kl}^{i,rs}(x)$$

Defining the following elementary integrals, analyzed in [16]

$$I_{kl}^{Q,p}(a,b,c) := \int_Q t_1^k t_2^l |at_1 + bt_2 + c|^{2p}\,dt_2\,dt_1, \quad Q \in \{\Delta, \square\} \tag{37}$$

we can identify

$$A^i_{kl}(x) = \int_Q \frac{\tilde{\varphi}_{kl}(t)}{|F_i(t) - x|}\, dt = \int_Q \frac{t_1^k t_2^l}{|a_i t_1 + b_i t_2 + x_i - x|}\, dt_2\, dt_1$$

$$= I^{Q,-\frac{1}{2}}_{kl}(a_i, b_i, x_i - x). \tag{38}$$

It remains to reduce the integral $B^{i,rs}_{kl}(x)$ to a linear combination of elementary integrals. We can compute

$$\left(\frac{\partial t_p}{\partial y_r}\right) = \left(\frac{\partial y_r}{\partial t_p}\right)^{-1} = \left(a_i|b_i|n_i\right)^{-1} = \frac{1}{a_i(b_i \times n_i)} \begin{pmatrix} b_i \times n_i \\ n_i \times a_i \\ a_i \times b_i \end{pmatrix}. \tag{39}$$

Therefore we obtain

$$\sum_{p=1}^{3} \frac{\partial t_p}{\partial y_r} \frac{\partial}{\partial t_p} = \frac{1}{a_i(b_i \times n_i)} \left((b_i \times n_i)_r \partial_{t_1} + (n_i \times a_i)_r \partial_{t_2} + (a_i \times b_i)_r \partial_{t_3}\right) \tag{40}$$

and consequently

$$B^{i,rs}_{kl}(x) = \int_Q \sum_{p=1}^{3} \frac{\partial t_p}{\partial y_r} \frac{\partial}{\partial t_p} \frac{(F_i(t) - x)_s}{|F_i(t) - x|} \tilde{\varphi}_{kl}(t)\, dt$$

$$= \int_Q \frac{\left((b_i \times n_i)_r \partial_{t_1} + (n_i \times a_i)_r \partial_{t_2} + (a_i \times b_i)_r \partial_{t_3}\right)}{a_i(b_i \times n_i)} \frac{(F_i(t) - x)_s}{|F_i(t) - x|} t_1^k t_2^l\, dt_2\, dt_1$$

$$=: \frac{(b_i \times n_i)_r}{a_i(b_i \times n_i)} C^{i,s}_{kl}(x) + \frac{(n_i \times a_i)_r}{a_i(b_i \times n_i)} D^{i,s}_{kl}(x) + \frac{(a_i \times b_i)_r}{a_i(b_i \times n_i)} E^{i,s}_{kl}(x).$$

For the last integral we obtain

$$E^{i,s}_{kl}(x) = \int_Q \partial_{t_3} \frac{(a_i t_1 + b_i t_2 + n_i t_3 + x_i - x)_s}{|a_i t_1 + b_i t_2 + n_i t_3 + x_i - x|} t_1^k t_2^l\, dt_2\, dt_1$$

$$= \int_Q \frac{(n_i)_s}{|a_i t_1 + b_i t_2 + x_i - x|} t_1^k t_2^l\, dt_2\, dt_1$$

$$\quad - \int_Q \frac{(a_i t_1 + b_i t_2 + x_i - x)_s (n_i(x_i - x))}{|a_i t_1 + b_i t_2 + x_i - x|^3} t_1^k t_2^l\, dt_2\, dt_1$$

$$= (n_i)_s I^{Q,-\frac{1}{2}}_{kl}(a_i, b_i, x_i - x) - (a_i)_s(n_i(x_i - x)) I^{Q,-\frac{3}{2}}_{k+1,l}(a_i, b_i, x_i - x)$$

$$\quad - (b_i)_s(n_i(x_i - x)) I^{Q,-\frac{3}{2}}_{k,l+1}(a_i, b_i, x_i - x)$$

$$\quad - (x_i - x)_s(n_i(x_i - x)) I^{Q,-\frac{3}{2}}_{k,l}(a_i, b_i, x_i - x).$$

The integrals $C^{i,s}_{kl}(x)$, $D^{i,s}_{kl}(x)$ can be treated by partial integration, but we have to distinguish between triangles and parallelograms. On parallelograms we simply obtain

$$C_{kl}^{i,s}(x) = \int_{\square} \partial_{t_1} \frac{(a_i t_1 + b_i t_2 + x_i - x)_s}{|a_i t_1 + b_i t_2 + x_i - x|} t_1^k t_2^l \, dt_2 \, dt_1$$

$$= \int_{-1}^{1} \frac{(a_i t_1 + b_i t_2 + x_i - x)_s}{|a_i t_1 + b_i t_2 + x_i - x|} t_1^k t_2^l \, dt_2 \Big|_{t_1=-1}^{t_1=1}$$

$$- k \int_{\square} \frac{(a_i t_1 + b_i t_2 + x_i - x)_s}{|a_i t_1 + b_i t_2 + x_i - x|} t_1^{k-1} t_2^l \, dt_2 \, dt_1$$

and on triangles there holds

$$C_{kl}^{i,s}(x) = \int_{\Delta} \partial_{t_1} \frac{(a_i t_1 + b_i t_2 + x_i - x)_s}{|a_i t_1 + b_i t_2 + x_i - x|} t_1^k t_2^l \, dt_2 \, dt_1$$

$$= \int_{0}^{1} t_1^k (1 - t_1)^l \frac{(a_i t_1 + b_i(1 - t_1) + x_i - x)_s}{|a_i t_1 + b_i(1 - t_1) + x_i - x|} \, dt_1$$

$$- \delta_{k,0} \int_{0}^{1} t_2^l \frac{(b_i t_2 + x_i - x)_s}{|b_i t_2 + x_i - x|} \, dt_2 - k \int_{\Delta} \frac{(a_i t_1 + b_i t_2 + x_i - x)_s}{|a_i t_1 + b_i t_2 + x_i - x|} t_1^{k-1} t_2^l \, dt_2 \, dt_1.$$

Double layer potential

Using the traction operator

$$(T\phi(y))_r = \lambda n_r \frac{\partial}{\partial y_t} \phi_t(y) + \mu n_t \frac{\partial}{\partial y_t} \phi_r(y) + \mu n_t \frac{\partial}{\partial y_r} \phi_t(y),$$

we can define the double layer potential operator by

$$\mathbf{K}\phi(x) := \int_{\Gamma} (T_y G(x,y))^t \phi(y) \, ds_y = \sum_{i=1}^{N} \sum_{r=1}^{3} \mathbf{e}_r \sum_{s=1}^{3} \sum_{kl} c_{kl}^s K_{kl}^{i,rs}(x) \qquad (41)$$

with

$$K_{kl}^{i,rs}(x) := \sum_{t=1}^{3} \left(\lambda n_{i,s} F_{kl}^{i,rtt}(x) + \mu n_{i,t} F_{kl}^{i,rst}(x) - \mu n_{i,t} F_{kl}^{i,rts}(x) \right) \qquad (42)$$

and

$$F_{kl}^{i,rst}(x) := \int_{\Gamma_i} \frac{\partial}{\partial y_t} G_{rs}(x,y) \varphi_{kl}^i(y) \, ds_y. \qquad (43)$$

We can decompose $F_{kl}^{i,rst}(x)$ as follows

$$F_{kl}^{i,rst}(x) := \frac{\lambda + 3\mu}{4\pi\mu(\lambda + 2\mu)} \delta_{rs} \int_{\Gamma_i} \frac{\partial}{\partial y_t} \frac{1}{|x - y|} \varphi_{kl}^i(y) \, ds_y$$

$$+ \frac{\lambda + \mu}{4\pi\mu(\lambda + 2\mu)} \int_{\Gamma_i} \frac{\partial}{\partial y_t} \frac{(x_r - y_r)(x_s - y_s)}{|x - y|^3} \varphi_{kl}^i(y) \, ds_y$$

$$= \frac{\lambda + 3\mu}{4\pi\mu(\lambda + 2\mu)} |\frac{\partial F_i}{\partial t}| \delta_{rs} \int_{Q} \sum_{p=1}^{3} \frac{\partial t_p}{\partial y_t} \frac{\partial}{\partial t_p} \frac{1}{|F_i(t) - x|} \tilde{\varphi}_{kl}(t) \, dt$$

$$+ \frac{\lambda + \mu}{4\pi\mu(\lambda + 2\mu)} |\frac{\partial F_i}{\partial t}| \int_{Q} \sum_{p=1}^{3} \frac{\partial t_p}{\partial y_t} \frac{\partial}{\partial t_p} \frac{(F_i(t) - x)_r (F_i(t) - x)_s}{|F_i(t) - x|^3} \tilde{\varphi}_{kl}(t) \, dt$$

$$=: \frac{\lambda + 3\mu}{4\pi\mu(\lambda + 2\mu)} \left|\frac{\partial F_i}{\partial t}\right| \delta_{rs} H_{kl}^{i,t}(x) + \frac{\lambda + \mu}{4\pi\mu(\lambda + 2\mu)} \left|\frac{\partial F_i}{\partial t}\right| J_{kl}^{i,rst}(x).$$

As before, we can represent the integrals $H_{kl}^{i,t}(x)$ and $J_{kl}^{i,rst}(x)$ in terms of the elementary integrals $I_{kl}^{Q,p}(a_i, b_i, x_i - x)$. We have

$$H_{kl}^{i,t}(x) = \int_Q \sum_{p=1}^{3} \frac{\partial t_p}{\partial y_t} \frac{\partial}{\partial t_p} \frac{1}{|F_i(t) - x|} \tilde{\varphi}_{kl}(t)\, dt$$

$$=: \frac{(b_i \times n_i)_t}{a_i(b_i \times n_i)} L_{kl}^i(x) + \frac{(n_i \times a_i)_t}{a_i(b_i \times n_i)} M_{kl}^i(x) + \frac{(a_i \times b_i)_t}{a_i(b_i \times n_i)} N_{kl}^i(x).$$

The last integral becomes

$$N_{kl}^i(x) = \int_Q \partial_{t_3} \frac{1}{|a_i t_1 + b_i t_2 + n_i t_3 + x_i - x|} t_1^k t_2^l\, dt_2\, dt_1$$

$$= - \int_Q \frac{n_i(x_i - x)}{|a_i t_1 + b_i t_2 + x_i - x|^3} t_1^k t_2^l\, dt_2\, dt_1 = -n_i(x_i - x) I_{k,l}^{Q,-\frac{3}{2}}(a_i, b_i, x_i - x).$$

The integrals $L_{kl}^i(x)$, $M_{kl}^i(x)$ can be treated like $C_{kl}^{i,s}(x)$, $D_{kl}^{i,s}(x)$ by partial integration and $J_{kl}^{i,rst}(x)$ is analyzed analogously.

Hypersingular integral operator

We implement the Galerkin matrix of the hypersingular integral operator via integration by parts which yields [9, 16]

$$\langle \mathbf{W}\phi, \psi \rangle = \int_\Gamma \int_\Gamma \frac{\mu}{2\pi} \frac{1}{|x - y|} \sum_{r,s=1}^{3} (\mathrm{curl}_\Gamma\, \phi_r(x))_s (\mathrm{curl}_\Gamma\, \psi_r(y))_s\, ds_y\, ds_x$$

$$+ \frac{\mu}{2\pi} \int_\Gamma \int_\Gamma \sum_{r,s,k,l,m,n=1}^{3} \varepsilon_{rsl}(\mathrm{curl}_\Gamma\, \phi_l(x))_s \frac{\delta_{rn}}{|x - y|} \varepsilon_{nkm}(\mathrm{curl}_\Gamma\, \psi_m)_k\, ds_y\, ds_x$$

$$- 4\mu^2 \int_\Gamma \int_\Gamma \sum_{r,s,k,l,m,n=1}^{3} \varepsilon_{rsl}(\mathrm{curl}_\Gamma\, \phi_l(x))_s G_{rn}(x, y) \varepsilon_{nkm}(\mathrm{curl}_\Gamma\, \psi_m)_k\, ds_y\, ds_x$$

$$- \int_\Gamma \int_\Gamma \frac{\mu}{4\pi} \frac{1}{|x - y|} \sum_{r,s=1}^{3} (\mathrm{curl}_\Gamma\, \phi_r(x))_r (\mathrm{curl}_\Gamma\, \psi_s(y))_s\, ds_y\, ds_x \qquad (44)$$

where $\mathrm{curl}_\Gamma\, u(x) = \mathbf{n}(x) \times \mathrm{grad}_\Gamma\, u(x)$, and ε_{ijk} is the total antisymmetric tensor ($\varepsilon_{123} = 1$). Using (44) the entries of the Galerkin matrix are computed analytically with the software package *maiprogs* [14].

4 Numerical Results

In this section we present numerical results of the above described Galerkin scheme for various examples. We perform h-, p- and hp-versions. Young's modulus (E-modulus) is $E = 2000$ and the Poisson number is $\nu = 0.3$.

For the computation of the error we use $\|\phi - \phi_N\|_W^2 = \|\phi\|_W^2 - \|\phi_N\|_W^2$
and $\|\mathbf{t} - \mathbf{t}_N\|_V^2 = \|\mathbf{t}\|_V^2 - \|\mathbf{t}_N\|_V^2$.

Example 1. For the Dirichlet problem of the Lamé equation with boundary
data $\mathbf{g}(x_1, x_2, x_3) = (-x_2, x_1, 0)$ in (7) on the square $\Gamma = [-1, 1]^2$ we know
the energy norm of the exact solution by extrapolation

$$\|\mathbf{t}\|_V = 115.0355908.$$

In Fig. 3 we present the numerical results for the Dirichlet problem. The
convergence rates which are given in Table 1, clearly confirm the exponentially
fast convergence of the hp-version with geometric mesh, which is expected due
to Theorem 4.

Fig. 3 shows clearly the exponentially fast convergence of the *hp*-version
on the geometric mesh with mesh grading parameter $\sigma = 0.17$. The parameter $\mu = 0.5$ describes the increase of the polynomial degree, namely
$(q, p), (q, p), (q, p+1), (q, p+1), (q, p+2), (q, p+2), \ldots$ in the x_2 direction and
correspondingly in the x_1 direction, for a geometric mesh consisting of rect-
angles only and refined towards the edges. Very good results are also obtained
for the *h*-version on an algebraically graded mesh towards the edges with mesh
grading parameter $\beta = 4.0$; this is in agreement with the theoretical results in
[26]. Also Fig. 3 and Table 1 show that the uniform *p*-version converges twice
as fast as the uniform *h*-version [3].

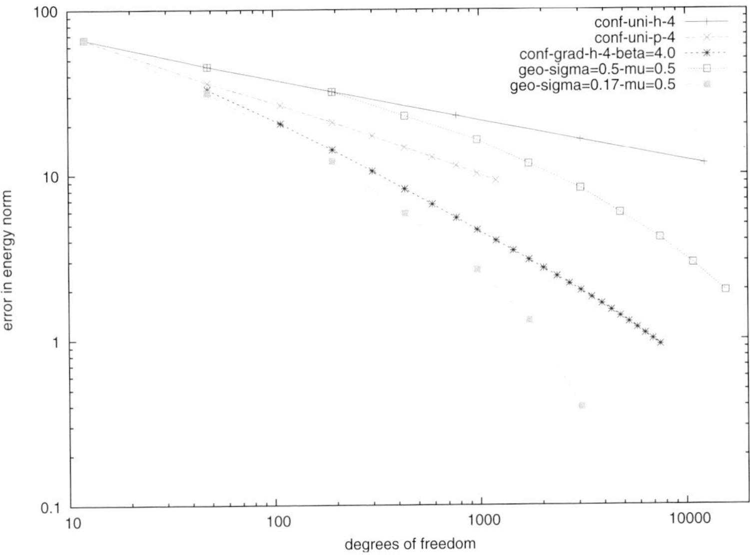

Fig. 3. Weakly singular integral equation (Lamé), Example 1.

Table 1. Convergence rates for the weakly singular integral equation on the Square.

N	$\|\mathbf{t}-\mathbf{t}_N\|_V$	α	p	N	$\|\mathbf{t}-\mathbf{t}_N\|_V$	α	N	$\|\mathbf{t}-\mathbf{t}_N\|_V$	α
h-Version, p=1			p-Version, 4 elements				hp-Version, $\sigma=0.17, \mu=0.5$		
12	65.977067		0	12	65.977067		12	65.977067	
48	45.338115	0.271	1	48	36.205111	0.433	48	31.511011	0.533
192	31.978059	0.252	2	108	26.548835	0.382	192	12.121016	0.689
768	22.804025	0.244	3	192	20.914871	0.415	432	5.8540817	0.897
3072	16.289194	0.243	4	300	17.265718	0.430	972	2.6642368	0.971
12228	11.618080	0.245	5	432	14.701526	0.441	1728	1.3123139	1.231
			6	588	12.801060	0.449	3072	0.3934324	2.094
			7	768	11.335587	0.455			
			8	972	10.170859	0.460			
			9	1200	9.2227497	0.464			
theoretically: 0.250			theoretically: 0.500						

Example 2. For the Neumann problem of the Lamé equation we consider the square $\Gamma = [-1,1]^2$ and choose $\mathbf{f} = (-x_2, x_1, 0)$ in (10). Via extrapolation we get $\|\phi\|_W = 0.04005011548$.

In Fig. 4 we present the numerical results for the Neumann problem. The convergence rates which are given in Table 2, clearly confirm the exponentially fast convergence of the hp-version with geometric mesh, which is expected due to Theorem 5.

Table 2. Convergence rates for the hypersingular integral equation on the square.

N	$\|\phi-\phi_N\|_W$	α	p	N	$\|\phi-\phi_N\|_W$	α	N	$\|\phi-\phi_N\|_W$	α
h-Version, p=1			p-Version, 4 elements				hp-Version, $\sigma=0.17, \mu=0.5$		
27	0.0258942		1	27	0.0258942		3	0.0400501	
147	0.0170821	0.245	2	147	0.0139794	0.364	27	0.0153835	0.435
675	0.0114749	0.261	3	363	0.0094512	0.433	147	0.0061827	0.538
2883	0.0078521	0.261	4	675	0.0071976	0.439	363	0.0035278	0.621
			5	1083	0.0058224	0.448	867	0.0012488	1.193
			6	1587	0.0048894	0.457	1587	0.0004945	1.532
			7	2187	0.0042117	0.465			
			8	2883	0.0037193	0.450			
theoretically: 0.250			theoretically: 0.500						

Fig. 4 shows clearly the exponentially fast convergence of the *hp*-version on the geometric mesh with $\sigma = 0.17$ and $\mu = 0.5$. Again we obtain very good results for the *h*-version on an algebraically graded mesh towards the edges with mesh grading parameter $\beta = 4.0$; which agrees with [26]. Also

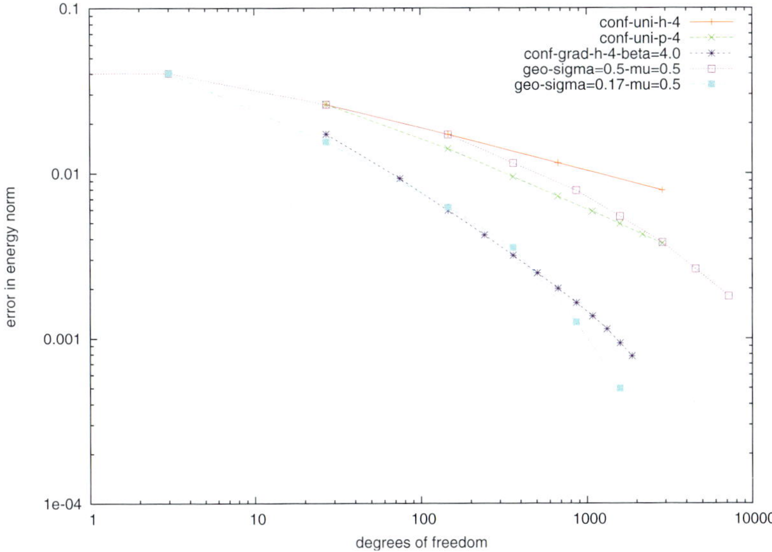

Fig. 4. Hypersingular integral equation (Lamé), Example 2.

Fig. 4 shows that the uniform p-version converges twice as fast as the uniform h-version [3].

References

1. S. Abou El-Seoud, V. J. Ervin, E. P. Stephan: An improved boundary element method for the charge density of a thin electrified plate in $\mathbb{R}^3$. Math. Meth. Appl. Sci. 13 (1990) 291–303.
2. I. Babuška, M. Suri: The optimal convergence rate of the p-version of the finite element method. SIAM J. Numer. Anal. 24 (1987) 750–776.
3. A. Bespalov, N. Heuer: The p-version of the boundary element method for hypersingular operators on piecewise plane open surfaces. Numer. Math. 100 (2005) 185–209.
4. M. Costabel: Boundary integral operators on Lipschitz domains: Elementary results. SIAM J. Math. Anal. 19 (1988) 613–626.
5. M. Costabel, E. P. Stephan: An improved boundary element Galerkin method for three dimensional crack problems. J. Integral Eqns. Operator Theory 10 (1987) 467–504.
6. W. Gui, I. Babuška: The h, p and h-p versions of the finite element method in 1 dimension, part iii: The adaptive h-p version. Numer. Math. 49 (1986) 659–683.
7. B. Q. Guo, I. Babuška: The h-p version of the finite element method, part i: The basic approximation results. Comp. Mech. 1 (1986) 21–41.
8. B. Q. Guo, N. Heuer, E. P. Stephan: The h-p version of the boundary element method for transmission problems with piecewise analytic data. SIAM J. Numer. Anal. 33 (1996) 789–808.

9. H. Han: The boundary integro-differential equations of three-dimensional neumann problem in linear elasticity. Numer. Math. 68 (1994) 269–281.

10. N. Heuer, M. Maischak, E. P. Stephan: Exponential convergence of the hp-version for the boundary element method on open surfaces. Numer. Math. 83 (1999) 641–666.

11. H. Holm, M. Maischak, E. P. Stephan: The *hp*-version of the boundary element method for the Helmholtz screen problems. Computing 57 (1996) 105–134.

12. M. Maischak: hp-Methoden für Randintegralgleichungen bei 3D-Problemen, Theorie und Implementierung. PhD thesis, Universität Hannover, 1995.

13. M. Maischak: FEM/BEM coupling methods for Signorini-type interface problems — error analysis, adaptivity, preconditioners. Habilitationsschrift, Universität Hannover, 2001.

14. M. Maischak: Manual of the software package maiprogs. Technical report Ifam48, Institut für Angewandte Mathematik, Universität Hannover, 2001.

15. M. Maischak: The hp-version boundary element Galerkin method for a hypersingular integral equation on open surfaces. Ifam preprint 59, 2003.

16. M. Maischak: Analytical evaluation of potentials and computation of the Galerkin integrals on triangles and parallelograms. Ifam preprint 87, Institut für Angewandte Mathematik, Universität Hannover, 2006.

17. M. Maischak, E. P. Stephan: The hp-version of the boundary element method in $\mathbb{R}^3$. The basic approximation results. Math. Meth. Appl. Sci. 20 (1997) 461–476.

18. M. Maischak, E. P. Stephan: The hp-version of the boundary element method in $\mathbb{R}^3$. Part II: approximation in countably normed spaces. Ifam preprint 43, Institut für Angewandte Mathematik, Universität Hannover, 2002. ftp://ftp.ifam.uni-hannover.de/pub/preprints/ifam43.ps.Z.

19. C. Schwab, M. Suri: The optimal p-version approximation of singularities on polyhedra in the boundary element method. SIAM J. Numer. Anal. 33 (1996) 729–759.

20. E. P. Stephan: A boundary integral equation method for three-dimensional crack problems in elasticity. Math. Meth. Appl. Sci. 8 (1986) 609–623.

21. E. P. Stephan: Boundary integral equations for screen problems in $\mathbb{R}^3$. J. Integral Eqns. and Operator Theory 10 (1987) 257–263.

22. E. P. Stephan: Improved Galerkin methods for integral equations on polygons and polyhedral surfaces. In: Japan-U.S. Symposium on boundary element methods, Tokyo, pp. 73–80, 1988.

23. E. P. Stephan: The h–p version of the boundary element method for solving 2– and 3–dimensional problems. Comp. Meth. Appl. Mech. Eng. 133 (1996) 183–208.

24. T. von Petersdorff: Randwertprobleme der Elastizitätstheorie für Polyeder – Singularitäten und Approximation mit Randelementmethoden. PhD thesis, Universität Darmstadt, 1989.

25. T. von Petersdorff, E. P. Stephan: Decompositions in edge and corner singularities for the solution of the Dirichlet problem of the Laplacian in a polyhedron. Math. Nachr. 149 (1990) 71–104.

26. T. von Petersdorff, E. P. Stephan: Regularity of mixed boundary value problems in $\mathbb{R}^3$ and boundary element methods on graded meshes. Math. Meth. Appl. Sci. 12 (1990) 229–249.

Sparse Convolution Quadrature for Time Domain Boundary Integral Formulations of the Wave Equation by Cutoff and Panel–Clustering

Wolfgang Hackbusch[1], Wendy Kress[1] and Stefan A. Sauter[2]

[1] Max–Planck–Institut für Mathematik in den Naturwissenschaften,
Inselstrasse 20–26, 04103 Leipzig, Germany
{wh,kress}@mis.mpg.de
[2] Institut für Mathematik, Universität Zürich,
Winterthurerstrasse 190, 8057 Zürich, Switzerland
stas@math.unizh.ch

Summary. We consider the wave equation in a time domain boundary integral formulation. To obtain a stable time discretization, we employ the convolution quadrature method in time, developed by Lubich. In space, a Galerkin boundary element method is considered. The resulting Galerkin matrices are fully populated and the computational complexity is proportional to $N \log^2 N M^2$, where M is the number of spatial unknowns and N is the number of time steps.

We present two ways of reducing these costs. The first is an a priori cutoff strategy, which allows to replace a substantial part of the matrices by 0. The second is a panel clustering approximation, which further reduces the storage and computational cost by approximating subblocks by low rank matrices.

1 Introduction

This paper is concerned with the numerical solution of the wave equation in an unbounded domain. Problems governed by the wave equation arise in many physical applications such as electromagnetic wave propagation or the computation of transient acoustic waves. When such problems are formulated in unbounded domains, the approach of *retarded potentials* allows a transformation of partial differential equations into space-time integral equations on the bounded surface of the scatterer.

Although this approach goes back to the early 1960s (cf. [11]) the development of fast numerical methods for integral equations in the field of hyperbolic problems is still in its infancies compared to the vast of fast methods for elliptic boundary integral equations (cf. [24] and references therein). Existing numerical discretisation methods include collocation methods with some stabilisation techniques (cf. [2, 3, 6, 7, 8, 22, 23]) and Laplace-Fourier methods coupled with Galerkin boundary elements in space (cf. [1, 5, 9, 12]). Numerical experiments can be found, e.g., in [13]. In [10], a fast version of the *marching-on-in-time* (MOT) method is presented which is based on a suitable plane wave expansion of the arising potential which reduces the storage and computational costs.

In this paper, we consider the convolution quadrature method for the time discretisation (cf. [18, 19, 20, 21]), and develop a panel-clustering method to obtain a data-sparse approximation of the underlying boundary integral equations. In [14], we have developed and analysed a simple cut-off strategy which reduces the number of entries in the system matrix which have to be computed while the rest is set to zero. The use of panel-clustering will further reduce the storage and computational complexity.

In [25, 26, 27] Lubich's convolution quadrature method is applied to problems such as viscoelastic and poroelastic continua.

2 Formulation of the Problem

We consider a scattering problem in an exterior domain. For this, let $\Omega \subset \mathbb{R}^3$ be an unbounded Lipschitz domain with boundary Γ. Let $\bar{u}$ be the solution to the wave equation

$$\partial_t^2 \bar{u} = \Delta \bar{u} + f \,, \text{ in } \Omega \times (0, T) \,,$$
$$\bar{u}(\cdot, 0) = u_0 \text{ in } \Omega \,,$$
$$\partial_t \bar{u}(\cdot, 0) = u_1 \text{ in } \Omega \,,$$
$$\bar{u} = 0 \text{ on } \Gamma \times (0, T) \,,$$

for some time interval $(0, T)$ and given data f, u_0 and u_1.

To formulate the differential equation as a boundary integral equation, we introduce an incident solution v and a diffracted solution u in the whole $\mathbb{R}^3$, with $\bar{u}|_\Omega = (u + v)|_\Omega$, where v solves the open space problem

$$\partial_t^2 v = \Delta v + f_p \text{ in } \mathbb{R}^3 \times (0, T) \,,$$
$$v(\cdot, 0) = u_{0p} \text{ in } \mathbb{R}^3 \,,$$
$$\partial_t v(\cdot, 0) = u_{1p} \text{ in } \mathbb{R}^3 \,,$$

where f_p, u_{ip} are prolongations of f and u_i to the whole $\mathbb{R}^3$, respectively. Given the solution to the above problem, v, u solves the homogeneous wave equation

$$\partial_t^2 u = \Delta u \text{ in } \Omega \times (0, T) \,, \tag{1a}$$
$$u(\cdot, 0) = \partial_t u(\cdot, 0) = 0 \text{ in } \Omega \,, \tag{1b}$$
$$u = g \text{ on } \Gamma \times (0, T) \,, \tag{1c}$$

where $g = -v|_{\Gamma \times (0, T)}$.

When considering a discretisation of the above partial differential equation on the unbounded domain Ω, one has to introduce an artificial boundary with additional boundary conditions. This is avoided by transforming the partial differential equation into a boundary integral equation. For this, we employ an ansatz as a *single layer potential*

$$u(x,t) = \int_0^t \int_\Gamma k(\|x-y\|, t-\tau)\phi(y,\tau)d\Gamma_y d\tau\,, \quad (x,t) \in \Omega \times (0,T)\,, \quad (2)$$

where $k(d,t)$ is the fundamental solution of the wave equation,

$$k(d,t) = \frac{\delta(t-d)}{4\pi d}\,, \tag{3}$$

$\delta(t)$ being the Dirac delta distribution. Inserting (2) into (1a), we see that the differential equation is satisfied. Also, the initial conditions are satisfied. An equation for the unknown density ϕ is obtained by taking the limit to the boundary. Since the single layer potential is continuous across the boundary, we obtain the following boundary integral equation for ϕ,

$$\int_0^t \int_\Gamma k(\|x-y\|, t-\tau)\phi(y,\tau)d\Gamma_y d\tau = g(x,t) \qquad \forall (x,t) \in \Gamma \times (0,T)\,. \quad (4)$$

Note that only the two-dimensional surface Γ is involved in this equation and not the three-dimensional domain Ω. This is one major advantage for the numerical solution process compared to finite element or finite volume methods.

3 Convolution Quadrature Method

Discretising (4) directly in space and time, e.g., with a Galerkin method in space and a collocation method in time, involves the treatment of the Dirac delta distribution. The resulting integration domains for a boundary element method are given by the intersection of the light cone (of finite width) with the triangles or quadrilaterals of the surface mesh which can be of quite general shape and, hence, numerical quadrature becomes rather complicated. In addition, care needs to be taken to obtain an unconditionally stable scheme.

The convolution quadrature approach for the time discretisation leads to an unconditionally stable scheme (see [20]). The resulting integration domains are just the boundary elements themselves. Furthermore, the approach allows a data-sparse approximation of the system matrix by panel-clustering.

To explain the convolution quadrature method, we consider a convolution of the form

$$(f \star g)(t) = \int_0^t f(t-\tau)g(\tau)d\tau\,, \quad t \geq 0\,. \tag{5}$$

Choosing a stepsize Δt, (5) can be approximated by a discrete convolution $(f \star_{\Delta t} g)(t_n)$ which will be based on the inverse Laplace transform

$$f(t) = \frac{1}{2\pi i} \int_{\sigma + i\mathbb{R}} \hat{f}(s) e^{st} ds$$

for some $\sigma > 0$. The inverse Laplace transform is defined if $\hat{f}$ is analytic and for $\operatorname{Re} s > \sigma$, $|\hat{f}(s)| \le c|s|^{-\mu}$ for some $c < \infty$ and $\mu > 0$. Inserting this representation of $f(t)$ into (5), we obtain

$$(f \star g)(t) = \frac{1}{2\pi i} \int_{\sigma + i\mathbb{R}} \hat{f}(s) y_g(s, t) ds \quad \text{with} \quad y_g(s, t) := \int_0^t e^{s(t-\tau)} g(\tau) d\tau .$$

Observe that the function $y_g(s, \cdot)$ satisfies the differential equation

$$\partial_t y(s, \cdot) = s y(s, \cdot) + g,$$

which can be approximated by a p-th order linear multistep method,

$$\sum_{j=0}^{k} \alpha_j y_{n+j-k}(s) = \Delta t \sum_{j=0}^{k} \beta_j \left(s y_{n+j-k}(s) + g((n+j-k)\Delta t) \right), \qquad (6)$$

with starting values $y_{-k}(s) = \ldots = y_{-1}(s) = 0$. We assume that sufficiently many time derivatives of g vanish at $t = 0$. Formally, a p-th order approximation of (5) is then given by

$$(f \star_{\Delta t} g)(t_n) = \frac{1}{2\pi i} \int_{\sigma + i\mathbb{R}} \hat{f}(s) y_n(s) ds . \qquad (7)$$

To see that (7) can be written as a discrete convolution, we multiply (6) by ζ^n for $|\zeta| < 1$ and $\frac{\operatorname{Re} \gamma(\zeta)}{\Delta t} > \sigma$ and sum over n to obtain

$$\sum_{n=0}^{\infty} y_n \zeta^n = \left(\frac{\gamma(\zeta)}{\Delta t} - s \right)^{-1} \sum_{n=0}^{\infty} g(n\Delta t) \zeta^n ,$$

with $\gamma(\zeta) := \frac{\sum_{j=0}^{k} \alpha_j \zeta^{k-j}}{\sum_{j=0}^{k} \beta_j \zeta^{k-j}}$. Doing the same for (7), we obtain

$$\sum_{n=0}^{\infty} (f \star_{\Delta t} g)(t_n) \zeta^n = \frac{1}{2\pi i} \int_{\sigma + i\mathbb{R}} \frac{\hat{f}(s)}{\frac{\gamma(\zeta)}{\Delta t} - s} ds \sum_{n=0}^{\infty} g(n\Delta t) \zeta^n$$

$$= \sum_{n=0}^{\infty} \hat{f}\left(\frac{\gamma(\zeta)}{\Delta t} \right) g(n\Delta t) \zeta^n ,$$

where we have employed Cauchy's integral formula in the last step. If we define $\omega_n^{\Delta t}$ by

$$\hat{f}\left(\frac{\gamma(\zeta)}{\Delta t}\right) = \sum_{n=0}^{\infty} \omega_n^{\Delta t}\zeta^n , \tag{8}$$

we have

$$\sum_{n=0}^{\infty}(f \star_{\Delta t} g)(t_n)\zeta^n = \sum_{n=0}^{\infty}\omega_n^{\Delta t}\zeta^n \sum_{m=0}^{\infty}g(m\Delta t)\zeta^m = \sum_{n=0}^{\infty}\left(\sum_{j=0}^{n}\omega_{n-j}^{\Delta t}g(j\Delta t)\right)\zeta^n .$$

Thus

$$(f \star_{\Delta t} g)(t_n) = \sum_{j=0}^{n}\omega_{n-j}^{\Delta t}g(j\Delta t) ,$$

which has the form of a discrete convolution.

4 Time Discretisation: Convolution Quadrature Method

In our case, the convolution coefficients are spatial boundary integral operators. The continuous convolution in (4) is approximated by the discrete convolution,

$$\sum_{j=0}^{n}\int_{\Gamma}\omega_{n-j}^{\Delta t}(\|x-y\|)\phi_{\Delta t}^{j}(y)d\Gamma_y = g(x,t_n) , \qquad n = 1,\dots,N, \quad x \in \Gamma, \tag{9}$$

where the convolution coefficients $\omega_n^{\Delta t}(d)$ are functions of $d = \|x-y\|$ determined by the power series (cf. (8)) of the Laplace transform

$$\hat{k}(d,s) = \frac{e^{-sd}}{4\pi d} ,$$

$$\hat{k}\left(d,\frac{\gamma(\zeta)}{\Delta t}\right) = \sum_{n=0}^{\infty}\omega_n^{\Delta t}(d)\zeta^n . \tag{10}$$

As a multistep method, we use the second order accurate, A-stable BDF2 method with

$$\gamma(\zeta) = \frac{1}{2}(\zeta^2 - 4\zeta + 3) .$$

The coefficients of the power series (10) can be obtained by the Taylor expansion of $\hat{k}(d,\frac{\gamma(\zeta)}{\Delta t})$ about $\zeta = 0$,

$$\omega_n^{\Delta t}(d) = \frac{1}{n!}\left.\frac{\partial^n \hat{k}(d,\frac{\gamma(\zeta)}{\Delta t})}{\partial\zeta^n}\right|_{\zeta=0} = \frac{1}{n!}\frac{1}{4\pi d}\left.\frac{\partial^n e^{-\frac{\gamma(\zeta)}{\Delta t}d}}{\partial\zeta^n}\right|_{\zeta=0} .$$

It can be shown that

$$\omega_n^{\Delta t}(d) = \frac{1}{n!}\frac{1}{4\pi d}\left(\frac{d}{2\Delta t}\right)^{n/2}e^{-\frac{3d}{2\Delta t}}H_n\left(\sqrt{\frac{2d}{\Delta t}}\right) , \tag{11}$$

where H_n are the Hermite polynomials.

5 Space Discretisation. Galerkin Boundary Element Methods

For the space discretisation, we employ a standard Galerkin boundary element method with piecewise constant or piecewise linear basis functions. Let $\mathcal{G}$ be a regular (in the sense of Ciarlet [4]) boundary element mesh on Γ consisting of shape regular, possibly curved triangles τ_i. Let $\mathbb{P}_0$ and $\mathbb{P}_1$ denote the space of constant and linear functions, respectively. We denote by

$$S_{-1,0} := \left\{ u \in L^\infty\left(\Gamma\right) \quad : \quad \forall \tau_i \in \mathcal{G} : u|_{\tau_i} \in \mathbb{P}_0 \right\}$$

the space of piecewise constant, discontinuous functions, and by

$$S_{0,1} := \left\{ u \in C^0\left(\Gamma\right) \quad : \quad \forall \tau_i \in \mathcal{G} : \left(u \circ \chi_i\right)|_{\tau_i} \in \mathbb{P}_1 \right\}$$

the space of continuous, piecewise linear functions, where χ_i denotes a regular mapping of the curved triangle τ_i to a planar reference triangle.

As a basis for $S_{-1,0}$ we choose

$$b_i(x) = \delta_{ij}, \text{ if } x \in \tau_j$$

and the basis for $S_{0,1}$ consists of the standard hat functions on the planar reference triangle, lifted to the surface Γ by the mapping χ_i. We generally refer to the boundary element space by S and its basis by $(b_i)_{i=1}^M$. The mesh width h is given by the maximum triangle diameter in $\mathcal{G}$.

For the Galerkin boundary element method, we replace $\phi_{\Delta t}^n$ in (9) by some $\phi_{\Delta t,h}^n \in S$ and impose the integral equation in a weak form. The fully discrete problem consists of finding $\phi_{\Delta t,h}^n \in S$, $n = 1, 2, \ldots, N$, of the form

$$\phi_{\Delta t,h}^n(y) = \sum_{i=1}^M \phi_{n,i} b_i(y),$$

such that

$$\sum_{j=0}^n \sum_{i=1}^M \phi_{j,i} \int_\Gamma \int_\Gamma \omega_{n-j}^{\Delta t}(\|x - y\|) b_i(y) b_k(x) d\Gamma_y d\Gamma_x = \int_\Gamma g(x, t_n) b_k(x) d\Gamma_x$$

$$(12)$$

for all $1 \leq k \leq M$ and $n = 1, \ldots, N$. This can be written as a linear system

$$\sum_{j=0}^n \mathbf{A}_{n-j} \phi_j = \mathbf{g}_n, \qquad n = 1, \ldots, N, \tag{13}$$

with the vectors $\phi_j = (\phi_{j,i})_{i=1}^M$ and the matrices

$$(\mathbf{A}_n)_{k,i} := \int_\Gamma \int_\Gamma \omega_n^{\Delta t}(\|x - y\|) b_i(y) b_k(x) d\Gamma_y d\Gamma_x,$$

and

$$(\mathbf{g}_n)_k = \int_\Gamma g(x, t_n) b_k(x) d\Gamma_x.$$

5.1 Efficient Algorithmic Realisation

Before we present a way to reduce the storage requirements, we take a look at the solution procedure. The problem to be solved is

$$\tilde{\phi}_n = \mathbf{A}_0^{-1}\left(\mathbf{g}_n - \sum_{i=0}^{n-1}\mathbf{A}_{n-i}\phi_i\right), \quad n = 0, 1, \ldots, N.$$

(14)

A straightforward way to solve (14) is to compute $\left(\mathbf{g}_n - \sum_{i=0}^{n-1}\mathbf{A}_{n-i}\phi_i\right)$ and then to solve the system for each n. The required work is however proportional to N^2. When using the following algorithm (cf. [16]) the computational costs are proportional to $N\log^2 N$. The procedure depends on a (small) control parameter r.

Algorithm 2 (Recursive solver for block triangular system)

    Comment: Main program
    begin
    solve_triangular$(0, N)$;
    end;
    Comment: The recursive subroutine **solve_triangular** is defined as follows.

    procedure solve_triangular $(a, b : \text{integer})$;
    begin
      if $b - a \le r - 1$ **then**
        for $n := a$ **to** b **do**

$$\phi_n := \mathbf{A}_0^{-1}\left(\mathbf{g}_n - \sum_{i=a}^{n-1}\mathbf{A}_{n-i}\phi_i\right)$$

(15)

      end
     else begin
      $m := \left\lceil \frac{b+a}{2} \right\rceil$;
      solve_triangular$(a, m - 1)$;
      for $n := m$ **to** b **do**

$$\mathbf{g}_n := \mathbf{g}_n - \sum_{i=a}^{m-1}\mathbf{A}_{n-i}\phi_i$$

(16)

      end;
      solve_triangular(m, b);
    end;
   end;

When using fast iterative methods, the computational costs for (15) are proportional to r^2 matrix vector multiplications. The special form of (16) allows the use of the discrete fast Fourier transform (see, e.g., [17]) and the updates of $\mathbf{g}$ can be done in $\mathcal{O}\left(M^2\left(b-a\right)\log\left(b-a\right)\right)$ operations. The **procedure solve_triangular** calls itself twice with half the dimension. The total computational cost sums up to $\mathcal{O}\left(M^2 N \log^2 N\right)$ (cf. [17]).

Remark 1. In the following, we will apply sparse approximation techniques to the matrices A_n. Further research will be concerned with a modification of the above algorithm making use of the sparse representation of the operator A_n. Note that already the use of (14) in combination with the fast evaluation of matrix vector products due to the sparse representation leads to a reduction of the overall complexity. The total computational cost sums up to $\mathcal{O}\left(M^{1+s}N\right)$ with $s < 1$.

6 Sparse Approximation of the Matrices $\mathbf{A}_n$ by Cutoff

6.1 Cutoff Strategy and Perturbation Analysis

The matrices $\mathbf{A}_n$ are full matrices. Thus, storage requirements and computational complexity for the solution of the fully discrete problem using fast iterative methods are proportional to M^2. However, a substantial part of the matrix consists of small entries and can be replaced by 0. To see this, we recall the definition of the convolution coefficients

$$\omega_n^{\Delta t}(d) = \frac{1}{n!}\frac{1}{4\pi d}\left(\frac{d}{2\Delta t}\right)^{n/2} e^{-\frac{3d}{2\Delta t}} H_n\left(\sqrt{\frac{2d}{\Delta t}}\right). \tag{17}$$

For $n = 0$, we have

$$\omega_0^{\Delta t}(d) = \frac{e^{-\frac{3}{2}\frac{d}{\Delta t}}}{4\pi d},$$

with a singularity at $d = 0$ and, for $n = 1$,

$$\omega_1^{\Delta t}(d) = \frac{1}{\Delta t}\frac{e^{-\frac{3}{2}\frac{d}{\Delta t}}}{2\pi}.$$

In Fig. 1, we plot $\omega_n^{\Delta t}(d)$ for $\Delta t = 1$ and different n. For general Δt, we have the relation

$$\omega_n^{\Delta t}(d) = \Delta t^{-1}\omega_n^1\left(\frac{d}{\Delta t}\right).$$

The convolution functions have their maximum near $d = t_n$. Away from this maximum, the coefficients decay fast. Using bounds for the Hermite polynomials, it can be shown (cf. [14]) that outside the interval

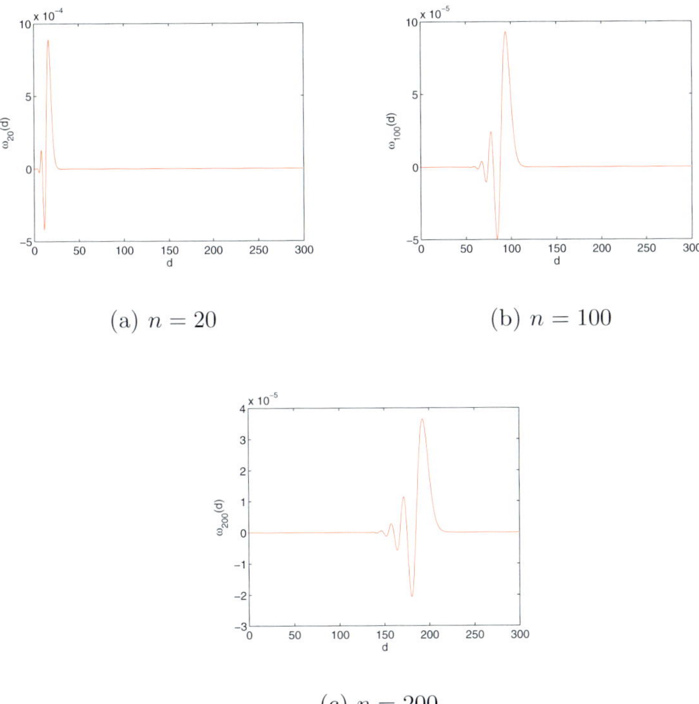

(a) $n = 20$

(b) $n = 100$

(c) $n = 200$

Fig. 1. The convolution weights $\omega_n^{\Delta t}(d)$ for $\Delta t = 1$ and different values of n.

$$I_{n,\varepsilon}^{\Delta t} := \left[t_n - 3\sqrt{\Delta t}\sqrt{t_n}|\log \varepsilon|, t_n + 3\sqrt{\Delta t}\sqrt{t_n}|\log \varepsilon| \right] \tag{18}$$

we have

$$|\omega_n^{\Delta t}(d)| \leq \frac{\varepsilon}{4\pi d} \quad \forall d \notin I_{n,\varepsilon}^{\Delta t}. \tag{19}$$

Given an error tolerance ε, we only consider those entries of $\mathbf{A}_n$, for which the possible values of $\|x - y\|$ lie inside $I_{n,\varepsilon}^{\Delta t}$. The remaining entries are set to zero. Let $\mathcal{P}_\varepsilon \subset \{1, \ldots, M\} \times \{1, \ldots, M\}$ be defined by

$$\mathcal{P}_\varepsilon := \left\{ (i,j) \, : \, \exists\, (x,y) \in \operatorname{supp} b_i \cap \operatorname{supp} b_j, \text{ s.t. } \|x - y\| \in I_{n,\varepsilon}^{\Delta t} \right\}. \tag{20}$$

This induces a sparse approximation $\tilde{\mathbf{A}}_n$ by

$$(\tilde{\mathbf{A}}_n)_{i,j} := \begin{cases} (\mathbf{A}_n)_{i,j} & \text{if } (i,j) \in \mathcal{P}_\varepsilon, \\ 0 & \text{otherwise.} \end{cases} \tag{21}$$

Instead of solving (13), we solve for an approximate solution $\tilde{\boldsymbol{\phi}}_j = \left(\tilde{\phi}_{j,i} \right)_{i=1}^{M}$,

$$\sum_{j=0}^{n} \tilde{\mathbf{A}}_{n-j} \tilde{\boldsymbol{\phi}}_j = \mathbf{g}_n \,, \qquad n = 1, \ldots, N \,, \tag{22}$$

and we have the approximate solution

$$\tilde{\phi}^n_{\Delta t,h}(y) := \sum_{i=1}^{M} \tilde{\phi}_{n,i} b_i(y) \,. \tag{23}$$

In [14], the following theorem is proven.

Theorem 1. *Let the exact solution $\phi(\cdot, t)$ of (4) be in $H^{m+1}(\Gamma)$ for any $t \in [0, T]$. There exists a constant $C > 0$ such that, for all cutoff parameters ε in (21) with $0 < \varepsilon < Ch\Delta t^3$, the solution $\tilde{\phi}_{\Delta t,h}$ in (23) exists and satisfies the error estimate*

$$\left\| \tilde{\phi}^n_{\Delta t,h} - \phi(\cdot, t_n) \right\|_{H^{-1/2}(\Gamma)} \leq C_g(T) \left(\varepsilon h^{-1} \Delta t^{-5} + \Delta t^2 + h^{m+3/2} \right) \,,$$

where C_g depends on the boundary data g.

Corollary 1. *Let the assumptions in Theorem 1 be satisfied. Let*

$$\Delta t^2 \sim h^{m+3/2} \,, \tag{24}$$

and choose

$$\varepsilon \sim h^{7m/2+25/4} \,.$$

Then the solution $\tilde{\phi}^n_{\Delta t,h}$ exists and converges with optimal rate

$$\left\| \tilde{\phi}^n_{\Delta t,h} - \phi(\cdot, t_n) \right\|_{H^{-1/2}(\Gamma)} \leq C_g(T) h^{m+3/2} \sim C_g(T) \Delta t^2 \,.$$

6.2 Storage Requirements

The approximation of the matrices $\mathbf{A}_n$ by sparse approximations $\tilde{\mathbf{A}}_n$ results in reduced storage requirements. To determine the storage requirements for the sparse matrices, assume that the dimension M of the boundary element space satisfies

$$c_1 h^{-2} \leq M \leq C_1 h^{-2}. \tag{25}$$

We further assume that there is a moderate constant C such that for any $1 \leq i \leq M$, the subset

$$\mathcal{P}_i := \{ j \in \{1, \ldots, M\} : (i, j) \in \mathcal{P}_\varepsilon \} \,,$$

with $\mathcal{P}_\varepsilon$ as in (20), satisfies

$$\sharp \mathcal{P}_i \leq C \max \left\{ 1, \frac{\sqrt{\Delta t}\, t_n^{3/2} \log M}{h^2} \right\} . \tag{26}$$

Table 1. Storage requirements for $\tilde{\mathbf{A}}_n$.

	$m = 0$	$m = 1$
$t_n = \mathcal{O}(\Delta t \log M)$	$CM^{1+\frac{1}{4}}\log^2 M$	CM
$t_n = \mathcal{O}(1)$	$Ct_n^{3/2}M^{1+\frac{13}{16}}\log M$	$Ct_n^{3/2}M^{1+\frac{11}{16}}\log M$

This assumption can be derived from the assumption that

$$ch^2 \leq \operatorname{supp} b_j \leq Ch^2$$

and that the area of

$$R_{i,n} := \left\{ y \in \Gamma \quad : \quad \exists x \in \operatorname{supp} b_i : \|x - y\| \in I_{n,\varepsilon}^{\Delta t} \right\}$$

satisfies $|R_{i,n}| \leq C\sqrt{\Delta t}\, t_n^{3/2} |\log(\varepsilon)|$ ($R_{i,n}$ is part of a ring with radius t_n and the same width as the interval $I_{n,\varepsilon}^{\Delta t}$.). Due to Corollary 1, $|\log \varepsilon| \sim \log M$.

With these assumptions, the number of nonzero matrix entries in $\tilde{\mathbf{A}}$ can be estimated by

$$\sum_{i=1}^{M} \sharp \mathcal{P}_i \leq CM \max\left\{ 1, \sqrt{\Delta t}\, t_n^{3/2} h^{-2} \log M \right\}.$$

Relation (24) allows to substitute $\sqrt{\Delta t}$ and the combination with (25) yields

Theorem 2. *The number of nonzero entries in the sparse approximation* $\tilde{\mathbf{A}}_n$ *is bounded from above by*

$$CM \max\left\{ 1, t_n^{3/2} M^{\frac{13}{16} - \frac{1}{8}m} \log M \right\}.$$

We distinguish between four cases: The case of piecewise constant and piecewise linear boundary elements ($m = 0$, and $m = 1$, respectively) and small and large n ($t_n = \mathcal{O}(\Delta t \log M)$ and $t_n = \mathcal{O}(1)$, respectively). The storage requirements for the different cases are summarised in Table 1. For small n, the storage requirements are significantly decreased. In Section 7, we present a method for further reducing the storage requirements even when $t_n > \mathcal{O}(\Delta t \log M)$.

7 Panel–Clustering

The panel-clustering method was developed in [15] for the data-sparse approximation of boundary integral operators which are related to elliptic boundary value problems. Since then, the field of sparse approximations of non-local operators has grown rapidly and nowadays advanced versions of the panel-clustering method are available and a large variety of alternative methods such as wavelet discretisations, multipole expansions, $\mathcal{H}$-matrices etc. exist.

However, these fast methods (with the exception of $\mathcal{H}$-matrices) are developed mostly for problems of elliptic type while the data-sparse approximation of retarded potentials is to our knowledge still in its infancies. In this section, we develop the panel-clustering method for retarded potentials.

7.1 The Algorithm

The panel-clustering can be applied as soon as $t_n > \mathcal{O}(\Delta t \, |\log \varepsilon|)$. (Note that for the first time steps the simple cutoff strategy reduces the computational complexity much more significantly than for the later time steps, see Table 1.)

For $t_n > \mathcal{O}(\Delta t \, |\log \varepsilon|)$, the matrices $\mathbf{A}_n$ in (13) are partitioned into sub-blocks $\mathbf{A}_n|_{s \times t}$ for some index set $s \times t \subset \{1, \ldots, M\} \times \{1, \ldots, M\}$. The sub-blocks are either replaced by zero, if the block entries are sufficiently small, or they are replaced by low rank matrices. To explain this approach in detail we first introduce the basic notation.

Let $\mathcal{I} := \{1, 2, \ldots, M\}$ denote the degrees of freedom for the space discretisation.

Definition 1 (Cluster). *A cluster t is a subset of $\mathcal{I}$. If t is a cluster, the corresponding subdomain of Γ is $\Gamma_t := \bigcup_{i \in t} \operatorname{supp}(b_i)$. The cluster box $Q_t \subset \mathbb{R}^3$ is the minimal axisparallel cuboid which contains Γ_t and the cluster size L_t is the maximal side length of Q_t.*

The clusters are collected in a hierarchical cluster tree $T_{\mathcal{I}}$.

Definition 2 (Cluster Tree). *A tree $T_{\mathcal{I}}$ is a cluster tree if the following conditions are satisfied.*

1. *The nodes in $T_{\mathcal{I}}$ are clusters.*
2. *The root of $T_{\mathcal{I}}$ is $\mathcal{I}$.*
3. *The leaves of $T_{\mathcal{I}}$ are the degrees of freedom, i.e., $\mathcal{L}(T_{\mathcal{I}}) = \mathcal{I}$ and the tree hierarchy is given by a father/son relation: For each interior node $t \in T_{\mathcal{I}} \setminus \mathcal{L}(T_{\mathcal{I}})$, the set $\operatorname{sons}(t)$ is the minimal subset in $T_{\mathcal{I}} \setminus \{t\}$ such that*

$$t = \bigcup_{s \in \operatorname{sons}(t)} s$$

holds. Vice versa, the father of any $s \in \operatorname{sons}(t)$ is t.

The standard construction of the cluster tree $T_{\mathcal{I}}$ is based on a recursive bisection of an axisparallel cuboid $\tilde{B}$ which contains Γ. The bisection of $\tilde{B}$ yields an auxiliary binary tree $T_{\tilde{B}}$. Then, the clusters in $T_{\mathcal{I}}$ are given by collecting, for any box $\tilde{B} \in T_{\tilde{B}}$, the indices $i \in \mathcal{I}$ which satisfy $\xi_i \in \tilde{B}$, where ξ_i denotes the nodal point for the i-th degree of freedom. Clusters in $T_{\mathcal{I}}$ which coincide with their father are removed from $T_{\mathcal{I}}$ and empty clusters are removed as well.

The kernel function $k(\|x - y\|, t)$ is approximated on $\Gamma_t \times \Gamma_s$, where (t, s) is a pair of clusters which satisfy the following condition. Recall the definition of the interval $I_{n,\varepsilon}^{\Delta t}$ as in (18).

Definition 3. *Let $\varepsilon > 0$ and $n > C\,|\log\varepsilon|$. Let $0 < \eta < 1$ be some control parameter. A pair of clusters $(t,s) \in T_{\mathcal{I}} \times T_{\mathcal{I}}$ is admissible at time step t_n if*

$$\forall\,(x,y) \in Q_t \times Q_s : \|x - y\| \notin I_{n,\varepsilon}^{\Delta t} \tag{27a}$$

or

$$\text{(27a) is violated and } \max\{L_t, L_s\} \le \eta\Delta t n^b. \tag{27b}$$

The power b in (27b) is a fixed number which is related to the accuracy of resulting discretisation.

A theoretical bound on b is $b \ge 1/4$ under the condition $n \ge C\,|\log\varepsilon|$. Numerical experiments indicate that the choice $b \approx 0.3$ also preserves the optimal convergence rates. This is shown in a forthcoming paper.

The following algorithm subdivides $\mathcal{I} \times \mathcal{I}$ into a matrix part P^{sparse}, corresponding to pairs of indices where the matrix has to be assembled in the conventional way, a zero part P^0 where the corresponding matrix entries are set to zero and a panel-clustering part P^{pc}, where the system matrix is approximated by panel-clustering. Note that the father/son relation of the cluster tree induces a father/son structure for pairs of clusters $\mathbf{b} = (c,s)$ by

$$\operatorname{sons}(\mathbf{b}) := \begin{cases} \operatorname{sons}(c) \times \operatorname{sons}(s) & \text{if } \operatorname{sons}(c) \ne \emptyset \text{ and } \operatorname{sons}(s) \ne \emptyset, \\ c \times \operatorname{sons}(s) & \text{if } \operatorname{sons}(c) = \emptyset \text{ and } \operatorname{sons}(s) \ne \emptyset, \\ \operatorname{sons}(c) \times s & \text{if } \operatorname{sons}(c) \ne \emptyset \text{ and } \operatorname{sons}(s) = \emptyset, \\ \emptyset & \text{if } \operatorname{sons}(c) \ne \emptyset \text{ and } \operatorname{sons}(s) \ne \emptyset. \end{cases}$$

Algorithm 3 *Let $n > C\,|\log\varepsilon|$. The minimal admissible block partitioning of $\mathcal{I} \times \mathcal{I}$ at time step t_n is obtained as the result of the procedure $\mathbf{divide}\big((\mathcal{I},\mathcal{I}),P^{\mathrm{sparse}},P^{\mathrm{pc}},P^0\big)$ defined by (cf. [15])*

    procedure divide $\big(\mathbf{b}, P^{\mathrm{sparse}}, P^{\mathrm{pc}}, P^0\big)$;
    begin
        if ($\mathbf{b}$ is non-admissible and $\operatorname{sons}(\mathbf{b}) = \emptyset$) **then** $P^{\mathrm{sparse}} := P^{\mathrm{sparse}} \cup \{\mathbf{b}\}$
        else if ($\mathbf{b}$ satisfies (27a)) **then** $P^0 := P^0 \cup \{\mathbf{b}\}$
        else if ($\mathbf{b}$ satisfies (27b)) **then** $P^{\mathrm{pc}} := P^{\mathrm{pc}} \cup \{\mathbf{b}\}$
        else for all $\widetilde{\mathbf{b}} \in \operatorname{sons}(\mathbf{b})$ **do divide** $\big(\widetilde{\mathbf{b}}, P^{\mathrm{sparse}}, P^{\mathrm{pc}}, P^0\big)$;
    end;

Remark 2. The set P^{sparse} is empty in most cases since the cluster sizes of the leaves satisfy

$$L_{\{i\}} = \mathcal{O}(h)$$

while relation (24) implies for the bound in (27b)

$$\eta\Delta t n^b = \mathcal{O}\left(\eta h^{m/2+3/4} n^b\right),$$

where $m = 0$ for constant and $m = 1$ for linear elements. Hence after a few time steps, $\eta\Delta t n^b \ge Ch$ and any pair $\mathbf{b}$ with $\operatorname{sons}(\mathbf{b}) = \emptyset$, i.e., $i,j \in \mathcal{I}$, satisfies (27a) or (27b).

Next, we explain the data sparse approximation on the blocks $\mathbf{b} = (c, s) \in P^{\mathrm{pc}}$. Since $\omega_n^{\Delta t}(\|\mathbf{x} - \mathbf{y}\|)$ is defined in $Q_c \times Q_s$ we may define its approximation by Čebyšev interpolation:

$$\omega_n^{\Delta t}(\|x - y\|) \approx \breve{\omega}_n^{\Delta t}(\|x - y\|) = \sum_{\mu,\nu \in \left(\mathbb{N}_{\leq q}\right)^3} \mathcal{L}_c^{(\mu)}(x)\mathcal{L}_s^{(\nu)}(y)\omega_n^{\Delta t}(\|x^\mu - y^\nu\|),$$

(28)

where $\mathcal{L}_c^{(\mu)}$ (resp. $\mathcal{L}_s^{(\nu)}$) are the tensorised versions of the q–th order Lagrange polynomials (properly scaled and translated to Q_c resp. Q_s) corresponding to the tensorised Čebyšev nodes x^μ for Q_c resp. y^ν for. Q_s.

The matrix $\mathbf{A}_n$ is the representation of the bilinear form $a_n : S \times S \to \mathbb{R}$,

$$a_n(\phi, \psi) := \int_\Gamma \int_\Gamma \omega_n^{\Delta t}(\|x - y\|)\phi(y)\psi(x)d\Gamma_y d\Gamma_x$$

with respect to the nodal basis $(b_i)_{i=1}^M$. We introduce the convention that, for any function $\varphi \in S$, the coefficient vector in the basis representation is denoted by $\boldsymbol{\varphi} = (\varphi_i)_{i=1}^M$, i.e., $\varphi = \sum_{i=1}^M \varphi_i b_i$.

The sparse approximation of a_n by our combined cutoff and panel-clustering strategy is given by

$$a_n(\phi, \psi) \approx \sum_{(i,j)\in P^{\mathrm{sparse}}} \psi_i\phi_j \left(\mathbf{A}_n^{\mathrm{sparse}}\right)_{i,j}$$

$$+ \sum_{\mathbf{b}=(\sigma,s)\in P^{\mathrm{pc}}} \sum_{\mu,\nu \in \left(\mathbb{N}_{\leq q}\right)^3} \left(\mathbf{S}_{\mathbf{b}}^{(n)}\right)_{\mu,\nu} J_\sigma^{(\mu)}(\psi) J_s^{(\nu)}(\phi),$$

with the sparse matrix part of $\mathbf{A}_n$

$$\left(\mathbf{A}_n^{\mathrm{sparse}}\right)_{i,j} := \begin{cases} \int_{\Gamma_{\{i\}}} \int_{\Gamma_{\{j\}}} \omega_n^{\Delta t}(\|x - y\|)b_j(y)\,b_i(x)\,d\Gamma_y d\Gamma_x & \text{if } (i,j) \in P^{\mathrm{sparse}}, \\ 0 & \text{otherwise}, \end{cases}$$

(29)

the *interaction matrix* $\mathbf{S}_{\mathbf{b}}^{(n)}$

$$\left(\mathbf{S}_{\mathbf{b}}^{(n)}\right)_{\mu,\nu} := \omega_n^{\Delta t}(\|x^\mu - y^\nu\|) \quad 0 \leq \mu_i, \nu_i \leq q,\ 1 \leq i \leq 3$$

and the *influence coefficients*

$$J_\sigma^{(\mu)}(\psi) := \sum_{i\in\sigma} \psi_i \int_{\Gamma_\sigma} \mathcal{L}_\sigma^{(\mu)}(x)b_i(x)\,d\Gamma_x, \quad 0 \leq \mu_i, \nu_i \leq q,\ 1 \leq i \leq 3.$$

The algorithmic realisation of the sparse matrix multiplication based on this approximation of the bilinear form and the recursive computation of the influence coefficients $J_\sigma^{(\mu)}(\psi)$ are structured as follows.

Phase 1: Computation and storage of the Galerkin operator

(a) Generate and store the cluster tree and the partitioning of $\mathcal{I} \times \mathcal{I}$ into P^{sparse}, P^{pc}, and P^0.

Introduce recursive tree levels $0 \le \ell \le \ell_{\max}$ by $T_{\mathcal{I}}(0) = \{\mathcal{I}\}$ and

$$T_{\mathcal{I}}(\ell + 1) := \{\sigma \in T_{\mathcal{I}} : \exists s \in T_{\mathcal{I}}(\ell) \text{ with "}\sigma \text{ is son of } s\text{"}\}.$$

Let $\ell_{\min}$ denote the minimal index such that (i) there exists $\sigma \in T_{\mathcal{I}}(\ell_{\min})$ with $L_\sigma \le \eta \Delta t n^b$ and (ii) for all $0 \le \ell < \ell_{\min}$ and $\sigma \in T_{\mathcal{I}}(\ell)$ there holds $L_\sigma > \eta \Delta t n^b$.

(b) Compute and store the nonzero entries of the matrix $\mathbf{A}_n^{\mathrm{sparse}}$.

(c) Compute and store the basis influence coefficients

$$J_{\{i\}}^{(\mu)}(b_i) := \int_{\mathrm{supp}(b_i)} \mathcal{L}_{\{i\}}^{(\mu)}(x)\, b_i(x)\, d\Gamma_x, \quad 1 \le i \le M, \ \mu \in (\mathbb{N}_{\le q})^3. \quad (30)$$

(d) Compute and store the interaction matrices $\mathbf{S}_{\mathbf{b}}^{(n)}$ for all $\mathbf{b} \in P^{\mathrm{pc}}$.

Phase 2: Evaluation of a matrix-vector multiplication $\varphi = \mathbf{A}_n \psi$

(a) For all $\sigma \in T_{\mathcal{I}}(\ell_{\max})$, for all $\mu \in (\mathbb{N}_{\le q})^3$ compute

$$J_\sigma^{(\mu)}(\psi) = \psi_i J_{\{i\}}^{(\mu)}(b_i).$$

For $\ell = \ell_{\max} - 1, \ell_{\max} - 2, \ldots, \ell_{\min}$, for all $\sigma \in T_{\mathcal{I}}(\ell)$ and all $\mu \in (\mathbb{N}_{\le q})^3$ compute

$$J_\sigma^{(\mu)}(\psi) = \sum_{s \in \mathrm{sons}(\sigma)} \sum_{\nu \in (\mathbb{N}_{\le q})^3} \gamma_{\mu,\nu,s} J_s^{(\nu)}(\psi) \quad \text{with} \quad \gamma_{\mu,\nu,s} := \mathcal{L}_\sigma^{(\mu)}(x_s^{(\nu)}).$$

(b) Let

$$T_{\mathcal{I}}^{\mathrm{pc}} := \{c \in T_{\mathcal{I}} \mid \exists s \in T_{\mathcal{I}} : (c, s) \in P^{\mathrm{pc}}\}$$

and, for $c \in T_{\mathcal{I}}^{\mathrm{pc}}$, let

$$P_{\mathrm{right}}^{\mathrm{pc}}(c) := \{s \in T_{\mathcal{I}} \mid (c, s) \in P^{\mathrm{pc}}\}.$$

For all $c \in T_{\mathcal{I}}^{\mathrm{pc}}$ and all $\mu \in (\mathbb{N}_{\le q})^3$ compute

$$R_c^{(\mu)}(\psi) := \sum_{s \in P_{\mathrm{right}}^{\mathrm{pc}}(c)} \sum_{\nu \in (\mathbb{N}_{\le q})^3} \left(\mathbf{S}_{\mathbf{b}}^{(n)}\right)_{\mu,\nu} J_s^{(\nu)}(\psi).$$

(c) For $\ell = \ell_{\min}, \ell_{\min} + 1, \ldots, \ell_{\max} - 1$, $\sigma \in T_{\mathcal{I}}(\ell)$, $s \in \mathrm{sons}(\sigma)$, and all $\nu \in (\mathbb{N}_{\le q})^3$ compute

$$R_s^{(\nu)}(\psi) := R_s^{(\nu)}(\psi) + \sum_{\mu \in (\mathbb{N}_{\le q})^3} \gamma_{\mu,\nu,s} R_\sigma^{(\mu)}(\psi).$$

For all $\{i\} \in T_{\mathcal{I}}\left(\ell_{\max}\right)$ do

$$\varphi_i := \sum_{\nu \in \left(\mathbb{N}_{\leq q}\right)^3} R_{\{i\}}^{(\nu)}\left(\psi\right) J_{\{i\}}^{(\nu)}\left(b_i\right).$$

(d) Evaluate (by taking into account the sparsity of $\mathbf{A}_n$)

$$\varphi := \varphi + \mathbf{A}_n^{\mathrm{sparse}}\psi.$$

7.2 Error Analysis

We proceed with the error analysis of the resulting perturbed Galerkin discretisation which leads to an a-priori choice of the interpolation order q such that the convergence rate of the unperturbed discretisation is preserved.

Standard estimates for tensorised Čebyšev-interpolation yield

$$\sup_{z \in Q_c - Q_s} \left| \omega_n^{\Delta t}(\|z\|) - \breve{\omega}_n^{\Delta t}(\|z\|) \right| \leq \tag{31}$$

$$C\frac{L^{q+1}\left(1 + \log^5 q\right)}{2^{2q+1}\left(q+1\right)!} \max_{i \in \{1,2,3\}} \sup_{z \in Q_c - Q_s} \left|\partial_{z_i}^{q+1}\omega\left(\|z\|\right)\right|,$$

where $C > 0$ is some constant independent of all parameters, L denotes the maximal side length of the boxes Q_c and Q_s and $Q_c - Q_s$ is the difference domain $\{x - y : (x, y) \in Q_c \times Q_s\}$.

Theorem 3. *For* $\mathbf{b} = (c, s) \in P^{\mathrm{pc}}$, *let* $(x, y) \in \Gamma_c \times \Gamma_s$ *and* $n \geq C\left|\log \varepsilon\right|$. *Assume that the partial derivatives of* $\omega_n^{\Delta t}(\|x - y\|)$ *satisfy*

$$\max_{1 \leq i \leq 3} \left|\partial_{z_i}^q \omega_n^{\Delta t}\left(\|z\|\right)\right| \leq c^q q! \|z\|^{-1}\left(\frac{1}{\Delta t n^b}\right)^q \quad \forall z \in Q_c - Q_s \tag{32a}$$

with b *as in Definition 3. Then*

$$\left|\omega_n^{\Delta t}(\|x - y\|) - \breve{\omega}_n^{\Delta t}(\|x - y\|)\right| \leq \frac{C_1}{\mathrm{dist}\left(Q_c, Q_s\right)}\left(C_2 \frac{L}{\Delta t n^b}\right)^{q+1} \tag{32b}$$

with L *as in* (31).

Note that in a forthcoming paper, the validity of assumption (32a) will be derived.

Theorem 4. *Let* $\varepsilon > 0$ *and* $n > C|\log^2 \varepsilon|$ *for some* C. *Let the assumptions of Theorem 3 be satisfied and the interpolation order chosen according to* $q \geq |\log \varepsilon| / \log 2$.

(a) *Let* $\mathbf{b} = (c, s) \in P^{\mathrm{pc}}$ *be admissible for some* $0 < \eta \leq \eta_0$ *and sufficiently small* $\eta_0 = \mathcal{O}\left(1\right)$. *Then*

$$\left|\omega_n^{\Delta t}(\|x - y\|) - \breve{\omega}_n^{\Delta t}(\|x - y\|)\right| \leq C\frac{\varepsilon}{\|x - y\|} \quad \forall (x, y) \in \Gamma_c \times \Gamma_s \tag{33a}$$

for some C *independent of* n *and* Δt.

(b) Let $\mathbf{b} = (c, s) \in P^0$. *Then*

$$\left|\omega_n^{\Delta t}(\|x - y\|)\right| \leq \frac{\varepsilon}{\|x - y\|} \quad \forall\, (x, y) \in \Gamma_c \times \Gamma_s. \tag{33b}$$

Proof. Assume that $(c, s) \in P^{\mathrm{pc}}$. Then, due to Condition (27b), we obtain from Theorem 3 the estimate

$$\left|\omega_n^{\Delta t}(\|x - y\|) - \breve{\omega}_n^{\Delta t}(\|x - y\|)\right| \leq \frac{C_1}{\mathrm{dist}\,(Q_c, Q_s)}\,(C_2\eta)^{q+1}\,.$$

The distance can be estimated by means of Condition (27b). For all $(x, y) \in Q_c \times Q_s$, there holds

$$\|x - y\| \leq \mathrm{dist}\,(Q_c, Q_s) + \sqrt{3}\,(L_c + L_s) \leq \mathrm{dist}\,(Q_c, Q_s) + 2\sqrt{3}\eta\Delta t n^b. \tag{34}$$

Because $(c, s) \in P^{\mathrm{pc}}$, Condition (27a) is violated and there exists $(x, y) \in Q_c \times Q_s$ such that $\|x - y\| \in I_{n,\varepsilon}^{\Delta t}$. Thus, by taking into account $n^b \leq n$, we obtain

$$\mathrm{dist}\,(Q_c, Q_s) \geq \|x - y\| - \sqrt{3}\,(L_c + L_s) \geq t_n - 3\sqrt{\Delta t}\sqrt{t_n}|\log\varepsilon| - 2\sqrt{3}\eta\Delta t n^b$$

$$= t_n\left(1 - 3\frac{|\log\varepsilon|}{\sqrt{n}} - 2\sqrt{3}\eta\right) \geq \frac{t_n}{10}$$

for $n > 15|\log^2\varepsilon|$ and $0 \leq \eta \leq \eta_0$ with $\eta_0 = \left(40\sqrt{3}\right)^{-1}$. Hence,

$$\mathrm{dist}\,(Q_c, Q_s) \geq \frac{t_n}{10} \geq 2\left(2\sqrt{3}\eta\Delta t n^b\right) \tag{35}$$

for all $0 \leq \eta \leq \eta_0$.

The combination of (34) and (35) yields

$$\frac{1}{\mathrm{dist}\,(Q_c, Q_s)} \leq \frac{3}{2\,\|x - y\|}$$

and

$$\left|\breve{\omega}_n^{\Delta t}(\|x - y\|) - \omega_n^{\Delta t}(\|x - y\|)\right| \leq \frac{3C_1}{2\,\|x - y\|}\,(C_2\eta)^{q+1}\,.$$

Finally, the condition $\eta_0 \leq (2C_2)^{-1}$ implies that the interpolation order

$$q \geq \frac{|\log\varepsilon|}{\log 2}$$

leads to an approximation which satisfies

$$\left|\breve{\omega}_n^{\Delta t}(\|x - y\|) - \omega_n^{\Delta t}(\|x - y\|)\right| \leq \frac{C_1\varepsilon}{2\,\|x - y\|}\,.$$

For $(c, s) \in P^0$, the assertion follows from (19). $\qquad\square$

In [14] an analysis of the perturbation error has been derived. Since it is only based on abstract approximations which satisfy an error estimate of type (33a) and (33b), we directly obtain a similar convergence theorem also for the panel clustering method. In the following, we denote by $\tilde{\phi}^n_{\Delta t,k} \in S$ the solution at time t_n of the Galerkin discretization with cutoff strategy and panel-clustering.

Theorem 5. *Let the assumption of Theorem 4 be satisfied. We assume that the exact solution $\phi(\cdot,t)$ is in $H^{m+1}(\Gamma)$ for any $t \in [0,T]$. Then there exists $C > 0$, such that for all cutoff parameters ε in (18) such that $0 < \varepsilon < Ch\Delta t^3$ and interpolation orders $q \geq |\log \varepsilon| / \log 2$, the solution $\tilde{\phi}_{\Delta t,h}$ with cutoff and panel-clustering satisfies the error estimate*

$$\left\| \tilde{\phi}^n_{\Delta t,h} - \phi(\cdot,t_n) \right\|_{H^{-1/2}(\Gamma)} \leq C_g(T) \left(\varepsilon h^{-1} \Delta t^{-5} + \Delta t^2 + h^{m+3/2} \right).$$

Corollary 2. *Let the assumptions of Theorem 5 be satisfied. Let $\Delta t \sim h^{m+3/2}$ and choose $\varepsilon \sim h^{7m/2+25/4}$. Then, the solution $\tilde{\phi}_{\Delta t,h}$ exists and converges with optimal rate*

$$\left\| \tilde{\phi}^n_{\Delta t,h} - \phi(\cdot,t_n) \right\|_{H^{-1/2}(\Gamma)} \leq C_g(T) h^{m+3/2} \sim C_g(T) \Delta t^2.$$

7.3 Complexity Estimates

In this subsection, we investigate the complexity of our data-sparse approximation of the wave discretisation. Since we will introduce numerical quadrature methods for approximating the integrals (29) and (30) (for possibly curved panels) in a forthcoming paper, we here restrict ourselves to the storage complexity of our data-sparse approximation scheme and discuss the computational complexity in a forthcoming paper. In this section, we always employ the theoretical value $1/4$ for the exponent b in (27b).

Sparse approximation of the system matrix $\tilde{\mathbf{A}}_n$.

To simplify the complexity analysis we assume that only the simple cutoff strategy and not the panel-clustering method is applied for the first time steps:

$$1 \leq n \leq C \max \left\{ \log M, M^{m-\frac{1}{2}} \right\}, \tag{36}$$

where the constant C depends only on the control parameter η. Note that the second argument in $\max\{\cdot,\cdot\}$ ensures that $P^{\mathrm{sparse}} = \emptyset$ and the matrix $\mathbf{A}^{\mathrm{sparse}}$ vanishes (cf. Remark 2). By using Theorem 2 and (24), the number of nonzero entries of $\tilde{\mathbf{A}}_n$ in this case is of order

$$M \max \left\{ M^{m-\frac{1}{2}} \log M, M^{\frac{1}{4}-\frac{1}{2}m} \log^{5/2} M \right\} = \begin{cases} M^{1+\frac{1}{4}} \log^{5/2} M & m = 0, \\ M^{1+\frac{1}{2}} \log M & m = 1, \end{cases}$$

where the leading constant in the $\mathcal{O}\left(\cdot\right)$-estimate depends only on η. Note that $\Delta t = \mathcal{O}\left(N^{-1}\right)$. Hence, relation (24) implies $N \sim M^{\frac{m}{4}+\frac{3}{8}}$ and allows to estimate the number of n's in (36) by

$$\max\left\{\log M, M^{m-\frac{1}{2}}\right\} \leq N \max\left\{M^{-\frac{m}{4}-\frac{3}{8}}\log M, M^{\frac{3}{4}m-\frac{7}{8}}\right\}.$$

Hence, the total cost for storing these matrices $\tilde{\mathbf{A}}_n$ is given by

$$\left(NM^{\frac{7}{8}+\frac{m}{2}}\log^{\kappa_m} M\right) \quad \text{with} \quad \kappa_m := \begin{cases} 7/2 & \text{if } m=0, \\ 1 & \text{if } m=1. \end{cases}$$

Basis influence coefficients.

The number of basis influence coefficients (cf. (30)) is bounded by

$$\mathcal{O}\left(M\log^3 M\right).$$

Since this step has to be computed and stored only once for all time steps the cost for this step (and the generation of the cluster tree) is negligible compared to the minimal cost $\mathcal{O}\left(NM\right)$ of the whole algorithm.

Influence matrices.

First, we compute the cardinality of P^{pc}. Note that the maximal diameter of a cluster $t \in T_{\mathcal{I}}$ satisfying condition (27b) is bounded by

$$L_t \leq \eta\Delta t n^b.$$

An assumption on the cluster tree and the geometric shape of the surface is that

$$\left|\left\{(x,y) \in \Gamma \times \Gamma \mid \|x-y\| \in I_{n,\varepsilon}^{\Delta t}\right\}\right| = \mathcal{O}\left(\sqrt{\Delta t}\, t_n^{3/2}\, |\log\varepsilon|\right),$$

where $|\omega|$ denotes the area measure of some $\omega \subset \Gamma \times \Gamma$. Hence, for sufficiently small Δt the number of pairs of clusters satisfying (27b) is bounded by

$$\mathcal{O}\left(\frac{\sqrt{\Delta t}\, t_n^{3/2}\, |\log\varepsilon|}{(\eta\Delta t n^b)^4}\right). \tag{37}$$

The storage requirements per matrix $\mathbf{S}_{\mathbf{b}}^{(n)}$ are given by $q^6 \sim |\log^6\varepsilon|$ and this leads to a storage complexity of

$$\mathcal{O}\left(\frac{n^{3/2-4b}\,|\log\varepsilon|^7}{\Delta t^2}\right). \tag{38}$$

Using the relations as in Corollary 2

$$\Delta t^2 \sim h^{m+3/2}, \qquad \varepsilon \sim h^{7m/2+25/4}, \qquad M = \mathcal{O}\left(h^{-2}\right)$$

Table 2. Storage requirements for the panel clustering approximation and sparse approximation.

	full matrix	cutoff	panel clustering+cutoff
$m = 0$	$\mathcal{O}\left(NM^2\right)$	$\mathcal{O}\left(NM^{1+\frac{13}{16}}\log M\right)$	$\mathcal{O}\left(NM^{1-\frac{1}{16}}\log^7 M\right)$
$m = 1$	$\mathcal{O}\left(NM^2\right)$	$\mathcal{O}\left(NM^{1+\frac{11}{16}}\log M\right)$	$\mathcal{O}\left(NM^{1+\frac{9}{16}}\log^7 M\right)$

we see that (38) is equivalent to (we here use $4b = 1$)

$$\mathcal{O}\left(n^{1/2}M^{m/2+3/4}\log^7 M\right).$$

To compute the total storage cost we sum over all $n \in \{0, 1, \ldots, N\}$ to obtain

$$\sum_{n=0}^{N} n^{\frac{1}{2}} M^{\frac{m}{2}+\frac{3}{4}} \log^7 M \leq C N^{\frac{3}{2}} M^{\frac{m}{2}+\frac{3}{4}} \log^7 M \leq C N M^{\frac{5m}{8}+\frac{15}{16}} \log^7 M$$

$$= C \left\{ \begin{array}{ll} NM^{\frac{15}{16}} \log^7 M & m = 0, \\ NM^{1+\frac{9}{16}} \log^7 M & m = 1. \end{array} \right.$$

Note that the storage cost for the temporary quantities in Phase 2 of the panel-clustering algorithm is proportional to $M \log^3 M$ and, hence, negligible compared to the other components of the algorithm.

The total storage requirements are summarised in Table 2. The table shows that the panel-clustering method combined with the cutoff strategy reduces the storage amount very significantly. For piecewise constant boundary elements we even get a storage complexity which behaves better than linearly, i.e., $\mathcal{O}\left(NM\right)$.

8 Conclusions

In this paper, we have followed the convolution quadrature approach by Lubich and combined it with Galerkin BEM for solving the retarded potential boundary integral formulation of the wave equation. The main goal was to develop fast and sparse algorithms for this purpose, i.e., a simple a-priori cutoff strategy where the number of matrix elements which have to be computed is substantially reduced and a significant portion of the matrix is replaced by zero. The panel-clustering method is applied to the remaining blocks which further reduces the computational costs.

In a forthcoming paper, we will introduce an efficient quadrature method and analyse the effect of these additional perturbations.

References

1. A. Bamberger, T. Ha-Duong: Formulation variationelle espace-temps pour le calcul par potentiel retardé d'une onde acoustique. Math. Meth. Appl. Sci. 8 (1986) 405–435, 598–608.
2. B. Birgisson, E. Siebrits, A. Pierce: Elastodynamic direct boundary element methods with enhanced numerical stability properties. Internat. J. Numer. Methods Engrg. 46 (1999) 871–888.
3. M. Bluck, S. Walker: Analysis of three-dimensional transient acoustic wave propagation using the boundary integral equation method. Internat. J. Numer. Methods Engrg. 39 (1996) 1419–1431.
4. P. Ciarlet: The finite element method for elliptic problems. North-Holland, 1987.
5. M. Costabel: Developments in boundary element methods for time-dependent problems. In: Problems and Methods in Mathematical Physics (L. Jentsch, F. Tröltzsch eds.), B.G. Teubner, Leipzig, pp. 17–32, 1994.
6. P. Davies: Numerical stability and convergence of approximations of retarded potential integral equations. SIAM J. Numer. Anal. 31 (1994) 856–875.
7. P. Davies: Averaging techniques for time marching schemes for retarded potential integral equations. Appl. Numer. Math. 23 (1997) 291–310.
8. P. Davies, D. Duncan: Stability and convergence of collocation schemes for retarded potential integral equations. SIAM J. Numer. Anal. 42 (2004) 1167–1188.
9. Y. Ding, A. Forestier, T. Ha-Duong: A Galerkin scheme for the time domain integral equation of acoustic scattering from a hard surface. J. Acoust. Soc. Am. 86 (1989) 1566–1572.
10. A. Ergin, B. Shanker, E. Michielssen: Fast analysis of transient acoustic wave scattering from rigid bodies using the multilevel plane wave time domain algorithm. J. Acoust. Soc. Am. 117 (2000) 1168–1178.
11. M. Friedman, R. Shaw: Diffraction of pulses by cylindrical obstacles of arbitrary cross section. J. Appl. Mech. 29 (1962) 40–46.
12. T. Ha-Duong: On retarded potential boundary integral equations and their discretization. In: Computational Methods in Wave Propagation, Vol. 31 (M. Ainsworth, P. Davies, D. Duncan, P. Martin, B. Rynne eds.), Heidelberg, Springer, pp. 301–336, 2003.
13. T. Ha-Duong, B. Ludwig, I. Terrasse: A Galerkin BEM for transient acoustic scattering by an absorbing obstacle. Internat. J. Numer. Methods Engrg. 57 (2003) 1845–1882.
14. W. Hackbusch, W. Kress, S. Sauter: Sparse convolution quadrature for time domain boundary integral formulations of the wave equation. Technical Report 116, Max-Planck-Institut, Leipzig, Germany, 2005.
15. W. Hackbusch, Z. Nowak: On the fast matrix multiplication in the boundary element method by panel-clustering. Numer. Math, 54 (1989) 463–491.
16. E. Hairer, C. Lubich, M. Schlichte: Fast numerical solution of nonlinear Volterra convolution equations. SIAM J. Sci. Stat. Comput. 6 (1985) 532–541.
17. P. Henrici: Fast Fourier methods in computational complex analysis. SIAM Review 21 (1979) 481–527.
18. C. Lubich: Convolution quadrature and discretized operational calculus I. Numer. Math. 52 (1988) 129–145.
19. C. Lubich: Convolution quadrature and discretized operational calculus II. Numer. Math. 52 (1988) 413–425.

134   W. Hackbusch, W. Kress, and S.A. Sauter

20. C. Lubich: On the multistep time discretization of linear initial-boundary value problems and their boundary integral equations. Numer. Math. 67 (1994) 365–389.
21. C. Lubich, R. Schneider: Time discretization of parabolic boundary integral equations. Numer. Math. 63 (1992) 455–481.
22. E. Miller: An overview of time-domain integral equations models in electromagnetics. J. of Electromagnetic Waves and Appl. 1 (1987) 269–293.
23. B. Rynne, P. Smith. Stability of time marching algorithms for the electric field integral equation. J. of Electromagnetic Waves and Appl. 4 (1990) 1181–1205.
24. S. Sauter, C. Schwab: Randelementmethoden. Analyse, Numerik und Implementierung schneller Algorithmen. B. G. Teubner, Stuttgart, Leipzig, Wiesbaden, 2004.
25. M. Schanz: Wave Propagation in Viscoelastic and Poroelastic Continua. A Boundary Element Approach. Lecture Notes in Applied and Computational Mechanics, Vol. 2, Springer, Heidelberg, 2001.
26. M. Schanz, H. Antes: Application of operational quadrature methods in time domain boundary element methods. Meccanica 32 (1997) 179–186.
27. M. Schanz, H. Antes, T. Rüberg: Convolution quadrature boundary element method for quasi-static visco- and poroelastic continua. Computers & Structures 83 (2005) 673–684.

Fast Multipole Methods and Applications

Günther Of

Institut für Numerische Mathematik, Technische Universität Graz,
Steyrergasse 30, 8010 Graz, Austria
of@tugraz.at

Summary. The symmetric formulation of boundary integral equations and the Galerkin boundary element method are considered to solve mixed boundary value problems of three–dimensional linear elastostatics. Fast boundary element techniques, like the fast multipole method, have to be used to overcome the quadratic complexity of standard boundary element methods. The fast methods provide a data sparse approximation of the fully populated matrices and reduce the computational costs and memory requirements from quadratic order to almost linear ones. Three different approaches to realize the boundary integral operators of linear elastostatics by the fast multipole method are described and numerical examples are given for one of these approaches.

1 Introduction

The Galerkin boundary element method for the symmetric formulation of boundary integral equations is an efficient and reliable tool to solve mixed boundary value problems in linear elastostatics by numerical simulations. This approach is based on a rigorous mathematical analysis. The related stability and error analysis can be found, for example, in [15, 16]. Mathematical books on boundary element methods are, for example, [17, 21, 35, 42].

As the involved boundary integral operators are non–local, standard boundary element techniques result in fully populated stiffness matrices. Therefore, standard boundary element methods are restricted to rather small problem sizes. Hence, fast boundary element methods have to be used for problems of engineering and industrial interest. There exist several fast boundary element methods reducing the memory requirements and the computational costs for a matrix times vector multiplication to almost linear complexity. Most of these methods rely on a clustering of the boundary elements. This leads to a block clustering of the considered matrix, too. Then low rank approximations are used for an appropriate approximation of the corresponding block matrices. The methods mainly differ in the construction and the realization of the low rank approximations. Among them there are the fast multipole method [7, 8, 34] and the panel clustering method [12] which both perform the low rank approximation by an approximation of the kernel by appropriate series expansions providing a separation of the variables. The panel clustering method uses the Taylor series expansion whereas the fast multipole method uses spherical harmonics. The adaptive cross approximation (ACA) method [1, 33] is an algebraic approach to construct low rank approximations. The $\mathcal{H}$–matrices [10] provide a complete arithmetic for the class of matrices with low rank approximations. The $\mathcal{H}^2$–matrices [11] use hierarchical basis functions for the low rank approximations. The wavelet approximation methods [4] construct special nested trial spaces which enable a sparse approximation of the matrix due to the rapid decay of the kernel.

An extensive overview is given in [25] for the large number of contributions to the fast multipole method. There exist several versions of the realization of the boundary integral operators by the fast multipole method in three–dimensional linear elastostatics. A fast multipole version based on the reformulation of the kernel with respect to the fast multipole method for the Laplacian is given in [6]. There, the kernels of the boundary integral operators of linear elastostatics are decomposed in terms depending on $|x - y|^{-1}$ and its derivatives. Then the fast multipole method for electrostatics problems is used as a black box. This approach leads to a rather large number of applications of the potential theoretic fast multipole method.

Starting from the kernel expansion of the fundamental solution of the Laplacian, a new multipole expansion together with the corresponding translations and conversions are derived in [46] for the fundamental solution of linear elastostatics. That leads to less applications of the fast multipole method, but the expansions and the operations get more costly. The authors do not make a clear statement in [46] whether their approach is faster than the approach presented in [6]. The same expansion in spherical harmonics is presented for the panel clustering method in linear elastostatics in [14].

In [32], a different approach based on Taylor series expansions, which is easier to adopt to other kernels, is used. This version of the fast multipole method lacks the translations of local expansions from the clusters to their sons and converts the multipole expansions to the clusters of the finest level. Therefore, the number of conversions is rather high.

We have presented a fast multipole method for linear elastostatics in [29]. The realization of the single layer potential is similar to the approach presented in [6] but guarantees the symmetry of the approximation of the Galerkin matrix of the single layer potential. Integration by parts is used to reduce the hypersingular operator, the double layer potential and its adjoint to double layer potentials of the Laplacian and to single layer potentials.

Here, we try to give an review of some approaches to realize a fast boundary element methods for three–dimensional linear elastostatics based on the fast multipole method. First, we describe the symmetric formulation and the considered Galerkin discretization in Sect. 2. In Sect. 3, the fast multipole method is introduced independent of the specific kernel expansions which are used in the approaches. These approaches are described and in parts compared in Sect. 4. Finally, several numerical examples are given in Sect. 5.

2 Symmetric Boundary Integral Formulation and Boundary Element Method

Let $\Omega \subset \mathbb{R}^3$ be a bounded, simply connected domain with a piecewise continuous Lipschitz boundary $\Gamma = \partial\Omega$, where the outer normal vector $n(x)$ is given for almost all $x \in \Gamma$. We consider a mixed boundary value problem of linear elastostatics, to determine the displacement field $u(x)$ for $x \in \Omega$,

$$
\begin{aligned}
-\operatorname{div} \sigma(u, x) &= 0 & &\text{for } x \in \Omega, \\
\gamma_0 u(x) &= g_D(x) & &\text{for } x \in \Gamma_D, \\
\gamma_1 u(x) &= g_N(x) & &\text{for } x \in \Gamma_N.
\end{aligned}
\tag{1}
$$

The boundary $\Gamma = \overline{\Gamma}_D \cup \overline{\Gamma}_N$ is decomposed in disjoint parts Γ_D and Γ_N. This decomposition may be given componentwise. To guarantee the unique solvability of the boundary value problem, we assume that the part with Dirichlet boundary conditions must not vanish in each component, i.e., meas $(\Gamma_{D,i}) > 0$. The stress tensor $\sigma(u)$ is related to the strain tensor $e(u)$ by Hooke's law

$$
\sigma(u) = \frac{E\nu}{(1+\nu)(1-2\nu)}\operatorname{tr} e(u)I + \frac{E}{(1+\nu)}e(u)
$$

with the Young modulus $E > 0$ and the Poisson ratio $\nu \in (-1, 1/2)$. For the case $\nu \to 1/2$, special techniques [40] have to be applied for the boundary element method. These techniques will not be addressed here. The strain tensor is defined by

$$
e(u) = \frac{1}{2}(\nabla u^{\top} + \nabla u).
$$

The trace operators are given by

$$
\begin{aligned}
\gamma_0 u(x) &:= \lim_{\Omega \ni \tilde{x} \to x \in \Gamma} u(\tilde{x}) & &\text{for almost all } x \in \Gamma, \\
\gamma_1 u(x) &:= \lim_{\Omega \ni \tilde{x} \to x \in \Gamma} [\sigma(u, \tilde{x})n(x)] & &\text{for almost all } x \in \Gamma.
\end{aligned}
$$

The solution of the mixed boundary value problem (1) is given by the representation formula

$$
u(x) = \int_{\Gamma} \gamma_{0,y} U^*(x, y)\gamma_1 u(y)ds_y - \int_{\Gamma} \left(\gamma_{1,y} U^*(x, y)\right)^{\top} \gamma_0 u(y)ds_y
\tag{2}
$$

for $x \in \Omega$. The fundamental solution of linear elastostatics is given by the Kelvin tensor

$$U_{ij}^*(x,y) = \frac{1}{8\pi}\frac{1}{E}\frac{1+\nu}{1-\nu}\left[\frac{(3-4\nu)}{|x-y|}\delta_{ij} + \frac{(x_i-y_i)(x_j-y_j)}{|x-y|^3}\right] \qquad (3)$$

for $i,j = 1,\ldots,3$. The application of the trace operators to the representation formula gives the boundary integral equation

$$\gamma_0 u(x) = \int_\Gamma U^*(x,y)\gamma_1 u(y)ds_y + \frac{1}{2}\gamma_0 u(x) - \int_{\Gamma\setminus\{x\}} T^*(x,y)\gamma_0 u(y)ds_y,$$

for almost all $x \in \Gamma$ with $T^*(x,y) = (\gamma_{1,y}U^*(x,y))^\top$, and the hypersingular boundary integral equation

$$\gamma_1 u(x) = \frac{1}{2}\gamma_1 u(x) + \int_{\Gamma\setminus\{x\}} \gamma_{1,x}U^*(x,y)\gamma_1 u(y)ds_y - \gamma_{1,x}\int_\Gamma T^*(x,y)\gamma_0 u(y)ds_y$$

for almost all $x \in \Gamma$, respectively. Both boundary integral equations together form a system of boundary integral equations

$$\begin{pmatrix}\gamma_0 u \\ \gamma_1 u\end{pmatrix} = \begin{pmatrix}\frac{1}{2}I - K & V \\ D & \frac{1}{2}I + K'\end{pmatrix}\begin{pmatrix}\gamma_0 u \\ \gamma_1 u\end{pmatrix}. \qquad (4)$$

In this representation, we use the standard notations for the boundary integral operators, in particular the single layer potential

$$(Vt)(x) = \int_\Gamma U^*(x,y)t(y)ds_y \qquad \text{for } x \in \Gamma,$$

the double layer potential

$$(Ku)(x) = \int_{\Gamma\setminus\{x\}} T^*(x,y)u(y)ds_y \qquad \text{for } x \in \Gamma,$$

its adjoint operator

$$(K't)(x) = \int_{\Gamma\setminus\{x\}} \gamma_{1,x}U^*(x,y)t(y)ds_y \qquad \text{for } x \in \Gamma,$$

and the hypersingular operator

$$(Du)(x) = -\gamma_{1,x}\int_\Gamma T^*(x,y)u(y)ds_y \qquad \text{for } x \in \Gamma.$$

Here and in what follows, t denotes the traction.

As the solution of the boundary value problem (1) is given by the representation formula (2), the complete Cauchy data $\gamma_0 u$ and $\gamma_1 u$ are sufficient for the evaluation of the solution $u(x)$ for $x \in \Omega$. It remains to determine $\gamma_0 u$ on Γ_N and $\gamma_1 u$ on Γ_D. First, suitable expansions $\widehat{g}_D \in H^{1/2}(\Gamma)$ and $\widehat{g}_N \in H^{-1/2}(\Gamma)$ of the given boundary data $g_D \in H^{1/2}(\Gamma_D)$ and $g_N \in H^{-1/2}(\Gamma_N)$ to the whole boundary Γ are chosen such that

$$\widehat{g}_D(x) = g_D(x) \quad \text{for } x \in \Gamma_D \quad \text{and} \quad \widehat{g}_N(x) = g_N(x) \quad \text{for } x \in \Gamma_N$$

hold. With the splitting of the Cauchy data into the known and the unknown parts,

$$\gamma_0 u(x) = \widehat{u}(x) + \widehat{g}_D(x) \quad \text{and} \quad \gamma_1 u(x) = \widehat{t}(x) + \widehat{g}_N(x),$$

only the functions $\widehat{u} \in \widetilde{H}^{1/2}(\Gamma_N)$ and $\widehat{t} \in \widetilde{H}^{-1/2}(\Gamma_D)$ have to be determined. The Sobolev space $\widetilde{H}^{1/2}(\Gamma_N)$ is the subset of functions in $H^{1/2}(\Gamma)$ with support on Γ_N. $\widetilde{H}^{-1/2}(\Gamma_D)$ is defined by duality of $H^{1/2}(\Gamma_D)$. The complete system (4) of the two boundary integral equations is used to determine the unknown functions $\widehat{u}$ and $\widehat{t}$. The use of the first boundary integral equation for $x \in \Gamma_D$ and of the hypersingular boundary integral equation for $x \in \Gamma_N$ ends up in the symmetric formulation [3, 39]:

$$(V\widehat{t})(x) - (K\widehat{u})(x) = (\frac{1}{2}I + K)\widehat{g}_D(x) - (V\widehat{g}_N)(x) \qquad \text{for } x \in \Gamma_D,$$

$$(K'\widehat{t})(x) + (D\widehat{u})(x) = (\frac{1}{2}I - K')\widehat{g}_N(x) - (D\widehat{g}_D)(x) \qquad \text{for } x \in \Gamma_N.$$

The equivalent variational formulation is given by:
Find $(\widehat{u}, \widehat{t}) \in \widetilde{H}^{1/2}(\Gamma_N) \times \widetilde{H}^{-1/2}(\Gamma_D)$, such that

$$a(\widehat{u}, \widehat{t}; v, \tau) = f(v, \tau) \qquad \text{for all } (v, \tau) \in \widetilde{H}^{1/2}(\Gamma_N) \times \widetilde{H}^{-1/2}(\Gamma_D) \qquad (5)$$

holds. The bilinear form is given by

$$a(\widehat{u}, \widehat{t}; v, \tau) = \langle V\widehat{t}, \tau \rangle_{\Gamma_D} - \langle K\widehat{u}, \tau \rangle_{\Gamma_D} + \langle K'\widehat{t}, v \rangle_{\Gamma_N} + \langle D\widehat{u}, v \rangle_{\Gamma_N},$$

and the linear form is defined by

$$f(v, \tau) = \langle (\frac{1}{2}I + K)\widehat{g}_D, \tau \rangle_{\Gamma_D} - \langle V\widehat{g}_N, \tau \rangle_{\Gamma_D}$$
$$+ \langle (\frac{1}{2}I - K')\widehat{g}_N(x), v \rangle_{\Gamma_N} - \langle D\widehat{g}_D, v \rangle_{\Gamma_N}.$$

The boundedness and the ellipticity of the bilinear form $a(\cdot; \cdot)$ on $\widetilde{H}^{1/2}(\Gamma_N) \times \widetilde{H}^{-1/2}(\Gamma_D)$ can be proofed by the boundedness of the operators and the ellipticity of the single layer potential V and of the hypersingular operator D on $\widetilde{H}^{-1/2}(\Gamma_D)$ and $\widetilde{H}^{1/2}(\Gamma_N)$. The unique solvability of the variational formulation (5) then follows by the Lemma of Lax–Milgram for the continuous linear form $f(\cdot)$.

Let the boundary $\Gamma = \partial\Omega$ be described by a union $\bigcup_{\ell=1}^{N} \overline{\tau}_\ell$ of plane triangles τ_ℓ with a local meshsize

$$h_\ell := \left(\int_{\tau_\ell} ds_x \right)^{1/2}.$$

The global meshsize is defined by

$$h := \max_{\ell=1,\dots,N} h_\ell.$$

Here, we consider a shape regular and quasi uniform boundary discretization for simplicity. We further assume that each boundary element τ_ℓ belongs either to $\Gamma_{D,i}$ or $\Gamma_{N,i}$ for each component $i = 1,\dots,3$. For each component $k = 1,\dots,3$, we use the finite–dimensional trial spaces

$$S_h^1(\Gamma_{N,k}) = \mathrm{span}\,\{\varphi_{i,k}\}_{i=1}^{M_{N,k}} \subset \widetilde{H}^{1/2}(\Gamma_{N,k})$$
$$S_h^0(\Gamma_{D,k}) = \mathrm{span}\,\{\psi_{j,k}\}_{j=1}^{N_{D,k}} \subset \widetilde{H}^{-1/2}(\Gamma_{D,k})$$

for the Galerkin discretization of the variational formulation (5) of the symmetric formulation. $S_h^1(\Gamma_N)$ is the space of the piecewise linear and continuous functions with support in Γ_N and is used for the approximation of the displacements u. The basis functions φ_i are the linear functions that are one in the node x_i of the boundary element mesh and zero in all other nodes. $S_h^0(\Gamma_D)$ denotes the space of piecewise constant functions with support in Γ_D and is used for the approximation of the tractions t. The basis functions ψ_j are one on the boundary element τ_j and zero on all others. N is the number of boundary elements and M is the number of nodes. An index restricts these numbers to the corresponding part of the boundary in the denoted component. For the componentwise trial functions

$$\widehat{u}_{h,k}(x) = \sum_{i=1}^{M_{N,k}} \widehat{u}_{i,k}\varphi_{i,k}(x) \quad \text{and} \quad \widehat{t}_{h,k}(x) = \sum_{j=1}^{N_{D,k}} \widehat{t}_{j,k}\psi_{j,k}(x),$$

we have to find the solution $(\widehat{u}_h, \widehat{t}_h) \in S_h^1(\Gamma_N) \times S_h^0(\Gamma_D)$ of the discrete Galerkin variational formulation

$$a(\widehat{u}_h, \widehat{t}_h; v_h, \tau_h) = f(v_h, \tau_h) \qquad \text{for all } (v_h, \tau_h) \in S_h^1(\Gamma_N) \times S_h^0(\Gamma_D). \tag{6}$$

It can be shown by means of Cea's lemma and the approximation properties of the trial spaces that the discrete variational formulation is uniquely solvable and that the following error estimate holds for the approximations $\widehat{u}_h$ and $\widehat{t}_h$ of the solutions $\widehat{u} \in \widetilde{H}^{s+1}(\Gamma_N)$ and $\widehat{t} \in \widetilde{H}_{\mathrm{pw}}^s(\Gamma_D)$ of the variational formulation (5):

$$\|\widehat{u} - \widehat{u}_h\|_{H^{1/2}(\Gamma)}^2 + \|\widehat{t} - \widehat{t}_h\|_{H^{-1/2}(\Gamma)}^2 \le ch^{2s+1}\left[\|\widehat{u}\|_{H^{s+1}(\Gamma)}^2 + \|\widehat{t}\|_{H_{\mathrm{pw}}^s(\Gamma)}^2\right] \tag{7}$$

for $-1/2 \le s \le 1$, in the case of suitable extensions $\widehat{g}_D \in H^{s+1}(\Gamma)$ and $\widehat{g}_N \in H_{\mathrm{pw}}^s(\Gamma)$ of the boundary data. Here, $H_{\mathrm{pw}}^s(\Gamma)$ denotes an appropriately defined piecewise Sobolev space, see [42]. In the case of a sufficiently smooth solution, i.e., $s = 1$, an optimal convergence rate of $3/2$ is obtained in the energy norm.

The equivalent system of linear equations of the discrete variational formulation (6) is given by

$$\begin{pmatrix} V_h & -K_h \\ K'_h & D_h \end{pmatrix} \begin{pmatrix} \widehat{\underline{t}} \\ \widehat{\underline{u}} \end{pmatrix} = \begin{pmatrix} \underline{f}^1 \\ \underline{f}^2 \end{pmatrix}. \tag{8}$$

$\widehat{\underline{t}} \in \mathbb{R}^{N_D}$ and $\widehat{\underline{u}} \in \mathbb{R}^{M_N}$ with $N_D = N_{D,1} + N_{D,2} + N_{D,3}$ and $M_N = M_{N,1} + M_{N,2} + M_{N,3}$ are the vectors of the coefficients $\widehat{t}_\ell$ and $\widehat{u}_j$ of the trial functions $\widehat{t}_h$ and $\widehat{u}_h$. The block matrices are given from the discretization of the corresponding boundary integral operators on the corresponding parts of the boundary for $i, j = 1, \ldots, 3$ by

$$V_h^{ij}[\ell, k] = \langle V\psi_{k,j}, \psi_{\ell,i} \rangle_{\Gamma_D} \qquad \text{for } \ell = 1, \ldots, N_{D,i}, k = 1, \ldots, N_{D,j},$$

$$K_h^{ij}[\ell, k] = \langle K\varphi_{k,j}, \psi_{\ell,i} \rangle_{\Gamma_D} \qquad \text{for } \ell = 1, \ldots, N_{D,i}, k = 1, \ldots, M_{N,j},$$

$$K'^{ij}_h[\ell, k] = \langle K'\psi_{k,j}, \varphi_{\ell,i} \rangle_{\Gamma_N} \qquad \text{for } \ell = 1, \ldots, M_{N,i}, k = 1, \ldots, N_{D,j},$$

$$D_h^{ij}[\ell, k] = \langle D\varphi_{k,j}, \varphi_{\ell,i} \rangle_{\Gamma_N} \qquad \text{for } \ell = 1, \ldots, M_{N,i}, k = 1, \ldots, M_{N,j}.$$

The vectors of the right hand side are given by

$$f^1_{\ell,i} = \langle (\frac{1}{2}I + K)\widehat{g}_D - V\widehat{g}_N, \psi_{\ell,i} \rangle_{\Gamma_D} \qquad \text{for } \ell = 1, \ldots, N_{D,i},$$

$$f^2_{k,i} = \langle (\frac{1}{2}I - K')\widehat{g}_N - D\widehat{g}_D, \varphi_{j,i} \rangle_{\Gamma_N} \qquad \text{for } k = 1, \ldots, M_{N,i}.$$

The matrix of the system of linear equations (8) is block skew symmetric and positive definite. Furthermore, all blocks are fully populated, i.e., the memory requirements and the effort for one matrix times vector multiplication is of order $\mathcal{O}(N_D^2 + M_N^2)$.

3 Fast Multipole Boundary Element Methods

In this section, we describe the realization of the matrix times vector multiplication $\underline{w} = A_h \underline{t}$ or componentwise,

$$w_\ell = \sum_{k=1}^{\widetilde{N}} A_h[\ell, k] t_k = \sum_{k=1}^{\widetilde{N}} \int_\Gamma (A\varphi_k)(x)\psi_\ell(x)t_k ds_x \qquad \text{for all } \ell = 1, \ldots, \widetilde{M}, \tag{9}$$

of some boundary integral operator

$$(At)(x) = \int_\Gamma \mathcal{Q}_x \mathcal{Q}_y k(x, y)t(y) ds_y$$

by the fast multipole method. $\mathcal{Q}_x$ and $\mathcal{Q}_y$ are some operators like linear combinations of partial derivatives operating on a kernel $k(x, y)$ with respect to x

and y, respectively. But these operators can also be just identities. $\{\varphi_k\}_{k=1}^{\widetilde{N}}$ is the set of trial functions and $\{\psi_\ell\}_{\ell=1}^{\widetilde{M}}$ is the set of test functions. These two sets might coincide. The basis functions φ_k and ψ_ℓ do not have to coincide with the definitions of Sect. 2. The main ingredients of the fast multipole method are the approximation of the kernel function by an appropriate series expansion and the use of a hierarchical structure to compute these expansions efficiently. We require that the kernel $k(x, y)$ is separable, i.e., an expansion

$$k(x, y) = \sum_{n=0}^{\infty} g_n(y) f_n(x)$$

exists with a separation of the variables x and y. Such an expansion can be given by a Taylor series expansion, for example. An approximation of the kernel is defined by truncating the infinite sum at a suitable chosen expansion degree $\tilde{p}$,

$$k_{\tilde{p}}(x, y) = \sum_{n=0}^{\tilde{p}} g_n(y) f_n(x). \tag{10}$$

If such a splitting of the kernel and the approximation (10) were valid for all x and y, the approximation of the matrix times vector multiplication (9) could be rewritten as

$$\widetilde{w}_\ell = \sum_{k=1}^{\widetilde{N}} t_k \int_\Gamma \int_\Gamma \mathcal{Q}_x \mathcal{Q}_y k_{\tilde{p}}(x, y) \varphi_k(y) ds_y \psi_\ell(x) ds_x$$

$$= \sum_{n=0}^{\tilde{p}} \int_\Gamma \mathcal{Q}_x f_n(x) \psi_\ell(x) ds_x \sum_{k=1}^{\widetilde{N}} t_k \int_\Gamma \mathcal{Q}_y g_n(y) \varphi_k(y) ds_y$$

and the total effort would be reduced to $\mathcal{O}(\tilde{p}(\widetilde{N} + \widetilde{M}))$, as the coefficients

$$\tilde{L}_n = \sum_{k=1}^{\widetilde{N}} t_k \int_\Gamma \mathcal{Q}_y g_n(y) \varphi_k(y) ds_y \qquad \text{for } n = 0, \ldots, \tilde{p}.$$

would be computed in $\mathcal{O}(\tilde{p}\widetilde{N})$ operations and the evaluation would take $\mathcal{O}(\tilde{p}\widetilde{M})$ operations.

But in general, the kernel approximation (10) is only valid for $|y| > d|x|$ with $d > 1$ and often an error estimate of the kind

$$|\mathcal{Q}_x \mathcal{Q}_y k(x, y) - \mathcal{Q}_x \mathcal{Q}_y k_{\tilde{p}}(x, y)| \leq c(\tilde{p}, d, |x|) \left(\frac{1}{d}\right)^{\tilde{p}+\varrho} \tag{11}$$

holds with some integer $\varrho \in \mathbb{Z}$. The constant $c(\tilde{p}, d, |x|)$ might be independent of $\tilde{p}$ or a polynomial in $\tilde{p}$ of low order. It also depends on d and $|x|$, but the error estimate is dominated by the exponential term $d^{-\tilde{p}-\varrho}$. Due to the restrictions on the validation of the expansion, the matrix times vector multiplication is separated into two parts, the nearfield part and the farfield part. The farfield $\mathrm{FF}(\ell)$ is the set of indices k, for which the supports of the test function ψ_ℓ and the trial functions φ_k are well separated and therefore it is suitable to apply the kernel approximation (10) due to the error estimate (11). The nearfield part $\mathrm{NF}(\ell)$ of the matrix times vector multiplication is realized as in standard boundary element methods [38]. An exact definition of nearfield and farfield will be given later. Now, the matrix times vector multiplication reads as

$$\widetilde{w}_\ell = \sum_{k \in \mathrm{NF}(\ell)} A_h[\ell, k] t_k + \sum_{n=0}^{\tilde{p}} M_n(O, \psi_\ell) \widetilde{L}_n(0, \mathrm{FF}(\ell)). \tag{12}$$

The coefficients

$$M_n(O, \psi_\ell) = \int_\Gamma \mathcal{Q}_x f_n(x) \psi_\ell(x) ds_x,$$

$$L_n(O, \varphi_k) = \int_\Gamma \mathcal{Q}_y g_n(y) \varphi_k(y) ds_y,$$

with reference to a local center O, can either be computed exactly, for example, in the case of spherical harmonics [23, 24], or can be approximated by the use of some numerical quadrature rule. If the coefficients

$$\widetilde{L}_n(O, \mathrm{FF}(\ell)) = \sum_{k \in \mathrm{FF}(\ell)} t_k L_n(O, k) \qquad \text{for } n = 0, \ldots, \tilde{p} \tag{13}$$

are known an efficient realization of the matrix times vector multiplication will be given by (12). These coefficients depend on the vector $\underline{t}$. Therefore, they have to be recalculated in each matrix times vector multiplication. As the coefficients $\widetilde{L}_n(O, \mathrm{FF}(\ell))$ depend on the farfield of the support of the basis function ψ_ℓ, they differ from each other in general and an efficient calculation is necessary.

The efficient computation of the coefficients $\widetilde{L}_n(O, \mathrm{FF}(\ell))$ in (13) will be described only very briefly. More detailed descriptions can be found, e.g., in [7, 8]. In order to do this computation efficiently, as much information as possible is shared when these coefficients are determined.

The second basic idea of the fast multipole method, the hierarchical structure is applied to compute these expansions. First, this hierarchical structure, called cluster tree, has to be build based on geometrical information. This structure can either be based on the boundary elements or on the supports of the basis functions φ_k and ψ_ℓ. The realization of a boundary integral operator might differ for these two approaches, since the nearfields and farfields differ from each other. In the latter approach, two cluster trees have to be built if the trial and test functions do not coincide.

Here, we describe the construction of the cluster tree based on the supports of the basis functions φ_k and ψ_ℓ. In the case of using boundary elements τ_i

for the construction of the cluster structures, this construction is almost the same. Then only one cluster tree has to be built, but the setup of the nearfield part of the matrix and the evaluation of the farfield part of the matrix times vector multiplication might need some more effort for the assembling. Here, the cluster tree is built from the top down based on the supports of the basis functions φ_k and ψ_ℓ. All trial functions $\{\varphi_k\}_{k=1}^{\widetilde{N}}$ are included in a box containing the original domain Ω. The cluster ω_1^0 of level 0 consists of all trial functions $\{\varphi_k\}_{k=1}^{\widetilde{N}}$ or the corresponding set of indices. The hierarchical structure is build recursively by the refinement of the box corresponding to a cluster ω_i^λ of the level λ into eight similar boxes. The trial functions $\{\varphi_k\}_{k=1}^{\widetilde{N}}$ are assigned to the boxes due to the centers of their supports. All trial functions, which are assigned to one of the refined boxes, form the cluster $\omega_j^{\lambda+1}$ of the finer level $\lambda + 1$ identified with the corresponding refined box. These clusters $\omega_j^{\lambda+1}$ are called the sons of the father cluster ω_i^λ. Empty boxes and the corresponding clusters containing no trial functions are neglected. This refinement is done until a minimal number of trial functions in the clusters is reached or until a maximal cluster level L is reached. Each of the trial functions φ_k is assigned to the cluster ω_i^L on the finest level L which contains the center of the support of φ_k. In this paper, we restrict our considerations to the case of a regular distribution of the boundary elements $\{\tau_k\}_{k=1}^N$ of a globally quasi uniform boundary element mesh. Nevertheless, the method can be extended to the adaptive case, see for example [2, 23].

Next, the second cluster tree with clusters σ_j^λ is build based on the supports of the test functions ψ_ℓ in the similar way. Depending on the choice of the two sets of basis functions, the two cluster trees might coincide.

We have used a more abstract definition of nearfield and farfield so far. Now, we can define these based on the cluster hierarchy. A cluster ω_i^λ is in the nearfield of a cluster σ_j^λ of the same level λ, if the condition

$$\text{dist}\,\{C(\omega_i^\lambda), C(\sigma_j^\lambda)\} \leq (d+1)\max\,\{r(\omega_i^\lambda), r(\sigma_j^\lambda)\} \tag{14}$$

holds for a parameter $d > 1$. $C(\omega_i^\lambda)$ denotes the center of the box identified with the cluster ω_i^λ, and $r(\omega_i^\lambda)$ is the corresponding radius of the cluster, i.e., $r(\omega_i^\lambda) = \sup_{x \in \omega_i^\lambda} |x - C(\omega_i^\lambda)|$. It is important for the multipole algorithm that the nearfield of a father cluster $\sigma_i^{\lambda-1}$ contains the nearfields of all its sons $\sigma_j^\lambda \subset \sigma_i^{\lambda-1}$. This definition of the nearfield and the farfield is transferred to the basis functions by their assignment to the leaves of the cluster tree:

$$\text{NF}(\ell) := \left\{k, 1 \leq k \leq \widetilde{N} \text{ and (14) holds for the cluster } \omega_i^L \text{ of } \varphi_k \right.$$

$$\left. \text{and } \sigma_j^L \text{ is the cluster of } \psi_\ell. \right\},$$

$$\text{FF}(\ell) := \{1, \ldots, N\} \setminus \text{NF}(\ell).$$

A symmetric definition of the nearfield helps to preserve the symmetry in the approximation of symmetric matrices, see for example [26, 30].

The efficient computation of the coefficients $\widetilde{L}_n(O, \mathrm{FF}(\ell))$ in (13) now uses this hierarchy. First, the coefficients

$$\widetilde{M}_n(C(\omega_j^L), \mathrm{P}(\omega_j^L)) = \sum_{k \in \omega_j^L} t_k \widehat{M}_n(C(\omega_j^L), \varphi_k) \tag{15}$$

are calculated for all clusters ω_j^L of the finest level L. $\mathrm{P}(\omega_j^\lambda) := \{k, \varphi_k \in \omega_j^\lambda\}$ is the set of all basis functions φ_k of the cluster ω_j^λ. The coefficients $\widehat{M}_n$ are given by

$$\widehat{M}_n(O, \varphi_k) := \int_\Gamma \mathcal{Q}_y f_n(y) \varphi_k(y) ds_y. \tag{16}$$

The coefficients $\widetilde{M}_n$ are now used to determine the multipole coefficients of the clusters on the coarser levels by a translation of the type

$$\widetilde{M}_n(C(\omega_j^\lambda), \mathrm{P}(\omega_j^\lambda)) = \sum_{\omega_i^{\lambda+1} \in \mathrm{sons}(\omega_j^\lambda)} \sum_s h_{n,s}^1 (\overrightarrow{C(\omega_j^\lambda)C(\omega_i^{\lambda+1})})$$
$$\cdot \widetilde{M}_{n-s}(C(\omega_i^{\lambda+1}), \mathrm{P}(\omega_i^{\lambda+1})) \tag{17}$$

with some coefficients $h_{n,s}^1$. From these multipole coefficients of a cluster ω_j^λ, the needed local coefficients of a second cluster σ_i^λ in the farfield of ω_j^λ can be calculated by a conversion of the type

$$\widetilde{L}_n(C(\sigma_i^\lambda), \mathrm{P}(\omega_j^\lambda)) = \sum_s h_{n,s}^2 (\overrightarrow{C(\omega_j^\lambda)C(\sigma_i^\lambda)}) \widetilde{M}_s(C(\omega_j^\lambda), \mathrm{P}(\omega_j^\lambda)) \tag{18}$$

with some coefficients $h_{n,s}^2$. These conversions are executed on the coarsest possible level, on which the admissibility condition (14) is fulfilled, i.e., for two clusters, which are in their mutual farfield, but their fathers are in their own mutual nearfields. These local coefficients are summed up for each cluster. Additionally, these coefficients are translated from each cluster σ_i^λ to its sons $\sigma_j^{\lambda+1}$ by

$$\widetilde{L}_n(C(\sigma_j^{\lambda+1}), \mathrm{FF}(\sigma_i^\lambda)) = \sum_s h_{n,s}^3 (\overrightarrow{C(\sigma_i^\lambda)C(\sigma_j^{\lambda+1})}) \widetilde{L}_s(C(\sigma_i^\lambda), \mathrm{FF}(\sigma_i^\lambda)) \tag{19}$$

with some coefficients $h_{n,s}^3$. The sum of all coefficients $\widetilde{L}_n(C(\sigma_j^L), \cdot)$ results in the local coefficients $\widetilde{L}_n(C(\sigma_j^L), \mathrm{FF}(\ell))$ needed for the matrix times vector multiplication (12). Here, σ_j^L is the cluster to which the test function ψ_ℓ is assigned. Now, the coefficients $\widetilde{L}_n(C(\sigma_j^L), \mathrm{FF}(\ell))$ are known for a fast evaluation of the farfield part in the matrix times vector product (12). Note that all the translations and conversions have to be executed in each matrix times vector multiplication, as the coefficients in (15) change for each vector $\underline{t}$. As we have described the fast multipole method for an abstract kernel expansion, we have to require that the corresponding translations and conversions exist.

A fixed expansion degree $\tilde{p}$ is not sufficient to guarantee the asymptotic convergence rate of the fast boundary element method, in general. Instead, the expansion degree $\tilde{p}$ has to be chosen proportional to $\log^2 N$, as shown for example in [30] in the case of spherical harmonics. Therefore the total effort of one matrix times vector multiplication is of order $\mathcal{O}(N \log^2 N)$. The memory requirements are also of order $\mathcal{O}(N \log^2 N)$. We have described the fast multipole method in its original version as given by [7, 8]. Several approaches have been made since then to increase the performance of the method. Especially, the translations and the conversions of the multipole and local expansions have been optimized. For example, the effort for these operations can be reduced by fast Fourier transforms [5] or an exponential representation [9]. But this speedup really pays off for larger expansion degrees, which might usually not be necessary in the case of a fast boundary element method for the Laplace equation and for linear elastostatics. As long as the expansion degree $\tilde{p}$ has to be chosen proportional to $\log^2 N$, the total effort of the fast multipole method is not of order $\mathcal{O}(N)$ but higher, since always $\mathcal{O}(N)$ local expansions of $\tilde{p} + 1$ coefficients have be evaluated.

A first approach to overcome this dependency of the expansion degree $\tilde{p}$ on the number of boundary elements has been made by [44] where the variable order approach of the panel clustering by [36] is transferred to the fast multipole method. In the case of boundary integral equations of the second kind with piecewise constant basis functions, one ends up with an $\mathcal{O}(N)$ algorithm. An approach that overcomes the dependency of the fast multipole method on the particular kernel expansion, which has to be derived for each differential operator separately, is given by [37].

4 Fast Boundary Element Methods for Linear Elastostatics

In this section, we try to show the differences of the approaches presented in [6], [29], and [46]. First, we consider the single layer potential of linear elastostatics

$$(Vt)(x) = \int_{\Gamma} U^*(x, y)t(y)ds_y \qquad \text{for } x \in \Gamma.$$

4.1 Realization of the Single Layer Potential as Linear Combination of the Kernel of the Laplacian and Its Derivatives

The fundamental solution $(U^*_{k\ell})_{\ell,k=1..3}$ of linear elastostatics

$$U^*_{k\ell}(x - y) = \frac{1 + \nu}{8\pi E(1 - \nu)} \left[(3 - 4\nu)\frac{\delta_{k\ell}}{|x - y|} + \frac{(x_k - y_k)(x_\ell - y_\ell)}{|x - y|^3} \right]$$

can be expressed by linear combinations of the kernel of the Laplacian and of its derivatives. In [6], the representation

$$U_{k\ell}^*(x - y) = \frac{1 + \nu}{2E(1 - \nu)} \frac{1}{4\pi} \left[(3 - 4\nu) \frac{\delta_{k\ell}}{|x - y|} - x_\ell \frac{\partial}{\partial x_k} \frac{1}{|x - y|} + \frac{\partial}{\partial x_k} \frac{y_\ell}{|x - y|} \right]$$

is chosen. A detailed analysis shows that the corresponding realization by the fast multipole methods requires four calls of the algorithm to compute the local expansions and seven evaluations of these local expansions in the case of a Galerkin matrix. But unfortunately, this realization of the Galerkin matrix V_h by the fast multipole method is not symmetric anymore, as a finite expansion degree has to be used.

4.2 Symmetric Realization of the Single Layer Potential as Linear Combination of the Kernel of the Laplacian and Its Derivatives

In the case of the Laplacian, the transposedness of the Galerkin matrices of the double layer potential and its adjoint operator can be preserved [26]. This gives the idea how to keep the symmetry of the Galerkin matrix of the single layer potential in linear elastostatics. The gradient terms are rewritten as

$$U_{k\ell}^*(x-y) = \frac{1 + \nu}{2E(1 - \nu)} \frac{1}{4\pi} \left[(3 - 4\nu) \frac{\delta_{k\ell}}{|x - y|} - x_\ell \frac{\partial}{\partial x_k} \frac{1}{|x - y|} - y_\ell \frac{\partial}{\partial y_k} \frac{1}{|x - y|} \right].$$

This representation preserves the symmetry within one block of the matrix. To guarantee the symmetry of the blocks, the expression of a block is added for interchanged indices k and ℓ and the sum is divided by two:

$$\begin{aligned}
U_{k\ell}^*(x - y) = \frac{1 + \nu}{2E(1 - \nu)} \frac{1}{4\pi} \Bigg[&(3 - 4\nu) \frac{\delta_{k\ell}}{|x - y|} \\
&- \frac{1}{2} x_\ell \frac{\partial}{\partial x_k} \frac{1}{|x - y|} - \frac{1}{2} y_\ell \frac{\partial}{\partial y_k} \frac{1}{|x - y|} \\
&- \frac{1}{2} x_k \frac{\partial}{\partial x_\ell} \frac{1}{|x - y|} - \frac{1}{2} y_k \frac{\partial}{\partial y_\ell} \frac{1}{|x - y|} \Bigg].
\end{aligned} \tag{20}$$

The realization of the single layer potential by this representation requires six calls of the fast multipole algorithm to compute local coefficients and nine evaluations [27]. The number of evaluations can be reduced to six by a more involved implementation of the fast multipole algorithm, which needs to store more local coefficients. This representation leads to a more expensive application of the single layer potential, but the preserved symmetry of the Galerkin matrix is often advantageous for iterative solvers.

The Kernel Expansion for the Laplacian by Reformulated Spherical Harmonics

The separation of the variables in the kernel of the Laplacian is done by a expansion in spherical harmonics, in general. For a simpler implementation and a fast realization, reformulated spherical harmonics [31, 45, 46] are used for the kernel expansion

$$\frac{1}{|x-y|} \approx \sum_{n=0}^{p} \sum_{m=-n}^{n} \overline{S_n^m(y)} R_n^m(x).$$
(21)

The reformulated spherical harmonics are given by

$$R_n^{\pm m}(x) = \frac{1}{(n+m)!} \frac{d^m}{du^m} P_n(u)\Big|_{u=\widehat{x}_3} (\widehat{x}_1 \pm i\widehat{x}_2)^m |x|^n,$$

$$S_n^{\pm m}(y) = (n-m)! \frac{d^m}{du^m} P_n(u)\Big|_{u=\widehat{y}_3} (\widehat{y}_1 \pm i\widehat{y}_2)^m \frac{1}{|y|^{n+1}}$$

in Cartesian coordinates for $m \geq 0$ and $\widehat{y}_i = y_i/|y|$. They can be computed efficiently by recursive schemes. $P_n(u)$ denote the Legendre polynomials. In the case of this expansion, the multipole coefficients (16) are computed by

$$\widehat{M}_n^m(O, \varphi_k) := \int_\Gamma \mathcal{Q}_y R_n^m(y)\varphi_k(y)ds_y$$

and form the multipole coefficients (15) of a cluster ω_j^L by

$$\widetilde{M}_n^m(C(\omega_j^L), \mathrm{P}(\omega_j^L)) = \sum_{k \in \omega_j^L} t_k \widehat{M}_n^m(C(\omega_j^L), \varphi_k).$$

The translation (17) of multipole coefficients reads as

$$\widetilde{M}_n^m(C(\omega_j^\lambda), \mathrm{P}(\omega_j^\lambda)) = \sum_{\omega_i^{\lambda+1} \in \mathrm{sons}(\omega_j^\lambda)} \sum_{s=0}^{n} \sum_{t=-s}^{s} R_s^t(\overrightarrow{C(\omega_j^\lambda)C(\omega_i^{\lambda+1})})$$
$$\cdot \widetilde{M}_{n-s}^{m-t}(C(\omega_i^{\lambda+1}), \mathrm{P}(\omega_i^{\lambda+1})).$$

The conversion (18) of multipole coefficients to local coefficients takes the form

$$\widetilde{L}_n^m(C(\sigma_i^\lambda), \mathrm{P}(\omega_j^\lambda)) = \sum_{s=0}^{\infty} \sum_{t=-s}^{s} (-1)^n \overline{S_{n+s}^{m+t}}(\overrightarrow{C(\omega_j^\lambda)C(\sigma_i^\lambda)}) \widetilde{M}_s^t(C(\omega_j^\lambda), \mathrm{P}(\omega_j^\lambda)),$$

while the translation (19) of local coefficients is executed by

$$\widetilde{L}_n^m(C(\sigma_j^{\lambda+1}), \mathrm{FF}(\sigma_i^\lambda)) = \sum_{s=n}^{p} \sum_{t=-s}^{s} R_{s-n}^{t-m}(\overrightarrow{C(\sigma_i^\lambda)C(\sigma_j^{\lambda+1})}) \widetilde{L}_s^t(C(\sigma_i^\lambda), \mathrm{FF}(\sigma_i^\lambda)).$$

With these conversion and translation formulae, all ingredients of the fast multipole method for the kernel of the Laplacian are now given. So the fast multipole method can be applied to the kernel of the Laplacian and to linear combinations of derivatives of this kernel. Also the single layer potential of linear elastostatics can be realized by these means, as described before in the approaches of [6] and [29].

4.3 Realization of the Boundary Integral Operators by an Expansion of the Kernel of Linear Elastostatics

In the approach of [46], the expansion of

$$|x - y| = \sum_{n=0}^{\infty} \sum_{m=-n}^{n} \left(\frac{\overline{S_n^m}(y)|x|^2 R_n^m(x)}{2n + 3} - \frac{|y|^2 \overline{S_n^m}(y) R_n^m(x)}{2n - 1} \right),$$

for $|x| < |y|$, is used to derive an expansion of the fundamental solution in linear elastostatics,

$$U_{k\ell}^*(x - y) = \frac{1}{8\pi\mu} \sum_{n=0}^{\infty} \sum_{m=-n}^{n} \left(F_{k\ell,n}^m(x)\overline{S_n^m}(y) + G_{k,n}^m(x)\overline{S_n^m}(y)y_\ell \right). \tag{22}$$

The coefficients $F_{k\ell,n}^m(x)$ and $G_{k,n}^m(x)$ are defined by

$$F_{k\ell,n}^m(x) = \frac{\lambda + 3\mu}{\lambda + 2\mu}\delta_{k\ell}R_n^m(x) - \frac{\lambda + \mu}{\lambda + 2\mu}x_\ell\frac{\partial}{\partial x_k}R_n^m(x),$$

$$G_{k,n}^m(x) = \frac{\lambda + \mu}{\lambda + 2\mu}\frac{\partial}{\partial x_k}R_n^m(x).$$

This expansion is used in [46] to realize the boundary integral operators of linear elastostatics. For the single layer potential, the matrix times vector product takes the form

$$\widetilde{w}_{\ell,i} = \sum_{j=1}^{3} \sum_{k\in\mathrm{NF}(\ell)} V_h^{ij}[\ell, k]t_{k,j} + \frac{1}{8\pi\mu} \sum_{n=0}^{p} \sum_{m=-n}^{n} \left(\sum_{j=1}^{3} F_{ij,n}^m(x)\widetilde{L}_{j,n}^{1,m}(\mathrm{FF}(\ell)) \right.$$

$$\left. + G_{i,n}^m(x)\widetilde{L}_n^{2,m}(\mathrm{FF}(\ell)) \right). \tag{23}$$

$\widetilde{w}_{\ell,i}$ denotes the ℓ–th entry of the vector $\widetilde{w}$ in the i–th coordinate. The computation of the local coefficients $\widetilde{L}_{j,n}^{1,m}(\mathrm{FF}(\ell))$ and $\widetilde{L}_n^{2,m}(\mathrm{FF}(\ell))$ is described next. The multipole coefficients

$$\widehat{M}_{k,n}^{1,m}(O, \varphi_i) = \int_\Gamma R_n^m(y)\varphi_{i,k}(y)ds_y \tag{24}$$

$$\widehat{M}_{k,n}^{2,m}(O, \varphi_i) = \int_\Gamma R_n^m(y)y_k\varphi_{i,k}(y)ds_y \tag{25}$$

have to be computed for each basis functions and for $k = 1,\ldots,3$, similar to (16). $\varphi_{i,k}$ denotes the k–th component of the trial function φ_i. These coefficients can be computed once in advance and be reused in each matrix times vector multiplication. Due to the more involved expansion series, four sets of coefficients have to be computed in this approach. In each matrix times vector multiplication the coefficients

$$\widetilde{M}_{k,n}^{1,m}(C(\omega_j^L), \mathrm{P}(\omega_j^L)) = \sum_{i \in \omega_j^L} t_{i,k} \widehat{M}_{k,n}^{1,m}(C(\omega_j^L), \varphi_i)$$

$$\widetilde{M}_n^{2,m}(C(\omega_j^L), \mathrm{P}(\omega_j^L)) = \sum_{i \in \omega_j^L} \sum_{\ell=1}^{3} t_{i,\ell} \widehat{M}_{\ell,n}^{2,m}(C(\omega_j^L), \varphi_i).$$

are calculated for all clusters ω_j^L of the finest level first as in (15). $t_{i,k}$ denotes the i–th entry of the vector $\underline{t}$ for the k–th component. The translation (17) of these multipole coefficients now looks like

$$\widetilde{M}_{k,n}^{1,m}(C(\omega_j^\lambda), \mathrm{P}(\omega_j^\lambda)) = \sum_{\omega_i^{\lambda+1} \in \mathrm{sons}(\omega_j^\lambda)} \sum_{s=0}^{n} \sum_{t=-s}^{s} R_s^t(z) \widetilde{M}_{k,n-s}^{1,m-t}(C(\omega_i^{\lambda+1}), \mathrm{P}(\omega_i^{\lambda+1})),$$

$$\widetilde{M}_n^{2,m}(C(\omega_j^\lambda), \mathrm{P}(\omega_j^\lambda)) = \sum_{\omega_i^{\lambda+1} \in \mathrm{sons}(\omega_j^\lambda)} \sum_{s=0}^{n} \sum_{t=-s}^{s} R_s^t(z) \left(\widetilde{M}_{n-s}^{2,m-t}(C(\omega_i^{\lambda+1}), \mathrm{P}(\omega_i^{\lambda+1})) \right.$$
$$\left. + \sum_{\ell=1}^{3} z_\ell \widetilde{M}_{\ell,n-s}^{1,m-t}(C(\omega_i^{\lambda+1}), \mathrm{P}(\omega_i^{\lambda+1})) \right)$$

where $z = \overrightarrow{C(\omega_j^\lambda)C(\omega_i^{\lambda+1})}$. The conversion (18) of multipole coefficients to local coefficients takes the form

$$\widetilde{L}_{\ell,n}^{1,m}(C(\sigma_i^\lambda), \mathrm{P}(\omega_j^\lambda)) = \sum_{s=0}^{\infty} \sum_{t=-s}^{s} (-1)^n \overline{S_{n+s}^{m+t}}(z) \widetilde{M}_{\ell,s}^{1,t}(C(\omega_j^\lambda), \mathrm{P}(\omega_j^\lambda)),$$

$$\widetilde{L}_n^{2,m}(C(\sigma_i^\lambda), \mathrm{P}(\omega_j^\lambda)) = \sum_{s=0}^{\infty} \sum_{t=-s}^{s} (-1)^n \overline{S_{n+s}^{m+t}}(z) \left(\widetilde{M}_s^{2,t}(C(\omega_j^\lambda), \mathrm{P}(\omega_j^\lambda)) \right.$$
$$\left. - \sum_{\ell=1}^{3} z_\ell \widetilde{M}_{\ell,s}^{1,t}(C(\omega_j^\lambda), \mathrm{P}(\omega_j^\lambda)) \right),$$

where $z = \overrightarrow{C(\omega_j^\lambda)C(\sigma_i^\lambda)}$, while the translation (19) of local coefficients is evaluated by

$$\widetilde{L}_{\ell,n}^{1,m}(C(\sigma_j^{\lambda+1}), \mathrm{FF}(\sigma_i^\lambda)) = \sum_{s=n}^{p} \sum_{t=-s}^{s} R_{s-n}^{t-m}(z) \widetilde{L}_{\ell,s}^{1,t}(C(\sigma_i^\lambda), \mathrm{FF}(\sigma_i^\lambda)),$$

$$\widetilde{L}_n^{2,m}(C(\sigma_j^{\lambda+1}), \mathrm{FF}(\sigma_i^\lambda)) = \sum_{s=n}^{p} \sum_{t=-s}^{s} R_{s-n}^{t-m}(z) \left(\widetilde{L}_s^{2,t}(C(\sigma_i^\lambda), \mathrm{FF}(\sigma_i^\lambda)) \right.$$
$$\left. - \sum_{\ell=1}^{3} z_\ell \widetilde{L}_{\ell,s}^{1,t}(C(\sigma_i^\lambda), \mathrm{FF}(\sigma_i^\lambda)) \right)$$

where $z = \overrightarrow{C(\sigma_i^\lambda)C(\sigma_j^{\lambda+1})}$.

In the case of the double layer potential K, its adjoint operator K' and the hypersingular operator D, operators $\mathcal{Q}_x$ and $\mathcal{Q}_y$, which are linear combinations of derivatives with respect to x and y, have to be applied to this expansion. In detail, the operator $\mathcal{Q}_y$ is applied in the computation of the multipole coefficients in (24) and (25), and the operator $\mathcal{Q}_x$ is applied in the evaluation of the expansion (23). Therefore, all boundary integral operators require the computation of four sets of coefficients $\widetilde{L}^{1,m}_{\ell,n}$ and $\widetilde{L}^{2,m}_{n}$. This corresponds to four calls of the fast multipole method, but the translations, the conversions and the evaluations get a little more involved.

4.4 Realization of the Double Layer Potential as Linear Combination of Derivatives of the Kernel of the Laplacian

In the approach of [6], the kernel $T^*(x,y) = (\gamma_{1,y}U^*(x,y))^\top$ of the double layer potential K is rewritten, in a similar way as the kernel of the single layer potential, as

$$T^*_{k\ell}(x,y) = \sum_{j=1}^{3} R_{k\ell j}\left[n_j(y)\frac{1}{|x-y|} \right] - \sum_{j=1}^{3} \frac{1}{8\pi(1-\nu)}\frac{\partial}{\partial x_k}\frac{\partial}{\partial x_j}\left[n_j(y)y_\ell \frac{1}{|x-y|} \right]$$

with an operator

$$R_{k\ell j} = \frac{1}{8\pi(1-\nu)}\left[(1-2\nu)\left(\delta_{\ell j}\frac{\partial}{\partial x_k} - \delta_{k\ell}\frac{\partial}{\partial x_j} \right) - 2(1-\nu)\delta_{kj}\frac{\partial}{\partial x_\ell} + x_\ell \frac{\partial}{\partial x_k}\frac{\partial}{\partial x_j} \right].$$

The realization of this representation by the fast multipole method ends up with twelve calls of the fast multipole algorithm of the Laplacian. Therefore, the effort of a application of the double layer potential K is more expensive than the corresponding realization of the single layer potential.

4.5 Realization of Boundary Integral Operators using Integration by Parts

In our approach [29], we use a representation of the double layer potential K of linear elastostatics by weakly singular boundary integral operators which can be derived by integration by parts [18]. The double layer potential K can be rewritten by

$$(Ku)(x) = \frac{1}{4\pi}\int_\Gamma u(y)\frac{\partial}{\partial n_y}\frac{1}{|x-y|}ds_y - \frac{1}{4\pi}\int_\Gamma \frac{1}{|x-y|}(\mathcal{M}u)(y)ds_y$$

$$+2\mu\left(V_E\left(\mathcal{M}u\right)\right)(x). \tag{26}$$

with $\mu = E/(2(1+\nu))$ and an operator $\mathcal{M}$, consisting of components of the surface curl,

$$\mathcal{M} = \begin{pmatrix} 0 & n_2\dfrac{\partial}{\partial x_1} - n_1\dfrac{\partial}{\partial x_2} & n_3\dfrac{\partial}{\partial x_1} - n_1\dfrac{\partial}{\partial x_3} \\[2ex] n_1\dfrac{\partial}{\partial x_2} - n_2\dfrac{\partial}{\partial x_1} & 0 & n_3\dfrac{\partial}{\partial x_2} - n_2\dfrac{\partial}{\partial x_3} \\[2ex] n_1\dfrac{\partial}{\partial x_3} - n_3\dfrac{\partial}{\partial x_1} & n_2\dfrac{\partial}{\partial x_3} - n_3\dfrac{\partial}{\partial x_2} & 0 \end{pmatrix}.$$

The representation (26) allows to realize the double layer potential of linear elastostatics by six calls of the fast multipole method and nine evaluations [27]. In a more involved implementation, this can be reduced to six evaluations again. Therefore the effort for an application of the double layer potential is comparable to the application of our realization of the single layer potential and not increased as in the approach of [6].

The representation (26) of the double layer potential is also used in the nearfield. Therefore, only weakly singular boundary integral operators have be be evaluated. The effort for the computation of the Galerkin weights is significantly reduced by the change from Cauchy singular to weakly singular integrals.

In the case of piecewise linear trial functions and plane triangles as boundary elements, $\mathcal{M}$ maps the linear basis functions to piecewise constant basis functions. Therefore, the already computed nearfield matrices of the single layer potentials can be reused.

The representation (26) of the double layer potential can be used to rewrite the bilinear form of its adjoint operator K' as

$$\langle K't, v\rangle_\Gamma = \langle K'_L t, v\rangle_\Gamma - \langle V_L t, \mathcal{M}v\rangle_\Gamma + 2\mu\langle Vt, \mathcal{M}v\rangle_\Gamma. \tag{27}$$

Here, V_L and K'_L denote the corresponding operators of the Laplacian which are applied componentwise. This is sufficient for the used Galerkin method. In this way, the bilinear form of the adjoint double layer potential K' can be realized by six calls of the fast multipole algorithm of the Laplacian.

Using piecewise constant trial functions and piecewise linear, continuous test functions, the already computed nearfield matrices can be reused for the realization of the bilinear form of the adjoint double layer potential.

As in the case of the Laplace equation, the bilinear form of the hypersingular operator can be transformed to bilinear forms of single layer potentials. Based on the representation (26) of the double layer potential K, integration by parts reduces the bilinear form of the hypersingular operator to [13]

$$\langle Du, v\rangle_\Gamma = \int_\Gamma \int_\Gamma \frac{\mu}{4\pi} \frac{1}{|x-y|} \left(\sum_{k=1}^{3} (\mathcal{M}_{k+2,k+1} v)(x) \cdot (\mathcal{M}_{k+2,k+1} u)(y) \right) ds_y ds_x$$

$$+ \int_\Gamma \int_\Gamma (\mathcal{M}v)^\top(x) \left(\frac{\mu}{2\pi} \frac{I}{|x-y|} - 4\mu^2 U^*(x,y) \right) (\mathcal{M}u)(y) ds_y ds_x$$

$$+ \int_\Gamma \int_\Gamma \sum_{i,j,k=1}^{3} (\mathcal{M}_{k,j} v_i)(x) \frac{\mu}{4\pi} \frac{1}{|x-y|} (\mathcal{M}_{k,i} u_j)(y) ds_y ds_x.$$

In the first line the indices 4 and 5 of the operator $\mathcal{M}$ have to be identified with 1 and 2, respectively.

Overall, it is sufficient to have a fast realization of the single layer potentials and the double layer potential of the Laplacian in our approach. Therefore, the effort for the computation of the Galerkin weights is reduced significantly. The approach based on integration by parts is not restricted to the fast multipole method but can also be used for other fast boundary element techniques.

By a detailed analysis [27], we have shown that the use of the fast multipole method as a fast boundary element method does not effect the main properties and the asymptotic error estimate of the boundary element method, summarized in the following theorem.

Theorem 1 ([27]). *Let $\widehat{t} \in \widetilde{H}^{\sigma}_{pw}(\Gamma_D)$ and $\widehat{u} \in \widetilde{H}^{\eta}(\Gamma_N)$ for $\sigma \in [0,1]$ and $\eta \in [1,2]$ be the unique solution of the variational problem (5). Let the discretization of the boundary be shape–regular and quasi–uniform. Let the expansion degree p of the multipole expansion (21) be proportional to $\log N$. Then the variational problem of the approximations of the boundary integral operators by the fast multipole method is uniquely solvable. For the approximate solutions $\widetilde{u}_h$ and $\widetilde{t}_h$ the following error estimate holds:*

$$\|\widehat{t} - \widetilde{t}_h\|^2_{H^{-1/2}(\Gamma)} + \|\widehat{u} - \widetilde{u}_h\|^2_{H^{1/2}(\Gamma)} \leq c \left(h^{2\sigma+1} \|\widehat{t}\|^2_{H^{\sigma}_{pw}(\Gamma)} + h^{2\eta-1} \|\widehat{u}\|^2_{H^{\eta}(\Gamma)} \right).$$

Similar results should hold for the other approaches to realize the boundary integral operators by a fast multipole method, as all approaches are based on the expansion in spherical harmonics. It seems to be an open question, how the expansion degrees have to be chosen optimally in each of the approaches. This optimal choice of the expansion degree has a big influence on the performance of the methods and has to be considered in comparisons between the methods.

As a fixed expansion degree is not sufficient to keep up with the asymptotic error estimate of a boundary element method, the expansion degree has to be adopted to the number of boundary elements like $\log N$. With a fixed expansion degree p over all levels in the cluster tree of the fast multipole method, the effort of the fast multipole method is always of order $\mathcal{O}(N \log^2 N)$, as for each boundary element an expansion with $\mathcal{O}(\log^2 N)$ coefficients has to be evaluated.

To overcome this logarithmic terms in the effort, an attempt has been made by [44] where the variable order approach of the panel clustering [36] is transferred to the fast multipole method. In the case of boundary integral equations of the second kind with piecewise constant basis functions, one ends up with an $\mathcal{O}(N)$ algorithm.

5 Numerical Examples

Finally, we show first some academic examples and then some examples of industrial interest. First, we compare our version [29] of the fast multipole method for linear elastostatics with a standard boundary element approach. The considered domain is the cuboid shown in Fig. 1.

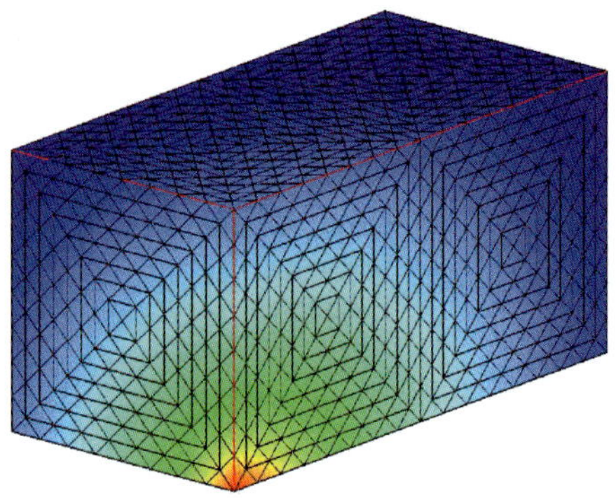

Fig. 1. Cuboid with 2560 boundary elements.

The left front side is the part with Dirichlet boundary conditions. The rest of the boundary has Neumann boundary conditions. The given boundary data are the traces of a chosen solution of the boundary value problem (1), which is given by a fundamental solution with the singularity outside of the domain. The system of linear equations (8) is solved as Schur complement system, with an iterative inversion of the matrix of the single layer potential in each iteration step. The results of this computations are given in Table 1.

Table 1. Comparison of standard and fast BEM.

L	N	M	dof	t_1	t_2	it	D-error	N-error
0	40	22	63	0	0	19	1.25e-3	7.23e-2
				0	0	19	1.25e-3	7.23e-2
1	160	82	255	2	0	24	2.86e-4	4.71e-2
				3	2	24	2.86e-4	4.71e-2
2	640	322	1035	37	5	27	5.15e-5	2.26e-2
				17	18	27	5.14e-5	2.26e-2
3	2560	1282	4179	609	82	30	1.06e-5	1.05e-2
				111	150	30	1.05e-5	1.05e-2
4	10240	5122	16803	(162 min)	(>23 min)			
				7.6 min	13.5 min	32	2.47e-6	5.06e-3
5	40960	20482	67395	(43.3 h)	(>6.6 h)			
				0.5 h	1.3 h	34	5.96e-7	2.50e-3

The first line of each refinement level L shows the data for the standard boundary element method, while the data in the second line refer to the fast boundary element method. N is the number of triangles and M denotes the number of nodes. Further, the number dof of degrees of freedom is given. t_1 and t_2 are the times in seconds for setting up and for solving the system of linear equations. It is the number of iteration steps needed in the Schur complement conjugate gradient method with a relative accuracy of 10^{-8}. Further, the errors of the computed Dirichlet and Neumann data are compared using the $L_2(\Gamma)$ norm. These experiments have been carried out on personal computer with an Intel Pentium 4 processor with 3.06 GHz and 1 GB of RAM. An artificial multilevel boundary element preconditioner [41] and the technique of operators of opposite order [28, 43] have been used as preconditioners for the single layer potential and the hypersingular operator, respectively.

The numbers of iterations are the same for the standard and the fast boundary element method. This is an evidence that the approximation of the system matrix is rather good and that important properties such as the symmetry are preserved. The standard method only works until the fourth refinement level due to the memory restrictions. Therefore, some values have been extrapolated and have been put in brackets. The numbers of iterations grow logarithmically as expected from theory [22]. There is some overhead in the fast boundary element method on the first refinement levels. But the fast multipole method speeds up the calculations on the higher levels significantly. The speedup is larger for setting up the system than for solving. This is typically for the fast multipole method. But here it is also caused by the special choice of the parameters of the fast multipole method for an optimal total time and by some implementation details. The factors of the increasing computational times for solving the Schur complement system are a little bit higher than expected for an algorithm of $\mathcal{O}(N \log^2 N)$. This is due to the increasing number of iterations in the inner iteration for the inversion of the single layer potential in the Schur complement conjugate gradient method. The errors of the Dirichlet and Neumann data match each other very well for the standard and the fast boundary element method. The convergence rates expected from theory are obtained.

In Table 2, a diagonal scaling, the artificial multilevel boundary element preconditioner [41] and an algebraic multigrid method [19, 20, 27] for the fast multipole method are compared as preconditioners of the single layer potential in a Dirichlet boundary value problem for the cuboid in Fig. 1. In the latter case, the algebraic multigrid preconditioner of the single layer potential of the Laplacian is applied componentwise, since its application is cheaper than the application of the operator of linear elastostatics and gives good iteration numbers.

Overall, six uniform refinement steps have been applied such that the finest grid consists of 163840 triangles and 491520 degrees of freedom. The computations have been executed on a personal computer with an AMD Opteron processor 146 with 2.0 GHz and 4 GB RAM. The iteration numbers of the

Table 2. Comparison of the preconditioners.

			scaling			ABPX			AMG		
L	N	dof	t_1	t_2	It	t_1	t_2	It	t_1	t_2	It
0	40	120	0	0	26	0	0	26	0	0	13
1	160	480	1	2	36	2	1	33	2	1	15
2	640	1920	5	13	51	5	10	40	5	4	16
3	2560	7680	18	93	70	19	58	44	21	25	16
4	10240	30720	75	680	92	76	370	50	88	160	17
5	40960	122880	365	6945	124	368	3080	55	457	1392	19
6	163840	491520	1749	55984	165	1750	20386	60	2325	9481	21

diagonal scaling grow quite fast. The iteration numbers of the artificial multilevel preconditioner increase logarithmically as predicted. As the costs for this preconditioner are very low, the reduced number of iterations results in a faster solving of the system. The algebraic multigrid method reduces the number of iterations once more. The application of the algebraic multigrid preconditioner is a lot more expensive than the artificial multilevel preconditioner, but nevertheless the computational times are reduced again. Therefore, the extra effort to set up the algebraic multilevel preconditioner is justifiable.

The first example of industrial interest is the stress analysis for a part of a press equipment and has been provided by W. Volk, M. Wagner and S. Wittig (BMW Research Center Munich). The two pictures in Fig. 2 show the deformed body under imposed deformations and stresses.

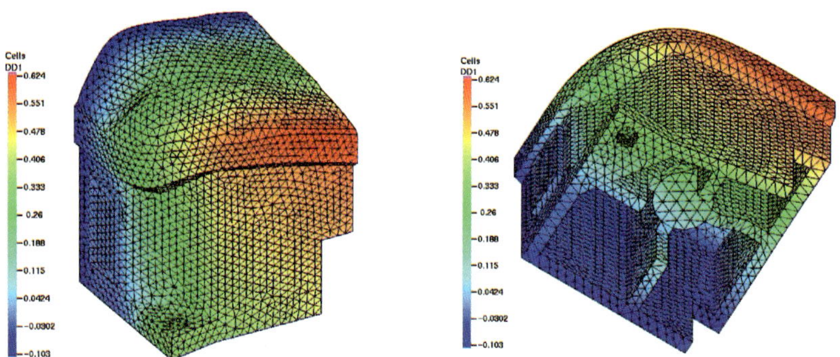

Fig. 2. Part of a press equipment.

The numbers N of boundary elements, the numbers of degrees of freedoms, the computational times for setting up and solving the system of linear equations and the numbers of iterations of the conjugate gradient method with a relative accuracy of 10^{-8} are given in Table 3. The press equipment is only fixed at a few points. Therefore, the block of the single layer potential is set up completely and inverted by a direct method. Than the complete system is solved as a Schur complement system on the Opteron computer.

Table 3. Computational times for the press equipment.

N	DOF(N)	DOF(D)	Setup	Solving	it
13144	182	19586	896	3398	343
52572	728	78426	3940	23623	372

The numbers of iterations might seem to be rather high on the first sight. But they are caused by the thin walls of the body. The slight increase of the numbers of iterations shows that the preconditioning of the hypersingular operator by the operator of opposite order [43] performs as expected.

The second example is a metal foam, see Fig. 3, provided by H. Andrä (Fraunhofer–Institut für Techno– und Wirtschaftsmathematik, Kaiserslautern). The left picture shows the body in the reference configuration, while the right picture shows the deformed body. The bottom side of the foam has been fixed and the top side has been pressed down by a given deformation in z–direction.

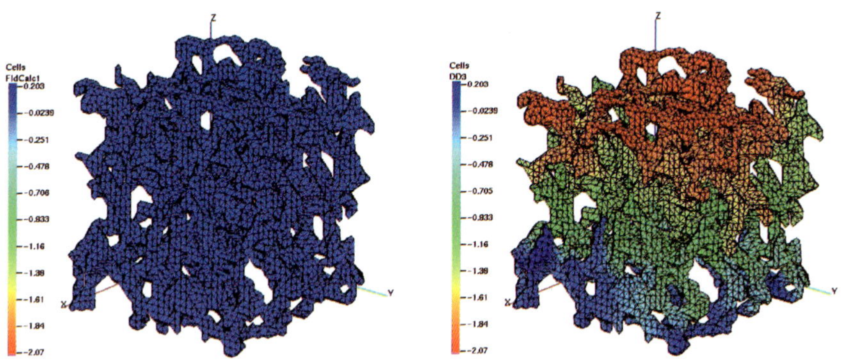

Fig. 3. Undeformed and deformed foam of metal.

The number of boundary elements, the number of nodes, the number of degrees of freedom, the number of iterations and the computational times are given in Table 4.

Table 4. Computational times for the foam.

N	M	DOF(N)	DOF(D)	Setup	Solving	It
28952	14152	396	41511	1730	9832	264

Several attempts of computations with commercial finite element software had been failed for this complex structure. But these computations were possible with the fast boundary element method. The preconditioners work well for this complex structure, too. The computations are rather costly for fast boundary element methods, as the boundary element mesh fills out the whole volume and the nearfields in the cluster tree are very large, consequently.

Overall, the realization of the boundary element method for linear elastostatics by the fast multipole method works very well and is applicable to complex structures of industrial interest.

Acknowledgment

This work has been supported by the German Research Foundation 'Deutsche Forschungsgemeinschaft (DFG)' under the grant SFB 404 'Multifield Problems in Continuum Mechanics' at the University of Stuttgart.

References

1. M. Bebendorf, S. Rjasanow: Adaptive low-rank approximation of collocation matrices. Computing 70 (2003) 1–24.
2. H. Cheng, L. Greengard, V. Rokhlin: A fast adaptive multipole algorithm in three dimensions. J. Comput. Phys. 155 (1999) 468–498.
3. M. Costabel: Symmetric methods for the coupling of finite elements and boundary elements. In: Boundary elements IX, Vol. 1, pp. 411–420. Comput. Mech., Southampton, 1987.
4. W. Dahmen, S. Prößdorf, R. Schneider: Wavelet approximation methods for pseudodifferential equations. II: Matrix compression and fast solution. Adv. Comput. Math. 1 (1993) 259–335.
5. W. D. Elliott, J. A. Board: Fast Fourier transform accelerated fast multipole algorithm. SIAM J. Sci. Comput. 17 (1996) 398–415.
6. Y. Fu, K. J. Klimkowski, G. J. Rodin, E. Berger, J. C. Browne, J. K. Singer, R. A. van de Geijn, K. S. Vemaganti: A fast solution method for three-dimensional many-particle problems of linear elasticity. Internat. J. Numer. Methods Engrg. 42 (1998) 1215–1229.
7. L. Greengard: The Rapid Evaluation of Potential Fields in Particle Systems. The MIT Press, Cambridge, MA, 1987.
8. L. Greengard, V. Rokhlin: A fast algorithm for particle simulations. J. Comput. Phys. 73 (1987) 325–348.
9. L. Greengard, V. Rokhlin: A new version of the fast multipole method for the Laplace equation in three dimensions. Acta Numerica 6 (1997) 229–269.
10. W. Hackbusch: A sparse matrix arithmetic based on $\mathcal{H}$-matrices. I: Introduction to $\mathcal{H}$-matrices. Computing 62 (1999) 89–108.

11. W. Hackbusch, B. Khoromskij, S. A. Sauter: On $\mathcal{H}^2$-matrices. In: Lectures on applied mathematics (H.-J. Bungartz et al. eds.), Proceedings of the symposium organized by the Sonderforschungsbereich 438 on the occasion of Karl-Heinz Hoffmann's 60th birthday, Munich, Germany, June 30-July 1, 1999, pp. 9–29, Springer, Berlin, 2000.

12. W. Hackbusch, Z. P. Nowak: On the fast matrix multiplication in the boundary element method by panel clustering. Numer. Math. 54 (1989) 463–491.

13. H. Han: The boundary integro-differential equations of three-dimensional Neumann problem in linear elasticity. Numer. Math. 68 (1994) 269–281.

14. K. Hayami, S. A. Sauter: A panel clustering method for 3-D elastostatics using spherical harmonics. In: Integral Methods in Science and Engineering (B. Bertram et al. eds.), Proceedings of the 5th International Conference, IMSE'98, Chapman Hall/CRC Res. Notes Math., Vol. 418, pp. 179–184, Chapman & Hall/CRC, Boca Raton, 2000.

15. G. C. Hsiao, W. L. Wendland: On a boundary integral method for some exterior problems in elasticity. Mat. Mech. Astron. 257 (1985) 31–60.

16. G. C. Hsiao, W. L. Wendland: Boundary element methods: foundation and error analysis. In: Encyclopedia of Computational Mechanics, Vol. I: Fundamentals (E. Stein, R. de Borst, T. Hughes eds.), pp. 339–373, John Wiley & Sons, 2004.

17. G. C. Hsiao, W. L. Wendland: Boundary Integral Equations: Variational Methods. Springer, Heidelberg, to be published.

18. V. D. Kupradze, T. G. Gegelia, M. O. Baseleisvili, T. V. Burculadze: Three-dimensional problems of the mathematical theory of elasticity and thermoelasticity. North-Holland Series in applied Mathematics and Mechanics, Vol. 25, North-Holland Publishing Company, Amsterdam, New York, Oxford, 1979.

19. U. Langer, D. Pusch: Data-sparse algebraic multigrid methods for large scale boundary element equations. Appl. Numer. Math. 54 (2005) 406–424.

20. U. Langer, D. Pusch, S. Reitzinger: Efficient preconditioners for boundary element matrices based on grey-box algebraic multigrid methods. Internat. J. Numer. Methods Engrg. 58 (2003) 1937–1953.

21. W. McLean: Strongly elliptic systems and boundary integral equations. Cambridge University Press, Cambridge, 2000.

22. W. McLean, O. Steinbach: Boundary element preconditioners for a hypersingular integral equation on an interval. Adv. Comput. Math. 11 (1999) 271–286.

23. K. Nabors, F. T. Korsmeyer, F. T. Leighton, J. White: Preconditioned, adaptive, multipole-accelerated iterative methods for three-dimensional first-kind integral equations of potential theory. SIAM J. Sci. Comput. 15 (1994) 713–735.

24. J. N. Newman: Distributions of sources and normal dipoles over a quadrilateral panel. J. Engrg. Math. 20 (1986) 113–126.

25. N. Nishimura: Fast multipole accelerated boundary integral equation methods. Appl. Mech. Rev. 55 (2002) 299–324.

26. G. Of: Die Multipolmethode für Randintegralgleichungen. Diplomarbeit, Universität Stuttgart, 2001.

27. G. Of: BETI–Gebietszerlegungsmethoden mit schnellen Randelementverfahren und Anwendungen. Doctoral thesis, Universtät Stuttgart, 2006.

28. G. Of, O. Steinbach: A fast multipole boundary element method for a modified hypersingular boundary integral equation. In: Analysis and simulation of multifield problems (W. L. Wendland et al. eds.), Selected papers of the international conference on multifield problems, Stuttgart, Germany, April 8-10, 2002. Lect. Notes Appl. Comput. Mech., Vol. 12, pp. 163–169, Springer, Berlin, 2003.

29. G. Of, O. Steinbach, W. L. Wendland: Applications of a fast multipole Galerkin boundary element method in linear elastostatics. Comput. Visual. Sci. 8 (2005) 201–209.

30. G. Of, O. Steinbach, W. L. Wendland: The fast multipole method for the symmetric boundary integral formulation. IMA J. Numer. Anal. 26 (2006) 272–296.

31. J. M. Perez-Jorda, W. Yang: A concise redefinition of the solid spherical harmonics and its use in the fast multipole methods. J. Chem. Phys. 104 (1996) 8003–8006.

32. V. Popov, H. Power: An $O(N)$ Taylor series multipole boundary element method for three-dimensional elasticity problems. Eng. Anal. Bound. Elem. 25 (2001) 7–18.

33. S. Rjasanow, O. Steinbach: The Fast Solution of Boundary Integral Equations. Springer Series in Mathematical and Analytical Technology with Applications to Engineering, Vol. 13, Springer, New York, 2006.

34. V. Rokhlin: Rapid solution of integral equations of classical potential theory. J. Comput. Phys. 60 (1985) 187–207.

35. S. A. Sauter, C. Schwab: Randelementmethoden. Analyse, Numerik und Implementierung schneller Algorithmen. B.G. Teubner, Stuttgart, Leipzig, Wiesbaden, 2004.

36. S. A. Sauter: Variable order panel clustering. Computing 64 (2000) 223–261.

37. G. Schmidlin, C. Lage, C. Schwab: Rapid solution of first kind boundary integral equations in $\mathbb{R}^3$. Eng. Anal. Bound. Elem. 27 (2003) 469–490.

38. C. Schwab, W. L. Wendland: On numerical cubatures of singular surface integrals in boundary element methods. Numer. Math. 62 (1992) 343–369.

39. S. Sirtori: General stress analysis method by means of integral equations and boundary elements. Meccanica 14 (1979) 210–218.

40. O. Steinbach: A robust boundary element method for nearly incompressible linear elasticity. Numer. Math. 95 (2003) 553–562.

41. O. Steinbach: Artificial multilevel boundary element preconditioners. Proc. Appl. Math. Mech. 3 (2003) 539–542.

42. O. Steinbach: Numerische Näherungsverfahren für elliptische Randwertprobleme. Finite Elemente und Randelemente. B.G. Teubner, Stuttgart, Leipzig, Wiesbaden, 2003.

43. O. Steinbach, W. L. Wendland: The construction of some efficient preconditioners in the boundary element method. Adv. Comput. Math. 9 (1998) 191–216.

44. J. Tausch: The variable order fast multipole method for boundary integral equations of the second kind. Computing 72 (2004) 267–291.

45. C. A. White, M. Head-Gordon: Derivation and efficient implementation of the fast multipole method. J. Chem. Phys. 101 (1994) 6593–6605.

46. K. Yoshida, N. Nishimura, S. Kobayashi: Application of fast multipole Galerkin boundary integral equation method to elastostatic crack problems in 3D. Internat. J. Numer. Methods Engrg. 50 (2001) 525–547.

A Fast Boundary Integral Equation Method for Elastodynamics in Time Domain and Its Parallelisation

Yoshihiro Otani[1], Toru Takahashi[2] and Naoshi Nishimura[3]

[1] Graduate School of Informatics, Kyoto University, Kyoto 606–8501, Japan
`otani@mbox.kudpc.kyoto-u.ac.jp`
[2] Earthquake Research Institute, University of Tokyo, Tokyo 113–0032, Japan
`ttaka@eri.u-tokyo.ac.jp`
[3] Graduate School of Informatics, Kyoto University, Kyoto 606–8501, Japan
`nchml@i.kyoto-u.ac.jp`

Summary. This paper discusses a time domain fast boundary integral equation method for three dimensional elastodynamics and its parallelisation for a large shared memory parallel computer. Some details of the PWTD (Plane Wave Time Domain) approach and an extension of the theory to the anisotropic case are presented. We then examine the parallelisation strategies of the code using OpenMP and MPI-OpenMP hybridisation. In the case of MPI-OpenMP hybrid parallelisation, a numerical example with more than one million spatial DOF is shown. It is concluded that the method is promising, its parallelisation with OpenMP is effective and that larger problem can be analysed with MPI-OpenMP hybrid parallelisation.

1 Introduction

When one solves wave problems in time domain with the conventional BIEM (Boundary Integral Equation Method), one will have an algorithm of the complexity of $O(N_s^2 N_t^2)$, where N_s and N_t are the spatial and time degrees of freedom (See Nishimura [4] for example). This complexity is considered expensive and has urged investigators to develop fast methods. Ergin et al. [1], for example, have proposed the so called Plane Wave Time Domain Algorithm (PWTD) which utilises the plane wave expansion of the fundamental solution and the hierarchical structure of the space-time in the scalar wave equation in 3D. Also, Lu et al. [3] have developed a PWTD algorithm for the 2 dimensional scalar wave equation, and shown that the complexity of this method is $O(N_s N_t \log N_s \log N_t)$.

Our group have been investigating fast algorithms for elastodynamics. In time domain, Takahashi et al. have extended the PWTD algorithm to elastodynamics in 2D and 3D and have solved problems of the spatial and temporal degrees of freedom of the orders of 10^4 and 10^2, respectively [5, 6].

In this paper we shall continue these investigations and report further progress made after the publication of these papers. Namely, we shall present a simplified derivation of the plane wave expansion of the fundamental solution for elastodynamics. With this derivation we shall obtain a PWTD formulation for general anisotropic elastodynamics.

We shall also investigate parallelisation of the algorithm for shared memory computers. In the area of high performance computing, the use of clusters of SMPs is becoming popular. In such platforms, each of the nodes is a shared memory parallel computer, in which all the CPUs share a large memory space. For shared memory parallel computers, OpenMP [8], which is reputed for its user friendliness, is considered to be the de facto standard for the thread-parallelism. In this investigation, we shall present an easy way of parallelising the fast time domain elastodynamics code with the help of OpenMP. We shall further seek an MPI [9]-OpenMP hybrid parallelisation of the algorithm, which enables an analysis of large scale problems with more than 1 million spatial degrees of freedom.

2 Formulation

2.1 Time Domain Elastodynamics in 3D and Integral Equations

In the following sections we shall outline the three dimensional fast boundary integral equation method for elastodynamics in time domain proposed by Takahashi et al. [6], to which the reader is referred for the details.

Let $D \subset \mathbb{R}^3$ be a domain which can be identified as the elastic body in question. We are now interested in finding the displacement field $u_i(\boldsymbol{x}, t)$ for $\boldsymbol{x} = (x_1, x_2, x_3) \in D, t \in (0, \infty)$ which satisfies

$$c_2^2 u_{i,jj}(\boldsymbol{x}, t) + (c_1^2 - c_2^2) u_{j,ij}(\boldsymbol{x}, t) + b_i(\boldsymbol{x}, t) = \ddot{u}_i(\boldsymbol{x}, t) \tag{1}$$

subject to initial conditions:

$$\boldsymbol{u}(\boldsymbol{x}, 0) = \boldsymbol{u}_0(\boldsymbol{x}), \quad \dot{\boldsymbol{u}}(\boldsymbol{x}, 0) = \boldsymbol{v}_0(\boldsymbol{x}) \quad \text{in } D \tag{2}$$

and boundary conditions:

$$\boldsymbol{u}(\boldsymbol{x}, t) = \bar{\boldsymbol{u}}(\boldsymbol{x}, t) \quad \text{on } S_1, \qquad \boldsymbol{t}(\boldsymbol{x}, t) = \bar{\boldsymbol{t}}(\boldsymbol{x}, t) \quad \text{on } S_2 \tag{3}$$

where a superposed $\bar{(\ )}$ indicates a function given on the boundary of D denoted by S, S_1 and S_2 are disjoint subsets of S such that $S_1 + S_2 = S$ holds, and the spatial derivatives with respect to x_i are denoted by $(\ )_{,i}$. Also, the initial displacement and velocity are denoted by $\boldsymbol{u}_0$ and $\boldsymbol{v}_0$, respectively, $\boldsymbol{t}$ is the surface traction defined by

$$t_i(\boldsymbol{x}, t) = C_{ijkl} n_j(\boldsymbol{x}) u_{k,l}(\boldsymbol{x}, t) \tag{4}$$

and $\boldsymbol{n}$ is the unit normal vector to S pointing into the complement of D. The symbol C_{ijkl} stands for the elasticity tensor which is written as

$$C_{ijkl} = \lambda \delta_{ij} \delta_{kl} + \mu(\delta_{ik}\delta_{jl} + \delta_{il}\delta_{jk}) \tag{5}$$

where λ and μ are Lamé constants, with which the velocities of P and S waves, denoted by c_1 and c_2, are written as

$$c_1 = \sqrt{\frac{\lambda + 2\mu}{\rho}}, \quad c_2 = \sqrt{\frac{\mu}{\rho}}. \tag{6}$$

where ρ is the density.

Assuming, for simplicity, that the initial displacement, velocity and body force vanish, we obtain the following BIE (Boundary Integral Equation) on $S \times (t > 0)$:

$$\frac{1}{2}u_i + \fint_S T_{ij}(\boldsymbol{x}, \boldsymbol{y}, t) * u_j(\boldsymbol{y}, t)dS_y = \int_S \Gamma_{ij}(\boldsymbol{x} - \boldsymbol{y}, t) * t_j(\boldsymbol{y}, t)dS_y \quad \boldsymbol{x} \in S \tag{7}$$

where $\fint$ stands for Cauchy's principal value and $*$ indicates the convolution with respect to time. Also, $\boldsymbol{\Gamma}$ and $\boldsymbol{T}$ are the fundamental solution and the double layer kernel for elastodynamics written as

$$\Gamma_{ij}(\boldsymbol{x}, t) = \frac{1}{4\pi\mu}\left[\frac{\delta(t - |\boldsymbol{x}|/c_2)}{|\boldsymbol{x}|}\delta_{ij} - c_2^2\partial_i\partial_j\left(\frac{(t - |\boldsymbol{x}|/c_2)_+}{|\boldsymbol{x}|} - \frac{(t - |\boldsymbol{x}|/c_1)_+}{|\boldsymbol{x}|}\right)\right] \tag{8}$$

$$T_{ij}(\boldsymbol{x}, \boldsymbol{y}, t) = C_{jlmn} n_l(\boldsymbol{y})\frac{\partial}{\partial y_n}\Gamma_{im}(\boldsymbol{x} - \boldsymbol{y}, t) \tag{9}$$

where δ is Dirac's delta, and $f_+ = (|f| + f)/2$.

2.2 Discretised BIE for Conventional Approach

We discretise (7) as follows: Discretise the boundary S with N_s planar triangles $S_I(I = 1, 2, \cdots, N_s)$ and take a spatial collocation point $\boldsymbol{x}_I$ at the centroid of each S_I. Divide the time into N_t time intervals $(t_{\alpha-1}, t_\alpha)$ having the constant length Δt. Take a time collocation point at the end point $(t_\alpha = \alpha\Delta t)$ of each interval. Denote the boundary densities at the collocation point, namely, the displacement $\boldsymbol{u}(\boldsymbol{x}_I, t_\alpha)$ and the traction $\boldsymbol{t}(\boldsymbol{x}_I, t_\alpha)$ by $\boldsymbol{u}_I^\alpha$ and $\boldsymbol{t}_I^\alpha$, respectively.

As the basis function, we use piecewise constant functions spatially and piecewise linear functions in time. Written explicitly, the time basis function takes the following form:

$$\Psi_\alpha(t) = \begin{cases} 0 & t < t_{\alpha-1}, t_{\alpha+1} \leq t \\ \frac{t-t_{\alpha-1}}{\Delta t} & t_{\alpha-1} \leq t < t_\alpha \\ \frac{t_{\alpha+1}-t}{\Delta t} & t_\alpha \leq t < t_{\alpha+1} \end{cases} \tag{10}$$

With these, the functions $u(x,t), t(x,t)$ are interpolated as:

$$u(x,t) = \sum_{\alpha=1} \Psi_\alpha(t) u(x,t_\alpha) \tag{11}$$

$$t(x,t) = \sum_{\alpha=1} \Psi_\alpha(t) t(x,t_\alpha) \tag{12}$$

with respect to time. As we write

$$W_{IJ}^{\alpha+\beta-1} = \frac{1}{2}\delta_{IJ}\delta_{\alpha\beta}I + \int_{t_{\beta-1}}^{\min(t_{\beta+1},t_\alpha)} \oint_{S_J} T(x_I, y, t_\alpha - s)\Psi_\beta(s)dS_y ds \tag{13}$$

$$U_{IJ}^{\alpha+\beta-1} = \int_{t_{\beta-1}}^{\min(t_{\beta+1},t_\alpha)} \int_{S_J} \Gamma(x_I - y, t_\alpha - s)\Psi_\beta(s)dS_y ds \tag{14}$$

we see that (7) allows the following discretisation:

$$\sum_{J=1}^{N_s} A_{IJ}^{(1)} a_J^\alpha = \sum_{J=1}^{N_s} B_{IJ}^{(1)} b_J^\alpha$$

$$- \sum_{\beta=1}^{\alpha-1} \sum_{J=1}^{N_s} \left(W_{IJ}^{(\alpha-\beta+1)} u_J^\beta - U_{IJ}^{(\alpha-\beta+1)} t_J^\beta \right) \tag{15}$$

where we have denoted unknown (known) quantities at the current time t_α on the element S_J as a_J^α (b_J^α) and the corresponding coefficients (either $W_{IJ}^{(1)}$ or $U_{IJ}^{(1)}$) by $A^{(1)IJ}$ ($B_{IJ}^{(1)}$). The system of equations in (15) has $3N_s$ unknowns. Since $A_{IJ}^{(1)}$ is sparse, one obtains its inverse with $O(N_s)$ operations. However, $U_{IJ}^{(\gamma)}$ and $W_{IJ}^{(\gamma)}$ in the 2nd term of (15) become dense as γ increases. Also, the number of matrix-vector multiplication required to obtain the RHS of this system of equations will be proportional to the number of time steps. Therefore, the number of operations required to obtain the solutions $a_I^{N_t}$ for N_t time steps is $O(N_s^2 N_t^2)$ with the conventional BIEM, which is quite expensive. It is thus seen that one needs a fast method of carrying out the mat-vec operations in the RHS of (15) in order to obtain a fast solution method for BIEMs in time domain. We are thus lead to the investigation of fast method of evaluating the RHS of (15) with the help of the plane wave expansion of the fundamental solution.

2.3 Plane Wave Expansion of the Fundamental Solution

We now consider an integral representation for the function given by

$$\Gamma_{ij}(\boldsymbol{x},t) - \Gamma'_{ij}(\boldsymbol{x},t)$$

where $\Gamma'_{ij}(\boldsymbol{x},t)$ is the anti-causal fundamental solution, or 'ghost', defined by $\Gamma'_{ij}(\boldsymbol{x},t) = \Gamma_{ij}(\boldsymbol{x},-t)$. As the limiting absorption principle tells, the causal (namely, the one which vanishes for $t < 0$) fundamental solution is obtained as

$$\Gamma_{ij}(\boldsymbol{x},t) = \mathcal{F}^{-1}\hat{\Gamma}_{ij}\big|_{\omega=\omega^+}, \quad \hat{\Gamma}_{ij} = [C_{ikjl}\xi_k\xi_l - \rho\omega^2\delta_{ij}]^{-1}$$

where ω^+ means that one takes the Fourier inverse transform assuming that the quantity ω has an infinitesimal positive imaginary part. Similarly, one shows that the following holds true:

$$\Gamma'_{ij}(\boldsymbol{x},t) = \mathcal{F}^{-1}\hat{\Gamma}_{ij}\big|_{\omega=\omega^-}$$

where ω^- has an obvious meaning. These results yield

$$\Gamma_{ij}(\boldsymbol{x},t) - \Gamma'_{ij}(\boldsymbol{x},t) = \frac{1}{2\pi}\mathcal{F}^{-1}_{\xi\to x}\int_p \hat{\Gamma}_{ij}e^{-i\omega t}d\omega$$

where p is a closed contour in the complex plane which includes in its interior all the poles of the integrand as a function of ω. See Fig. 1. After some manipulation, one has the following plane wave expansion for the fundamental solution:

$$\Gamma_{ij}(\boldsymbol{x},t) - \Gamma'_{ij}(\boldsymbol{x},t)$$
$$= -\frac{\partial t}{8\pi^2}\int_{S_k}\left[\frac{k_ik_j}{\rho c_1^3}\delta(t - \boldsymbol{x}\cdot\boldsymbol{k}/c_1) + \frac{k_pk_q e_{pik}e_{qjk}}{\rho c_2^3}\delta(t - \boldsymbol{x}\cdot\boldsymbol{k}/c_2)\right]dS_k.$$
$$(16)$$

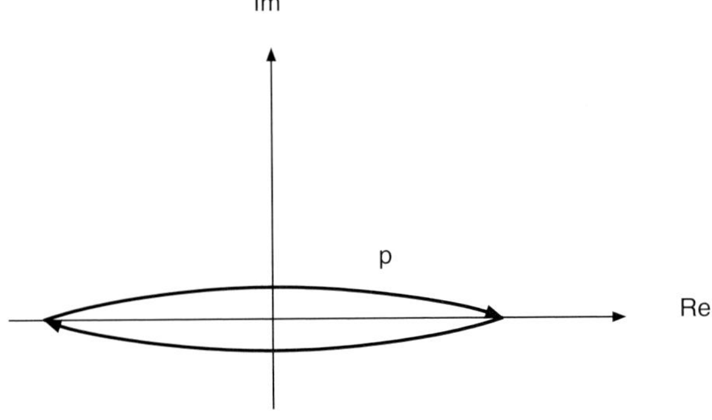

Fig. 1. Path.

Since the ghost has no physical meaning, one has to make sure that this term does not pollute the numerical solution. As a matter of fact, the technique of sectioning the time interval, to be presented later, guarantees that the ghost has no effect on the numerical solution.

Finally, we notice that Takahashi et al. [6] obtained the same result using somewhat more involved calculation, while the present version requires just residue calculus. This simplification makes it possible to use the same method in obtaining similar results for the general anisotropic case, as we shall see later.

2.4 Evaluation of Potentials with the Plane Wave Expansion

We now construct a fast method of evaluating the layer potentials in (7) with the help of (16).

We consider 2 spherical regions S_s and S_o, both with the radius of R, and denote the origin of each of them by s and o, respectively. Also, we shall denote the distance between the centres $|o - s|$ by R_c, which are assumed to satisfy $R_c > 2R$. A part of S_s, denoted by S_0, is then considered. We now evaluate the layer potentials due to the densities t and u distributed on $S_0 \times (0, t]$ and observed at the collocation point (x, t) $(x \in S_o, t \in (0, \infty))$ using (16). The densities u and t are divided into sub-densities u^z and t^z which have supports only in the time interval $(T_1^z, T_2^z]$ $(T_{1,2}^z < t)$. If

$$R_c - 2R \geq c_1(T_2^z - T_1^z) \tag{17}$$

holds, the potential due to known densities t^z and u^z on $S_0 \times (T_1^z, T_2^z]$ is expressed as [6]:

$$\int_{S_0} \left(T_{ij}(x, y, t) * u_j^z(y, t) - \Gamma_{ij}(x - y, t) * t_j^z(y, t) \right) dS_y$$

$$= -\frac{\partial_t}{8\pi^2} \int_{S_k} \left[k_i \delta(t - (x - s) \cdot k/c_1) * O^z(s, t, k) \right.$$

$$\left. + e_{pik} k_p \delta(t - (x - s) \cdot k/c_2) * O_k^z(s, t, k) \right] dS_k. \tag{18}$$

Namely, the effect of the ghost is absent if (17) holds. In this statement O^z and $O_k^z (k = 1, 2, 3)$ are the outgoing rays defined by

$$O^z(s, t, k) = \int_{S_0} \left(\frac{C_{jlnm} n_l k_m k_n}{\rho c_1^4} \dot{u}_j^z(y, t - (s - y) \cdot k/c_1) \right.$$

$$\left. - \frac{k_j}{\rho c_1^3} t_j^z(y, t - (s - y) \cdot k/c_1) \right) dS_y \tag{19}$$

$$O_k^z(s, t, k) = \int_{S_0} \left(\frac{C_{jlnm} n_l e_{qmk} k_q k_n}{\rho c_2^4} \dot{u}_j^z(y, t - (s - y) \cdot k/c_2) \right.$$

$$\left. - \frac{e_{qjk} k_q}{\rho c_2^3} t_j^z(y, t - (s - y) \cdot k/c_2) \right) dS_y. \tag{20}$$

Equation (18) can also be computed as follows:

$$\int_{S_0} \left(T_{ij}(\boldsymbol{x}, \boldsymbol{y}, t) * u_j^z(\boldsymbol{y}, t) - \Gamma_{ij}(\boldsymbol{x} - \boldsymbol{y}, t) * t_j^z(\boldsymbol{y}, t) \right) dS_y$$

$$= -\frac{1}{8\pi^2} \int_{S_k} \left[k_i \delta(t - (\boldsymbol{x} - \boldsymbol{o}) \cdot \boldsymbol{k}/c_1) * I^z(\boldsymbol{o}, t, \boldsymbol{k}) \right.$$

$$\left. + e_{pik} k_p \delta(t - (\boldsymbol{x} - \boldsymbol{o}) \cdot \boldsymbol{k}/c_2 * I_k^z(\boldsymbol{o}, t, \boldsymbol{k})) \right] dS_k \qquad (21)$$

where I^z and $I_k^z (k = 1, 2, 3)$ are incoming rays defined by

$$I_{(i)}^z(\boldsymbol{o}, t, \boldsymbol{k}) = \partial_t \delta(t - (\boldsymbol{o} - \boldsymbol{s}) \cdot \boldsymbol{k}/c_{1(2)}) * O_{(i)}^z(\boldsymbol{s}, t, \boldsymbol{k}). \qquad (22)$$

One can shift the origin of the outgoing and incoming rays via

$$O_{(i)}^z(\boldsymbol{s}, t, \boldsymbol{k}) = \delta(t - (\boldsymbol{s} - \boldsymbol{s}') \cdot \boldsymbol{k}/c_{1(2)}) * O_{(i)}^z(\boldsymbol{s}', t, \boldsymbol{k}), \qquad (23)$$

$$I_{(i)}^z(\boldsymbol{s}, t, \boldsymbol{k}) = \delta(t - (\boldsymbol{s} - \boldsymbol{s}') \cdot \boldsymbol{k}/c_{1(2)}) * I_{(i)}^z(\boldsymbol{s}', t, \boldsymbol{k}). \qquad (24)$$

3 Fast Algorithm

We now construct a fast method of evaluating the layer potentials in the RHS of (15) with the help of the plane wave expansion introduced in Section 2.4. In this algorithm, we evaluate the contributions to the potentials at a collocation point from far elements via the plane wave expansion and those from nearby elements via conventional approaches.

3.1 Cells and Time Intervals

We first divide the spatial domain under consideration into hierarchical cell structure. For this purpose we take a cube which contains the whole boundary S and call it the cell of the level 0. We then divide it into 8 equal cubes called level 1 cells by bisecting the edges of the level 0 cell. We further subdivide the level l cell into 8 cubes, from which those containing boundary elements are kept as the level $l + 1$ cells. We continue this subdivision until the cell contains less than a certain number of elements. The lowest level number is denoted by l_{max}. A childless cell is called a leaf.

We next introduce the concept of neighbourhood in the following manner: Let the coordinates of the centre of level l cells C and C' be C_i and C_i', respectively, and the edge length of a level l cell be $L^{(l)}$. The cells C and C' are said to be close if

$$|C_i - C_i'| < (\beta + 1)L^{(l)} \quad i = 1, 2, 3 \qquad (25)$$

holds, and called far otherwise, where β is a number. In the present investigation we put $\beta = 1$. Namely, two cells are said to be far if there are more than 1 cells between them.

In the evaluation of the effects from far cells, we use the plane wave expansion, which is applicable only when (17) holds. Therefore, we introduce a set of time intervals of a constant length $M\Delta t$, the zth one of which is given by $((z-1)M\Delta t, zM\Delta t](z = 1, 2, \cdots)$. The densities $\boldsymbol{u}$ and $\boldsymbol{t}$ are then divided into $\boldsymbol{u}^z$ and $\boldsymbol{t}^z$ according to these intervals. Since the discretised densities are interpolated with interpolation functions having supports of the lengths $2p_t$, the support of the zth density, i.e. $(T_1^z, t_2^z]$, is given by

$$T_1^z = ((z-1)M + 1 - p_t)\Delta t \tag{26}$$

$$T_2^z = (zM + p_t)\Delta t. \tag{27}$$

In order to evaluate effects from far cells with the plane wave expantion, one takes the M for each level as

$$M^{(l_{max})} = \frac{\beta L^{(l_{max})}}{c_1 \Delta t} - 2p_t + 1 \tag{28}$$

$$M^{(l)} = 2M^{(l+1)} \quad l = 2, 3, \cdots, l_{max} - 1 \tag{29}$$

considering (17). With this $M^{(l)}$ the time interval $z^{(l)}$ for each level l is defined and thus $T_1^{z^{(l)}}$ and $T_2^{z^{(l)}}$ are determined by (26) and (27).

3.2 Outline of the Fast Algorithm

We now outline the fast method based on the plane wave expansion:

1. Evaluation of the layer potentials in a cell C due to densities on $S \times (0, t_{\alpha-1}]$:
   Let the current time be $t_\alpha = \alpha \Delta t$. We then evaluate the layer potential produced by known densities on $S \times (0, t_{\alpha-1}]$ by separating it into the contributions from the near and far cells.
   - Contribution from the near cells can be evaluated directly with the conventional method.
   - Contribution from far cells is evaluated as follows:
     Let the current time t_α be in the $z_\alpha^{(l)}$th time interval in level l. Then the outgoing and incoming rays $O_p^{z^{(l)}}$ and $I_p^{z^{(l)}}$ due to boundary densities $\boldsymbol{u}^{z^{(l)}}$ and $\boldsymbol{t}^{z^{(l)}}$ which have supports in the passed time intervals $z^{(l)} = 1, 2, \cdots, z_\alpha^{(l)} - 1$ are known, and their effects have already been evaluated via (21), as we shall see later in 3 b), and stored. Therefore all it takes is to recall the stored values for the potential. Notice that the contributions of the densities in the current time interval from far cells will not reach C at the current time and, hence, do not have to be considered.

2. Determination of the unknown boundary densities at the current time:
   The RHS of (15) has already been obtained with the help of the procedures described above in 1. Hence one solves (15) with any iterative method to determine the unknowns at the current time.

3. Computation of the outgoing and incoming rays:
   We compute the outgoing and incoming rays produced by the densities in the current time interval $z_\alpha^{(l)}$.

   a) upward pass

   We compute the outgoing ray for the current time interval $z_\alpha^{(l)}$ at the centre of each cell, starting from the lowest level l_{max} and going upward to the level 2. The highest level in this algorithm is 2 since the cells in levels 1 and 0 are all neighbours which means that no combination of cells in these levels allow the use of the plane wave expansion. This upward pass is carried out only in the level l where the current time step number α is a multiple of $M^{(l)}$. In leaves we use (19) and (20) to obtain the outgoing rays. For parent cells we add outgoing rays of the children after shifting the origin using (23).

   b) downward pass

   We compute incoming rays for the current time interval $z_\alpha^{(l)}$ at the centre of each cell starting from level 2 and going downward along the tree structure of cells. But, one carries this out only for levels where the current time step number α is a multiple of $M^{(l)}$. From the outgoing rays of far cells one computes the incoming rays of a cell C with (22). When C is a parent, we use (24) to shift the incoming ray of C to the centres of its children. When C is childless, we use (21) to compute the potentials due to far cells at the collocation points $(\boldsymbol{x}, t_{\alpha'})(\alpha' = \alpha + 1, \cdots, N_t)$ in C.

   As regrads the third argument $\boldsymbol{k}$ of the incoming and outgoing rays, which represents a vector on the unit sphere, we use $(2K^{(l)}+1)(K^{(l)}+1)$ samples in the level l. The number $K^{(l)}$ is taken in a way that it satisfies $K^{(l)} = 2K^{(l+1)}$.

4. Update
   Increment the time step number by 1.

4 Remarks on PWTD Algorithm

4.1 Detail of the Downward Pass

The PWTD algorithm is quite complicated. Particularly complicated is the downward pass which is considered to be worth the efforts to elaborate.

We first note that the layer potentials in a cell C produced by cells far from C can be computed in the zth time interval as follows:

$$-\frac{1}{8\pi^2} \int_{S_k} [k_i \delta(t - (\boldsymbol{x} - \boldsymbol{o}) \cdot \boldsymbol{k}/c_1) * \mathcal{L}^z(\boldsymbol{o}, t, \boldsymbol{k})$$
$$+ e_{pik} k_p \delta(t - (\boldsymbol{x} - \boldsymbol{o}) \cdot \boldsymbol{k}/c_2) * \mathcal{L}_k^z(\boldsymbol{o}, t, \boldsymbol{k})] \, dS_k \qquad (30)$$

where $\boldsymbol{x}$ is a point in C and $\boldsymbol{o}$ is the centroid of C. In this expression, $\mathcal{L}$ is defined by

$$\mathcal{L}^z_{(k)}(t) = \sum_{\zeta=1}^{z-1} I^\zeta_{(k)}(t) \qquad t \in \left[(z-1)M\Delta t - \frac{\sqrt{3}L}{2c}, zM\Delta t + \frac{\sqrt{3}L}{2c} \right] \tag{31}$$

and will be called the 'accumulated incominig ray'.

Note that the accumulated incoming ray is defined in an interval for t obtained by adding the maximum time shift in (30) (i.e. $\frac{\sqrt{3}L}{2c}$) before and after the zth time interval. As a matter of fact, we need the accumulated incominig ray for

$$t \in \left[t_1 - \frac{\sqrt{3}L}{2c}, t_2 + \frac{\sqrt{3}L}{2c} \right] \tag{32}$$

in order to compute potentials for $t \in (t_1, t_2)$ everywhere in a cell C having the side length of L, where c is either c_1 or c_2. Also, note that rays of any type defined at $\boldsymbol{x}$ for $t \in (t_1, t_2)$ will shift into another at $\boldsymbol{y}$ for

$$t \in \left[t_1 - \frac{|\boldsymbol{x} - \boldsymbol{y}|}{c}, t_2 + \frac{|\boldsymbol{x} - \boldsymbol{y}|}{c} \right]. \tag{33}$$

We now design the algorithm in a way that the calculation of the accumulated incominig ray $\mathcal{L}^{z^{(l)}+1}_{(k)}$ is completed at the end of the $z^{(l)}$th time interval in C. To this end we consider what information will have to be passed to the child cell and what information will have to be stored before the next $((z^{(l)}+1)$th) time interval. We notice that the end point of the $z^{(l)}$th time interval at the level l is also an end point of the $z'^{(l+1)}$th time interval for a certain $z'^{(l+1)}$ at the level $(l+1)$. Also, the mid point of the $(z^{(l)}+1)$th time interval at the level l is the end point of the $(z'^{(l+1)}+1)$th time interval at the level $(l+1)$. Therefore, if a parent C shifts its accumulated incominig ray $\mathcal{L}^{z^{(l)}+1}_{(k)}$ to the children at the end of $z^{(l)}$th time interval, this will feed the children with the incoming rays required to compute $\mathcal{L}^{z'^{(l+1)}+1}_{(k)}$ and $\mathcal{L}^{z'^{(l+1)}+2}_{(k)}$. The part of the accumulated incominig ray of C shifted to the children is then cleared and the rest of the accumulated incominig ray which is required for the next and later time intervals is stored. This is the outline of the downward pass.

We next describe further details of the algorithm. Let the current time be $t_\alpha = z^{(l)}M^{(l)}\Delta t$. We first compute the incoming rays from the interaction list of C. Namely, for a cell C' not in the neighbourhood of C but in the neighbourhood of C's parent, we use (22) to convert the outgoing rays of C' to the incoming rays for C. This process produces the incoming rays in the interval given by (see (32) and (33))

$$\Bigg(((z^{(l)}-1)M^{(l)}-M_*^{(l)})\Delta t - \frac{(2\beta+1)\sqrt{3}L^{(l)}}{c},$$

$$(z^{(l)}M^{(l)}+M_*^{(l)})\Delta t + \frac{(2\beta+1)\sqrt{3}L^{(l)}}{c}\Bigg]. \tag{34}$$

where

$$M_*^{(l)} = \frac{\sqrt{3}L^{(l)}}{2c\Delta t} + p_t$$

and c stands for one of $c_{1,2}$. Since one needs the accumulated incominig ray only for the time in the $(z^{(l)}+1)$th time interval and after, we need to store the accumulated incominig ray only in the time interval given by (see (32))

$$\left[(\alpha+1)\Delta t - \frac{\sqrt{3}L^{(l)}}{2c}, (\alpha+M_*^{(l)})\Delta t + \frac{(2\beta+1)\sqrt{3}L^{(l)}}{c}\right]. \tag{35}$$

Hence, we compute and add only the part of the incoming ray in the time interval of (35) to the accumulated incominig ray for C. The part of the stored accumulated incominig ray in the interval given by

$$\left[(\alpha+1)\Delta t - \frac{\sqrt{3}L^{(l)}}{2c}, (\alpha+M^{(l)})\Delta t + \frac{\sqrt{3}L^{(l)}}{2c}\right] \tag{36}$$

will be needed in the $(z^{(l)}+1)$th time interval. Indeed, this part will be needed in order to evaluate potentials in the next time interval if C is a leaf cell (see (32)) or will have to be passed to the children because of the requirement of the algorithm. In the latter case, i.e. when the cell C under consideration has a child, we shift only the part of the accumulated incominig ray in (36) with (24) and pass it to the child, and then clear the part of the accumulated incominig ray for C in (36). Once this is done, the first part of the length $M^{(l)}\Delta t$ of the accumulated incoming ray of C stored for the interval in (35) will not be referred to later and, hence, need not be stored any longer. The rest of the accumulated incominig ray, however, is kept and the incominig ray to be computed in the next time interval will be added to this.

When the cell C is a leaf, we compute the potentials from the accumulated incominig ray in the interval in (36) using (30). Since the first part of the accumulated incoming ray of the length $M^{(l)}\Delta t$ in the interval in (35) will never be referred to in later time intervals we will not store it. The rest of the accumulated incominig ray is kept and to this we shall add the incominig ray to be computed in the next time interval.

The above operation is carried out downward from level 2 to level l_{max}, but only at the relevant levels, where the current time step is a multiple of $M^{(l)}$.

Notice that the process of shifting the accumulated incominig ray in (36) in the level l to the children in the level $(l+1)$ uses (24), which gives the time shift of $\frac{\sqrt{3}L^{(l)}}{4c}$ at the maximum. Hence the length of the accumulated incominig ray to be sampled at the level $(l+1)$ by the children when a parent shifts its accumulated incominig ray to its children is given by

$$\left[(\alpha+1)\Delta t - \frac{\sqrt{3}L^{(l+1)}}{2c}, (\alpha+M^{(l)})\Delta t + \frac{3\sqrt{3}L^{(l)}}{4c}\right] \tag{37}$$

Since $M_*^{(l+1)} = \frac{\sqrt{3}L^{(l+1)}}{2c\Delta t} + p_t$, $L^{(l)} = 2L^{(l+1)}$ and $M^{(l)} \leq \frac{\beta L^{(l)}}{c\Delta t}$, we see that the interval in (37) is included in the interval in (35) (with $l = l+1$). Hence the amount of memory required for the accumulated incoming ray is determined by (35).

We remark that the length of the interval given by (35) does not depend on the number of the time steps N_t. We also remark that the amount of memory required for the accumulated incoming ray is always the same since the first $M^{(l)}$ time steps of the accumulated incoming ray are removed and the same number of steps are added every time one updates the accumulated incoming ray, i.e. every $M^{(l)}$ time steps.

See Fig. 2 which summarises the detail of the downward pass.

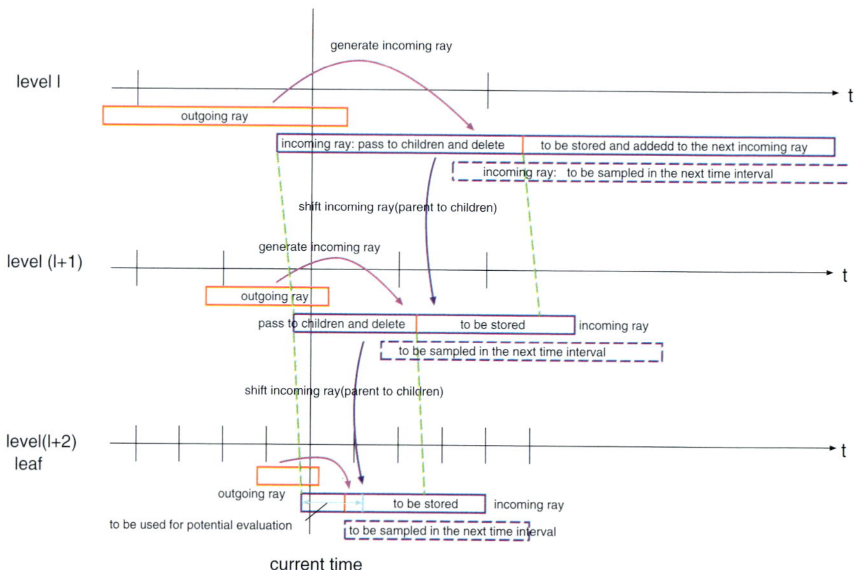

Fig. 2. Detail of the downward pass.

4.2 Anisotropy

The PWTD approach is somewhat exceptional in the fast methods of the FMM type in that all the details of the formulation can be derived from the Fourier analysis. This suggests that it may be possible to derive PWTD formulations for general anisotropic elastodynamics. In the rest of this section we shall show that this is true.

We first derive the plane wave expansion formula for the fundamental solution $\boldsymbol{\Gamma}$, which, in the case of general anisotropy, can be written as

$$\Gamma_{ij}(\boldsymbol{x},t) = \mathcal{F}^{-1}(A^{-1})_{ij}, \tag{38}$$

where

$$A_{ij} = C_{ikjl}\xi_k\xi_l - \rho\omega^2\delta_{ij}, \tag{39}$$

δ_{ij} is Kronecker's delta and ξ_i and ω are the Fourier parameters for the spatial variable x_i and time t, with which the Fourier transform and its inverse transform are written as:

$$\mathcal{F}f = \int\int e^{-i\xi_i x_i + i\omega t} f(x,t)dxdt$$

$$\mathcal{F}^{-1}\hat{f} = \frac{1}{(2\pi)^4}\int\int e^{i\xi_i x_i - i\omega t}\hat{f}(\xi,\omega)d\xi d\omega.$$

The elasticity tensor C_{ijkl} now has 21 independent components.

We introduce the following sextic equation for ω

$$\det(C_{ikjl}\xi_k\xi_l - \rho\omega^2\delta_{ij}) = 0$$

for a fixed ξ. Obviously, the solution to this equation can be written as

$$\pm c_p(\boldsymbol{k})|\xi|, \quad (p = 1,2,3) \tag{40}$$

where $c_p^2(\boldsymbol{k})$ $(p = 1,2,3)$ are the eigenvalues of the matrix

$$\frac{C_{ikjl}}{\rho}k_k k_l,$$

and $\boldsymbol{k} = \xi/|\xi|$. We shall assume that the eigenvalues are distinct and denote the corresponding eigenvector by v_i^p.

As in the isotropic case, we obtain

$$\Gamma_{ij}(\boldsymbol{x},t) - \Gamma'_{ij}(\boldsymbol{x},t) = \frac{1}{2\pi}\mathcal{F}^{-1}_{\xi\to x}\int_p \hat{\Gamma}_{ij}e^{-i\omega t}d\omega$$

where Γ'_{ij} is the 'ghost' and p is the contour in Fig. 1. By evaluating this integral, one obtains the following:

$$\Gamma_{ij}(\boldsymbol{x},t) - \Gamma'_{ij}(\boldsymbol{x},t)$$

$$= -\frac{\partial_t}{8\pi^2\rho} \int_{S_k} \sum_{p=1}^{3} \left[\frac{v_i^p(k)v_j^p(k)}{\rho c_p^3(k)} \delta(t - \boldsymbol{x} \cdot \boldsymbol{k}/c_p(k)) \right] dS_k. \tag{41}$$

This formula gives the plane wave expansion of the fundamental solution.

The rest of the analysis goes parallel to the isotropic case. Indeed, one has the local expansion formula

$$\int_{S_0} \left(T_{ij}(\boldsymbol{x},\boldsymbol{y},t) * u_j^z(\boldsymbol{y},t) - \Gamma_{ij}(\boldsymbol{x}-\boldsymbol{y},t) * t_j^z(\boldsymbol{y},t) \right) dS_y$$

$$= -\frac{1}{8\pi^2} \sum_p \int_{S_k} v_i^p(\boldsymbol{k})\delta(t - (\boldsymbol{x}-\boldsymbol{o}) \cdot \boldsymbol{k}/c_p(\boldsymbol{k})) * I^{zp}(\boldsymbol{o},t,\boldsymbol{k})dS_k.$$

where the incoming ray $I^{zp}(\boldsymbol{o},t,\boldsymbol{k})$ is related to the outgoing ray $O^{zp}(\boldsymbol{s},t,\boldsymbol{k})$ by

$$I^{zp}(\boldsymbol{o},t,\boldsymbol{k}) = \partial_t\delta(t - (\boldsymbol{o}-\boldsymbol{s}) \cdot \boldsymbol{k}/c_p(\boldsymbol{k})) * O^{zp}(\boldsymbol{s},t,\boldsymbol{k})$$

while the outgoing ray is defined by

$$O^{zp}(\boldsymbol{s},t,\boldsymbol{k}) = \frac{1}{\rho} \int_{S_0} \left(\frac{C_{jsqr}n_q(\boldsymbol{y})k_s v_j^p(\boldsymbol{k})}{c_p^4} \dot{u}_r^z(\boldsymbol{y},t) - \frac{v_j^p(\boldsymbol{k})t_j^z(\boldsymbol{y},t)}{c_p^3(\boldsymbol{k})} \right)$$
$$* \delta(t - (\boldsymbol{s}-\boldsymbol{y}) \cdot \boldsymbol{k}/c_p(\boldsymbol{k}))dS_y.$$

One can shift the origin of the outgoing and incoming rays via

$$O^{zp}(\boldsymbol{s},t,\boldsymbol{k}) = \delta(t - (\boldsymbol{s}-\boldsymbol{s}') \cdot \boldsymbol{k}/c_p(\boldsymbol{k})) * O^{zp}(\boldsymbol{s}',t,\boldsymbol{k}),$$
$$I^{zp}(\boldsymbol{o},t,\boldsymbol{k}) = \delta(t - (\boldsymbol{o}-\boldsymbol{o}') \cdot \boldsymbol{k}/c_p(\boldsymbol{k})) * I^{zp}(\boldsymbol{s}',t,\boldsymbol{k}).$$

So far, no numerical attempts have been made with the PWTD approach presented above.

5 Parallelisation

In this section we shall consider parallelisation of the fast BIEM discussed so far on shared memory computers.

Use of parallel algorithms has become quite common thanks to the success of the Beowulf PC clusters. MPI is the most popular API for parallelisation on such platforms. In the area of high performance computing, further enhancement of parallel performance is pursued with the use of clusters of SMPs. In such platforms, each of the nodes is a shared memory parallel computer, in which all the CPUs share a large memory space. For shared memory parallel computers, OpenMP is the de facto standard for the thread-parallelism. Parallelisation with OpenMP is remarkably simple and one can possibly achieve equally high parallel performance as one would with MPI. Unfortunately, shared memory parallel computers have been accessible only to limited users so far. However, the situation is changing because of the recent development of multi-core processors. In view of this, we shall discuss parallelisation of the PWTD approach with OpenMP and MPI-OpenMP hybridisation.

5.1 Parallelisation with OpenMP

An advantage of OpenMP is that one can easily parallelise only the 'hot spots' of the code. In this paper, we shall parallelise only the following parts:

- direct computation of the coefficient matrix
- upward pass
- downward pass

For the solution of the linear equation we use non-parallelised GMRES, although its parallelisation is easy. This is because the solution of the linear equation requires only a marginal proportion of the total CPU time thanks to the sparsity and good conditioning of the coefficient matrix.

Direct Computation of the Coefficient Matrix

The part of the code related to the direct evaluation of the components of the coefficient matrix (influence coefficients) has been parallelised as shown below in a Fortran like pseudocode:

```
do level=lowest level, 2, -1
!$OMP PARALLEL DO
  do cells of the relevant level
    if (the relevant cell is a leaf)  then
      compute the influence coefficients from the cells
      near the relevant one.
    endif
  enddo
!$OMP END PARALLEL DO
enddo
```

Upward Pass

The upward pass has been parallelised as follows:

```
do level=lowest level, 2, -1
  if(current time step number is a multiple of M)
!$OMP PARALLEL DO
```

```
  do cells of the relevant level
    if (the relevant cell is a leaf)  then
      generate outgoing rays from the boundary densities
      within the leaf.
    else
      shift outgoing rays from the children
    endif
  enddo
!$OMP END PARALLEL DO
  endif
enddo
```

Note that one may pass outgoing rays from children to the parent in a do loop either for the child cells or that for the parents. In the present inplementation we use the latter since one may otherwise have conflicting write to a memory corresponding to the outgoing ray of the same parent from different children belonging to different threads.

Downward Pass

Still easier is the parallelisation of the downward path. Indeed, the only change one has to make in the sequential code is to add OpenMP directives as indicated in the following pseudocode:

```
do level=2, lowest level
  if(if the current time step number is a multiple of M)
!$OMP PARALLEL DO
    do cells of the relevant level

      compute incoming rays from the outgoing rays of cells
      in the interaction list

      if(the relevant cell is a leaf) then
        compute potentials at collocation points in the cell
        from the accumulated incoming ray
      else
        shift accumulated incoming rays to the children
      endif
    enddo
!$OMP END PARALLEL DO
  endif
enddo
```

5.2 MPI-OpenMP Hybridisation

In order to solve still larger problems, one may want to use more memory than could be accessed from one shared memory machine. In such cases, MPI-

OpenMP hybridisation will be useful. The basic idea is to use the domain decomposition and assign each of subdomains to an MPI process, which is further thread-parallelised with OpenMP. A relatively simple MPI code will be sufficient to this end, as we shall see.

The detail of our hybrid implementation goes as follows.

In the process-parallelisation via MPI, we decompose the boundary S into subdomains and assign each of them to a MPI process. The domain decomposition is done according to the cell division at level 2. Since one has 64 level 2 cells, one process will cover $64/m$ cells where m is the number of processes. Each process will store only relevant incoming and outgoing rays, and influence coefficients for collocation points belonging to its own domain. As for the iterative solver we use a sequential version of GMRES.

For the downward pass, it may happen that some of the cells in the interaction list of a cell belonging to a certain process may not belong to the same process. In such cases one will have to use inter-process communication. In order to reduce the amount of communication, we list up the cases where inter-process communications are necessary, and do all the required data transfers at the beginning of the procedures for each level in the downward pass. In this manner one obtains the RHS of (15) distributed among processes. One finally gathers them to complete the computation of the RHS of (15).

Since the influence coefficients are computed only for the collocation points belonging to the process, the coefficient matrix on the RHS of (15) are divided row-wise and stored separately to the memory accessible to a process. Therefore, one will have to use inter-process communication to obtain a matrix-vector product when GMRES requires one.

We note that the process-parallelisation with MPI discussed above can be used on its own, but a more sophisticated coding will be preferable if one wants to solve very large problems with MPI alone (See [7] for such an attempt). However, one can use the hybrid of the present MPI implementation with the thread-parallelisation via OpenMP in the solution of very large problmes, as we shall see. We now sketch the MPI-OpenMP hybridisation referring to the OpenMP pseudocode in section 5.1. As has been stated, each process computes and stores influence coefficients for collocation points in its domain, and outgoing and incoming rays related to cells in its domain. These computations within a process are further thread-parallelised with OpenMP. Therefore, the loops for cells in the pseudocode in section 5.1 have to be changed to those for cells belonging to the process.

6 Numerical Calculation

We now examine the performance of the parallelisation via OpenMP. We also present examples of large scale problems solved with MPI-OpenMP hybrid parallelisation.

6.1 Computer

We have used Fujitsu HPC2500 of the Academic Center for Computing and Media Studies of Kyoto University for all the computations. This machine consists of 11 computational nodes, connected by optical interconnects. One node has 128 CPUs (SPARC64V(1.56GHz)), and a shared memory of 512GB.

6.2 Numerical Examples

We solve an interior problem for a parallelepiped and an exterior problem for a system of spherical cavities in order to test the performance of the OpenMP-parallelisation. In the former problem we also examine the accuracy of the solution. In the latter we test the performance of the process-parallelisation with MPI and that of the MPI-OpenMP hybrid parallelisation.

Interior Problem for a Parallelepiped

We consider an interior problem for a parallelepiped whose space diagonal connects two points given by $(0, 0, 0)$ and $(1.40, 0.68, 0.80)$. We have discretised

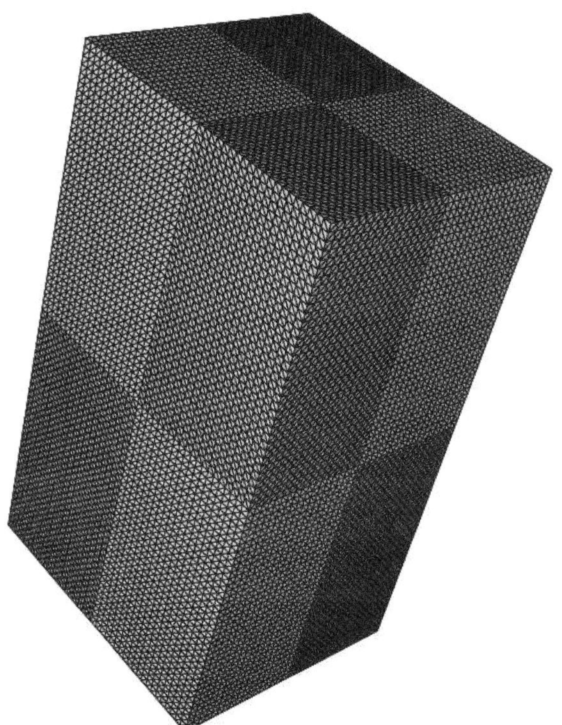

Fig. 3. Boundary elements.

the boundary with 26,160 triangular elements (78,480 DOF), as shown in Fig. 3. Also an appropreate non-dimentionalisation has been made, which gives $\rho = 1$, $c_1 = 1$ and $c_2 = 1/\sqrt{2}$. As the boundary condition we give the traction associated with the following displacement field, which satisfies the governing equation of elastodynamics:

$$\boldsymbol{u}(\boldsymbol{x}, t) = \boldsymbol{d}\left[1 - \mathrm{Cos}\frac{2\pi}{\varLambda}\left(t - \frac{\boldsymbol{d}\cdot\boldsymbol{x}}{c_1}\right)\right] \tag{42}$$

where $\boldsymbol{d}$ stands for the direction of the wave propagation which is set equal to $\boldsymbol{d} = (0,0,1)$ and $\varLambda = 1.0$ is a constant. The definition of the 'Cos' function is as follows:

$$\mathrm{Cos}\, x = \begin{cases} \cos x & 0 \leq x \leq 2\pi \\ 1 & x < 0, x > 2\pi \end{cases} \tag{43}$$

The time increment $\varDelta t$ is 0.01 and the number of time steps is $N_t = 200$. Also, the maximum number of elements in a leaf is 150 and the number of incoming and outgoing ray sampling points on the unit sphere in level 2 is chosen to be $K^{(2)} = 56$.

In Fig. 4 we have shown the relation between the thread numbers (number of CPUs) and the computational time in the OpenMP thread-parallelisation. Fig. 5 plots the speed up defined by (computational time when one has just 1 thread)/(computational time with n threads) vs the number of threads. Ideally, the speed up is equal to the number of threads, which is also plotted in the same figure with a dotted line.

The error of the numerical solution relative to the exact one in (42) has been obtained at each time step, whose maximum was 3.28%.

As is seen the computational time is minimum with 64 threads while the computation with 128 threads took slightly longer than the 64 thread case. One of the reasons for this saturation is the following: From the point of view of the algorithm, level 2 is the most costly one because both the number of sampling points for the incoming and outgoing rays and the lengths of these rays are the maximum at this level. However, the maximim number of level 2 cells is 64, and the use of more than 64 threads for this level is not considered effective. This explains the sudden deterioration of the efficiency beyond 64 threads. From the point of view of hardware, one may mention the conflict of memory access as a possible reason of the saturation of the efficiency beyond 64 threads.

Spherical Cavities

We next consider the scattering of a plane incident wave by a three dimensional array of spherical cavities in an infinite space.

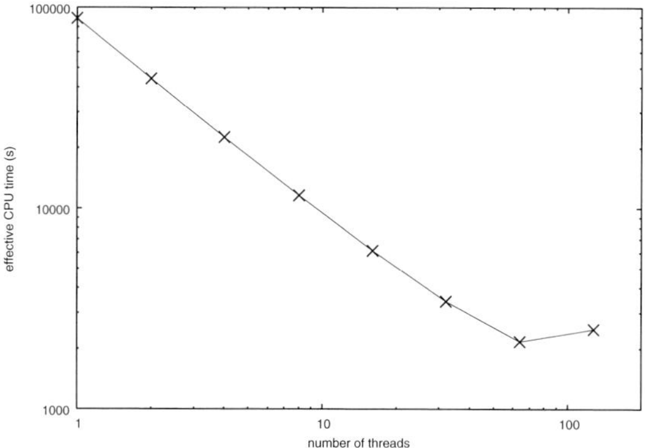

Fig. 4. Computational time for the OpenMP parallelised code (parallelepiped).

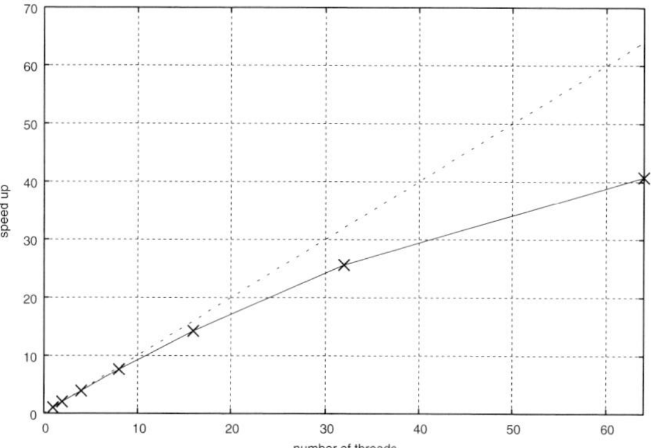

Fig. 5. Speedup with OpenMP (parallelepiped).

In the first example we have a $4 \times 4 \times 4$ array of spherical cavities in x_1, x_2 and x_3 directions. The distance between the neighbouring cavities are $2a = 0.15$ and the radius of each sphere is $0.9a = 0.0675$. Each of the spheres are subdivided with 320 triangular boundary elements, and the total number of elements is 20,480 (61,440 DOF). See Fig. 6.

The material constants are chosen to be the same as in the parallelepiped case in the previous subsection. The time increment is chosen as $\Delta t = 0.01$, and the number of the time steps is $N_t = 200$. The maximum number of elements in a leaf is 100 and the number of sampling points of rays on the unit sphere for the level 2 is $K^{(2)} = 24$.

The incident wave $\boldsymbol{u}^I$ is given by

$$
\boldsymbol{u}^I(\boldsymbol{x}, t) = \boldsymbol{d} \left[1 - \mathrm{Cos}\frac{2\pi}{\Lambda} \left(t - \frac{\boldsymbol{d} \cdot \boldsymbol{x}}{c_1} \right) \right]
\tag{44}
$$

with the incident direction of $\boldsymbol{d} = (0, 0, 1)$ and $\Lambda = 4a = 0.3$.

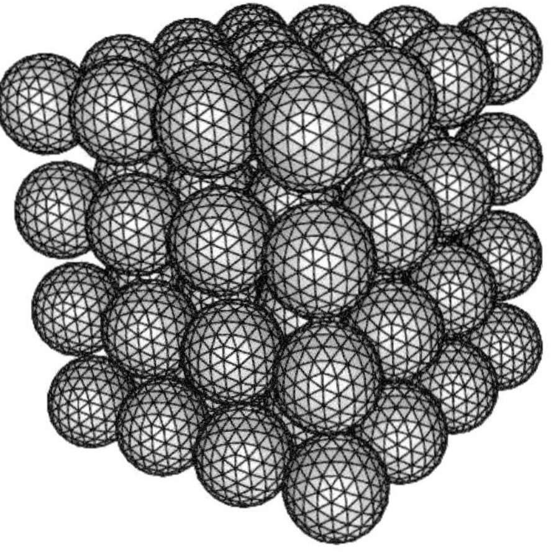

Fig. 6. Boundary elements for spheres.

Fig. 7 shows the relation between the number of threads and the computational time with the OpenMP thread-parallelisation. We also plot the computational time for the MPI process-parallelised code vs the number of processes for the purpose of comparison. The increase of computational time with 128 thread seen in this figure can be explained similarly as in the previous section 6.2. This figure also shows that the OpenMP version of parallel code is slightly faster than the MPI version. Of course, the computational time depends on the detail of the coding and it may not be very appropreate to generalise this timing results. Fig. 8 plots the speedup vs the number of threads in the OpenMP thread-parallelised code. Table 1 gives the computational time for the MPI-OpenMP hybrid parallelisation. This table shows that the case with 1×64 MPI-OpenMP parallelisation is the fastest, followed by the 8×8 case.

From these results we can say that OpenMP provides a much less painful way to achieve almost the same level of the parallel efficiency as that with MPI. We can thus conclude that the parallelisation of FMM with OpenMP is practical.

Finally, we remark that we have solved the same problem with $K^{(2)} = 42$ and compared the result with the $K^{(2)} = 24$ analysis. The results were close

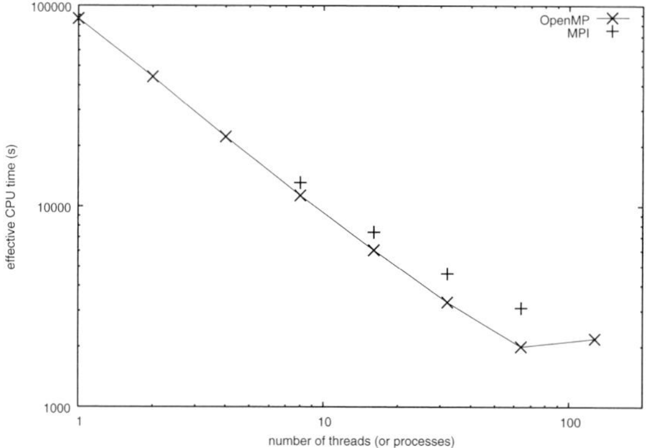

Fig. 7. Computational time with OpenMP (spherical cavities).

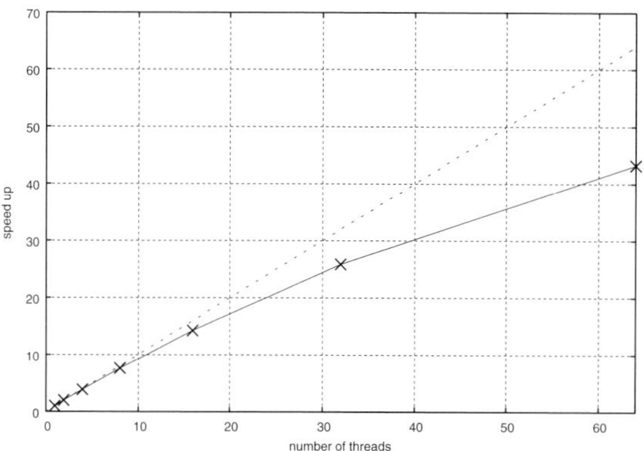

Fig. 8. Speedup with OpenMP (spherical cavities).

Table 1. Computational time with MPI-OpenMP hybrid parallelisation (spherical cavities).

number of processes	number of threads	computational time (s)
1	64	2053
2	32	2379
4	16	2356
8	8	2301
16	4	2603
32	2	2956
64	1	2947

to each other, the average difference of the these solutions (displacement on the boundary) being less than 1%.

Large Scale Problem

We finally present an analysis with $12 \times 12 \times 8$ cavities in $x_1 \times x_2 \times x_3$ directions. The number of boundary elements is 368,640 and the total DOF is 1,105,290.

The constants and parameters are chosen as follows: $\rho = 1$, $c_1 = 1$ and $c_2 = 1/\sqrt{2}$. Also, the time increment is $\Delta t = 0.01$, the number of time steps is $N_t = 200$, the maximum number of elements in a leaf is 100 and $K^{(2)} = 80$. In this problem we have used a hybrid parallelisation with 8 processes and 32 threads, thus using 256 CPUs in total. An analysis of this size is possible only with the hybrid parallelisation.

Fig. 9 shows the obtained boundary displacements for $t = 200\Delta t$. The accuracy of the analysis, however, is unknown. The reqired computational time was 10 hours 47 minutes. The memory used was about 18.8GB per process and about 150.4GB in total. Since HPC2500 has 512GB of memory per node, it is possible to solve still larger problems, at least theoretically.

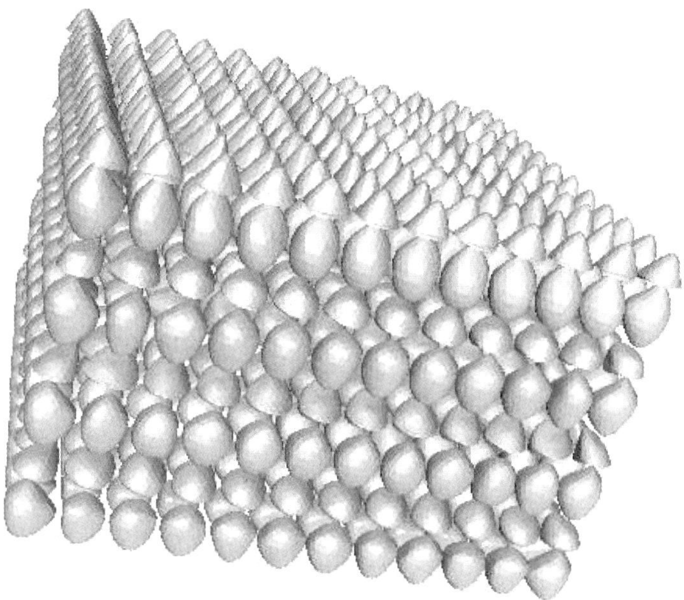

Fig. 9. Deformation of $12 \times 12 \times 8$ cavities. $t = 200\Delta t$.

7 Concluding Remarks

- In this paper, we have seen that the PWTD algorithm can be applied successfully to time domain elastodynamic problems in 3D.
- We have shown that it is easy to parallelise the fast BIEM using PWTD algorithm for 3 dimensional elastodynamics in time domain with OpenMP and shared memory parallel computers. We have also seen that the parallel efficiency of the code is satisfactory.
- MPI-OpenMP hybrid implementation is seen to be effective in large scale problems. With this implementation, we could solve three dimensional time domain elastodynamic problems with more than 1 million spatial degrees of freedom. It is possible to solve still larger problems considering the computational resources available.
- With the techniques presented in this paper one can now solve really large problem in time domain. However, we have to remember that the stability of the scheme is still an open issue. In this connection, we point out that Ergin et al. [2] proposed the use of the combination of the ordinary and traction boundary integral equations (i.e. Burton-Miller type integral equations) in order to avoid corrupse of the time domain BIEM for exterior problems for the wave equation. Use of similar approaches in elastodynamics may be an interesting research subject in the future.
- The formulation for the anisotropic case suggests that one may use only 3 types of rays rather than 4 as in the present implementation even in the isotropic case. So far, no attempts of implementing the 3 ray formulation in the isotropic case have been made. A possible benefit of the present 4 ray formulation may be found in the easier implementation for the interpolation and anterpolation [4] of various rays.

References

1. A. A. Ergin, B. Shanker, E. Michielssen: Fast evaluation of three-dimensional transient wave fields using diagonal translation operators. J. Comput. Phys. 146 (1988) 157–180.
2. A. A. Ergin, B. Shanker, E. Michielssen: Analysis of transient wave scattering from rigid bodies using a Burton-Miller approach. J. Acoust. Soc. Am. 106 (1999) 2396–2404.
3. M. Lu, J. Wang, A. A. Ergin, E. Michielssen: Fast evaluation of two dimensional transient wave fields. J. Comput. Phys. 158 (2000) 161–185.
4. N. Nishimura: Fast multipole accelerated boundary integral equation methods. Appl. Mech. Rev. 55 (2002) 299–324.
5. T. Takahashi, N. Nishimura, S. Kobayashi: Fast boundary integral equation method for elastodynamic problems in 2D in time domain. Trans JSME A67-661 (2001) 1409–1416 (in Japanese).

6. T. Takahashi, N. Nishimura, S. Kobayashi: A fast BIEM for three-dimensional elastodynamics in time domain, Eng. Anal. Boundary Elements 27 (2003) 491–506. (see vol. 28 (2004) 165–180 for Erratum to "A fast BIEM for three-dimensional elastodynamics in time domain")
7. L. Ying, G. Biros, D. Zorin, H. Langston: A new parallel kernel-independent fast multipole method, Proc. 16th ACM/IEEE Conference on Supercomputing, 2003.
   http://mrl.nyu.edu/~harper/kifmm3d/documentation/papers/sc03-fmm.pdf
8. http://www.openmp.org/
9. http://www-unix.mcs.anl.gov/mpi/

FM–BEM and Topological Derivative Applied to Acoustic Inverse Scattering

Marc Bonnet and Nicolas Nemitz

Solid Mechanics Laboratory (UMR CNRS 7649), Ecole Polytechnique,
91128 Palaiseau, France
{bonnet,nemitz}@lms.polytechnique.fr

Summary. This study is set in the framework of inverse scattering of scalar (e.g. acoustic) waves. A qualitative probing technique based on the distribution of topological sensitivity of the cost functional associated with the inverse problem with respect to the nucleation of an infinitesimally-small hard obstacle is formulated. The sensitivity distribution is expressed as a bilinear formula involving the free field and an adjoint field associated with the cost function. These fields are computed by means of a boundary element formulation accelerated by the Fast Multipole method. A computationally fast approach for performing a global preliminary search based on the available overspecified boundary data is thus defined. Its usefulness is demonstrated through results of numerical experiments on the qualitative identification of a hard obstacle in a bounded acoustic domain, for configurations featuring $O(10^5)$ nodal unknowns and $O(10^6)$ sampling points.

1 Introduction

Defect identificaton problems are often solved by minimization of a cost function featuring the experimental data and (if available) prior information. Such cost functions are non-convex and exhibit local minima. Despite that fact, traditional iterative minimization or equation-solving methods are usually preferred to global search techniques such as evolutionary algorithms due to the prohibitive computational cost of solving large numbers of forward wave scattering problems. To perform optimally, gradient-based iterative algorithms are used in conjunction with shape sensitivity techniques see e.g. [15, 18, 20].

Still, the stand-alone use of gradient-type minimization for such purposes is not satisfactory for its success is strongly dependent on a reliable prior information about the geometry of the hidden object. This has prompted the development of 'sampling' or 'probe' non-iterative methods, which may be used in isolation or as a preliminary step for choosing adequate initial guesses to be used in subsequent standard optimization schemes. Such methods are surveyed in a recent review article [21] and include the linear sampling method [3, 17], not pursued here, and the application of topological sensitivity, which is the subject of this article. The concept of topological sensitivity consists in quantifying the sensitivity of a cost function with respect to the creation of a scatterer $B_\varepsilon(\boldsymbol{x}_\mathrm{s})$ of small characteristic radius ε and given location $\boldsymbol{x}_\mathrm{s}$, as a function of $\boldsymbol{x}_\mathrm{s}$. It appeared first in [6, 22] in connection with topological optimization of mechanical structures, allowing to define algorithms where "excess" material is iteratively removed until a satisfactory shape and topology is reached [9]. More recently, other investigations have studied the topological sensitivity as a sampling tool for inverse scattering problems, in the context of identification of cavities in 3D semi-infinite and infinite elastic media [13] and in elastic 3D bounded bodies [2], and of elastic inclusions [14] (see also [8] for 2D elastostatics and [7] for 2D linear acoustics).

The distribution of topological sensitivity can be expressed in terms of a bilinear formula involving the free field and an adjoint field associated with the cost function. However, the computational cost of solving the forward and adjoint problems and evaluating the topological sensitivity distribution on a fine sampling grid increases rapidly with the non-dimensional wavenumber. The purpose of this article is to propose the topological sensitivity field computed by means of the Fast Multipole BEM (FM-BEM) [11, 12, 19] as the basis of a computationally fast tool for probing acoustic media for hidden hard obstacles on the basis of overdetermined boundary data, within the model framework of forward scattering problems governed by the scalar Helmholtz equation. To that end, the FM-BEM is in particular applied to evaluate in a fast way the integral representation formulae expressing the free and adjoint fields at a large number of sampling points points inside the medium.

This article is organized as follows. After some preliminaries concerning the forward and inverse problems of interest (Section 2), the concept of topological sensitivity is presented in Section 3. The FM-BEM approach for the scalar Helmholtz equation is then summarized in Section 4. Finally, results of numerical experiments on qualitative scatterer identification using computed distributions of topological sensitivity are presented in Section 5, for configurations featuring $O(10^5)$ nodal BE unknowns and $O(10^6)$ sampling points.

2 Forward and Inverse Problems

This article is concerned with the identification of rigid obstacles embedded in acoustic media. The generic acoustic scattering problem of interest is defined as follows. Let Ω denote a three-dimensional open domain, either bounded or unbounded, with a sufficiently regular boundary S and filled with an acoustic medium characterized by wave velocity c and mass density ρ; this configuration will be referred to as the reference (i.e. obstacle-free) medium. Let $B^\star$ denote a rigid scatterer (or a set thereof) bounded by the closed surface $\Gamma^\star$, so that $\Omega^\star = \Omega \backslash \bar{B}^\star$ is the acoustic region surrounding the scatterer. Steady-state excitations on S with angular frequency ω generate an acoustic pressure field

$u^\star$ in the acoustic domain $\Omega^\star$, governed by the following set of field equations and boundary conditions (collectively denoted by $\mathcal{P}(B^\star)$ for later reference):

$$\mathcal{P}(B^\star): \qquad \begin{aligned} (\Delta + k^2)u^\star &= 0 && (\text{in } \Omega^\star), \\ p[u^\star] &= p^{\mathrm{D}} && (\text{on } S), \\ p[u^\star] &= 0 && (\text{on } \Gamma^\star), \end{aligned} \qquad (1)$$

where $k = \omega/c$ is the wavenumber, $w \to p[w] \equiv w_{,n} = \boldsymbol{\nabla} w \cdot \boldsymbol{n}$ denotes the normal derivative operator, $\boldsymbol{n}$ is the normal on $S \cup \Gamma^\star$ outward to $\Omega^\star$, and p^{D} is the prescribed Neumann data over S (other types of boundary conditions may be considered as well). It is assumed that ω is not an eigenfrequency of any of the boundary-value problems arising in the ensuing developments.

In the inverse scattering problem of interest, an unknown obstacle B^{true}, of boundary Γ^{true}, is to be identified. The corresponding exact acoustic field u^{true} is then governed by problem $\mathcal{P}(B^{\mathrm{true}})$. With reference to problem $\mathcal{P}(B^{\mathrm{true}})$, supplementary information is needed for the identification of B^{true}. Here, measured values u^{obs} of acoustic pressure are assumed to be available over the measurement surface $S^{\mathrm{obs}} \subset S$. Ideally (i.e. assuming that the physics is exactly described by the chosen linear acoustics setting and that no measurement errors are present), u^{obs} is the trace of u^{true} on S^{obs}. The identification of B^{true} may then be formulated in terms of the minimization of a cost function. Generic cost function of format

$$\mathcal{J}(\Omega^\star) = \int_{S^{\mathrm{obs}}} \varphi\big(u_{\mathrm{R}}^\star(\boldsymbol{\xi}), u_{\mathrm{I}}^\star(\boldsymbol{\xi}), \boldsymbol{\xi} \big) \, \mathrm{d}\Gamma \qquad (2)$$

are considered, where $u^\star$ is the boundary trace of the solution to the forward problem $\mathcal{P}(B^\star)$ for an assumed obstacle configuration $B^\star$, the subscripts 'R' and 'I' being used to indicate the real and imaginary parts of a complex quantity (i.e. $w_{\mathrm{R}} = \mathrm{Re}(w)$ and $w_{\mathrm{I}} = \mathrm{Im}(w)$). For instance, the output least-squares cost function associated to measurement u^{obs} on S^{obs}, commonly used for such purposes, corresponds to

$$\varphi(w_{\mathrm{R}}, w_{\mathrm{I}}, \boldsymbol{\xi}) = \frac{1}{2}\big| w(\boldsymbol{\xi}) - u^{\mathrm{obs}}(\boldsymbol{\xi}) \big|^2, \qquad (3)$$

The minimization of such cost functions can be performed using many methods, all of which are iterative and need repeated evaluations of $\mathcal{J}(\Omega^\star)$. Traditional gradient-based minimization may converge within a moderate number of evaluations of $\mathcal{J}(\Omega^\star)$ if the trial surface $\Gamma^\star$ can be described in terms of a few geometrical parameters, but reach a local minimum which depends on the choice of initial guess. Global search techniques, e.g. evolutionary algorithms [16] or sampling methods based on the Metropolis algorithm [25], perform a global search (i.e. identify absolute and/or multiple minima), but at the cost of very large numbers of cost functions evaluations. In this article, the topological sensitivity is proposed as a tool for performing a qualitative global search at a computational cost which is far below that entailed by a true global optimization technique. The results may then (for example) be used as initial guesses in subsequent minimization-based inversion procedures.

3 Topological Sensitivity of the Cost Function

3.1 Notations

Let $B_\varepsilon(\boldsymbol{x}_\mathrm{s}) = \boldsymbol{x}_\mathrm{s} + \varepsilon\mathcal{B}$, where $\mathcal{B} \subset \mathbb{R}^3$ is a fixed bounded open set with boundary $\mathcal{S}$ and volume $|\mathcal{B}|$ containing the origin, define the region of space occupied by a hard obstacle of (small) size $\varepsilon > 0$ containing a fixed sampling point $\boldsymbol{x}_\mathrm{s}$. It is convenient to introduce the scaled position vector $\bar{\boldsymbol{\xi}}$ defined by

$$\boldsymbol{\xi} = \boldsymbol{x}_\mathrm{s} + \varepsilon\bar{\boldsymbol{\xi}} \qquad (\boldsymbol{\xi} \in B_\varepsilon, \bar{\boldsymbol{\xi}} \in \mathcal{B}) \tag{4}$$

In particular, this mapping recasts integrals over B_ε and Γ_ε into integrals over $\mathcal{B}$ and $\mathcal{S}$, respectively, and transforms the differential volume and area elements according to

$$\mathrm{d}V_\xi = \varepsilon^3 \, \mathrm{d}\bar{V}_{\bar{\xi}} \ (\boldsymbol{\xi} \in B_\varepsilon, \bar{\boldsymbol{\xi}} \in \mathcal{B}), \quad \mathrm{d}\Gamma_\xi = \varepsilon^2 \, \mathrm{d}\bar{\Gamma}_{\bar{\xi}} \ (\boldsymbol{\xi} \in \Gamma_\varepsilon, \bar{\boldsymbol{\xi}} \in \mathcal{S}) \tag{5}$$

Without loss of generality, $\boldsymbol{x}_\mathrm{s}$ can be chosen as the center of B_ε, i.e. such that

$$\int_{B_\varepsilon} (\boldsymbol{\xi} - \boldsymbol{x}_\mathrm{s}) \, \mathrm{d}V_\xi = \mathbf{0}, \ \text{ i.e. } \ \int_{\mathcal{B}} \bar{\boldsymbol{\xi}} \, \mathrm{d}\bar{V}_{\bar{\xi}} = \mathbf{0}. \tag{6}$$

Let $u^\star = u^\varepsilon(\boldsymbol{\xi}; \boldsymbol{x}_\mathrm{s})$ denote the solution to the scattering problem $\mathcal{P}\big(B_\varepsilon(\boldsymbol{x}_\mathrm{s})\big)$ defined by (1), where $\Omega^\star = \Omega_\varepsilon(\boldsymbol{x}_\mathrm{s}) = \Omega \setminus \overline{B_\varepsilon}(\boldsymbol{x}_\mathrm{s})$ and $\overline{B_\varepsilon}(\boldsymbol{x}_\mathrm{s})$ is the closure of $B_\varepsilon(\boldsymbol{x}_\mathrm{s})$. Further, let $J(\varepsilon; \boldsymbol{x}_\mathrm{s})$ be defined by

$$J(\varepsilon; \boldsymbol{x}_\mathrm{s}) = \mathcal{J}\big(\Omega_\varepsilon(\boldsymbol{x}_\mathrm{s})\big) = \int_{S^{\mathrm{obs}}} \varphi\big(u_\mathrm{R}^\varepsilon(\boldsymbol{\xi}), u_\mathrm{I}^\varepsilon(\boldsymbol{\xi}), \boldsymbol{\xi}\big) \, \mathrm{d}\Gamma, \tag{7}$$

For convenience, explicit references to $\boldsymbol{x}_\mathrm{s}$ will often be omitted in the sequel, e.g. by writing $J(\varepsilon)$ or $u^\varepsilon(\boldsymbol{\xi})$ instead of $J(\varepsilon; \boldsymbol{x}_\mathrm{s})$ or $u^\varepsilon(\boldsymbol{\xi}; \boldsymbol{x}_\mathrm{s})$.

The evaluation of $J(\varepsilon)$ entails solving for u^ε the forward problem $\mathcal{P}(B_\varepsilon)$. It is convenient, and customary, to decompose u^ε according to

$$u^\varepsilon = u + v^\varepsilon, \tag{8}$$

where u, the *free field* defined as the response of the obstacle-free (reference) medium Ω due to the given excitation p^D, solves

$$\begin{aligned} (\Delta + k^2)u &= 0 &&(\text{in } \Omega), \\ p[u] &= p^\mathrm{D} &&(\text{on } S), \end{aligned} \tag{9}$$

while v^ε, the *scattered field*, solves

$$\begin{aligned} (\Delta + k^2)v^\varepsilon &= 0 &&(\text{in } \Omega_\varepsilon), \\ p[v^\varepsilon] &= 0 &&(\text{on } S), \\ p[v^\varepsilon] &= -p[u] &&(\text{on } \Gamma_\varepsilon). \end{aligned} \tag{10}$$

3.2 Expansion of $J(\varepsilon)$

To establish the topological sensitivity of $J(\varepsilon)$, one starts with the expansion

$$J(\varepsilon) = J(0) + \int_{S^{\mathrm{obs}}} \mathrm{Re}\big[\,\varphi_{,u}\, v^{\varepsilon}\,\big]\, \mathrm{d}\Gamma + o\big(\,|v^{\varepsilon}|_{S^{\mathrm{obs}}}\,\big), \tag{11}$$

with

$$\varphi_{,u} \equiv \Big(\frac{\partial\varphi}{\partial u_{\mathrm{R}}^{\varepsilon}} - \mathrm{i}\,\frac{\partial\varphi}{\partial u_{\mathrm{I}}^{\varepsilon}}\Big)\Big|_{u^{\varepsilon}=u}. \tag{12}$$

Let the *adjoint field* $\hat{u}$ be defined by

$$\begin{aligned}
(\Delta + k^2)\hat{u} &= 0 && (\text{in } \Omega)\,, \\
p[\hat{u}] &= \varphi_{\mathrm{N}} && (\text{on } S^{\mathrm{obs}})\,, \\
p[\hat{u}] &= 0 && (\text{on } S \setminus S^{\mathrm{obs}}).
\end{aligned} \tag{13}$$

Then, the reciprocity identity (i.e. third Green's formula) applied to the states $\hat{u}$ and v^{ε} over the domain Ω_{ε} leads, by virtue of the boundary conditions in (10) and (13), to the identity

$$\int_{S^{\mathrm{obs}}} \varphi_{,u}\, v^{\varepsilon}\, \mathrm{d}\Gamma + \int_{\Gamma_{\varepsilon}} p[\hat{u}]v^{\varepsilon}\, \mathrm{d}\Gamma + \int_{\Gamma_{\varepsilon}} \hat{u}p[u]\, \mathrm{d}\Gamma = 0 \tag{14}$$

As a result, the integral in the r.h.s. of (11) is converted into integrals over the vanishing cavity surface. Besides, since both u and $\hat{u}$ are also defined inside B_{ε}, the last integral in (14) can be recast into a domain integral over B_{ε} by means of the divergence formula. Expansion (11) then takes the form

$$J(\varepsilon) = J(0) + \mathrm{Re}\Big\{\int_{B_{\varepsilon}} \big[\,\boldsymbol{\nabla} u \cdot \boldsymbol{\nabla}\hat{u} - k^2 u\hat{u}\,\big]\, \mathrm{d}V - \int_{\Gamma_{\varepsilon}} v^{\varepsilon} p[\hat{u}]\, \mathrm{d}\Gamma\Big\}$$
$$+ o\big(|v^{\varepsilon}|_{S^{\mathrm{obs}}}\big), \tag{15}$$

The first integral in (15) features a density whose definition does not depend on ε, and its expansion about $\varepsilon = 0$ can therefore be obtained by simply using the scaled coordinates (4), (5) and expanding $\big[\,\boldsymbol{\nabla} u \cdot \boldsymbol{\nabla}\hat{u} - k^2 u\hat{u}\,\big](\boldsymbol{x}_{\mathrm{s}} + \varepsilon\bar{\boldsymbol{\xi}})$. In contrast, the second integral of (15) features the scattered field v^{ε}, which depends on ε. Its asymptotic behaviour must then be obtained from that of v^{ε} on Γ_{ε} (taking into account the fact that Γ_{ε} also depends on ε). This step is based on exploiting an integral equation reformulation of equations (10).

3.3 Governing Integral Equation Formulation for the Scattered Field

The governing problem (10) for the scattered field $v^{\varepsilon} = v^{\varepsilon}(\cdot\,;\boldsymbol{x}_{\mathrm{s}})$ can be recast as the boundary integral equation [1, 4]:

$$\frac{1}{2}v^{\varepsilon}(\boldsymbol{x}) + \int_{\Gamma_{\varepsilon}} \mathcal{H}(\boldsymbol{x},\boldsymbol{\xi};k)v^{\varepsilon}(\boldsymbol{\xi})\, \mathrm{d}\Gamma_{\xi} = -\int_{\Gamma_{\varepsilon}} \mathcal{G}(\boldsymbol{x},\boldsymbol{\xi};k)p[u](\boldsymbol{\xi})\, \mathrm{d}\Gamma_{\xi} \quad (\boldsymbol{x} \in \Gamma_{\varepsilon}), \tag{16}$$

where the Green's function $\mathcal{G}(\boldsymbol{x},\boldsymbol{\xi},k)$ is defined by

$$(\Delta_\xi + k^2)\mathcal{G}(\boldsymbol{x},\boldsymbol{\xi},k) + \delta(\boldsymbol{\xi} - \boldsymbol{x}) = 0 \quad (\boldsymbol{\xi} \in \Omega), \quad \mathcal{H}(\boldsymbol{x},\boldsymbol{\xi},k) = 0 \quad (\boldsymbol{\xi} \in S), \quad (17)$$

and $\mathcal{H}(\boldsymbol{x},\boldsymbol{\xi};k) = \boldsymbol{\nabla}\mathcal{G}(\boldsymbol{x},\boldsymbol{\xi},k) \cdot \boldsymbol{n}(\boldsymbol{\xi})$ is the normal derivative of $\mathcal{G}(\boldsymbol{x},\boldsymbol{\xi},k)$ (the nabla symbol $\boldsymbol{\nabla}$, when used in front of such kernel, conventionally indicates a gradient with respect to the second argument $\boldsymbol{\xi}$). Moreover, the free and adjoint fields have the explicit expressions

$$u(\boldsymbol{x}) = \int_S \mathcal{G}(\boldsymbol{x},\boldsymbol{\xi};k)p^{\mathrm{D}}(\boldsymbol{\xi})\,\mathrm{d}\Gamma_\xi\,, \qquad \hat{u}(\boldsymbol{x}) = \int_S \mathcal{G}(\boldsymbol{x},\boldsymbol{\xi};k)\varphi_{,u}(\boldsymbol{\xi})\,\mathrm{d}\Gamma_\xi. \quad (18)$$

It is convenient for the present purposes to split $(\mathcal{G},\mathcal{H})$ according to:

$$\begin{aligned}
\mathcal{G}(\boldsymbol{x},\boldsymbol{\xi};k) &= G(\boldsymbol{x},\boldsymbol{\xi};k) + G_{\mathrm{C}}(\boldsymbol{x},\boldsymbol{\xi};k) \\
\mathcal{H}(\boldsymbol{x},\boldsymbol{\xi};k) &= H(\boldsymbol{x},\boldsymbol{\xi};k) + H_{\mathrm{C}}(\boldsymbol{x},\boldsymbol{\xi};k),
\end{aligned} \quad (19)$$

where (G,H) is the well-known singular free-space fundamental solution for the Helmholtz equation, given by

$$G(\boldsymbol{x},\boldsymbol{\xi};k) = \frac{1}{4\pi r}e^{\mathrm{i}kr}\,, \qquad H(\boldsymbol{x},\boldsymbol{\xi};k) = [\boldsymbol{r} \cdot \boldsymbol{n}(\boldsymbol{\xi})]\frac{\mathrm{i}kr - 1}{4\pi r^3}e^{\mathrm{i}kr}, \quad (20)$$

with $\boldsymbol{r} = \boldsymbol{\xi} - \boldsymbol{x}$ and $r = |\boldsymbol{\xi} - \boldsymbol{x}| = |\boldsymbol{r}|$, and the complementary part $(G_{\mathrm{C}}, H_{\mathrm{C}})$ is not singular at $\boldsymbol{\xi} = \boldsymbol{x}$. On using decomposition (19), equation (16) becomes

$$\begin{aligned}
\frac{1}{2}f(\boldsymbol{x}) &+ \int_{\Gamma_\varepsilon} H(\boldsymbol{x},\boldsymbol{\xi};k)v^\varepsilon(\boldsymbol{\xi})\,\mathrm{d}\Gamma_\xi + \int_{\Gamma_\varepsilon} H_{\mathrm{C}}(\boldsymbol{x},\boldsymbol{\xi};k)v^\varepsilon(\boldsymbol{\xi})\,\mathrm{d}\Gamma_\xi \\
&= -\int_{\Gamma_\varepsilon} G(\boldsymbol{x},\boldsymbol{\xi};k)p[u](\boldsymbol{\xi})\,\mathrm{d}\Gamma_\xi - \int_{\Gamma_\varepsilon} G_{\mathrm{C}}(\boldsymbol{x},\boldsymbol{\xi};k)p[u](\boldsymbol{\xi})\,\mathrm{d}\Gamma_\xi \quad (\boldsymbol{x} \in \Gamma_\varepsilon). \quad (21)
\end{aligned}$$

3.4 Leading Asymptotic Contribution to the Scattered Field

To study the asymptotic behaviour of integral equation (21) as $\varepsilon \to 0$, it is useful to introduce further scaled geometric quantities:

$$\boldsymbol{x} = \varepsilon\bar{\boldsymbol{x}}\,, \quad \boldsymbol{r} = \varepsilon\bar{\boldsymbol{r}}\,, \quad r = \varepsilon\bar{r} \quad (\boldsymbol{x},\boldsymbol{\xi} \in \Gamma_\varepsilon; \bar{\boldsymbol{x}},\bar{\boldsymbol{\xi}} \in \mathcal{S}) \quad (22)$$

in addition to definition (4) of $\bar{\boldsymbol{\xi}}$. The leading contributions as $\varepsilon \to 0$ to the fundamental kernels featured in equation (21) are

$$\begin{aligned}
G(\boldsymbol{x},\boldsymbol{\xi};k) &= \varepsilon^{-1}G(\bar{\boldsymbol{x}},\bar{\boldsymbol{\xi}}) + O(1) \\
H(\boldsymbol{x},\boldsymbol{\xi};k) &= \varepsilon^{-2}H(\bar{\boldsymbol{x}},\bar{\boldsymbol{\xi}}) + O(1)
\end{aligned} \quad (\boldsymbol{x},\boldsymbol{\xi} \in \Gamma_\varepsilon), \quad (23)$$

for the singular kernels (G,H) defined by (20), where

$$G(\bar{\boldsymbol{x}},\bar{\boldsymbol{\xi}}) = \frac{1}{4\pi\bar{r}}\,, \qquad H(\bar{\boldsymbol{x}},\bar{\boldsymbol{\xi}}) = -\frac{\bar{\boldsymbol{r}} \cdot \boldsymbol{n}(\bar{\boldsymbol{\xi}})}{4\pi\bar{r}^3} \quad (24)$$

are the fundamental kernels for the Laplace equation, and

$$G_C(\boldsymbol{x}, \boldsymbol{\xi}; k) = G_C(\boldsymbol{x}_s, \boldsymbol{x}_s; k) + O(\varepsilon)$$
$$H_C(\boldsymbol{x}, \boldsymbol{\xi}; k) = H_C(\boldsymbol{x}_s, \boldsymbol{x}_s; k) + O(\varepsilon) \qquad (\boldsymbol{x}, \boldsymbol{\xi} \in \Gamma_\varepsilon). \tag{25}$$

for the nonsingular kernels (G_C, H_C).

On performing the coordinate transformations (4), (22) and using estimates (23), (25) together with (5b), one finds that

$$\int_{\Gamma_\varepsilon} H_C(\boldsymbol{x}, \boldsymbol{\xi}; k) v^\varepsilon(\boldsymbol{\xi}) \, \mathrm{d}\Gamma_\xi = O(\varepsilon^2 |v^\varepsilon|)$$
$$\int_{\Gamma_\varepsilon} G_C(\boldsymbol{x}, \boldsymbol{\xi}; k) p[u](\boldsymbol{\xi}) \, \mathrm{d}\Gamma_\xi = O(\varepsilon^3). \tag{26}$$

The limiting form of integral equation (21) as $\varepsilon \to 0$, retaining only the leading contributions, is hence found to be

$$\bar{\mathcal{L}}_{\mathcal{S}}\left[v^\varepsilon\left(\boldsymbol{x}_s + \varepsilon\bar{\boldsymbol{\xi}}\right)\right](\bar{\boldsymbol{x}}) = -\varepsilon \boldsymbol{\nabla} u(\boldsymbol{x}_s) \cdot \, ints \frac{1}{4\pi\bar{r}} \boldsymbol{n}(\bar{\boldsymbol{\xi}}) \, \mathrm{d}\bar{\Gamma}_{\bar{\xi}} + o(\varepsilon) \tag{27}$$

where $\bar{\mathcal{L}}_{\mathcal{S}}$, defined by

$$\left[\bar{\mathcal{L}}_{\mathcal{S}} f\right](\bar{\boldsymbol{x}}) = \frac{1}{2} f(\bar{\boldsymbol{x}}) + \int_{\mathcal{S}} H(\bar{\boldsymbol{x}}, \bar{\boldsymbol{\xi}}) f(\bar{\boldsymbol{\xi}}) \, \mathrm{d}\bar{\Gamma}_{\bar{\xi}} \qquad (\bar{\boldsymbol{x}} \in \mathcal{S}), \tag{28}$$

is in fact the governing integral operator associated with exterior Neumann problems for the Laplace equation in the normalized domain $\mathbb{R}^3 \setminus \bar{\mathcal{B}}$. Equation (27) indicates that the scattered field is of order $O(\varepsilon)$ on Γ_ε:

$$v^\varepsilon(\boldsymbol{\xi}) = v^\varepsilon\left(\boldsymbol{x}_s + \varepsilon\bar{\boldsymbol{\xi}}\right) = \varepsilon \boldsymbol{\nabla} u(\boldsymbol{x}_s) \cdot \mathcal{V}(\bar{\boldsymbol{\xi}}) + o(\varepsilon) \qquad (\boldsymbol{\xi} \in \Gamma_\varepsilon, \bar{\boldsymbol{\xi}} \in \mathcal{S}) \tag{29}$$

where the vector function $\mathcal{V}(\bar{\boldsymbol{\xi}})$ solves the exterior Laplace problem

$$\left[\bar{\mathcal{L}}_{\mathcal{S}} \mathcal{V}\right](\bar{\boldsymbol{x}}) = -\boldsymbol{\nabla} u(\boldsymbol{x}_s) \cdot \int_{\mathcal{S}} \frac{1}{4\pi\bar{r}} \boldsymbol{n}(\bar{\boldsymbol{\xi}}) \, \mathrm{d}\bar{\Gamma}_{\bar{\xi}} \tag{30}$$

i.e.

$$\begin{aligned}
\Delta \mathcal{V} &= 0 & (\bar{\boldsymbol{\xi}} \in \mathbb{R}^3 \setminus \bar{\mathcal{B}}), \\
\boldsymbol{\nabla}_\xi \mathcal{V}.\boldsymbol{n} &= -\boldsymbol{n} & (\bar{\boldsymbol{\xi}} \in \mathcal{S}), \\
|\boldsymbol{\xi}|^2 \mathcal{V} &= O(1) & (|\bar{\boldsymbol{\xi}}| \to +\infty)
\end{aligned} \tag{31}$$

It is important to note that $\mathcal{V}(\bar{\boldsymbol{\xi}})$ does not depend on the sampling point $\boldsymbol{x}_s$, and hence needs to be computed only once.

3.5 Topological Derivative

On substituting (29) into (15) and taking (5) into account, one finally arrives at the following expansion of $J(\varepsilon)$:

$$J(\varepsilon) = J(0) + \varepsilon^3 \mathcal{T}(\boldsymbol{x}_\mathrm{s}) + o(\varepsilon^3) \tag{32}$$

where $\mathcal{T}(\boldsymbol{x}_\mathrm{s})$, the *topological derivative* of $J(\varepsilon)$, is given in terms of the free and adjoint fields by

$$\mathcal{T}(\boldsymbol{x}_\mathrm{s}) = \mathrm{Re}\big\{\boldsymbol{\nabla}\hat{u} \cdot \mathcal{A}(\mathcal{S}) \cdot \boldsymbol{\nabla}u - |\mathcal{B}|\, k^2 \hat{u}u\big\}(\boldsymbol{x}_\mathrm{s}) \tag{33}$$

and with the second-order tensor $\mathcal{A}(\mathcal{S})$ defined by

$$\mathcal{A}(\mathcal{S}) = |\mathcal{B}|\, \boldsymbol{I} - \int_S \big[\, \mathcal{V}(\bar{\boldsymbol{\xi}}) \otimes \boldsymbol{n}(\bar{\boldsymbol{\xi}}) \,\big]\, \mathrm{d}\bar{\Gamma}_{\bar{\xi}} \tag{34}$$

$$= |\mathcal{B}|\, \boldsymbol{I} - \int_S \big[\, \boldsymbol{n}(\bar{\boldsymbol{\xi}}) \otimes \mathcal{V}(\bar{\boldsymbol{\xi}}) \,\big]\, \mathrm{d}\bar{\Gamma}_{\bar{\xi}}$$

where the second equality (i.e. the fact that $\mathcal{A}(\mathcal{S})$ is symmetric) easily stems from the third Green's identity and the definition (31) of $\mathcal{V}$.

For arbitrary surfaces $\mathcal{S}$ which are sufficiently regular for integral equation (30) to be mathematically meaningful (this includes surfaces with edges and corners, e.g. box-shaped scatterers, but precludes infinitely-thin screens), the vector density $\mathcal{V}$ may be found by e.g. solving numerically three sets of BEM equations for exterior Laplace problems, a computationally modest task.

For the simplest case of a rigid spherical obstacle (where $\mathcal{B}$ is the unit sphere, $|\mathcal{B}| = 4\pi/3$, and on which $\boldsymbol{n}(\bar{\boldsymbol{\xi}}) = -\bar{\boldsymbol{\xi}}$) one easily finds by analytical means that $\mathcal{V}(\bar{\boldsymbol{\xi}}) = \bar{\boldsymbol{\xi}}/2$. Then, (34) is readily found to be given by

$$\mathcal{A}(\mathcal{S}) = 2\pi \boldsymbol{I}. \tag{35}$$

3.6 Practical Computation of Topological Sensitivity

The developments thus far are based on the Green's function $\mathcal{G}$ defined by (17), and lead to almost explicit formulae for $\mathcal{T}(\boldsymbol{x}_\mathrm{s})$, their only non-explicit component being the auxiliary density $\mathcal{V}$, which must be computed numerically except for simple shapes of the trial scatterer $\mathcal{B}$.

In practice, this explicit character is retained only for geometries Ω such that the corresponding Green's function is known analytically. Such cases correspond with geometrically simple configurations, e.g. the acoustic half-space. For configurations where the Green's function is not available, the free and adjoint fields and the nonsingular kernels may be sought as solutions of boundary integral equations [1, 4]. The free and adjoint fields, defined by (9) and (13), satisfy the well-known integral identities

$$c(\boldsymbol{x})u(\boldsymbol{x}) + \int_S H(\boldsymbol{x},\boldsymbol{\xi};k)u(\boldsymbol{\xi})\, \mathrm{d}\Gamma_\xi = \int_S G(\boldsymbol{x},\boldsymbol{\xi};k)p^\mathrm{D}(\boldsymbol{\xi})\, \mathrm{d}\Gamma_\xi \tag{36}$$

$$c(\boldsymbol{x})\hat{u}(\boldsymbol{x}) + \int_S H(\boldsymbol{x},\boldsymbol{\xi};k)\hat{u}(\boldsymbol{\xi})\, \mathrm{d}\Gamma_\xi = -\int_{S^\mathrm{obs}} G(\boldsymbol{x},\boldsymbol{\xi};k)\varphi_\mathrm{N}(\boldsymbol{\xi})\, \mathrm{d}\Gamma_\xi \tag{37}$$

which provide integral equations for $\boldsymbol{x} \in S$ (with $c(\boldsymbol{x}) = 1/2$, except at points $\boldsymbol{x}$ where S is only piecewise smooth, such as edges or corners, for which $c(\boldsymbol{x})$ is also known) and integral representation formulae for $\boldsymbol{x} \in \Omega$ (with $c(\boldsymbol{x}) = 1$). Differentiation of the latter under the integral sign provide integral representations of $\nabla u(\boldsymbol{x})$ and $\nabla u(\boldsymbol{x}_\mathrm{s})$ in Ω.

Hence, evaluation of the topological sensitivity field as given by (33) entails the solution of integral equations (36) and (37) for u and $\hat{u}$ on S, followed by an evaluation of $u(\boldsymbol{x}_\mathrm{s})$, $\nabla u(\boldsymbol{x}_\mathrm{s})$ and $\hat{u}(\boldsymbol{x}_\mathrm{s})$, $\nabla u(\boldsymbol{x}_\mathrm{s})$ by means of (36) and (37) used as integral representations. The first step may involve large numbers of unknowns if the diameter of Ω spans more than a few wavelengths. Moreover, the present objective being to comprehensively explore a 3-D region of space for hidden scatterers by examining the distribution of $\mathcal{T}(\boldsymbol{x}_\mathrm{s})$, formula (33) is to be evaluated at a large number of sampling points $\boldsymbol{x}_\mathrm{s}$. Both steps can then be considerably accelerated using the FM-BEM.

4 The Fast Multipole Method for Helmholtz Equation

4.1 BEM Discretization

Equations (36) and (37) are in this article solved by means of the simplest BEM discretization, which employs flat triangular boundary elements with straight edges and piecewise-linear C^0 interpolation of u and $\hat{u}$ (other choices, e.g. C^0 quadratic interpolation, would of course have been possible). All numerical results presented hereinafter have been obtained on that basis. The primary unknowns are the values of u or $\hat{u}$ at the mesh nodes, i.e. at all the element vertices. Equations (36) and (37) are collocated at all mesh nodes. All singular element integrals associated with the kernel $H(\boldsymbol{x}, \boldsymbol{\xi}; k)$ are zero because of the piecewise-flat geometry representation, which of course simplifies the implementation. Denoting by N the total number of nodes (and hence of unknowns), this procedure gives rise to the linear systems of equations

$$[A]\{u\} = \{b\} \tag{38}$$

$$[A]\{\hat{u}\} = \{\hat{b}\} \tag{39}$$

where the N-vectors $\{u\}$ and $\{\hat{u}\}$ collect all nodal values of u and $\hat{u}$. The discussion to follow will focus on solving system (38), the adjoint system (39) being of course solved in exactly the same way with $\{b\}$ replaced by $\{\hat{b}\}$.

As the problem size N grows, direct solvers become impractical or infeasible with respect to both computing time and storage, mainly due to the fully-populated nature of the BEM matrix $[A]$, and iterative solvers are used instead. Since $[A]$ is a non-symmetric, invertible matrix (except when k is a eigenvalue for Ω and homogeneous Neumann BCs), the iterative solution technique most often used for systems such as (38) is the generalized minimal residual (GMRES) algorithm (see e.g. [10]), which is applicable to general invertible square matrices. Such algorithms are based on matrix-vector evaluations, and therefore do not require actual assembly and storage of $[A]$. The GMRES algorithm requires repeated evaluations of the *residual*

$$\{b\} - [A]\{u\} \tag{40}$$

where $\{u\}$ is a given solution candidate, which is updated at each GMRES iteration. Hence, one needs to compute (discretized versions of) the double-layer and single-layer potentials featured in the left-hand and right-hand sides, respectively, of (36) and (37) for known densities.

Traditional BE methods lead to a $O(N^2)$ computational cost for each residual evaluation, because element integrals computed for a collocation point cannot be reused for another collocation point. By adopting the Fast Multipole boundary element method (FM-BEM), each residual can be computed within a $O(N \mathrm{Log} N)$ time. The implementation used here, concisely described in the remainder of this section, follows Darve [5] and Sylvand [23, 24].

4.2 Expansion of the Fundamental Solution

The starting point for the FM-BEM is the following representation of the full-space fundamental solution (20):

$$G(\boldsymbol{x}, \boldsymbol{\xi}; k) = \lim_{p \to \infty} \frac{\mathrm{i}k}{4\pi} \int_{\hat{S}} e^{-\mathrm{i}k(\hat{\boldsymbol{s}} \cdot \tilde{\boldsymbol{x}})} T_p(\hat{\boldsymbol{s}}, \boldsymbol{r}_0) e^{\mathrm{i}k(\hat{\boldsymbol{s}} \cdot \tilde{\boldsymbol{\xi}})} \, \mathrm{d}\Gamma_{\hat{\boldsymbol{s}}} \tag{41}$$

where $\hat{S} = \{\hat{\boldsymbol{s}}, |\hat{\boldsymbol{s}}| = 1\}$ is the unit sphere, the position vector $\boldsymbol{r} = \boldsymbol{\xi} - \boldsymbol{x}$ has been decomposed as

$$\boldsymbol{r} = (\boldsymbol{\xi}_0 - \boldsymbol{x}_0) + (\boldsymbol{\xi} - \boldsymbol{\xi}_0) - (\boldsymbol{x} - \boldsymbol{x}_0) = \boldsymbol{r}_0 + \tilde{\boldsymbol{\xi}} - \tilde{\boldsymbol{x}} \tag{42}$$

in terms of two poles $\boldsymbol{x}_0$ and $\boldsymbol{\xi}_0$, and with the transfer function $T_p(\hat{\boldsymbol{s}}, \boldsymbol{r}_0)$ defined by

$$T_p(\hat{\boldsymbol{s}}, \boldsymbol{r}_0) = \frac{(-\mathrm{i})^n}{4\pi} \sum_{n=0}^{p} (-1)^n (2n+1) h_n^{(1)}(k r_0) P_n(\hat{\boldsymbol{s}} \cdot \hat{\boldsymbol{r}}_0) \tag{43}$$

In (43), the $h_n^{(1)}$ and P_n are respectively the spherical Hankel functions of first kind and the Legendre polynomials, and $r_0 \equiv |\boldsymbol{\xi}_0 - \boldsymbol{x}_0|$. Moreover, for any vector $\boldsymbol{z} \in \mathbb{R}^3$, a hat symbol indicates the corresponding unit vector, i.e. $\hat{\boldsymbol{z}} = \boldsymbol{z}/|\boldsymbol{z}|$. Representation (41) holds under the condition

$$|r_0| > |\tilde{\boldsymbol{\xi}} - \tilde{\boldsymbol{x}}| \tag{44}$$

The poles $\boldsymbol{x}_0$ and $\boldsymbol{\xi}_0$ are actually meant to be chosen close to $\boldsymbol{\xi}$ and $\boldsymbol{x}$, respectively, so as to satisfy the stronger condition

$$|\boldsymbol{\xi} - \boldsymbol{x}_0| > |\boldsymbol{x} - \boldsymbol{x}_0| \quad \text{and} \quad |\boldsymbol{x} - \boldsymbol{\xi}_0| > |\boldsymbol{\xi} - \boldsymbol{\xi}_0| \tag{45}$$

The representation of $H(\boldsymbol{x}, \boldsymbol{\xi}; k)$ obtained by differentiating (41) is then

$$H(\boldsymbol{x}, \boldsymbol{\xi}; k) = \lim_{p \to \infty} \frac{-k^2}{4\pi} \int_{\hat{S}} e^{-\mathrm{i}k(\hat{\boldsymbol{s}} \cdot \tilde{\boldsymbol{x}})} T_p(\hat{\boldsymbol{s}}, \boldsymbol{r}_0) \big[\hat{\boldsymbol{s}} \cdot \boldsymbol{n}(\boldsymbol{\xi}) \big] e^{\mathrm{i}k(\hat{\boldsymbol{s}} \cdot \tilde{\boldsymbol{\xi}})} \, \mathrm{d}\Gamma_{\hat{s}} \qquad (46)$$

In practice, representations (41) and (46) are approximated by (i) using the transfer function $T_p(\hat{\boldsymbol{s}}, \boldsymbol{r}_0)$ with a finite index p and (ii) replacing the integral over the unit sphere with a quadrature rule with Q points $\hat{\boldsymbol{s}}_q$ and weights w_q. So, one replaces (41) and (46) with the approximations

$$G(\boldsymbol{x}, \boldsymbol{\xi}; k) \approx \frac{\mathrm{i}k}{4\pi} \sum_{q=1}^{Q} w_q e^{-\mathrm{i}k(\hat{\boldsymbol{s}}_q \cdot \tilde{\boldsymbol{x}})} T_p(\hat{\boldsymbol{s}}_q, \boldsymbol{r}_0) e^{\mathrm{i}k(\hat{\boldsymbol{s}}_q \cdot \tilde{\boldsymbol{\xi}})} \qquad (47)$$

$$H(\boldsymbol{x}, \boldsymbol{\xi}; k) \approx \frac{\mathrm{i}k}{4\pi} \sum_{q=1}^{Q} w_q e^{-\mathrm{i}k(\hat{\boldsymbol{s}}_q \cdot \tilde{\boldsymbol{x}})} T_p(\hat{\boldsymbol{s}}_q, \boldsymbol{r}_0) \big[\hat{\boldsymbol{s}}_q \cdot \boldsymbol{n}(\boldsymbol{\xi}) \big] e^{\mathrm{i}k(\hat{\boldsymbol{s}}_q \cdot \tilde{\boldsymbol{\xi}})} \qquad (48)$$

The choice of points $\hat{\boldsymbol{s}}_q$ and weights w_q, and their number Q, depends on the truncation order p used in the transfer function T_p. On parameterizing unit vectors $\hat{\boldsymbol{s}} \in \hat{S}$ using spherical angular coordinates (θ, ϕ), a commonly used choice [23, 5] consists in using the $Q(p) = (p+1)(2p+1)$ quadrature points of the form $\hat{\boldsymbol{s}}_q = (\theta_a, \phi_b)$, where $\cos\theta_a$ $(1 \leq a \leq p+1)$ are the abscissae for the $(p+1)$-point Gauss-Legendre 1-D quadrature rule over $[-1, 1]$, $\phi_b = 2\pi b/(2p+1)$ $(1 \leq b \leq 2p+1)$ are uniformly-spaced abscissae on $[0, 2\pi]$. The associated weights are $w_q = 2\pi w_a^\theta/(2p+1)$, where w_a^θ are the Gauss-Legendre weights for the $(p+1)$-point rule.

Now, let S_x and S_ξ denote two disjoint portions of S, and let the poles $\boldsymbol{x}_0$ and $\boldsymbol{\xi}_0$ be chosen close to S_x and S_ξ, respectively, in such a way that (45) holds for any $\boldsymbol{x} \in S_x$ and $\boldsymbol{\xi} \in S_\xi$. Consider the computation of

$$I(\boldsymbol{x}) = \int_{S_\xi} G(\boldsymbol{x}, \boldsymbol{\xi}; k) v(\boldsymbol{\xi}) \, \mathrm{d}\Gamma_\xi \qquad (\boldsymbol{x} \in S_x)$$

for a given density $v(\boldsymbol{\xi})$, which is a typical contribution to the evaluation of the residual (40). On substituting (47), one obtains

$$I(\boldsymbol{x}) \approx \frac{\mathrm{i}k}{4\pi} \sum_{q=1}^{Q(p)} w_q e^{-\mathrm{i}k(\hat{\boldsymbol{s}}_q \cdot \tilde{\boldsymbol{x}})} T_p(\hat{\boldsymbol{s}}_q, \boldsymbol{r}_0) \int_{S_\xi} e^{\mathrm{i}k(\hat{\boldsymbol{s}}_q \cdot \tilde{\boldsymbol{\xi}})} v(\boldsymbol{\xi}) \, \mathrm{d}\Gamma_\xi \qquad (\boldsymbol{x} \in S_x) \quad (49)$$

So, the *same* integral over S_ξ can be re-used for *all* collocation points $\boldsymbol{x} \in S_x$. Moreover, for a chosen portion S_ξ, this is true for any portion S_x and associated pole $\boldsymbol{x}_0$ such that condition (45) holds. Computations of the form (49) can be decomposed into three stages: (i) compute for each quadrature point of $\hat{S}$ the multipole moment $\mathcal{R}(\hat{\boldsymbol{s}}_q; \boldsymbol{\xi}_0)$:

$$\mathcal{R}(\hat{\boldsymbol{s}}_q; \boldsymbol{\xi}_0) = \int_{S_\xi} e^{\mathrm{i}k(\hat{\boldsymbol{s}}_q \cdot \tilde{\boldsymbol{\xi}})} v(\boldsymbol{\xi}) \, \mathrm{d}\Gamma_\xi \qquad (50)$$

(ii) multiply $\mathcal{R}(\hat{\boldsymbol{s}}_q; \boldsymbol{\xi}_0)$ by the transfer function $T_p(\hat{\boldsymbol{s}}_q, \boldsymbol{r}_0)$, to obtain local expansion coefficients $\mathcal{L}(\hat{\boldsymbol{s}}_q; \boldsymbol{x}_0)$ at $\boldsymbol{x}_0$:

$$\mathcal{L}(\hat{\boldsymbol{s}}_q; \boldsymbol{x}_0) = T_p(\hat{\boldsymbol{s}}_q, \boldsymbol{\xi}_0 - \boldsymbol{x}_0)\mathcal{R}(\hat{\boldsymbol{s}}_q; \boldsymbol{\xi}_0) \tag{51}$$

(iii) for all $\boldsymbol{x} \in S_x$, transfer $\mathcal{L}(\hat{\boldsymbol{s}}_q; \boldsymbol{x}_0)$ locally from $\boldsymbol{x}_0$ to $\boldsymbol{x}$ and perform the numerical quadrature over $\hat{S}$, to obtain (an approximation of) $I(\boldsymbol{x})$:

$$I(\boldsymbol{x}) \approx \frac{\mathrm{i}k}{4\pi} \sum_{q=1}^{Q(p)} w_q e^{-\mathrm{i}k(\hat{\boldsymbol{s}}_q \cdot \tilde{\boldsymbol{x}})} \mathcal{L}(\hat{\boldsymbol{s}}_q; \boldsymbol{x}_0) \tag{52}$$

The one-level fast multipole method consists in partitioning the spatial region containing S into cubic cells of identical sizes whose vertices lie on a regular cubic grid. Each pair (S_x, S_ξ) is such that $S_x = S \cap C_x$ and $S_\xi = S \cap C_\xi$, where (C_x, C_ξ) is any pair of *disjoint* cubic cells. The poles $\boldsymbol{x}_0$ and $\boldsymbol{\xi}_0$ are the respective cell centroids. The one-level FM-BEM has a complexity of $O(N^{3/2})$ per iteration for equations of type (38), which is of course better than the $O(N^2)$ complexity of traditional BEM, but not optimal. Further acceleration is provided by using the multi-level fast multipole method, where the size of clusters S_x, S_ξ depends of their distance.

4.3 Multilevel FM-BEM Algorithm

To exploit optimally the acceleration afforded by (49), a hierarchical oct-tree structure of elements is introduced. For that purpose, a cube containing the boundary S, called 'level-0 cell', is divided into eight cubes (level-1 cells), each of which is divided in the same fashion, and so on. A level-ℓ cell is divided into level-$(\ell+1)$ cells unless it contains less than a preset (relatively small) number E of boundary elements (such cells are termed *leaves*). A noteworthy feature of the FM-BEM applied to Helmholtz-type equations is that to achieve the same accuracy in approximations (47), (48) at all levels, the truncation parameter p is level-dependent. A often-used formula [5] for the adjustment of p is of the form

$$p(\ell) = \sqrt{3}kd(\ell) + C\mathrm{Log}_{10}(\sqrt{3}kd(\ell) + \pi) \tag{53}$$

where $d(\ell)$ is the size of a cubic ℓ-level cell (so, $d(\ell+1) = 2d(\ell)$ and $p(\ell+1)$ is roughly $2p(\ell)$) and C is a constant. As a result, the set of quadrature points $\hat{\boldsymbol{s}}_q$ on $\hat{S}$ is also level-dependent: each level necessitates a distinct set $\hat{\boldsymbol{s}}_q^\ell$ of quadrature points and associated weights. For the present implementation, values of C such that $2 \le C \le 8$ were found to provide a acceptable compromise between accuracy and cost, and $C = 4$ was actually used.

The FM-BEM algorithm then consists of:

- An *upward pass* where multipole moments (50) are first computed for the lowest-level cells and then recursively aggregated by moving upward in the tree until level 2 (for which there are $4 \times 4 \times 4$ cells) is reached. Letting $C(\boldsymbol{\xi}_0)$ denote the set of children of a given ℓ-level cell $C(\boldsymbol{\xi}_0)$, i.e. of $(\ell + 1)$-level cells $C(\boldsymbol{\xi}_0')$ contained in $C(\boldsymbol{\xi}_0)$, this operation relies on the identity

$$\mathcal{R}(\hat{\boldsymbol{s}}_q^{\ell+1}; \boldsymbol{\xi}_0) = \sum_{C(\boldsymbol{y}-0) \in \mathcal{C}(\boldsymbol{x}_0)} e^{\mathrm{i}k(\hat{\boldsymbol{s}}_q^{\ell+1} \cdot (\boldsymbol{\xi}_0' - \boldsymbol{\xi}_0))} \mathcal{R}(\hat{\boldsymbol{s}}_q^{\ell+1}; \boldsymbol{\xi}_0') \tag{54}$$

for shifting the origin from the center $\boldsymbol{\xi}_0'$ of a level-$(\ell+1)$ cell to the center $\boldsymbol{\xi}_0$ of a level-ℓ cell in the the contribution of a given cell.

Then, it is necessary to interpolate, i.e. compute $\mathcal{R}(\hat{\boldsymbol{s}}_q^{\ell}; \boldsymbol{\xi}_0)$ at the quadrature points $\hat{\boldsymbol{s}}_q^{\ell}$ from the previously determined values $\mathcal{R}(\hat{\boldsymbol{s}}_q^{\ell+1}; \boldsymbol{\xi}_0)$. The procedure used follows [23, 5] and is not detailed here.

- A *downward pass* where the coefficients of local expansions are first computed at level $\ell = 2$ and then evaluated at lower-level cells by tracing the tree structure downwards. This operation relies on the identity

$$\mathcal{L}(\hat{\boldsymbol{s}}_q^{\ell}; \boldsymbol{x}_0') = e^{-\mathrm{i}k(\hat{\boldsymbol{s}}_q^{\ell+1} \cdot (\boldsymbol{x}_0 - \boldsymbol{x}_0'))} \mathcal{L}(\hat{\boldsymbol{s}}_q^{\ell}; \boldsymbol{x}_0) \tag{55}$$

for shifting the origin from the center $\boldsymbol{x}_0$ of a level-(ℓ) cell to the center $\boldsymbol{x}_0'$ of a level-$(\ell+1)$ cell. This operation is not performed at the root level, i.e. when $\ell + 1 = 2$.

Then, the contributions of all level-$(\ell+1)$ cells belonging to the *interaction list* $\mathcal{I}(\boldsymbol{x}_0)$ of the level-$\ell + 1$ cell $C(\boldsymbol{x}_0)$ (i.e. all such cells which are non-adjacent to $C(\boldsymbol{x}_0)$ but whose father is adjacent to the father of $C(\boldsymbol{x}_0)$) are aggregated:

$$\sum_{C(\boldsymbol{y}-0) \in \mathcal{I}(\boldsymbol{x}_0)} T_p(\hat{\boldsymbol{s}}_q^{\ell}, \boldsymbol{y} - 0 - \boldsymbol{x}_0) \mathcal{R}(\hat{\boldsymbol{s}}_q^{\ell}; \boldsymbol{y} - 0)$$

and the result is added to $\mathcal{L}(\hat{\boldsymbol{s}}_q^{\ell}; \boldsymbol{x}_0')$ given by (55)

Then, the values $\mathcal{L}(\hat{\boldsymbol{s}}_q^{\ell}; \boldsymbol{x}_0')$ are converted to values at the quadrature points $\hat{\boldsymbol{s}}_q^{\ell+1}$ by a 'reverse interpolation', or 'anterpolation', procedure.

- When the lowest level is reached, all quadratures of the form (52) are finally performed, where $\boldsymbol{x}_0$ is the centers of a leaf cell, thus evaluating all far-field contributions to the residual at all collocation points.

Moreover, for all leaf cells $C(\boldsymbol{x}_0)$ and all collocation points $\boldsymbol{x} \in C(\boldsymbol{x}_0)$, the near-field contributions are computed by evaluating using traditional integration methods the element contributions for all elements located in $C(\boldsymbol{x}_0)$ and all cells of same level adjacent to $C(\boldsymbol{x}_0)$.

The computation of integral representation formulae for $u(\boldsymbol{x}_\mathrm{s})$, $\boldsymbol{\nabla} u(\boldsymbol{x}_\mathrm{s})$ and $\hat{u}(\boldsymbol{x}_\mathrm{s})$, $\boldsymbol{\nabla} u(\boldsymbol{x}_\mathrm{s})$ at all chosen sampling points follows the same approach, with collocation points $\boldsymbol{x} \in S$ replaced with sampling points $\boldsymbol{x}_\mathrm{s} \in \Omega$. For sampling points lying in leaf cells not adjacent to any same-level cell intersecting S, the integral representations result from far-field interactions (i.e. fast-multipole contributions) only. Besides, all multipole moments used in this step are those corresponding to the solution of (38) or (39), i.e. those evaluated at the last iteration of the GMRES solution algorithm

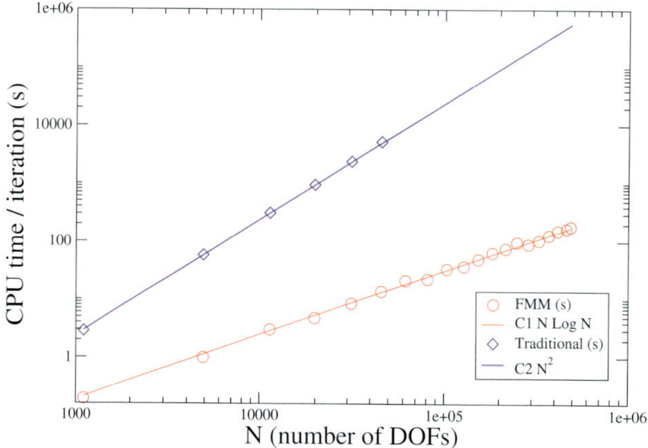

Fig. 1. CPU time for one evaluation of residual (40) as a function of the number N of nodal unknowns, with matrix-vector products computed using classical BEM techniques (lozenges, theoretical complexity $O(N^2)$) or the FM-BEM (circles, theoretical complexity $O(N\mathrm{Log}N)$).

4.4 Numerical Verification of Theoretical Complexity

To check that the theoretical complexity of the FM-BEM is achieved and illustrate the computational advantage brought by the FM-BEM over the conventional BEM, the simple situation of a spherical acoustic domain subjected on its surface to a uniform normal velocity, is considered. BEM solutions for this problem have been computed for a sequence of meshes with decreasing element size. For each mesh, the prescribed frequency is selected so that the number of nodes per wavelength is (approximately) the same for all meshes. Figure 1 shows the CPU time used for one evaluation of residual (40), using either classical BEM integration techniques (with theoretical complexity $O(N^2)$) or the FM-BEM (with theoretical complexity $N\mathrm{Log}N$), as a function of the number N of nodal unknowns. Functions of the form $C_1 N^2$ and $C_2 N\mathrm{Log}N$ closest to the actual recorded values of $\mathrm{CPU}(N)$ are also plotted on the same graph. The theoretical complexity for both the classical BEM and the FM-BEM are very well verified by the actual CPU times. The FM-BEM, as expected, performs much better for large problems.

5 Preliminary Identification via Topological Sensitivity: Numerical Examples

To illustrate the approach described in Sections 3 and 4, the following configuration has been considered: the bounded acoustic domain is the cube defined by $\Omega(L) = \{\, |\xi_i| \leq L \ (i = 1, 2, 3)\,\}$, with $L = 8a$ or $L = 16a$ in terms of a

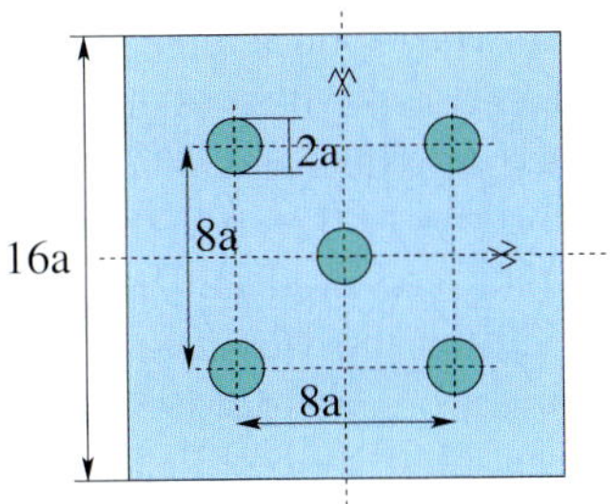

Fig. 2. Pattern of excitation surfaces S_q on each face of external boundary S.

reference length a. A simulated testing configuration is based on 30 experiments, each of which consists in applying a uniform excitation over a small region S_q ($1 \leq q \leq 30$) of S and recording the acoustic pressure over the whole boundary S (i.e. at all BE mesh nodes), so that $S^{\mathrm{obs}} = S$. The acoustic excitation is such that the wavelength is $\lambda = 3a$. Each of the six faces of the cubical domain Ω supports five excitation surfaces S_q, each a disk of radius a, arranged as depicted on Fig. 2. The cost function for the inverse problem is defined by

$$\mathcal{J}(\Omega^\star) = \frac{1}{2} \sum_{q=1}^{30} \int_S |u^\star - u_q^{\mathrm{obs}}|^2 \, \mathrm{d}\Gamma_\xi \tag{56}$$

where u_q^{obs} is the data obtained for the q-th applied excitation, with $u_q^{\mathrm{obs}} = u^{\mathrm{true}}$ in the absence of data noise. The centroid $\boldsymbol{x}^{\mathrm{true}}$ of true scatterer B^{true} to be identified is located at $\boldsymbol{x}^{\mathrm{true}} = (2a, 3a, 2a)$.

To facilitate the graphical interpretation, a thresholded variant $\hat{\mathcal{T}}(\boldsymbol{x}_{\mathrm{s}})$ of $\mathcal{T}(\boldsymbol{x}_{\mathrm{s}})$ is introduced according to

$$\hat{\mathcal{T}}(\boldsymbol{x}_{\mathrm{s}}) = \begin{cases} \mathcal{T}(\boldsymbol{x}_{\mathrm{s}}), & \mathcal{T} \leq C\,\mathcal{T}_{\mathrm{min}} \\ 0, & \mathcal{T} > C\,\mathcal{T}_{\mathrm{min}} \end{cases} \qquad \text{with } C = 0.25. \tag{57}$$

The BE meshes used for computing the free field u, the adjoint field $\hat{u}$ and the simulated data u_q^{true} are made of three-noded flat triangular elements, arranged in a regular mesh with approximately 15 nodes per wavelength. For the purposes of computing the simulated error-free data u^{true} for each synthetic experiment, BE meshes of Γ^{true} have been set up as well. Table 1 indicates the numbers of nodes and elements supported by the BE meshes.

Table 1. Number of element and DOFs supported by the BE meshes.

Cube size	Cube		Obstacle		Total	
	Elements	nodes	Elements	nodes	Elements	nodes
$2L = 16a$	76800	38402	336	170	77136	38572
$2L = 32a$	307200	153602	336	170	307536	153772

First, the identification of one spherical scatterer is considered, for three cases with increasing scatterer radii $0.2a$, $0.4a$ and $0.8a$. The field $\mathcal{T}$ has been computed for each case on the basis of error-free synthetic data, over a sampling grid $\mathcal{S}$ made of $100 \times 100 \times 100$ sampling points located on the vertices of a regular cubic grid, centered at the origin and with grid spacing $\Delta x_{\mathrm{s}} = 16a/101$, uniformly filling the whole acoustic domain bounded by $\mathcal{S}$. Figures 3, 4 and 5 show, for each scatterer configuration considered in turn, the distribution of the thresholded topological sensitivity $\hat{\mathcal{T}}$ defined by (57) in the three coordinate planes containing the true scatterer centroid $\boldsymbol{x}^{\mathrm{true}}$. In all cases, $\hat{\mathcal{T}}$ (and hence $\mathcal{T}$) is seen to attain its lowest values in zones corresponding to, or close to, the actual true scatterer location. It should however be mentioned that $\mathcal{T}$ has been observed to also achieve low values in zones located near edges or corners of $\mathcal{S}$, where there is no scatterer. To emphasize and illustrate this remark, Fig. 6 shows the iso-surfaces of the field $\mathcal{T}$ corresponding to $\mathcal{T} = 0.55\mathcal{T}_{\mathrm{min}}$, computed on the entire search grid $\mathcal{S}$ (Fig. 6a) and on truncated search grids such that $\{\, |\xi_i| \leq 6.5a\ (i = 1, 2, 3) \,\}$ (Fig. 6b) and $\{\, |\xi_i| \leq 5a\ (i = 1, 2, 3) \,\}$ (Fig. 6c), i.e. in which sampling points situated at distances less than $1.5a$ and $3a$, respectively, from S have been taken out of $\mathcal{S}$. Figure 6c shows that, at least in the central region $\{\, |\xi_i| \leq 5a\ (i = 1, 2, 3) \,\}$ of the acoustic domain, low negative values of $\mathcal{T}$ occur only in a small region which is consistent with the actual location of the scatterer.

Then, the effect of data errors is considered for the spherical scatterer of radius $0.8a$, by using synthetic data u^{obs} in the form $u^{\mathrm{obs}} = u^{\mathrm{true}}(1 + \eta)$, where η are random numbers with zero mean and uniform distribution over the interval $[-0.1, 0.1]$. Figure 7 shows the distribution of $\hat{\mathcal{T}}$ in the three coordinate planes containing $\boldsymbol{x}^{\mathrm{true}}$. A comparison between Figs. 5 and 7 reveals that the distribution of $\hat{\mathcal{T}}$ is only marginally affected by the data noise.

The examples shown thus far illustrate the capability of $\mathcal{T}$, here defined on the basis of an asymptotic analysis involving vanishing *spherical* obstacles, to identify the location of spherical obstacles of finite extent. Now, the identification of a non-spherical, box-shaped obstacle whose sides are aligned along the coordinale axes and of finite size $0.8a \times 0.8a \times 1.6a$ and whose centroid is still $\boldsymbol{x}^{\mathrm{true}} = (2a, 3a, 2a)$, is considered. Figure 8 shows the distribution of $\hat{\mathcal{T}}$ in the three coordinate planes containing $\boldsymbol{x}^{\mathrm{true}}$. The true obstacle is again satisfactorily located.

A last example illustrates the case of a larger acoustic domain $\Omega(16a)$, instead of $\Omega(8a)$ considered up to this point, with the same wavelength $\lambda = 3a$ as before. The 'true' scatterer (again a sphere of radius $0.8a$) is still located at $\boldsymbol{x}^{\mathrm{true}} = (2a, 3a, 2a)$, and hence is located at a larger distance (expressed in wavelengths) from the measurement surfaces. The sampling grid $\mathcal{S}$ is now made of $150 \times 150 \times 150$ regularly-spaced sampling points, with a grid spacing now of $\Delta x_{\mathrm{s}} = 32a/151$. Figure 9 shows the distribution of $\hat{\mathcal{T}}$ in the three coordinate planes containing $\boldsymbol{x}^{\mathrm{true}}$. Presumably as a result of greater remoteness (and hence lower identifiability) of the scatterer, these distributions show, in addition to the correct one, secondary spatial zones where the presence of a

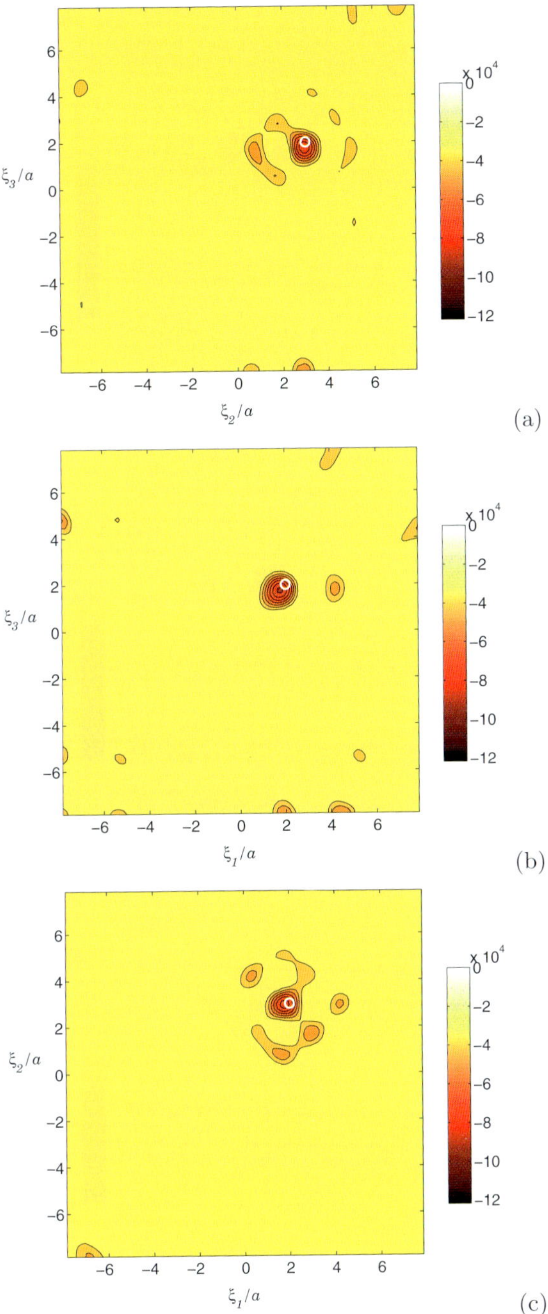

Fig. 3. Identification of spherical hard scatterer of radius $0.2a$: distribution of thresholded topological sensitivity $\hat{\mathcal{T}}(\boldsymbol{x}_s)$ for sampling points $\boldsymbol{x}_s$ in coordinate planes $\xi_1 = x_1^{\text{true}}$ (**a**), $x_2 = x_2^{\text{true}}$ (**b**) and $x_3 = x_3^{\text{true}}$ (**c**).

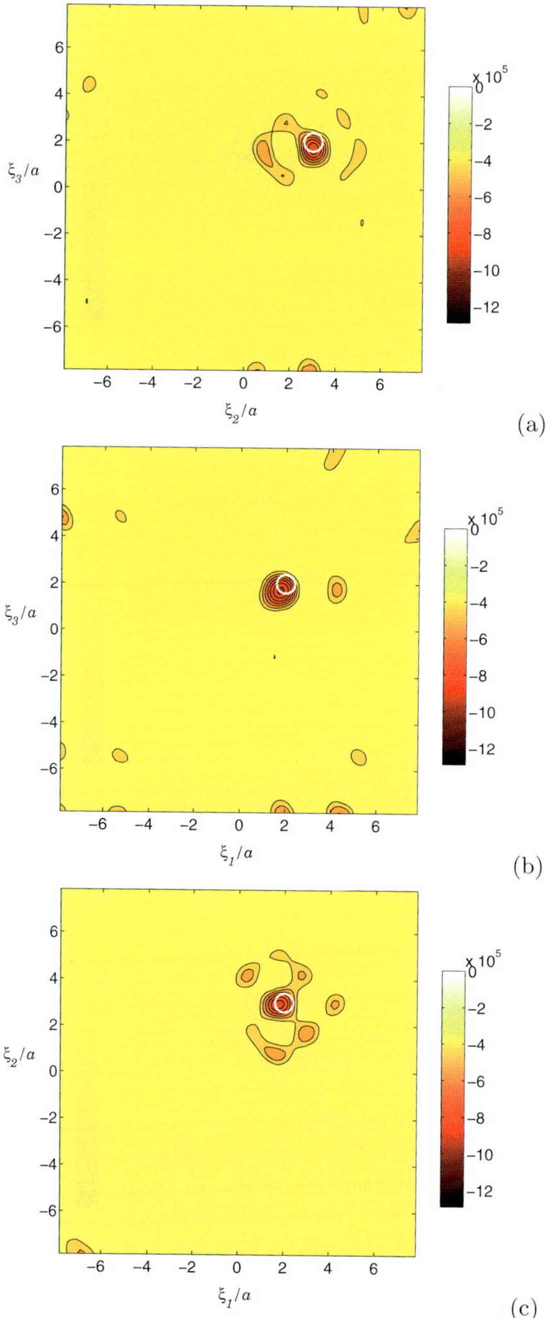

Fig. 4. Identification of spherical hard scatterer of radius $0.4a$: distribution of thresholded topological sensitivity $\hat{\mathcal{T}}(\boldsymbol{x}_{\mathrm{s}})$ for sampling points $\boldsymbol{x}_{\mathrm{s}}$ in coordinate planes $\xi_1 = x_1^{\mathrm{true}}$ (**a**), $x_2 = x_2^{\mathrm{true}}$ (**b**) and $x_3 = x_3^{\mathrm{true}}$ (**c**).

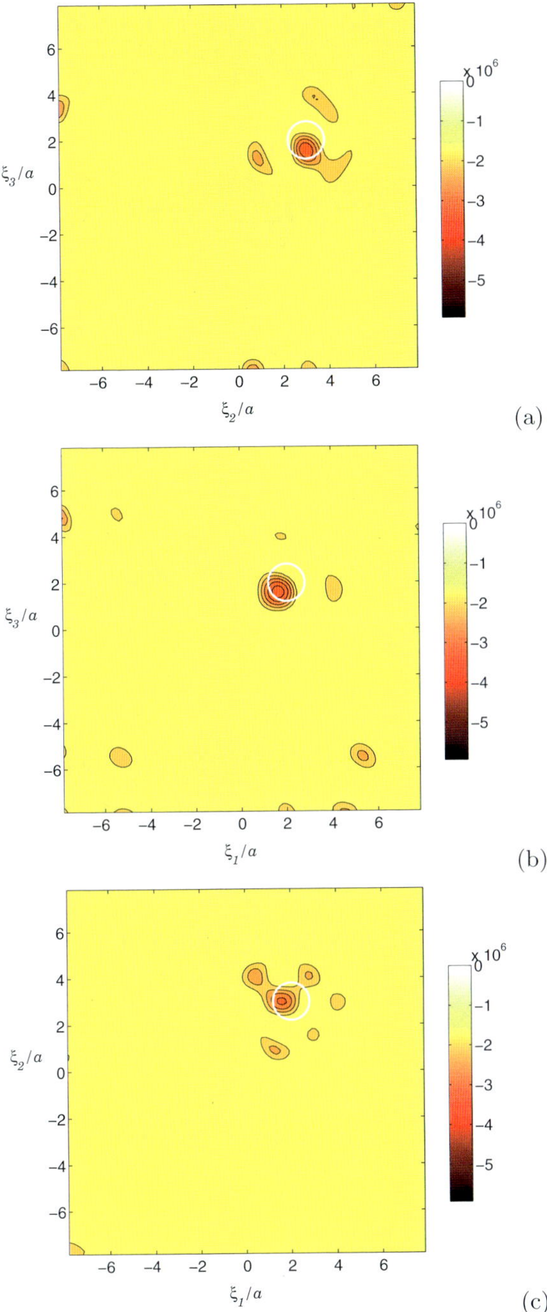

Fig. 5. Identification of spherical hard scatterer of radius $0.8a$: distribution of thresholded topological sensitivity $\hat{\mathcal{T}}(\boldsymbol{x}_{\mathrm{s}})$ for sampling points $\boldsymbol{x}_{\mathrm{s}}$ in coordinate planes $\xi_1 = x_1^{\mathrm{true}}$ (**a**), $x_2 = x_2^{\mathrm{true}}$ (**b**) and $x_3 = x_3^{\mathrm{true}}$ (**c**).

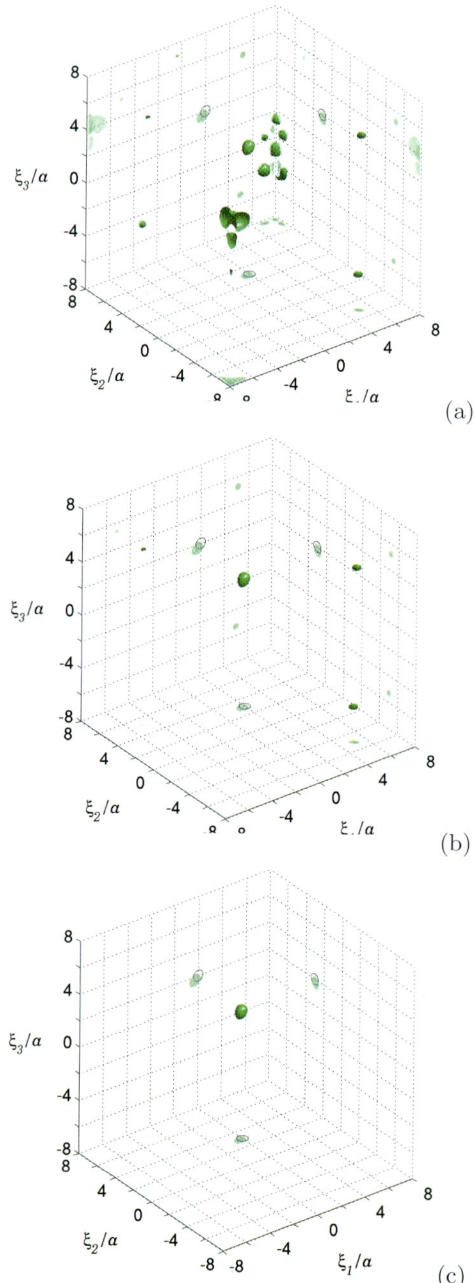

Fig. 6. Identification of spherical hard scatterer of radius $0.4a$: iso-surfaces of $\mathcal{T}(\boldsymbol{x}_\mathrm{s})$ for $\mathcal{T} = 0.55\mathcal{T}_\mathrm{min}$, computed on the entire search grid $\mathcal{S}$ (**a**) and on truncated search grids such that $\{\,|\xi_i| \leq 6.5a\ (i = 1, 2, 3)\,\}$ (**b**) and $\{\,|\xi_i| \leq 5a\ (i = 1, 2, 3)\,\}$ (**c**). Values of $\mathcal{T}(\boldsymbol{x}_\mathrm{s})$ lower than the iso-value are inside the iso-surface.

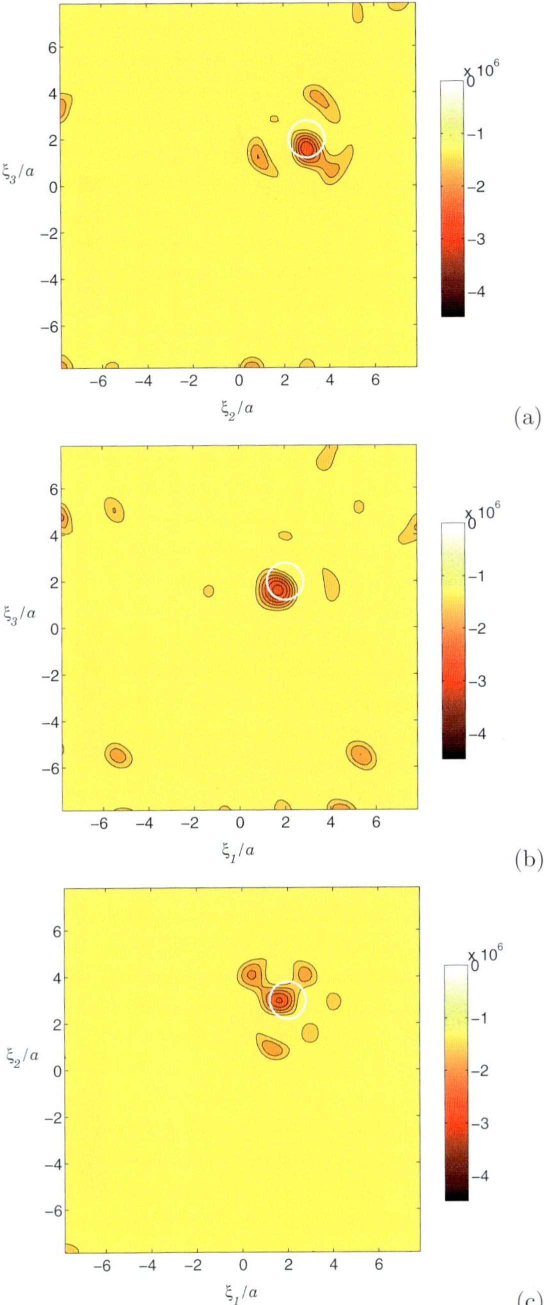

Fig. 7. Identification of spherical hard scatterer of radius $0.4a$, synthetic data with 10% noise: distribution of thresholded topological sensitivity $\hat{\mathcal{T}}(\boldsymbol{x}_\mathrm{s})$ for sampling points $\boldsymbol{x}_\mathrm{s}$ in coordinate planes $\xi_1 = x_1^{\mathrm{true}}$ (**a**), $x_2 = x_2^{\mathrm{true}}$ (**b**) and $x_3 = x_3^{\mathrm{true}}$ (**c**).

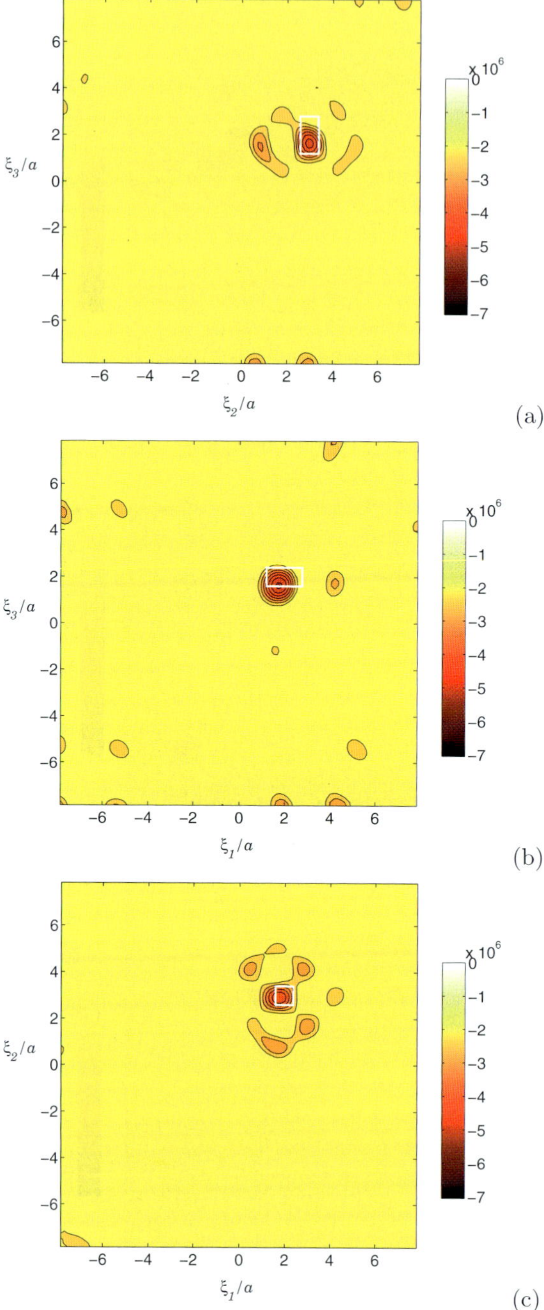

Fig. 8. Identification of box-shaped scatterer: distribution of thresholded topological sensitivity $\hat{\mathcal{T}}(\boldsymbol{x}_{\mathrm{s}})$ for sampling points $\boldsymbol{x}_{\mathrm{s}}$ in coordinate planes $\xi_1 = x_1^{\mathrm{true}}$ (**a**), $x_2 = x_2^{\mathrm{true}}$ (**b**) and $x_3 = x_3^{\mathrm{true}}$ (**c**).

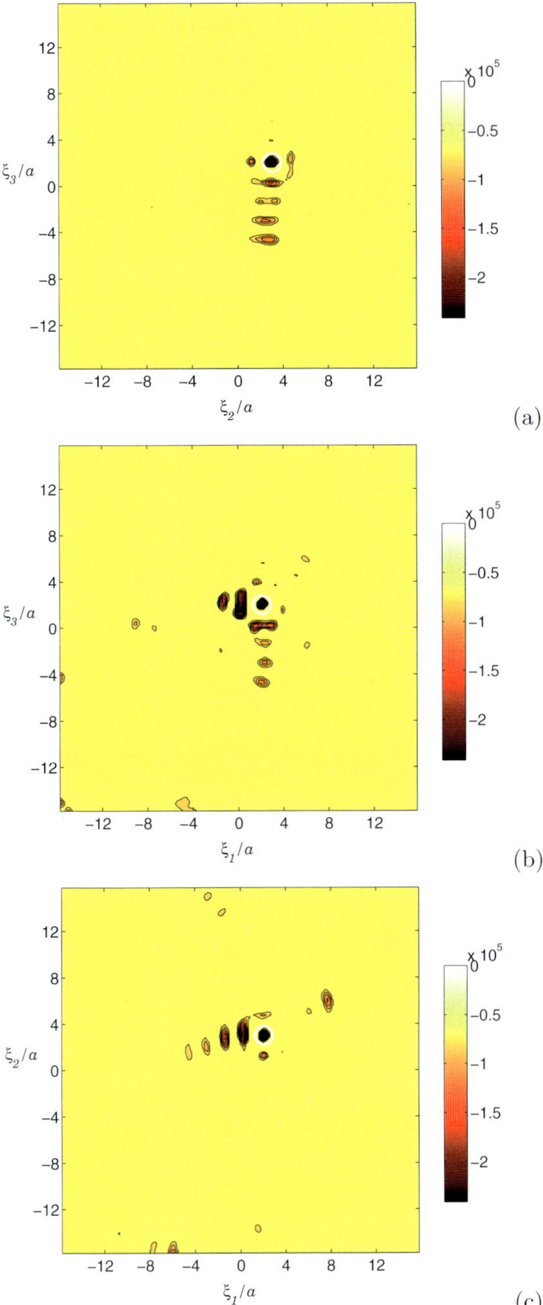

Fig. 9. Identification of spherical hard scatterer of radius $0.4a$ in large domain: distribution of thresholded topological sensitivity $\hat{\mathcal{T}}(\boldsymbol{x}_\mathrm{s})$ for sampling points $\boldsymbol{x}_\mathrm{s}$ in coordinate planes $\xi_1 = x_1^{\mathrm{true}}$ (**a**), $x_2 = x_2^{\mathrm{true}}$ (**b**) and $x_3 = x_3^{\mathrm{true}}$ (**c**).

small scatterer is also consistent with the data. Still, the lowest values of $\hat{T}$ furnish a reasonable indication of the true obstacle location, as seen on the 3-D plots of iso-surfaces $T = 0.55T_{\mathrm{min}}$ and $T = 0.7T_{\mathrm{min}}$ of Fig. 10 for the truncated grid defined by $\{\,|\xi_i| \leq 14a\ (i = 1, 2, 3)\,\}$. Again, one notes that values of T close to its minimum T_{min} occur only in the vicinity of the correct obstacle location (excluding, as before, regions close to the external surface S).

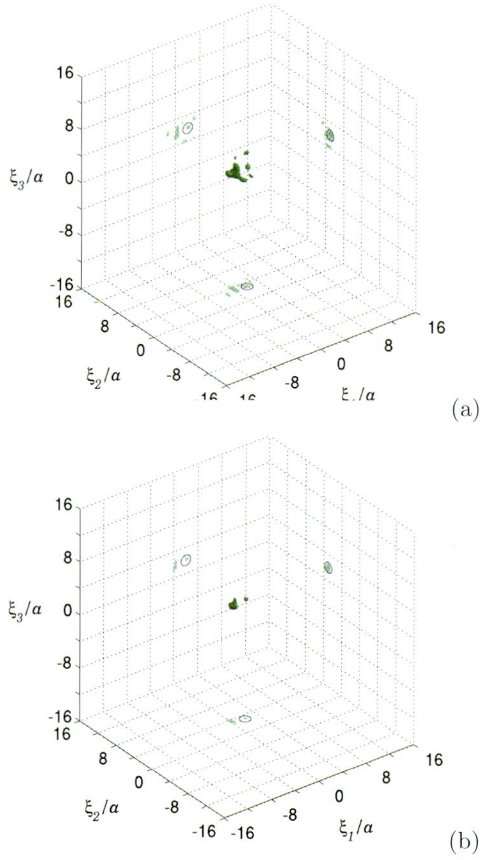

Fig. 10. Identification of spherical hard scatterer of radius $0.8a$ embedded in large domain $\Omega(16a)$: iso-surfaces of (a) $T(\boldsymbol{x}_{\mathrm{s}})$ for $T = 0.55T_{\mathrm{min}}$ and (b) $T = 0.7T_{\mathrm{min}}$ computed on the truncated search grid such that $\{\,|\xi_i| \leq 14a\ (i = 1, 2, 3)\,\}$. Values of $T(\boldsymbol{x}_{\mathrm{s}})$ lower than the iso-value are inside the iso-surface.

Typical CPU times and GMRES iteration counts are provided in Table 2. It is interesting to observe that the overall CPU times for $\Omega(16a)$, which involves roughly 4 times as many nodal unknowns as $\Omega(8a)$, are about 5 times higher than those for $\Omega(8a)$ (while the expected ratio for traditional BEM would be $4^3 = 64$), and that the GMRES iteration counts are only fractionally higher for $\Omega(16a)$. All computations have been performed on a Linux PC computer with one 3 GHz processor.

Table 2. CPU times and (in parentheses) GMRES iteration count for computing the true, free and adjoint solutions on the boundary, and CPU times for computing the topological sensitivity over the whole sampling grid $\mathcal{S}$.

	u^{true} on $S \cup \Gamma^{\text{true}}$	u on S	$\hat{u}$ on S	$\mathcal{T}$ on $\mathcal{S}$
$2L = 16a$	1444s (435)	969s (282)	1163s (342)	852s
$2L = 32a$	6461s (439)	5615s (388)	6818s (476)	1860s

6 Conclusion

In this article, a computationally fast qualitative technique for probing acoustic media for hidden hard obstacles on the basis of overdetermined boundary data, based on the computation via the FM-BEM of the distribution of topological sensitivity of the cost functional associated with the inverse problem, has been presented. Its usefulness has been demonstrated through results of numerical experiments on the qualitative identification of a hard obstacle in a bounded acoustic domain, for configurations featuring $O(10^5)$ nodal unknowns and $O(10^6)$ sampling points, resulting in overall computing times of a few hours on a 3 GHz PC computer. There is ample scope for increasing these computational sizes. Besides, the proposed approach can be developed for many other physical models, e.g. elastodynamics or electromagnetic waves.

References

1. M. Bonnet: Boundary Integral Equations Methods for Solids and Fluids. John Wiley and Sons, 1999.
2. M. Bonnet, B. B. Guzina: Sounding of finite solid bodies by way of topological derivative. Internat. J. Numer. Methods Engrg. 61 (2004) 2344–2373.
3. D. Colton, A. Kirsch: A simple method for solving inverse scattering problems in the resonance region. Inverse Problems 12 (1996) 383–393.
4. D. Colton, R. Kress: Integral Equation Method in Scattering Theory. John Wiley and Sons, 1983.
5. E. Darve: The fast multipole method: numerical implementation. J. Comp. Phys. 160 (2000) 195–240.

6. H. A. Eschenauer, V. V. Kobelev, A. Schumacher: Bubble method for topology and shape optimization of structures. Structural Optimization 8 (1994) 42–51.

7. G. R. Feijóo: A new method in inverse scattering based on the topological derivative. Inverse Problems 20 (2004) 1819–1840.

8. R. Gallego, G. Rus: Identification of cracks and cavities using the topological sensitivity boundsary integral equation. Comp. Mech. 33 (2004) 154–163.

9. S. Garreau, P. Guillaume, M. Masmoudi: The topological asymptotic for PDE systems: the elasticity case. SIAM J. Control Opt. 39 (2001) 1756–1778.

10. A. Greenbaum: Iterative methods for solving linear systems. SIAM, Philadelphia, 1997.

11. L. Greengard, V. Rokhlin: A fast algorithm for particle simulations. J. Comp. Phys. 73 (1987) 325–348.

12. N. A. Gumerov, R. Duraiswami: Fast multipole methods for the Helmholtz equation in three dimensions. Elsevier Series in Electromagnetism, 2005.

13. B. B. Guzina, M. Bonnet: Topological derivative for the inverse scattering of elastic waves. Quart. J. Mech. Appl. Math. 57 (2004) 161–179.

14. B. B. Guzina, I. Chikichev: From imaging to material identification: a generalized concept of topological sensitivity. In: Proceedings of ECCOMAS 2004 (P. Neittanmäki, et al. eds.).

15. A. Kirsch: The domain derivative and two applications in inverse scattering theory. Inverse Problems 9 (1993) 81–96.

16. Z. Michalewicz, D. B. Fogel: How to solve it: modern heuristics. Springer, 2004.

17. S. Nintcheu Fata, B. B. Guzina: A linear sampling method for near-field inverse problems in elastodynamics. Inverse Problems 20 (2004) 713–736.

18. S. Nintcheu Fata, B. B. Guzina, M. Bonnet: A computational basis for elastodynamic cavity identification in a semi-infinite solid. Comp. Mech. 32 (2003) 370–380.

19. N. Nishimura: Fast multipole accelerated boundary integral equation methods. Appl. Mech. Rev. 55 (2002) 299–324.

20. R. Potthast: On the convergence of a new Newton-type method in inverse scattering. Inverse Problems 17 (2001) 1419–1434.

21. R. Potthast: A survey on sampling and probe methods for inverse problems. Inverse Problems 22 (2006) R1–R47.

22. A. Schumacher: Topologieoptimierung von Bauteilstrukturen unter Verwendung von Lochpositionierungskriterien. Ph.D. thesis, Universität Siegen, 1995.

23. G. Sylvand: La méthode multipôle rapide en électromagnétisme: performances, parallélisation, applications. Ph.D. thesis, Ecole Nationale des Ponts et Chaussées, Noisy le Grand, 2002.

24. G. Sylvand: Complex industrial computations in electromagnetism using the fast multipole method. In: Mathematical and numerical methods aspects of wave propagation (G. Cohen, E. Heikkola, P. Joly, P. Neittaanmäki eds.), pp. 657–662, Springer, 2003.

25. A. Tarantola: Inverse problem theory and methods for model parameter estimation. SIAM, 2005.

Boundary Element Methods for Eddy Current Computation

Ralf Hiptmair

Seminar für Angewandte Mathematik, ETH Zürich, 8092 Zürich, Switzerland
hiptmair@sam.math.ethz.ch

Summary. This paper studies numerical methods for time-harmonic eddy current problems in the case of homogeneous, isotropic, and linear materials. It provides a survey of approaches that entirely rely on boundary integral equations and their conforming Galerkin discretization. Starting point are both **E**- and **H**-based strong formulation, for which issues of gauging and topological constraints on the existence of potentials are discussed.

Direct boundary integral equations and the so-called symmetric coupling of the integral equations corresponding to the conductor and the non-conducting regions are employed. They give rise to coupled variational problems that are elliptic in suitable trace spaces. This implies quasi-optimal convergence of conforming Galerkin boundary element methods, which make use of div_Γ-conforming trial spaces for surface currents.

1 Introduction

A great deal of electromagnetic field problems faced in an industrial context fall into the category of eddy current problems. This applies, for instance, for problems of inductive heating, magnetomechanical valves, and the computation of inductances of bulky conductors in power electronics.

The typical setting of eddy current problems involves a bounded conducting region Ω_c and its complement $\Omega_e := \mathbb{R}^3 \setminus \bar{\Omega}_c$, the non-conducting air region. Usually, Ω_e is supposed to have the electromagnetic properties of empty space ($\epsilon = \epsilon_0$, $\mu = \mu_0$), whereas Ω_c might be filled with some "complex" conducting material. In this paper we restrict our attention to the case of a simple, linear, homogeneous, and isotropic conductor characterized by a constant conductivity $\sigma > 0$ and permeability $\mu_c > 0$. This can be a reasonable approximation for a non-ferromagnetic material like aluminum.

In eddy current simulations the shape of the conductor is usually provided in some CAD format. Therefore, we can take for granted that the surface of Ω_c is piecewise smooth and consists of a few curved faces. In mathematical terms, Ω_c is a curvilinear Lipschitz polyhedron in the sense of [30, Sect. 1]. All the developments of this paper refer to such a geometric setting.

We restrict ourselves to time harmomic current excitation with angular frequency $\omega > 0$. Hence, thanks to the assumed linearity of all materials involved, temporal Fourier transform allows reduction to pure spatial boundary value problems for the unknown complex amplitudes (phasors) of the electromagnetic fields. Two common types of exciting alternating currents will be taken into account:

1. The total current in a loop of the conductor is prescribed (non-local inductive excitation, [42, Sect. 5]). Here, by "loop" we mean a connected component of Ω_c, whose first Betti number is equal to 1. Homeomorphic images of a torus are typical examples, see Fig. 1 (left)
2. A driving force on charge carriers is modelled by a compactly supported generator current $\mathbf{j}_s$, which has to be divergence free everywhere [42, Sect. 3]. The case $\operatorname{supp}\mathbf{j}_s \cap \Omega_c cl = \emptyset$ describes excitation through a stranded inductor coil or antenna (inductive coupling), whereas $\operatorname{supp}\mathbf{j}_s \cap \Omega_c \neq \emptyset$ models wires feeding a current into Ω_c (galvanic coupling, see Fig. 1, right).

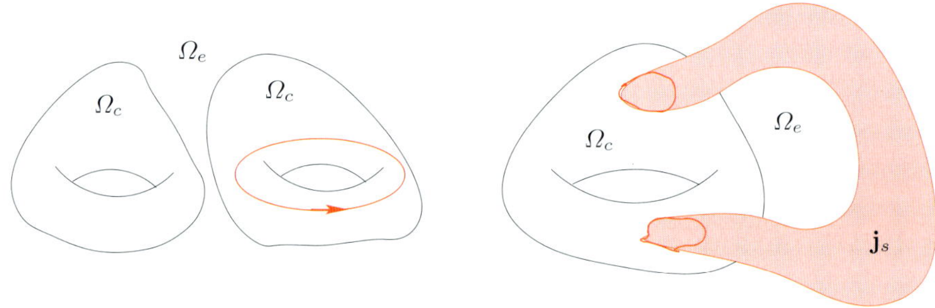

Fig. 1. Current excitations: prescribed total current in a conducting loop (left), generator current $\mathbf{j}_s$ (right). Note that $\mathbf{j}_s$ must be continued inside Ω_c in order to ensure $\operatorname{div}\mathbf{j}_s = 0$!

The goal of the numerical simulation may be the approximate computation of the total Ohmic losses in the conductors, and of the electromagnetic forces acting on the conductor. This entails discretizing the field equations and, in particular, coping with the *unbounded* part Ω_e of the generic computational domain $\mathbb{R}^3$. The standard approach is the finite element method [38], in which artificial homogeneous boundary conditions for the fields are imposed "sufficiently" far away from the conductor. This is justified by the decay properties of the fields, though it may be difficult to fix a viable cut-off distance a priori (see [5] for an adaptive procedure). After meshing the resulting bounded computational domain, the finite element discretization can proceed in the standard fashion. However, in case of a delicate shape of Ω_c, suitable finite element meshes may comprise a prohibitively large number of elements in Ω_e.

Boundary element methods (BEM) applied to the field equations steer clear of these difficulties, since they are based on integral equations posed on the surface $\Gamma := \partial\Omega$. These are only available for homogeneous equations with constant coefficients, but this is just the setting we take for granted (both in Ω_c and Ω_e). Consequently, boundary element methods that rely on a triangulation of Γ alone become an option and will be the focus of this presentation.

A central issue is how to couple the boundary integral equations associated with Ω_c and Ω_e. The basic coupling is provided by the transmission conditions for the electric and magnetic fields, more precisely, their tangential continuity. This still leaves many options, most of which lead to variational problems lacking useful structural properties.

The coupling challenge was first addressed in the context of linking domain based variational formulations with integral equations, a prerequisite for coupling finite elements (FEM) with boundary elements. In this context a breakthrough was achieved when M. Costabel in [27] introduced the so-called *symmetric coupling* by using the integral equations in the form of the Calderón projectors. This idea has been successfully extended to computational electromagnetism in [39, 40, 48].

Representations of Poincare-Steklov operators derived from Calderón projectors also guide the derivation of variational formulations involving only boundary integral equations on an interface [31, 63, 20], see [60, 50] for an application to domain decomposition. Here we aim to adapt these ideas to eddy current models. It turns out that surprising new aspects come into play, related to the issues of *gauging* and *topological obstructions*.

This paper deals with integral equations in variational form and their Galerkin discretization by means of boundary elements. We do not discuss "details" of implementation like computation of matrix entries [58, Ch. 5], matrix compression [58, Ch. 7], and boundary approximation, however important these topics are for a viable code. Instead we refer to the theses [54] and [10] for further information and numerical examples. We also gloss over the issue of how to construct fast iterative solvers for the resulting linear systems of equation. The reader may consult [39, Sect. 9] and [15, 23, 61].

2 Eddy Current Model

The behavior of an electromagnetic field is governed by Maxwell's equations. Instead of using these, in special situations simplified *quasistatic models* supply sufficiently accurate approximations to the true fields [33]. One of them is the eddy current model, representing a magneto-quasistatic approximation to Maxwell's equations in the sense that the electric field energy is neglected.

This model is reasonably accurate for *slowly varying* fields, for which the change in magnetic field energy is dominant [3, 33]. "Slowly varying", means that

$$L \sqrt{\epsilon_0 \, \mu_0} \, \omega \ll 1 \,, \tag{1}$$

where L is the characteristic size of the region of interest: Ω_c has to be small compared to the wavelength of electromagnetic waves, which makes it possible to ignore wave propagation. There is a second condition for the validity of the eddy current approximation, requiring that the typical time-scale is long compared to the relaxation time for space charges, that is, the conductivity must be large enough so that

$$\omega \frac{\epsilon_0}{\sigma} \ll 1 \,. \tag{2}$$

This implies that no space charges need to be taken into account. We point out that (1) and (2) provide a "rule of thumb", but ignore the impact of geometry: in the presence of thin slots or gaps the eddy current approximation might become invalid locally [8, Ch. 8].

Formally, the eddy current model arises from Maxwell's equations by dropping the displacement current $\mathbf{D}$. In the frequency domain the eddy current model for complex field amplitudes (for the electric field $\mathbf{E}$ and the magnetic field $\mathbf{H}$) reads

$$\mathbf{curl}\,\mathbf{E} = -i\omega\mu\mathbf{H} \quad \text{in } \mathbb{R}^3\,, \quad \mathbf{curl}\,\mathbf{H} = \begin{cases} \sigma\mathbf{E} & \text{in } \Omega_c \\ \mathbf{j}_s & \text{in } \Omega_e \end{cases}. \tag{3}$$

According to the aforementioned assumptions, the permeability μ is constant $\equiv \mu_c$ in Ω_c and equal to μ_0 in the air region Ω_e. The conductivity σ is constant in Ω_c and set to zero in Ω_e. The first equation is called Faraday's law, the second (reduced) Ampere's law. These equations have to be supplemented by the decay conditions

$$\mathbf{H}(\mathbf{x}) = O(|\mathbf{x}|^{-1}), \quad \mathbf{E}(\mathbf{x}) = O(|\mathbf{x}|^{-1}) \quad \text{uniformly for } |\mathbf{x}| \to \infty \,. \tag{4}$$

Switching from the full Maxwell's equations to the eddy current equations obviously involves a breach of the symmetry between electric and magnetic quantities. As a first consequence, we cannot expect a solution for $\mathbf{E}$ to be unique, because it can be altered by any gradient supported in Ω_e and will still satisfy the equations (3). The solution for $\mathbf{H}$ will not be affected. This reflects the fact that in a magnetoquasistatic model $\mathbf{E}$ is relegated to the role of a "fictitious quantity". Imposing the constraints

$$\operatorname{div}\mathbf{E} = 0 \quad \text{in } \Omega_e \quad \text{and} \quad \int_{\Gamma_k} \mathbf{E} \cdot \mathbf{n}\, dS = 0 \,, \tag{5}$$

where Γ_k, $k = 1, \ldots, L$, are the connected components of Γ, will restore uniqueness of the solution for $\mathbf{E}$. Thus, one can single out a physically meaningful electric field in Ω_e [1]. However, this is rather a gauging procedure, i.e.

the selection of a representative from an equivalence class of meaningful fields [44], than part of the generic eddy current model. When devising a numerical scheme, we should target $\mathbf{H}$ as principal variable.

How can there be a role of the electric field in a magneto-quasistatic context? To understand this, recall that Faraday's law in strong form involves $\operatorname{div}(\mu\mathbf{H}) = 0$ everywhere. This makes it possible to introduce a magnetic vector potential $\mathbf{A}$ such that $\operatorname{\mathbf{curl}}\mathbf{A} = \mu\mathbf{H}$ and to express $\mathbf{E}$ via a scalar potential Ψ as $\mathbf{E} = -\operatorname{\mathbf{grad}}\Psi - i\mathbf{A}$. We have ample freedom to perform gauging and use it to set $\Psi = 0$. Thus, $\mathbf{E}$ turns out to be a scaled magnetic vector potential in disguise. I endorse this view as the proper reading of $\mathbf{E}$ in (3).

A second consequence of the magneto-quasistatic model reduction is the partial decoupling of electric and magnetic field in Ω_e. In fact, knowing $\mathbf{H}$ on Γ, we can solve a div-$\operatorname{\mathbf{curl}}$ boundary value problem to obtain $\mathbf{H}$ and then, in light of (5), another div-$\operatorname{\mathbf{curl}}$ problem will yield $\mathbf{E}$. Conversely, within the conductor, (3) permits the elimination of either $\mathbf{H}$ and $\mathbf{E}$, which leads to a second-order boundary value problem. The bottom line is, that in Ω_c and Ω_e we encounter elliptic systems of PDEs of different character. This will have profound consequences for the statement of transmission problems, see Sect. 5.

We finish this introduction to the eddy current model by explaining how to incorporate current excitation through *offset fields* $\mathbf{E}_s$ and $\mathbf{H}_s$. We demand

$$\operatorname{\mathbf{curl}}\operatorname{\mathbf{curl}}\mathbf{E}_s = -i\omega\mu_0\mathbf{j}_s \ , \qquad \operatorname{\mathbf{curl}}\mathbf{H}_s = \mathbf{j}_s \ ,$$
$$\operatorname{div}\mathbf{E}_s = 0 \ , \qquad\qquad \operatorname{div}\mathbf{H}_s = 0 \ , \qquad \text{in } \Omega_e \ . \qquad (6)$$

Such fields can be computed by evalutating the Newton potentials

$$\mathbf{E}_s(\mathbf{x}) = -i\omega\frac{\mu_0}{4\pi}\int_{\mathbb{R}^3}\frac{\mathbf{j}_s(\mathbf{y})}{|\mathbf{x}-\mathbf{y}|}\,\mathrm{d}\mathbf{y} \ , \quad \mathbf{H}_s(\mathbf{x}) = \frac{1}{i\omega\mu_0}\operatorname{\mathbf{curl}}\mathbf{E}_s \ , \qquad (7)$$

provided that $\mathbf{j}_s$ has vanishing divergence everywhere in $\mathbb{R}^3$. In the case of thin wires represented by line currents, (7) amounts to the well-known Biot-Savart formula.

The requirements (6) impliy for the *reaction fields* $\mathbf{E}_r := \mathbf{E} - \mathbf{E}_s$, $\mathbf{H}_r := \mathbf{H} - \mathbf{H}_s$ that

$$\operatorname{\mathbf{curl}}\operatorname{\mathbf{curl}}\mathbf{E}_r = 0 \quad , \quad \operatorname{\mathbf{curl}}\mathbf{H}_r = 0 \quad \text{in } \Omega_e \ . \qquad (8)$$

In Ω_c we retain the original phasors $\mathbf{E}$, $\mathbf{H}$, often referrred to as *total fields*. By using offset fields the spatially distributed excitation $\mathbf{j}_s$ can be converted into an inhomogeneous jump condition across Γ for the fields. Spatial source terms are no longer present, which greatly facilitates the implementation of boundary element methods. The treatment of an excitation through a total loop current will be postponed until discretization is discussed in Sects. 7.2, 8.3.

3 Spaces and Traces

All developments in this paper will be consistently set in a variational framework. The Hilbert spaces, on which the variational approach rests, have a very concrete physical meaning as spaces of fields with finite energy. Let $\Omega \subset \mathbb{R}^3$ be a generic domain, not necessarily bounded. The natural Hilbert space for magnetic fields with finite total energy on Ω is

$$\boldsymbol{H}(\mathbf{curl}; \Omega) := \{ \mathbf{V} \in \boldsymbol{L}^2(\Omega), \mathbf{curl}\, \mathbf{V} \in \boldsymbol{L}^2(\Omega) \} \,,$$

equipped with the graph norm (cf. [36, Ch. 1]). In the context of the eddy current model the energy associated with the electric field is measured only by its **curl**. Of course, also the mean dissipated energy has to be finite, which entails square integrability over Ω_c, but in Ω_e the L^2-norm of the field need not be bounded. Therefore, weighted Beppo-Levi type spaces (cf. [35])

$$\boldsymbol{W}(\mathbf{curl}, \Omega) := \{ \frac{\mathbf{V}(\mathbf{x})}{\sqrt{1 + |\mathbf{x}|^2}} \in \boldsymbol{L}^2(\Omega), \mathbf{curl}\, \mathbf{V} \in \boldsymbol{L}^2(\Omega) \}$$

are the proper choice for **E**. The property that their energy only depends on certain derivatives is characteristic for potentials. For them weighted spaces have to be used, for instance the standard Beppo Levi space (cf. [53, Sect. 2.5.4])

$$W^1(\Omega) := \{ \frac{\Phi(\mathbf{x})}{\sqrt{1 + |\mathbf{x}|^2}} \in L^2(\Omega), \mathbf{grad}\, \Phi \in \boldsymbol{L}^2(\Omega) \} \,.$$

For each of the above spaces, the restrictions to Ω of smooth functions that are compactly supported in $\mathbb{R}^3$ form dense subsets.

Thanks to this density property we may wonder how to extend certain restrictions of smooth functions onto boundaries to continuous and surjective trace mappings. Now, assume that the boundary $\partial\Omega$ is compact and endowed with an exterior unit normal vectorfield $\mathbf{n} \in L^\infty(\partial\Omega)$. The pointwise restriction of functions in $C^\infty(\bar{\Omega})$ spawns the standard trace $\gamma : W^1(\Omega) \mapsto H^{\frac{1}{2}}(\partial\Omega)$. However, the relevant traces for electromagnetic fields are tangential traces of vectorfields. We can distinguish between the tangential components trace $\gamma_{\mathbf{t}}$ for $\mathbf{U} \in \boldsymbol{C}^\infty(\bar{\Omega})$ defined by $(\gamma_{\mathbf{t}}\mathbf{U})(\mathbf{x}) = \mathbf{n}(\mathbf{x}) \times (\mathbf{U}(\mathbf{x}) \times \mathbf{n}(\mathbf{x}))$ for almost all $\mathbf{x} \in \partial\Omega$, and the twisted tangential trace $(\gamma_\times \mathbf{U})(\mathbf{x}) := \mathbf{U}(\mathbf{x}) \times \mathbf{n}(\mathbf{x})$.

In eddy current computations we usually face non-smooth surfaces. This profoundly affects the smoothness of restrictions, in particular of tangential traces. Just keep in mind that even for smooth vectorfields their tangential traces will feature discontinuities at ridges and corners of $\partial\Omega$. Therefore it takes sophisticated techniques to devise meaningful tangential trace operators on the function spaces. For domains with piecewise smooth boundary they were developed in [16, 17, 12, 18]. These papers and, in particular [13], should be consulted as main references.

Before we tackle $\boldsymbol{W}(\mathbf{curl}, \Omega)$, we remind (see [16, Prop. 1.7]) that on piecewise smooth boundaries spaces $\boldsymbol{H}^{\frac{1}{2}}_{\|}(\Gamma)$ and $\boldsymbol{H}^{\frac{1}{2}}_{\perp}(\Gamma)$ can be introduced so that the tangential traces become continuous and surjective operators $\gamma_{\mathbf{t}} : \boldsymbol{H}^1(\Omega) \mapsto \boldsymbol{H}^{\frac{1}{2}}_{\|}(\Gamma), \gamma_\times : \boldsymbol{H}^1(\Omega) \mapsto \boldsymbol{H}^{\frac{1}{2}}_{\perp}(\Gamma)$. Sloppily speaking, $\boldsymbol{H}^{\frac{1}{2}}_{\|}(\Gamma)$

contains the tangential surface vectorfields that are in $\boldsymbol{H}^{\frac{1}{2}}(\partial\Omega)$ for each smooth component of $\partial\Omega$ and feature a suitable "tangential continuity" across the edges. A corresponding "normal continuity" is satisfied by surface vectorfields in $\boldsymbol{H}_{\perp}^{\frac{1}{2}}(\Gamma)$. The associated dual spaces will be denoted by $\boldsymbol{H}_{\parallel}^{-\frac{1}{2}}(\Gamma)$ and $\boldsymbol{H}_{\perp}^{-\frac{1}{2}}(\Gamma)$, respectively.

Armed with these spaces and the density of smooth functions, the integration by parts formula

$$\int_{\Omega} \mathbf{curl}\,\mathbf{V}\cdot\mathbf{U} - \mathbf{V}\cdot\mathbf{curl}\,\mathbf{U}\,\mathrm{d}\mathbf{x} = \int_{\partial\Omega} \gamma_{\times}\mathbf{U}\cdot\gamma_{\mathbf{t}}\mathbf{V}\,\mathrm{d}S \qquad (9)$$

is the key to establishing trace theorems for $\boldsymbol{W}(\mathbf{curl},\Omega)$. Recall that the surface divergence operator div_{Γ} is the $\boldsymbol{L}^2(\partial\Omega)$-adjoint of the surface gradient $\mathbf{grad}_{\Gamma}$. By rotating tangential surface vectorfields by $\frac{\pi}{2}$, we get the same relationship between the scalar valued surface rotation curl_{Γ} and the tangent vector valued $\mathbf{curl}_{\Gamma}$. Using, first, $\mathbf{V}\in\boldsymbol{H}^1(\Omega)$, and, secondly, $\mathbf{V}\in\mathbf{grad}\,H^2(\Omega)$, we learn from (9) that

$$\gamma_{\mathbf{t}}: \boldsymbol{H}(\mathbf{curl};\Omega)\mapsto\boldsymbol{H}_{\perp}^{-\frac{1}{2}}(\mathbf{curl}_{\Gamma},\Gamma)\,,$$

$$\gamma_{\times}: \boldsymbol{H}(\mathbf{curl};\Omega)\mapsto\boldsymbol{H}_{\parallel}^{-\frac{1}{2}}(\mathrm{div}_{\Gamma},\Gamma)\,,$$

are continuous trace mappings. Here, we used the notations

$$\boldsymbol{H}_{\perp}^{-\frac{1}{2}}(\mathbf{curl}_{\Gamma},\Gamma) := \{\mathbf{v}\in\boldsymbol{H}_{\perp}^{-\frac{1}{2}}(\Gamma),\,\mathrm{curl}_{\Gamma}\,\mathbf{v}\in H^{-\frac{1}{2}}(\partial\Omega)\}\,,$$

$$\boldsymbol{H}_{\parallel}^{-\frac{1}{2}}(\mathrm{div}_{\Gamma},\Gamma) := \{\lambda\in\boldsymbol{H}_{\parallel}^{-\frac{1}{2}}(\Gamma),\,\mathrm{div}_{\Gamma}\,\lambda\in H^{-\frac{1}{2}}(\partial\Omega)\}\,,$$

for spaces of tangential traces. Moreover, according to Thm. 2.7 and Thm. 2.8 in [16], they are surjective, too. Thus, we have found the right tangential trace spaces for $\boldsymbol{H}(\mathbf{curl};\Omega)$. By (9) the spaces $\boldsymbol{H}_{\parallel}^{-\frac{1}{2}}(\mathrm{div}_{\Gamma},\Gamma)$ and $\boldsymbol{H}_{\perp}^{-\frac{1}{2}}(\mathbf{curl}_{\Gamma},\Gamma)$ can be seen to be dual to each other (see [17, Sect. 4]). The sesqui-linear duality pairing will be denoted by $\langle\cdot,\cdot\rangle_{\tau}$. Moreover, the rotation mapping $\mathrm{R}\mathbf{v}:=\mathbf{v}\times\mathbf{n}$ can be extended to an isometry between the two spaces.

Integration by parts permits us to introduce several important weakly defined traces: The weak normal trace $\gamma_{\mathbf{n}}$ is defined for vectorfields $\mathbf{U}\in\boldsymbol{H}(\mathrm{div};\Omega):=\{\mathbf{V}\in\boldsymbol{L}^2(\Omega),\,\mathrm{div}\,\mathbf{V}\in L^2(\Omega)\}$ by

$$\langle\gamma_{\mathbf{n}}\mathbf{U},\gamma\Phi\rangle_{1/2,\Gamma} = \int_{\Omega}\mathrm{div}\,\mathbf{U}\,\overline{\Phi} + \mathbf{U}\cdot\mathbf{grad}\,\overline{\Phi}\,\mathrm{d}\mathbf{x}\quad\forall\Phi\in H^1(\Omega)\,,$$

with $\langle\cdot,\cdot\rangle_{1/2,\Gamma}$ as duality pairing between $H^{-\frac{1}{2}}(\partial\Omega)$ and $H^{\frac{1}{2}}(\partial\Omega)$. The mapping $\gamma_{\mathbf{n}}:\boldsymbol{H}(\mathrm{div};\Omega)\mapsto H^{-\frac{1}{2}}(\partial\Omega)$ is continuous and surjective, and an extension of the normal components trace $\gamma_{\mathbf{n}}\mathbf{U}(\mathbf{x}):=\mathbf{U}(\mathbf{x})\cdot\mathbf{n}(\mathbf{x})$. Thus, the conormal trace $\partial_{\mathbf{n}}:=\gamma_{\mathbf{n}}\circ\mathbf{grad}$ is continuous and surjective from $H(\Delta,\Omega):=\{\Phi\in W^1(\Omega),\,\Delta\Phi\in\boldsymbol{L}^2(\Omega)\}$ onto $H^{-\frac{1}{2}}(\partial\Omega)$.

Against the backdrop of boundary value problems for the Laplacian $-\Delta$, the trace operator $\gamma : H^1(\Omega) \mapsto H^{\frac{1}{2}}(\Gamma)$ can be called "Dirichlet trace", whereas $\partial_\mathbf{n}$ provides the "Neumann trace". For $\Psi \in H(\Delta, \Omega)$ and $\Phi \in H^1(\Omega)$, they are linked by another Green's formula

$$\langle \partial_\mathbf{n} \Psi, \gamma\Phi \rangle_{1/2,\Gamma} = \int_\Omega \Delta\Psi\,\overline{\Phi} + \mathbf{grad}\,\Psi \cdot \mathbf{grad}\,\overline{\Phi}\,\mathrm{dx}\ . \tag{10}$$

The eddy current equations prominently feature the operator $\mathbf{curl\,curl}$ and we may wonder about suitable Dirichlet- and Neumann traces. Since the energy space associated with $\mathbf{curl\,curl}$ is $\boldsymbol{H}(\mathbf{curl};\Omega)$, the previous discussion reveals that $\gamma_\mathbf{t}$ can be used as Dirichlet trace. In view of (10) a $\mathbf{curl\,curl}$-counterpart γ_N of $\partial_\mathbf{n}$ can be defined for

$$\mathbf{U} \in \boldsymbol{W}(\mathbf{curl}^2, \Omega) := \{\mathbf{V} \in \boldsymbol{W}(\mathbf{curl}, \Omega), \mathbf{curl\,curl}\,\mathbf{V} \in \boldsymbol{L}^2(\Omega)\}$$

by demanding that for all $\mathbf{V} \in \boldsymbol{H}(\mathbf{curl};\Omega)$

$$\langle \gamma_N \mathbf{U}, \gamma_\mathbf{t} \mathbf{V} \rangle_\tau = \int_\Omega \mathbf{curl}\,\mathbf{U} \cdot \mathbf{curl}\,\overline{\mathbf{V}} - \mathbf{curl\,curl}\,\mathbf{U} \cdot \overline{\mathbf{V}}\,dx\ . \tag{11}$$

The trace γ_N furnishes a continuous and surjective mapping

$$\gamma_N : \boldsymbol{W}(\mathbf{curl}^2, \Omega) \mapsto \boldsymbol{H}_{||}^{-\frac{1}{2}}(\mathrm{div}_\Gamma, \Gamma) \quad (\text{cf. [39, Lemma 3.3]}),$$

which can be regarded as an extension of the restriction $(\gamma_N \mathbf{U})(\mathbf{x}) := \mathbf{curl}\,\mathbf{U}(\mathbf{x}) \times \mathbf{n}(\mathbf{x})$, $\mathbf{x} \in \partial\Omega$, for smooth $\mathbf{U}$.

We mention two commuting relationships between traces that are elementary for smooth functions and, by extension, carry over to the trace operators in Sobolev spaces:

$$\mathbf{grad}_\Gamma \circ \gamma = \gamma_\mathbf{t} \circ \mathbf{grad} \quad \text{on } W^1(\Omega)\ , \tag{12}$$

$$\gamma_\mathbf{n} \circ \mathbf{curl} = \mathrm{curl}_\Gamma \circ \gamma_\mathbf{t} = \mathrm{div}_\Gamma \circ \gamma_\times \quad \text{on } \boldsymbol{W}(\mathbf{curl}, \Omega)\ , \tag{13}$$

where equality is in the sense of the trace spaces $\boldsymbol{H}_\perp^{-\frac{1}{2}}(\mathbf{curl}_\Gamma, \Gamma)$ and $H^{-\frac{1}{2}}(\Gamma)$, respectively.

Integration by parts also shows that a vectorfield in $\boldsymbol{C}^\infty(\Omega_c cl) \cap \boldsymbol{C}^\infty(\overline{\Omega}_e)$ must feature tangential continuity in order to be contained in $\boldsymbol{W}(\mathbf{curl}, \mathbb{R}^3)$. Thus, both $\mathbf{E}$ and $\mathbf{H}$ can only belong to $\boldsymbol{W}(\mathbf{curl}, \mathbb{R}^3)$, if the following *transmission conditions* hold across $\Gamma := \partial\Omega_c$

$$[\gamma_\mathbf{t} \mathbf{E}]_\Gamma = 0 \quad \text{and} \quad [\gamma_\mathbf{t} \mathbf{H}]_\Gamma = 0\ . \tag{14}$$

Here, the "jump" $[\cdot]_\Gamma$ designates the difference of the values of a trace from Ω_e ("exterior") and from Ω_c ("interior"). We also stick to the convention that exterior traces will be labeled by a superscript $+$, whereas traces from Ω_c bear a superscript $-$.

4 Topological Prerequisites

Topological considerations come into play, when one wants to represent irrotational vectorfields on manifolds through gradients of scalar potentials. This is only possible, if the first cohomology group of the manifold is trivial [59, Ch. 6]. Otherwise, *cuts* have to be used to take care of irrotational vectorfields that are no gradients [8, Sect. 8.3.4], see Fig. 2.

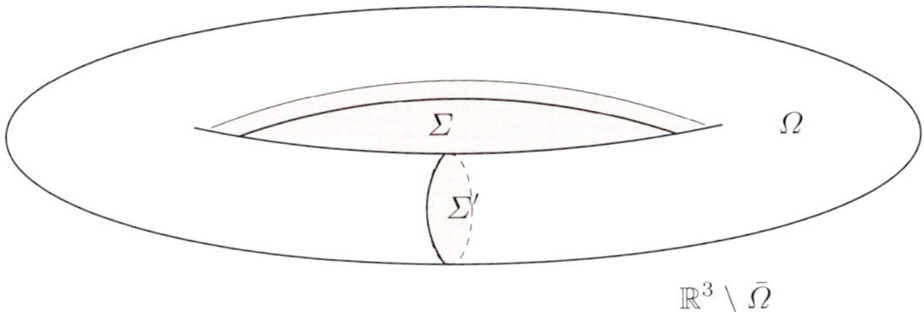

Fig. 2. Cut Σ' for the torus and cut Σ for its complement in $\mathbb{R}^3$.

Theorem 1. *For every domain $\Omega \subset \mathbb{R}^3$ with piecewise smooth boundary there exist piecewise smooth orientable embedded surfaces $\Sigma_1, \ldots, \Sigma_N \subset \bar{\Omega}$ (cuts), where N agrees with the first Betti number of Ω, such that*

- *the Σ_k, $k = 1, \ldots, N$, are mutually disjoint.*
- *the first cohomology group $H^1(\Omega', \mathbb{Z})$ of $\Omega' := \Omega \setminus (\Sigma_1 \cup \ldots \cup \Sigma_N)$ is trivial.*
- *Ω' is a generalized Lipschitz domain in the sense of [29], that is, when "seen from one side" its boundary $\partial \Omega'$ is Lipschitz continuous.*

Proof. The theorem is proved in [45]. $\square$

In the sequel we are going to equip Ω_e with a set of cuts $\Sigma_1, \ldots, \Sigma_N$, according to Thm. 1. Each Σ_k has an orientation that translates into a crossing direction and thus we can distinguish between an "upper" surface Σ_k^+ and a "lower" surface Σ_k^-. Both surfaces are equipped with unit normal vectorfields $\mathbf{n}_k^+, \mathbf{n}_k^-$ pointing "away from Σ_k" into the interior of $\Omega' := \Omega_e \setminus (\Sigma_1 \cup \ldots \cup \Sigma_N)$. We fix $\mathbf{n}_{|\Sigma_k} := \mathbf{n}_k^+$ so that it agrees with the crossing direction.

The statement of Thm. 1 implies

$$\mathbf{V} \in \boldsymbol{H}(\mathbf{curl}; \Omega'), \ \mathbf{curl}\,\mathbf{V} = 0 \quad \Rightarrow \quad \exists \Phi \in H^1(\Omega') : \ \mathbf{V} = \mathbf{grad}\,\Phi .$$

It is even possible to characterize low dimensional spaces of vectorfields that fill the gap between $\mathrm{Ker}(\mathbf{curl}) \cap \boldsymbol{H}(\mathbf{curl}; \Omega_e)$ and $\mathbf{grad}\,H^1(\Omega_e)$. To that end, consider functions $\eta_k \in H^1(\Omega_e \setminus \Sigma_k)$, $k = 1, \ldots, N$, with $[\eta_k]_{\Sigma_k} = 1$. Here, $[\cdot]_S$ denotes the jump of some function across the externally oriented surface S, i.e. the difference of its value on the "+-side" and the "−-side".

Theorem 2. *Using the notations introduced above, we have the representation*

$$\mathrm{Ker}(\mathbf{curl}) \cap \boldsymbol{H}(\mathbf{curl};\Omega_e) = \mathbf{grad}\, H^1(\Omega_e) + \mathrm{Span}\left\{\widetilde{\mathbf{grad}}\,\eta_1, \ldots, \widetilde{\mathbf{grad}}\,\eta_N\right\},$$

where $\widetilde{\mathbf{grad}}\,\eta_k \in \boldsymbol{L}^2(\Omega_e)$ *is the gradient of* η_k *on* $\Omega_e \setminus \Sigma_k$.

Proof. Compare Sect. 3 in [4].   □

From Thm. 2 we learn that

$$\mathrm{Ker}(\mathbf{curl}) \cap \boldsymbol{H}(\mathbf{curl};\Omega_e) = \widetilde{\mathbf{grad}}\, H^1_\Sigma(\Omega_e),\tag{15}$$

with $H^1_\Sigma(\Omega_e) := \{\varphi \in H^1(\Omega'), [\varphi]_{\Sigma_k} = \mathrm{const.}, 1 \le k \le N\}$.

Thm. 1 may also be applied to Ω_c yielding N cutting surfaces $\widehat{\Sigma}_1, \ldots, \widehat{\Sigma}_N$, since the first Betti numbers of Ω_c and Ω_e agree. The boundaries $\sigma_1, \ldots, \sigma_N$, $\widehat{\sigma}_1, \ldots, \widehat{\sigma}_N$ of Σ_k and $\widehat{\Sigma}_k$, $k = 1, \ldots, N$, respectively, represent a basis of the homology group $H_1(\Gamma, \mathbb{Z})$, see Fig. 3. In analogy to Thm. 2 we find that

$$\mathrm{Ker}(\mathrm{div}_\Gamma) \cap \boldsymbol{H}(\mathrm{div};\Gamma) = \mathbf{curl}_\Gamma\, H^1(\Gamma) + \mathrm{Span}\left\{\mathbf{g}^1, \ldots, \mathbf{g}^N, \widehat{\mathbf{g}}^1, \ldots, \widehat{\mathbf{g}}^N\right\},\tag{16}$$

where $\mathbf{g}^k$ is the vectorial surface rotation $\mathbf{curl}_\Gamma\, \varphi$ of some $\varphi \in H^1(\Gamma \setminus \sigma_k)$ that has a jump of constant height 1 across σ_i. The $\widehat{\mathbf{g}}^k$ are constructed analogously with respect to $\widehat{\sigma}_k$.

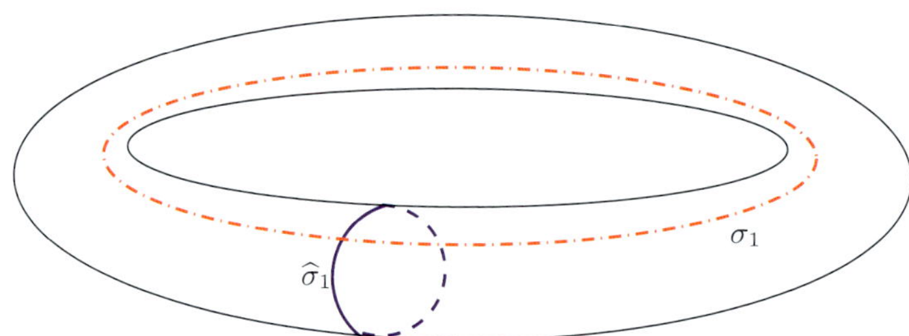

Fig. 3. Fundamental cycles σ_1 and $\widehat{\sigma}_1$ for the surface of the torus, a domain with first Betti number $= 1$.

We remark that if Γ is equipped with some non-degenerate triangulation Γ_h (rendering it a cellular complex) the boundaries of interior and exterior cutting surfaces can be chosen such that they agree with edge cycles of Γ_h. Further, it is possible to pick piecewise smooth Lipschitz surfaces as related cuts. Such a choice of cuts will be a tacit assumption, whenever a triangulation of Γ has been fixed.

Remark 1. Please be aware that it is not the purpose of cuts to render Ω' simply connected, i.e., to ensure that it has a trivial first *homotopy* group. This is easily seen in the case of knotted geometries.

5 Variational Formulations and Transmission Problems

Two fundamentally different approaches to a variational formulation of (3) are conceivable. They can be distinguished by which equation is preserved in strong form and which is taken into account only in weak form [7]. This distinction parallels the primal and dual variational principles known from second order elliptic boundary value problems [11, Ch. 1]. A discusion for the full Maxwell's equations in frequency domain is given in [38, Sect. 2.3].

The first approach involves Faraday's law in strong form. It is used to replace $\mathbf{H}$ in Ampere's law and the latter is multiplied by a test function in $\boldsymbol{H}(\mathbf{curl}; \mathbb{R}^3)$ and subjected to integration by parts according to (9). This results in the following "$\mathbf{E}$-based" variational problem (cf. [7, Sect. 3], [56], and [57]): Seek $\mathbf{E} \in \boldsymbol{W}(\mathbf{curl}, \mathbb{R}^3)$ such that for all $\mathbf{V} \in \boldsymbol{W}(\mathbf{curl}, \mathbb{R}^3)$

$$\left(\tfrac{1}{\mu} \mathbf{curl}\,\mathbf{E}, \mathbf{curl}\,\mathbf{V} \right)_{\boldsymbol{L}^2(\mathbb{R}^3)} + i\omega\, (\sigma\mathbf{E}, \mathbf{V})_{\boldsymbol{L}^2(\Omega_c)} = -i\omega\, (\mathbf{j}_s, \mathbf{V})_{\boldsymbol{L}^2(\mathbb{R}^3)} \ . \qquad (17)$$

Theorem 3. *The variational problem* (17) *has a unique solution for* $\mathbf{H} := -\tfrac{1}{i\omega\mu}\,\mathbf{curl}\,\mathbf{E} \in \boldsymbol{H}(\mathbf{curl}; \mathbb{R}^3)$. *If it is posed on the constrained space*

$$\mathcal{W} := \{\mathbf{V} \in \boldsymbol{W}(\mathbf{curl}, \mathbb{R}^3),\ \mathrm{div}\,\mathbf{V} = 0 \ \text{in}\ \Omega_e,\ \int_{\Gamma_k} \gamma_{\mathbf{n}}\mathbf{V}\,dS = 0,\ k = 1, \ldots, L \} \ ,$$

a unique solution $\mathbf{E} \in \mathcal{W}$ *exists. Here* Γ_k, $k = 1, \ldots, L$, *stand for the connected components of* Γ.

Proof. The reader is referred to [3, Sect. 3] and [39, Sect. 2]. $\quad\square$

A crucial observation is that (17) is equivalent to a *transmission problem*. To state it, we first appeal to the transmission conditions (14). Secondly, testing (17) with fields compactly supported in Ω_c or Ω_e, and making use of the offset fields from (6), we get

$$\mathbf{curl}\,\mathbf{curl}\,\mathbf{E} + i\omega\mu_c\sigma\mathbf{E} = 0 \quad \text{in}\ \Omega_c \ , \qquad (18)$$

$$\mathrm{div}\,\mathbf{E}_r = 0 \quad , \quad \mathbf{curl}\,\mathbf{curl}\,\mathbf{E}_r = 0 \quad \text{in}\ \Omega_e \ ,$$

$$\gamma_{\mathbf{t}}^+\mathbf{E}_r - \gamma_{\mathbf{t}}^-\mathbf{E} = -\gamma_{\mathbf{t}}^+\mathbf{E}_s \quad , \quad \frac{1}{\mu_0}\gamma_N^+\mathbf{E}_r - \frac{1}{\mu_c}\gamma_N^-\mathbf{E} = -\frac{1}{\mu_0}\gamma_N^+\mathbf{E}_s \ .$$

Here we have skimped on the full "gauge conditions" (5), that is $\mathbf{E}_r \in \mathcal{W}$, for the reaction field $\mathbf{E}_r$ in Ω_e.

The second option for a variational formulation is to keep Ampere's law strongly, leading to "$\mathbf{H}$-based" formulations. Then, we have to use the trial space $\mathbf{H}_s + \mathcal{V}$ with

$$\mathcal{V} := \{\mathbf{V} \in \boldsymbol{H}(\mathbf{curl}; \mathbb{R}^3),\ \mathbf{curl}\,\mathbf{V} = 0 \ \text{in}\ \Omega_e\}$$

for $\mathbf{H}$. Remember that the offset field $\mathbf{H}_s$ is to satisfy $\mathbf{curl}\,\mathbf{H}_s = \mathbf{j}_s$ and $\mathrm{div}\,\mathbf{H}_s = 0$ in Ω_e. Now, testing the first equation of (3) with a compactly supported $\mathbf{V} \in \mathcal{V}$, employing integration by parts on a ball with sufficiently large radius, and using the second equation inside Ω_c, we obtain: Seek $\mathbf{H} \in \mathcal{V} + \mathbf{H}_s$ such that

$$\left(\sigma^{-1}\operatorname{\mathbf{curl}}\mathbf{H}, \operatorname{\mathbf{curl}}\mathbf{V}\right)_{\boldsymbol{L}^2(\Omega_c)} + i\omega\left(\mu\mathbf{H}, \mathbf{V}\right)_{\boldsymbol{L}^2(\mathbb{R}^3)} = 0 \quad \forall \mathbf{V} \in \mathcal{V}. \qquad (19)$$

For a more detailed presentation of the considerations leading to (19) the reader is referred to [9], [7, Sect. 2], and [8, Ch. 8]. Existence and uniqueness of solutions of (19) immediately follow from the Lax-Milgram lemma.

Straight from (19) we infer $\operatorname{div}(\mu_r\mathbf{H}) = 0$ in all of $\mathbb{R}^3$. This involves the normal continuity of $\mu_r\mathbf{H}$ across Γ. We are led to a transmission problem for the total magnetic field inside Ω_c and the reaction field $\mathbf{H}_r$ outside:

$$\operatorname{\mathbf{curl}}\sigma^{-1}\operatorname{\mathbf{curl}}\mathbf{H} + i\omega\mu_c\mathbf{H} = 0 \quad \text{in } \Omega_c ,$$
$$\operatorname{\mathbf{curl}}\mathbf{H} = 0 \quad, \quad \operatorname{div}\mathbf{H} = 0 \quad \text{in } \Omega_e , \qquad (20)$$
$$\mu_0\gamma_{\mathbf{n}}^+\mathbf{H}_r - \mu_c\gamma_{\mathbf{n}}^-\mathbf{H} = -\mu_0\gamma_{\mathbf{n}}^+\mathbf{H}_s \quad, \quad \gamma_{\mathbf{t}}^+\mathbf{H}_r - \gamma_{\mathbf{t}}^-\mathbf{H} = \gamma_{\mathbf{t}}\mathbf{H}_s \quad \text{on } \Gamma .$$

However, if Ω_c has non-vanishing first Betti number, then there is no unique solution of (5) [55, 34]. To see this please notice that thanks to Thm. 2 the path integrals $f_k(\mathbf{H}) := \int_{\widehat{\sigma}_k}\mathbf{H}\cdot d\mathbf{s}$ supply continuous functionals on $\mathcal{V}$. They do not vanish, because plugging in an extension to $\boldsymbol{W}(\operatorname{\mathbf{curl}}, \mathbb{R}^3)$ of $\widetilde{\operatorname{\mathbf{grad}}}\,\eta_k$ results in 1. Next, consider the variational problem (19) posed over $\mathcal{V}$, but with f_k as non-homogeneous right hand side. A unique non-zero solution $\mathbf{H}_k \in \mathcal{V}$ exists. From $f_k(\operatorname{\mathbf{grad}}\Phi) = 0$ for all $\Phi \in W^1(\mathbb{R})$ we conclude that still $\operatorname{div}(\mu_r\mathbf{H}_k) = 0$. Hence, $\mathbf{H}_k$ satisfies all the transmission conditions of (5). Testing with smooth vectorfields that are compactly supported in Ω_c establishes the first equation of (5). Summing up, (5) may have non-zero solutions, even if $\mathbf{H}_s = 0$.

These considerations refute the equivalence of (5) and (19). The bottom line is that in general the $\mathbf{H}$-based model does not allow a formulation as transmission problem, unless some extra coupling conditions that, however, fail to involve traces on Γ only, are taken into account. These additional conditions are formulated and investigated in [1] (see also [47]). They turn out to be an integral version of Faraday's law with respect to cuts.

A third class of variational formulations, the hybrid methods, combines primal and dual variational principles, one kind applied in Ω_c the other in Ω_e. An extensive discussion with finite elements in mind is given in [2]. The first option is to work "$\mathbf{H}$-based" inside Ω_c and "$\mathbf{E}$-based" in Ω_e: these formulations can be nicely combined into a transmission problem

$$\operatorname{\mathbf{curl}}\operatorname{\mathbf{curl}}\mathbf{H} + i\omega\mu_c\sigma\mathbf{H} = 0 \quad \text{in } \Omega_c ,$$
$$\operatorname{\mathbf{curl}}\operatorname{\mathbf{curl}}\mathbf{E}_r = 0 \quad, \quad \operatorname{div}\mathbf{E}_r = 0 \quad \text{in } \Omega_e , \qquad (21)$$
$$\gamma_{\times}^+\mathbf{E}_r - \frac{1}{\sigma}\gamma_N^-\mathbf{H} = -\gamma_{\times}^+\mathbf{E}_s \quad, \quad -\frac{1}{i\omega\mu_0}\gamma_N^+\mathbf{E} - \gamma_{\times}^-\mathbf{H} = \gamma_{\mathbf{t}}^+\mathbf{H}_s \quad \text{on } \Gamma .$$

Alternatively, Faraday's law can be used in strong form in Ω_c, and Ampere's law is tested with $\mathbf{V} \in \boldsymbol{H}(\operatorname{\mathbf{curl}}; \mathbb{R}^3)$, but integration by parts is performed on Ω_c only. Therefore, boundary terms have to be retained in the variational equation

$$\left(\frac{1}{\mu_c}\operatorname{\mathbf{curl}}\mathbf{E},\operatorname{\mathbf{curl}}\mathbf{V}\right)_{L^2(\Omega_c)} + i\omega\left(\sigma\mathbf{E},\mathbf{V}\right)_{L^2(\Omega_c)} - \left\langle\frac{1}{\mu_c}\gamma_N^-\mathbf{E},\gamma_{\mathbf{t}}^-\mathbf{V}\right\rangle_\tau = 0$$

for $\mathbf{V}\in\boldsymbol{H}(\operatorname{\mathbf{curl}};\Omega_c)$. In Ω_e Ampere's law is incorporated strongly by zeroing in on $\mathbf{H}\in\mathbf{H}_s+\mathcal{V}$. Faraday's law is tested with compactly supported irrotational fields only, and subsequently we integrate by parts. We end up with

$$\left\langle\gamma_\times^+\mathbf{V},\gamma_{\mathbf{t}}^+\mathbf{E}\right\rangle_\tau + i\omega\left(\mu_0\mathbf{H},\mathbf{V}\right)_{L^2(\Omega_e)} = 0 \quad \forall\mathbf{V}\in\mathcal{V}\,.$$

Both variational problems are linked through the transmission conditions, which enable us to replace $\frac{1}{\mu_c}\gamma_N^-\mathbf{E}$ by $-i\omega\gamma_\times^+\mathbf{H}$ in the boundary terms. This results in the variational problem [48]: Seek $\mathbf{E}\in\boldsymbol{H}(\operatorname{\mathbf{curl}};\Omega_c)$, $\mathbf{H}\in\mathbf{H}_s+\mathcal{V}$ such that for all $\mathbf{W}\in\boldsymbol{H}(\operatorname{\mathbf{curl}};\Omega_c)$, $\mathbf{V}\in\mathcal{V}$

$$\left(\tfrac{1}{\mu_c}\operatorname{\mathbf{curl}}\mathbf{E},\operatorname{\mathbf{curl}}\mathbf{W}\right)_{L^2(\Omega_c)} + i\omega\left(\sigma\mathbf{E},\mathbf{W}\right)_{L^2(\Omega_c)} + i\omega\left\langle\gamma_\times\mathbf{H},\gamma_{\mathbf{t}}\mathbf{W}\right\rangle_\tau = 0\,,$$
$$i\omega\left\langle\gamma_\times\mathbf{V},\gamma_{\mathbf{t}}\mathbf{E}\right\rangle_\tau \qquad\qquad - \omega^2\overline{\left(\mu_0\mathbf{H},\mathbf{V}\right)_{L^2(\Omega_e)}} = 0\,.$$

$$(22)$$

Theorem 4. *The bilinear form associated with the variational problem* (22) *is* $\boldsymbol{H}(\operatorname{\mathbf{curl}};\Omega_c)\times\mathcal{V}$-*elliptic.*

Proof. Setting $\mathbf{W}:=\mathbf{E}$, $\mathbf{V}:=\mathbf{H}$, and subtracting both equations makes the "off-diagonal" terms cancel. $\square$

Similarly as in the case of the $\mathbf{H}$-based model, an equivalent transmission problem is also elusive for the variational problem (22).

In the sequel we are going to focus on the pure $\mathbf{E}/\mathbf{H}$-based formulations (18) and (19), respectively.

6 Boundary Integral Operators

The theory of boundary integral operators for strongly elliptic partial differential operators of second order is well developed [52, 28, 58]. Here, we summarize some of the results as a guidance for developing a similar theory for boundary integral operators for second-order partial differential equations involving the **curl**-operator. The relevance of this for the transmission problem (18) and the variational problem (19) is evident.

The starting point is a representation formula, the famous Green's representation formula for solutions of the homogeneous Helmholtz equation. It relies on the scalar single layer potential

$$\Psi_V^0(\varphi)(\mathbf{x}) := \int_\Gamma G_\kappa(\mathbf{x},\mathbf{y})\,\varphi(\mathbf{y})\,\mathrm{d}\mathbf{y} \quad \mathbf{x} \notin \Gamma\,, \quad \varphi \in H^{-\frac{1}{2}}(\Gamma)\,, \qquad (23)$$

and the scalar double layer potential

$$\Psi_K^0(v)(\mathbf{x}) := \int_\Gamma \frac{\partial}{\partial \mathbf{n}(\mathbf{y})} G_\kappa(\mathbf{x},\mathbf{y})\,v(\mathbf{y})\,\mathrm{d}\mathbf{y} \quad \mathbf{x} \notin \Gamma\,, \quad v \in H^{\frac{1}{2}}(\Gamma)\,, \qquad (24)$$

both based on the Helmholtz kernel [52, Ch. 9]

$$G_\kappa(\mathbf{x},\mathbf{y}) := \frac{\exp(-\kappa|\mathbf{x}-\mathbf{y}|)}{4\pi|\mathbf{x}-\mathbf{y}|}\,, \quad \mathbf{x} \neq \mathbf{y}\,.$$

The potentials owe their significance to the following result [52, Thm. 6.10], [58, Thm. 3.1.6]:

Theorem 5. *Assume $\Re\kappa \geq 0$. Any distribution $U \in H^1_{\mathrm{loc}}(\Omega_c \cup \Omega_e)$ with $-\Delta U + \kappa^2 U = 0$ in $\Omega_c \cup \Omega_e$ and $|U(x)| = O(|x|^{-1})$ for $|x| \to \infty$ can be represented as*

$$U(\mathbf{x}) = -\Psi_V^\kappa([\partial_\mathbf{n} U]_\Gamma) + \Psi_K^\kappa([\gamma U]_\Gamma)\,, \quad \mathbf{x} \notin \Gamma\,.$$

It is well known [52, Thm. 6.11] that the potentials Ψ_V^κ and Ψ_K^κ provide continuous mappings

$$\Psi_V^\kappa : H^{-\frac{1}{2}}(\Gamma) \mapsto H^1_{\mathrm{loc}}(\mathbb{R}^3) \cap H(\Delta, \Omega_c \cup \Omega_e) \qquad (25)$$

$$\Psi_K^\kappa : H^{\frac{1}{2}}(\Gamma) \mapsto H(\Delta, \Omega_c \cup \Omega_e)\,. \qquad (26)$$

In fact, $(-\Delta + \kappa^2)\Psi_V^\kappa = (-\Delta + \kappa^2)\Psi_K^\kappa = 0$ away from Γ [58, Thm. 3.1.1]. We also recall the fundamental *jump relations* for the potentials

$$[\gamma\Psi_V^\kappa(\varphi)]_\Gamma = 0\,, \qquad [\partial_\mathbf{n}\Psi_V^\kappa(\varphi)]_\Gamma = -\varphi\,, \qquad \varphi \in H^{-\frac{1}{2}}(\Gamma)\,, \qquad (27)$$

$$[\gamma\Psi_K^\kappa(v))]_\Gamma = v\,, \qquad [\partial_\mathbf{n}\Psi_K^\kappa(v)]_\Gamma = 0 \qquad v \in H^{\frac{1}{2}}(\Gamma)\,. \qquad (28)$$

The mapping properties (25) and (26) of the potentials ensure that the *boundary integral operators*

$$\begin{aligned}
\mathsf{V}^\kappa &:= \gamma\Psi_V^\kappa & &: H^{-\frac{1}{2}}(\Gamma) \mapsto H^{\frac{1}{2}}(\Gamma)\,,\\
\mathsf{K}^\kappa &:= \tfrac{1}{2}(\gamma^- + \gamma^+)\Psi_K^\kappa & &: H^{\frac{1}{2}}(\Gamma) \mapsto H^{\frac{1}{2}}(\Gamma)\,,\\
\mathsf{K}^{\kappa,*} &:= \tfrac{1}{2}(\partial_\mathbf{n}^- + \partial_\mathbf{n}^+)\Psi_V^\kappa & &: H^{-\frac{1}{2}}(\Gamma) \mapsto H^{-\frac{1}{2}}(\Gamma)\,,\\
\mathsf{D}^\kappa &:= -\partial_\mathbf{n}\Psi_K^\kappa & &: H^{\frac{1}{2}}(\Gamma) \mapsto H^{-\frac{1}{2}}(\Gamma)\,.
\end{aligned} \qquad (29)$$

are well defined and continous [58, Sect. 3.1.2]. Moreover, the single layer boundary integral operator V^κ and hypersingular boundary integral operator D^κ are elliptic in the following sense, see [52, Thms. 7.6,7.8]

$$|\langle\varphi, \mathsf{V}^0\varphi\rangle_{1/2,\Gamma}| \geq c\|\varphi\|^2_{H^{-\frac{1}{2}}(\Gamma)} \quad \forall\varphi \in H^{-\frac{1}{2}}(\Gamma)\,, \qquad (30)$$

$$|\langle\mathsf{D}^0 v, v\rangle_{1/2,\Gamma}| \geq c\|v\|^2_{H^{\frac{1}{2}}(\Gamma)/\mathbb{R}} \quad \forall v \in H^{\frac{1}{2}}(\Gamma)\,, \qquad (31)$$

with constants $c > 0$ depending on Γ only.

Now we attempt to develop similar representation formulas and boundary integral operators related to the differential operator $\mathbf{curl\,curl} + \kappa^2$. It is our first objective to derive a boundary integral representation formula for distributions satisfying the homogeneous equation $\mathbf{curl\,curl\,U} + \kappa^2\mathbf{U} = 0$ in $\Omega_c \cup \Omega_e$. In order to handle transmission conditions in the calculus of distributions, we introduce *currents*, that is, distributions supported on Γ. For a function $\varphi \in H^{-\frac{1}{2}}(\Gamma)$, a tangential vector-field $\xi \in \boldsymbol{H}_{\perp}^{-1}(\Gamma)$, and test functions $\Phi \in \mathcal{D}(\mathbb{R}^3)$, $\boldsymbol{\Phi} \in \boldsymbol{\mathcal{D}}(\mathbb{R}^3) := (\mathcal{D}(\mathbb{R}^3))^3$, we define

$$(\varphi\delta_\Gamma)(\Phi) := \langle \varphi, \gamma\Phi\rangle_{1/2,\Gamma} \,, \quad (\xi\delta_\Gamma)(\boldsymbol{\Phi}) := \langle \xi, \gamma_{\mathbf{t}}\boldsymbol{\Phi}\rangle_\tau = \langle \xi, \gamma\boldsymbol{\Phi}\rangle_{-1,\Gamma} \,.$$

Now, in the sense of distributions, integration by parts yields, cf. [14, Sect. 2.3],

$$\begin{aligned}
&\text{for } \mathbf{U} \in \boldsymbol{H}_{\mathrm{loc}}(\mathrm{div}; \Omega_c \cup \Omega_e) &:& \quad \mathrm{div}\,\mathbf{U} = \mathrm{div}\,\mathbf{U}_{|\Omega_c\cup\Omega_e} + [\gamma_\mathbf{n}\mathbf{U}]_\Gamma\,\delta_\Gamma \,, \\
&\text{for } \mathbf{U} \in \boldsymbol{H}_{\mathrm{loc}}(\mathbf{curl}; \Omega_c \cup \Omega_e) &:& \quad \mathbf{curl}\,\mathbf{U} = \mathbf{curl}\,\mathbf{U}_{|\Omega_c\cup\Omega_e} - [\gamma_\times\mathbf{U}]_\Gamma\,\delta_\Gamma \,, \\
&\text{for } \xi \in \boldsymbol{H}_{\|}^{-\frac{1}{2}}(\mathrm{div}_\Gamma, \Gamma) &:& \quad \mathrm{div}(\xi\,\delta_\Gamma) = (\mathrm{div}_\Gamma\,\xi)\,\delta_\Gamma \,.
\end{aligned}$$

For notational simplicity, we introduce the average $\{\gamma\}_\Gamma = \frac{1}{2}(\gamma^+ + \gamma^-)$ for some trace operator γ. Remember that the superscripts $-$ and $+$ tag traces onto Γ from Ω_c and Ω_e respectively.

Now let $\mathbf{U}$ satisfy $\mathbf{curl\,curl\,U} + \kappa^2\mathbf{U} = 0$ along with $\mathrm{div}\,\mathbf{U} = 0$ in $\Omega_c\cup\Omega_e$. Then the following identity holds in the sense of distributions,

$$\begin{aligned}
-\delta\mathbf{U} + \kappa^2\mathbf{U} &= \mathbf{curl\,curl\,U} - \mathbf{grad}\,\mathrm{div}\,\mathbf{U} + \kappa^2\mathbf{U} \\
&= \mathbf{curl}\left(\mathbf{curl}\,\mathbf{U}_{|\Omega_c\cup\Omega_e} - [\gamma_\times\mathbf{U}]_\Gamma\,\delta_\Gamma\right) - \mathbf{grad}\left([\gamma_\mathbf{n}\mathbf{U}]_\Gamma\,\delta_\Gamma\right) + \kappa^2\mathbf{U} \\
&= \mathbf{curl\,curl}\,\mathbf{U}_{|\Omega_c\cup\Omega_e} - [\gamma_N\mathbf{U}]_\Gamma\,\delta_\Gamma - \mathbf{curl}([\gamma_\times\mathbf{U}]_\Gamma\,\delta_\Gamma) - \\
&\qquad - \mathbf{grad}([\gamma_\mathbf{n}\mathbf{U}]_\Gamma\,\delta_\Gamma) + \kappa^2\mathbf{U} \\
&= -[\gamma_N\mathbf{U}]_\Gamma\,\delta_\Gamma - \mathbf{curl}([\gamma_\times\mathbf{U}]_\Gamma\,\delta_\Gamma) - \mathbf{grad}([\gamma_\mathbf{n}\mathbf{U}]_\Gamma\,\delta_\Gamma) \,.
\end{aligned}$$

We know from [26, Theorem 6.7] that the Cartesian components of $\mathbf{U}$ will satisfy decay conditions and the scalar Helmholtz equation in $\Omega_c \cup \Omega_e$. Using the results from [52, Ch. 9], we can apply component-wise convolution with the outgoing fundamental solution G_κ for the operator $-\Delta + \kappa^2$. We find that almost everywhere in $\mathbb{R}^3$ the components of $\mathbf{U} = (u_1, u_2, u_3)^T$ satisfy

$$\begin{aligned}
u_j(\mathbf{x}) = &- ([\gamma_N\mathbf{U}]_\Gamma\,\delta_\Gamma)(G_\kappa(\mathbf{x} - \cdot)\mathbf{e}_j) - ([\gamma_\times\mathbf{U}]_\Gamma\,\delta_\Gamma)(\mathbf{curl}_\mathbf{y}(G_\kappa(\mathbf{x} - \cdot)\mathbf{e}_j)) + \\
&+ ([\gamma_\mathbf{n}\mathbf{U}]_\Gamma\,\delta_\Gamma)(\mathrm{div}(G_\kappa(\mathbf{x} - \cdot)\mathbf{e}_j)) \,, \quad j = 1, 2, 3 \,.
\end{aligned}$$

Using $\mathbf{grad}_\mathbf{x}\,G_\kappa(\mathbf{x} - \mathbf{y}) = -\mathbf{grad}_\mathbf{y}\,G_\kappa(\mathbf{x} - \mathbf{y})$, we arrive at the famous Stratton–Chu representation formula for the electric field in $\Omega_c \cup \Omega_e$ [62], cf. [26, Sect. 6.2], [53, Sect. 5.5], [21, Ch. 3, Sect. 1.3.2], and [19, Sect. 4]

Theorem 6. *If, for $\kappa \in \mathbb{C}$, $\Re\kappa \geq 0$, a distribution $\mathbf{U} \in \boldsymbol{H}_{\mathrm{loc}}(\mathbf{curl}; \Omega_c \cup \Omega_e)$ satisfies $\mathbf{curl\,curl\,U} + \kappa^2\mathbf{U} = 0$ and $\mathrm{div}\,\mathbf{U} = 0$ in $\Omega_c\cup\Omega_e$, along with the decay condition $|\mathbf{U}(\mathbf{x})| = O(|\mathbf{x}|^{-1})$ for $|\mathbf{x}| \to \infty$, then it possess the representation*

$$\mathbf{U} = -\boldsymbol{\Psi}_A^\kappa([\gamma_N\mathbf{U}]_\Gamma) - \boldsymbol{\Psi}_M^\kappa([\gamma_\mathbf{t}\mathbf{U}]_\Gamma) - \mathbf{grad}\,\boldsymbol{\Psi}_V^\kappa([\gamma_\mathbf{n}\mathbf{U}]_\Gamma) \,.$$

Here, we used the notations $\boldsymbol{\Psi}_A^\kappa$ for the the vectorial single layer potential

$$\boldsymbol{\Psi}_A^\kappa(\lambda)(\mathbf{x}) := \int_\Gamma G_\kappa(\mathbf{x},\mathbf{y})\lambda(\mathbf{y})\,dS(\mathbf{y}) \quad \mathbf{x} \notin \Gamma \,,$$

and $\boldsymbol{\Psi}_M^\kappa$ for the "Maxwell double layer potential"

$$\boldsymbol{\Psi}_M^\kappa(\mathbf{v}) := \mathbf{curl}\,\boldsymbol{\Psi}_A^\kappa(\mathsf{R}\mathbf{v}) \,.$$

From the representation formula it is clear that the potentials have the following mapping properties, see [39, Sect. 5]:

Theorem 7. *The potential mappings*

$$\boldsymbol{\Psi}_A^\kappa : \boldsymbol{H}_\|^{-\frac{1}{2}}(\mathrm{div}_\Gamma, \Gamma) \mapsto \boldsymbol{W}^1(\mathbb{R}^3) \cap \boldsymbol{W}(\mathbf{curl}^2, \Omega_c \cup \Omega_e)\,,$$

$$\boldsymbol{\Psi}_M^\kappa : \boldsymbol{H}_\perp^{-\frac{1}{2}}(\mathbf{curl}_\Gamma, \Gamma) \mapsto \boldsymbol{W}(\mathbf{curl}^2, \Omega_c \cup \Omega_e)\,,$$

are continuous.

We remark that any distribution complying with the assumptions of the theorem actually behaves like $|\mathbf{U}(\mathbf{x})| = O(|\mathbf{x}|^{-2})$ for $|\mathbf{x}| \to \infty$, see [3, Prop. 3.1].

In light of Thm. 7, the representation formula of Thm. 6 allows to deduce *jump relations*. For formal derivations please consult [39, Sect. 5], or [55], [53, Thm. 5.5.1], and [26, Thm. 6.11] for smooth boundaries.

Theorem 8. *The potentials satisfy the jump relations*

$$\begin{aligned}
[\gamma_{\mathbf{t}}\boldsymbol{\Psi}_A^\kappa]_\Gamma &= 0 & , && [\gamma_N\boldsymbol{\Psi}_A^\kappa]_\Gamma &= -Id\,, \\
[\gamma_{\mathbf{t}}\boldsymbol{\Psi}_M^\kappa]_\Gamma &= -Id & , && [\gamma_N\boldsymbol{\Psi}_M^\kappa]_\Gamma &= 0\,, \\
[\gamma_{\mathbf{n}}\boldsymbol{\Psi}_A^\kappa]_\Gamma &= 0 & , && [\gamma_{\mathbf{n}}\boldsymbol{\Psi}_M^\kappa]_\Gamma &= 0\,.
\end{aligned}$$

If $\kappa \neq 0$, then, by virtue of (13),

$$\gamma_{\mathbf{n}}^{\pm}\mathbf{U} = -\frac{1}{\kappa^2}\gamma_{\mathbf{n}}^{\pm}\,\mathbf{curl\,curl}\,\mathbf{U} = -\frac{1}{\kappa^2}\,\mathrm{div}_\Gamma(\gamma_N^{\pm}\mathbf{U})\,.$$

This permits us to rewrite the representation formula of Thm. 6 for the case $\kappa \neq 0$:

$$\mathbf{U} = -\boldsymbol{\Psi}_A^\kappa([\gamma_N\mathbf{U}]_\Gamma) - \boldsymbol{\Psi}_M^\kappa([\gamma_{\mathbf{t}}\mathbf{U}]_\Gamma) + \frac{1}{\kappa^2}\,\mathbf{grad}\,\Psi_V^\kappa(\mathrm{div}_\Gamma([\gamma_N\mathbf{U}]_\Gamma))\,. \quad (32)$$

After introducing the "Maxwell single layer potential"

$$\boldsymbol{\Psi}_S^\kappa(\mu) := \boldsymbol{\Psi}_A^\kappa(\mu) - \frac{1}{\kappa^2}\,\mathbf{grad}\,\Psi_V^\kappa(\mathrm{div}_\Gamma\,\mu)\,, \quad \mu \in \boldsymbol{H}_\|^{-\frac{1}{2}}(\mathrm{div}_\Gamma, \Gamma)\,, \quad (33)$$

the formula (32) becomes a perfect analogue to the representation formula of
Thm. 5:

$$\mathbf{U} = -\boldsymbol{\Psi}^{\kappa}_{S}([\gamma_N\mathbf{U}]_{\Gamma}) - \boldsymbol{\Psi}^{\kappa}_{M}([\gamma_{\mathbf{t}}\mathbf{U}]_{\Gamma}) \,. \tag{34}$$

Again, the analoguous roles of γ and $\gamma_{\mathbf{t}}$ as "Dirichlet traces" and $\partial_{\mathbf{n}}$ and γ_N
as "Neumann traces" become apparent, cf. Sect. 3 and [19, Sect. 3].

For $\kappa = 0$, the jump of the normal trace cannot be elimintated from
the Stratton-Chu representation formula. This stark difference between the
situations $\kappa \neq 0$ and $\kappa = 0$ can be blamed on the divergence constraint, which
is redundant for $\kappa \neq 0$, but becomes essential, if κ vanishes. This profoundly
changes the characteristics of the differential operator and in the latter case
we have to deal with $\gamma_N\mathbf{U}$ and $\gamma_{\mathbf{n}}\mathbf{U}$ as "Neumann data".

As above we introduce boundary integral operators by taking different
traces of potentials. Their continuity properties can be directly inferred from
those of the potentials, see Thm. 7, and those of the trace operators.

Theorem 9. *For $\kappa \neq 0$ the boundary integral operators*

$$
\begin{aligned}
\mathsf{A}^{\kappa} &:= \gamma_{\mathbf{t}}\boldsymbol{\Psi}^{\kappa}_{S} &&: \boldsymbol{H}^{-\frac{1}{2}}_{\|}(\mathrm{div}_{\Gamma},\Gamma) \mapsto \boldsymbol{H}^{-\frac{1}{2}}_{\perp}(\mathbf{curl}_{\Gamma},\Gamma) \,, \\
\mathsf{B}^{\kappa} &:= \tfrac{1}{2}(\gamma^{-}_{N} + \gamma^{+}_{N})\boldsymbol{\Psi}^{\kappa}_{S} &&: \boldsymbol{H}^{-\frac{1}{2}}_{\|}(\mathrm{div}_{\Gamma},\Gamma) \mapsto \boldsymbol{H}^{-\frac{1}{2}}_{\|}(\mathrm{div}_{\Gamma},\Gamma) \,, \\
\mathsf{C}^{\kappa} &:= \tfrac{1}{2}(\gamma^{-}_{\mathbf{t}} + \gamma^{+}_{\mathbf{t}})\boldsymbol{\Psi}^{\kappa}_{M} &&: \boldsymbol{H}^{-\frac{1}{2}}_{\perp}(\mathbf{curl}_{\Gamma},\Gamma) \mapsto \boldsymbol{H}^{-\frac{1}{2}}_{\perp}(\mathbf{curl}_{\Gamma},\Gamma) \,, \\
\mathsf{N}^{\kappa} &:= \gamma_{N}\boldsymbol{\Psi}^{\kappa}_{M} &&: \boldsymbol{H}^{-\frac{1}{2}}_{\perp}(\mathbf{curl}_{\Gamma},\Gamma) \mapsto \boldsymbol{H}^{-\frac{1}{2}}_{\|}(\mathrm{div}_{\Gamma},\Gamma) \,,
\end{aligned}
$$

are well defined and continuous. The same holds for

$$
\begin{aligned}
\mathsf{A}^{0} &:= \gamma_{\mathbf{t}}\boldsymbol{\Psi}^{0}_{A} &&: \boldsymbol{H}^{-\frac{1}{2}}_{\|}(\Gamma) \mapsto \boldsymbol{H}^{\frac{1}{2}}_{\|}(\Gamma) \,, \\
\mathsf{B}^{0} &:= \tfrac{1}{2}(\gamma^{-}_{N} + \gamma^{+}_{N})\boldsymbol{\Psi}^{0}_{A} &&: \boldsymbol{H}^{-\frac{1}{2}}_{\|}(\mathrm{div}_{\Gamma},\Gamma) \mapsto \boldsymbol{H}^{-\frac{1}{2}}_{\|}(\mathrm{div}_{\Gamma},\Gamma) \,.
\end{aligned}
$$

We know that the double layer boundary integral operators K^{κ} and $\mathsf{K}^{\kappa,*}$
are adjoints with respect to the sesquilinear duality pairing $\langle\cdot,\cdot\rangle_{1/2,\Gamma}$ [52,
Thm. 6.17]. A similar property is enjoyed by their counterparts
$mathsf{B}^{\kappa}$ and C^{κ}:

Theorem 10. *If $\kappa \neq 0$, the boundary integral operators B^{κ} and C^{κ} satisfy*

$$\langle\mathsf{B}^{\kappa}\mu,\mathbf{v}\rangle_{\tau} = -\langle\mu,\mathsf{C}^{\kappa}\mathbf{v}\rangle_{\tau} \quad \forall\mu \in \boldsymbol{H}^{-\frac{1}{2}}_{\|}(\mathrm{div}_{\Gamma},\Gamma),\, \mathbf{v} \in \boldsymbol{H}^{-\frac{1}{2}}_{\perp}(\mathbf{curl}_{\Gamma},\Gamma) \,.$$

The same relationship holds in the case $\kappa = 0$, if μ is restricted to

$$\boldsymbol{H}^{-\frac{1}{2}}_{\|}(\mathrm{div}_{\Gamma}\,0,\Gamma) := \{\boldsymbol{\eta} \in \boldsymbol{H}^{-\frac{1}{2}}_{\|}(\mathrm{div}_{\Gamma},\Gamma) : \mathrm{div}_{\Gamma}\,\boldsymbol{\eta} = 0\} \,.$$

Proof. We appeal to the relationship, see [39, Lemma 5.2] or [51, Lemma 2.3],

$$\mathrm{div}\,\boldsymbol{\Psi}^{\kappa}_{A}(\boldsymbol{\eta}) = \boldsymbol{\Psi}^{\kappa}_{V}(\mathrm{div}_{\Gamma}\,\boldsymbol{\eta}), \quad \boldsymbol{\eta} \in \boldsymbol{H}^{-\frac{1}{2}}_{\|}(\mathrm{div}_{\Gamma},\Gamma)$$

to conclude

$$(\mathbf{curl}\,\mathbf{curl}\,+\kappa^2 Id)\boldsymbol{\Psi}_S^\kappa(\mu) = 0 \quad \text{for } \kappa \neq 0\,,$$

$$(\mathbf{curl}\,\mathbf{curl}\,+\kappa^2 Id)\boldsymbol{\Psi}_A^\kappa(\mu) = 0 \quad \text{for } \mu \in \boldsymbol{H}_{||}^{-\frac{1}{2}}(\mathrm{div}_\Gamma\, 0, \Gamma)\,.$$

We use these relationships together with the integration by parts formula (11): pick any $\mathbf{v} \in \boldsymbol{H}_\perp^{-\frac{1}{2}}(\mathbf{curl}_\Gamma, \Gamma)$, $\mu \in \boldsymbol{H}_{||}^{-\frac{1}{2}}(\mathrm{div}_\Gamma, \Gamma)$ ($\mu \in \boldsymbol{H}_{||}^{-\frac{1}{2}}(\mathrm{div}_\Gamma\, 0, \Gamma)$, if $\kappa = 0$) and set $\mathbf{V} = \boldsymbol{\Psi}_M^\kappa(\mathbf{v})$ and $\mathbf{U} = \boldsymbol{\Psi}_S^\kappa(\mu)$ ($\mathbf{U} = \boldsymbol{\Psi}_A^0(\mu)$, if $\kappa = 0$). Then,

$$\begin{aligned}
\left\langle \gamma_N^+ \mathbf{U}, \gamma_\mathbf{t}^+ \mathbf{V} \right\rangle_\tau &= -\int_{\Omega_e} \mathbf{curl}\,\mathbf{U} \cdot \mathbf{curl}\,\overline{\mathbf{V}} - \mathbf{curl}\,\mathbf{curl}\,\mathbf{U} \cdot \overline{\mathbf{V}}\,\mathrm{dx} \\
&= -\int_{\Omega_e} \mathbf{curl}\,\mathbf{U} \cdot \mathbf{curl}\,\overline{\mathbf{V}} - \mathbf{U} \cdot \mathbf{curl}\,\mathbf{curl}\,\overline{\mathbf{V}}\,\mathrm{dx} = \left\langle \gamma_N^+ \overline{\mathbf{V}}, \gamma_\mathbf{t}^+ \overline{\mathbf{U}} \right\rangle_\tau \\
&= \quad \left\langle \gamma_N^- \overline{\mathbf{V}}, \gamma_\mathbf{t}^- \overline{\mathbf{U}} \right\rangle_\tau\,, \text{ by jump conditions of Thm. 8} \\
&= \int_{\Omega_c} \mathbf{curl}\,\mathbf{U} \cdot \mathbf{curl}\,\overline{\mathbf{V}} - \mathbf{U} \cdot \mathbf{curl}\,\mathbf{curl}\,\overline{\mathbf{V}}\,\mathrm{dx} \\
&= \int_{\Omega_c} \mathbf{curl}\,\mathbf{U} \cdot \mathbf{curl}\,\overline{\mathbf{V}} - \mathbf{curl}\,\mathbf{curl}\,\mathbf{U} \cdot \overline{\mathbf{V}}\,\mathrm{dx} = \left\langle \gamma_N^- \mathbf{U}, \gamma_\mathbf{t}^- \mathbf{V} \right\rangle_\tau
\end{aligned}$$

We remark that "boundary terms at ∞" can be discarded due to the decay $O(|\mathbf{x}|^{-2})$ for $|\mathbf{x}| \to \infty$ of both fields. Thus, using the other set of jump relations from Thm. 8, we have obtained

$$\begin{aligned}
\left\langle \mathsf{B}^\kappa(\mu), \mathbf{v} \right\rangle_\tau &= -\frac{1}{2}\left\langle \gamma_N^+ \mathbf{U} + \gamma_N^- \mathbf{U}, \gamma_\mathbf{t}^+ \mathbf{V} - \gamma_\mathbf{t}^- \mathbf{V} \right\rangle_\tau \\
&= -\frac{1}{2}\left\langle \gamma_N^- \mathbf{U} - \gamma_N^+ \mathbf{U}, \gamma_\mathbf{t}^+ \mathbf{V} + \gamma_\mathbf{t}^- \mathbf{V} \right\rangle_\tau = -\left\langle \mu, \mathsf{C}^\kappa(\mathbf{v}) \right\rangle_\tau\,,
\end{aligned}$$

which finishes the proof. $\square$

Ellipticity estimates corresponding to or extending (30) and (31) are available, too:

Theorem 11. *If $\Re\kappa^2 \geq 0$ and $\Im\kappa^2 \geq 0$, the following estimates hold true for all $\forall \mu \in \boldsymbol{H}_{||}^{-\frac{1}{2}}(\mathrm{div}_\Gamma\, 0, \Gamma)$ and $\mathbf{v} \in \boldsymbol{H}_\perp^{-\frac{1}{2}}(\mathbf{curl}_\Gamma, \Gamma)$*

$$\Im\{\langle \mu, \mathsf{A}^\kappa \mu \rangle_\tau\} \geq 0\,, \quad \Re\{\langle \mathsf{N}^\kappa \mathbf{v}, \mathbf{v} \rangle_\tau\} \geq 0\,, \quad \Im\{\langle \mathsf{N}^\kappa \mathbf{v}, \mathbf{v} \rangle_\tau\} \geq 0\,.$$

Moreover, with $c > 0$ that may depend on Γ and κ,

$$\Re\{\langle \mu, \mathsf{A}^\kappa \mu \rangle_\tau\} \geq c\,\|\mu\|_{\boldsymbol{H}_{||}^{-\frac{1}{2}}(\mathrm{div}_\Gamma, \Gamma)}^2\,, \quad |\langle \mathsf{N}^\kappa \mathbf{v}, \mathbf{v} \rangle_\tau| \geq c\,\|\mathbf{v}\|_{\boldsymbol{H}_\perp^{-\frac{1}{2}}(\mathbf{curl}_\Gamma, \Gamma)}^2\,.$$

Proof. As in the proof of Thm. 10, we rely on the integration by parts formula (11) and jump relations from Thm. 8 to get (for the case $\kappa \neq 0$)

$$\langle \mu, \mathsf{A}^\kappa \mu\rangle_\tau = -\left\langle [\gamma_N \boldsymbol{\Psi}_S^\kappa(\mu)]_\Gamma, \gamma_\mathbf{t} \boldsymbol{\Psi}_S^\kappa(\mu)\right\rangle_\tau$$

$$= \left\langle \gamma_N^- \boldsymbol{\Psi}_S^\kappa(\mu), \gamma_\mathbf{t}^- \boldsymbol{\Psi}_S^\kappa(\mu)\right\rangle_\tau - \left\langle \gamma_N^+ \boldsymbol{\Psi}_S^\kappa(\mu), \gamma_\mathbf{t}^+ \boldsymbol{\Psi}_S^\kappa(\mu)\right\rangle_\tau$$

$$= \int_{\mathbb{R}^3 \setminus \Gamma} |\,\mathbf{curl}\,\boldsymbol{\Psi}_S^\kappa(\mu)|^2 - \mathbf{curl}\,\mathbf{curl}\,\boldsymbol{\Psi}_S^\kappa(\mu) \cdot \overline{\boldsymbol{\Psi}_S^\kappa(\mu)}\, d\mathbf{x}$$

$$= \int_{\mathbb{R}^3 \setminus \Gamma} |\,\mathbf{curl}\,\boldsymbol{\Psi}_S^\kappa(\mu)|^2 + \kappa^{-2} |\,\mathbf{curl}\,\mathbf{curl}\,\boldsymbol{\Psi}_S^\kappa(\mu)|^2\, d\mathbf{x}\,.$$

If $\kappa = 0$ we replace $\boldsymbol{\Psi}_S^\kappa$ with $\boldsymbol{\Psi}_A^0$, for which we know $\mathbf{curl}\,\mathbf{curl}\,\boldsymbol{\Psi}_A^0(\mu) = 0$, if $\mathrm{div}_\Gamma\, \mu = 0$.

This identity can be combined with the continuity of the trace γ_N: with a constant $c = c(\Gamma) > 0$

$$\|\mu\|_{\boldsymbol{H}_{\|}^{-\frac{1}{2}}(\mathrm{div}_\Gamma, \Gamma)} = \left\|[\gamma_N \boldsymbol{\Psi}_S^\kappa(\mu)]_\Gamma\right\|_{\boldsymbol{H}_{\|}^{-\frac{1}{2}}(\mathrm{div}_\Gamma, \Gamma)}$$

$$\leq c \left(\|\mathbf{curl}\,\boldsymbol{\Psi}_S^\kappa(\mu)\|_{\boldsymbol{L}^2(\Omega_c \cup \Omega_e)} + |\mathbf{curl}\,\mathbf{curl}\,\boldsymbol{\Psi}_S^\kappa(\mu)|_{\boldsymbol{L}^2(\Omega_c \cup \Omega_e)} \right)\,.$$

Similar arguments show ellipticity for N^κ:

$$\langle \mathsf{N}^\kappa \mathbf{v}, \mathbf{v}\rangle_\tau = -\left\langle \gamma_N \boldsymbol{\Psi}_M^\kappa(\mathbf{v}), [\gamma_\mathbf{t} \boldsymbol{\Psi}_M^\kappa(\mathbf{v})]_\Gamma\right\rangle_\tau$$

$$= \int_{\mathbb{R}^3 \setminus \Gamma} |\,\mathbf{curl}\,\boldsymbol{\Psi}_M^\kappa(\mathbf{v})|^2 - \mathbf{curl}\,\mathbf{curl}\,\boldsymbol{\Psi}_M^\kappa \cdot \overline{\boldsymbol{\Psi}_M^\kappa(\mathbf{v})}\, d\mathbf{x}$$

$$= \int_{\mathbb{R}^3 \setminus \Gamma} |\,\mathbf{curl}\,\boldsymbol{\Psi}_M^\kappa(\mathbf{v})|^2 + \kappa^2 |\boldsymbol{\Psi}_M^\kappa(\mathbf{v})|^2\, d\mathbf{x}$$

$$\geq c \,\|\boldsymbol{\Psi}_M^\kappa(\mathbf{v})\|_{\boldsymbol{H}(\mathbf{curl}; \Omega_c \cup \Omega_e)}^2\,.$$

Now we have to make use of the continuity of the tangential trace $\gamma_\mathbf{t}$: for $c > 0$ independent of $\mathbf{v}$,

$$\|\mathbf{v}\|_{\boldsymbol{H}_\perp^{-\frac{1}{2}}(\mathrm{curl}_\Gamma, \Gamma)} = \left\|[\gamma_\mathbf{t} \boldsymbol{\Psi}_M^\kappa(\mathbf{v})]_\Gamma\right\|_{\boldsymbol{H}_\perp^{-\frac{1}{2}}(\mathrm{curl}_\Gamma, \Gamma)} \leq c \,\|\boldsymbol{\Psi}_M^\kappa(\mathbf{v})\|_{\boldsymbol{H}(\mathbf{curl}; \Omega_c \cup \Omega_e)}\,.$$

$\square$

The same arguments confirm the following estimates for the scalar single layer potential boundary integral operator based on the Helmholtz kernel:

$$\Re\{\langle \varphi, \mathsf{V}^\kappa \varphi\rangle_{1/2, \Gamma}\} \geq c \,\|\varphi\|_{H^{-\frac{1}{2}}(\Gamma)}^2 \,, \quad \Im\{\langle \varphi, \mathsf{V}^\kappa \varphi\rangle_{1/2, \Gamma}\} \geq 0\,, \qquad (35)$$

for all $\varphi \in H^{-\frac{1}{2}}(\Gamma)$.

7 E-Based Model

Now we discuss the steps leading to a symmetrically coupled boundary element formulation for the transmission problem (18).

232   R. Hiptmair

7.1 Coupled Problem

Now, let $(\mathbf{E}, \mathbf{E}_r)$ stand for the solution of the transmission problem (18) in Ω_c and Ω_e, respectively. Suitable trace operators can be applied to the representation formulas and this procedure yields the *Calderon identities*. From (34) we get

$$
\begin{aligned}
\gamma_{\mathbf{t}}^{-}\mathbf{E} &= \mathsf{A}^{\kappa}(\gamma_N^{-}\mathbf{E}) + (\tfrac{1}{2}Id + \mathsf{C}^{\kappa})(\gamma_{\mathbf{t}}^{-}\mathbf{E}) , \\
\gamma_N^{-}\mathbf{E} &= (\tfrac{1}{2}Id + \\
mathsf{B}^{\kappa})(\gamma_N^{-}\mathbf{E}) &+ \mathsf{N}^{\kappa}(\gamma_{\mathbf{t}}^{-}\mathbf{E}) ,
\end{aligned}
\tag{36}
$$

where $\kappa = \tfrac{1}{2}\sqrt{2}(1+i)\sqrt{\omega\sigma\mu_c}$. Thanks to Thm. 6 we have

$$
\begin{aligned}
\gamma_{\mathbf{t}}^{+}\mathbf{E}_r &= -\mathsf{A}^{0}(\gamma_N^{+}\mathbf{E}_r) + (\tfrac{1}{2}Id - \mathsf{C}^{0})(\gamma_{\mathbf{t}}^{+}\mathbf{E}_r) - \mathbf{grad}_\Gamma \mathsf{V}^{0}(\gamma_{\mathbf{n}}^{+}\mathbf{E}_r) , \\
\gamma_N^{+}\mathbf{E}_r &= (\tfrac{1}{2}Id - \mathsf{B}^{0})(\gamma_N^{+}\mathbf{E}_r) - \mathsf{N}^{0}(\gamma_{\mathbf{t}}^{+}\mathbf{E}_r) , \\
\gamma_{\mathbf{n}}^{+}\mathbf{E}_r &= -\gamma_{\mathbf{n}}^{+}\boldsymbol{\Psi}_A^{0}(\gamma_N^{+}\mathbf{E}_r) - \gamma_{\mathbf{n}}^{+}\boldsymbol{\Psi}_M^{0}(\gamma_{\mathbf{t}}^{+}\mathbf{E}_r) + (\tfrac{1}{2}Id - \mathsf{K}^{0})(\gamma_{\mathbf{n}}^{+}\mathbf{E}_r) .
\end{aligned}
\tag{37}
$$

The boundary data for any solution of the interior/exterior $\mathbf{E}$-based eddy current equations will fulfill (36) and (37), respectively.

The gist of the symmetric coupling approach according to Costabel [27] is to use all of the equations of the Calderon identities in conjunction with the transmission conditions. However, here we have to grapple with a mismatch of interior and exterior boundary data due to the presence of $\gamma_{\mathbf{n}}^{+}\mathbf{E}_r$ in (37). A remedy is motivated by the observation

$$
\mathbf{curl\,curl}\,\mathbf{E}_r = 0 \quad \text{in } \Omega_e \quad \Rightarrow \quad \mathrm{div}_\Gamma(\gamma_N^{+}\mathbf{E}_r) = 0 ,
$$

which is an immediate consequence of the identity (13). We observe that $\gamma_N^{+}\mathbf{E}_r$ has to be sought in the space $\boldsymbol{H}_{\|}^{-\frac{1}{2}}(\mathrm{div}_\Gamma 0, \Gamma)$!

By the transmission condition for γ_N and the fact that $\mathbf{curl\,curl}\,\mathbf{E}_s = 0$ in a neighborhood of Γ, $\gamma_N^{-}\mathbf{E}$ has to be div_Γ-free, as well. Hence, we can restrict our attention to boundary data $\gamma_N^{-}\mathbf{E}, \gamma_N^{+}\mathbf{E}_r$ in $\boldsymbol{H}_{\|}^{-\frac{1}{2}}(\mathrm{div}_\Gamma 0, \Gamma)$ throughout. Recalling the dualities, this is a proper test space for those equations of the Calderon identities that are set in $\boldsymbol{H}_{\perp}^{-\frac{1}{2}}(\mathbf{curl}_\Gamma, \Gamma)$. Since div_Γ is the $\boldsymbol{L}^2(\Gamma)$-adjoint of $\mathbf{grad}_\Gamma$, we find

$$
\mu \in \boldsymbol{H}_{\|}^{-\frac{1}{2}}(\mathrm{div}_\Gamma 0, \Gamma) \quad \Rightarrow \quad \langle \mu, \mathbf{grad}_\Gamma \varphi \rangle_\tau = 0 \quad \forall \varphi \in H^{\frac{1}{2}}(\Gamma) .
$$

This makes the undesirable terms disappear, when switching to a weak form of the top equations in the Calderon identities (36) and (37)! For all $\mu \in \boldsymbol{H}_{\|}^{-\frac{1}{2}}(\mathrm{div}_\Gamma 0, \Gamma)\boldsymbol{H}_{\|}^{-\frac{1}{2}}(\mathrm{div}_\Gamma 0, \Gamma)$ we obtain

$$
\begin{aligned}
\langle \mu, \gamma_{\mathbf{t}}^{-}\mathbf{E} \rangle_\tau &= \langle \mu, \mathsf{A}^{\kappa}(\gamma_N^{-}\mathbf{E}) \rangle_\tau + \langle \mu, (\tfrac{1}{2}Id + \mathsf{C}^{\kappa})\gamma_{\mathbf{t}}^{-}\mathbf{E} \rangle_\tau , \\
\langle \mu, \gamma_{\mathbf{t}}^{+}\mathbf{E}_r \rangle_\tau &= \langle \mu, -\mathsf{A}^{0}(\gamma_N^{+}\mathbf{E}_r) \rangle_\tau + \langle \mu, (\tfrac{1}{2}Id - \mathsf{C}^{0})\gamma_{\mathbf{t}}^{+}\mathbf{E}_r \rangle_\tau .
\end{aligned}
$$

From the transmission conditions we know $\gamma_{\mathbf{t}}^{+}\mathbf{E}_r - \gamma_{\mathbf{t}}^{-}\mathbf{E} = -\gamma_{\mathbf{t}}^{+}\mathbf{E}_s$. Thus, subtracting the above equations leads to

$$-\left\langle \mu, \mathsf{A}^0(\gamma_N^{+}\mathbf{E}_r) + \mathsf{A}^{\kappa}(\gamma_N^{-}\mathbf{E})\right\rangle_{\tau} - \left\langle \mu, \mathsf{C}^0(\gamma_{\mathbf{t}}^{+}\mathbf{E}_r) + \mathsf{C}^{\kappa}(\gamma_{\mathbf{t}}^{-}\mathbf{E}_r)\right\rangle_{\tau}$$
$$= -\tfrac{1}{2}\left\langle \mu, \gamma_{\mathbf{t}}^{+}\mathbf{E}_s\right\rangle_{\tau}$$

for all $\mu \in \boldsymbol{H}_{\|}^{-\frac{1}{2}}(\mathrm{div}_{\Gamma}\, 0, \Gamma)$. From the transmission condition $\frac{1}{\mu_0}\gamma_N^{+}\mathbf{E}_r - \frac{1}{\mu_r}\gamma_N^{-}\mathbf{E} = -\frac{1}{\mu_0}\gamma_N^{+}\mathbf{E}_s$ and the second equations of the Calderon identities we directly conclude

$$\frac{1}{\mu_0}(\tfrac{1}{2}Id - \mathsf{B}^0)\gamma_N^{+}\mathbf{E}_r - \frac{1}{\mu_0}\mathsf{N}^0(\gamma_{\mathbf{t}}^{+}\mathbf{E}_r) - \frac{1}{\mu_c}(\tfrac{1}{2}Id + \mathsf{B}^{\kappa})\gamma_N^{-}\mathbf{E} - \frac{1}{\mu_c}\mathsf{N}^{\kappa}(\gamma_{\mathbf{t}}^{-}\mathbf{E})$$
$$= -\frac{1}{2\mu_0}\gamma_N^{+}\mathbf{E}_s\ .$$

As final unknown quantities we introduce the tangential trace of the electric field $\mathbf{u} := \gamma_{\mathbf{t}}^{-}\mathbf{E} \in \boldsymbol{H}_{\perp}^{-\frac{1}{2}}(\mathbf{curl}_{\Gamma}, \Gamma)$ and the tangential trace of the magnetic field $\lambda := \frac{1}{\mu_c}\gamma_N^{-}\mathbf{E} \in \boldsymbol{H}_{\|}^{-\frac{1}{2}}(\mathrm{div}_{\Gamma}\, 0, \Gamma)$. The latter is also known as *equivalent surface current*. The transmission conditions enable us to express the exterior traces in these unknowns. We end up with the coupled variational problem: Seek $\mathbf{u} \in \boldsymbol{H}_{\perp}^{-\frac{1}{2}}(\mathbf{curl}_{\Gamma}, \Gamma)$, $\lambda \in \boldsymbol{H}_{\|}^{-\frac{1}{2}}(\mathrm{div}_{\Gamma}\, 0, \Gamma)$ such that

$$\boxed{\begin{aligned}\left\langle (\tfrac{1}{\mu_0}\mathsf{N}^0 + \tfrac{1}{\mu_c}\mathsf{N}^{\kappa})\mathbf{u}, \mathbf{v}\right\rangle_{\tau} &+ \left\langle (\mathsf{B}^0 + \mathsf{B}^{\kappa})\lambda, \mathbf{v}\right\rangle_{\tau} &= f(\mathbf{v})\, ,\\ \left\langle \mu, (\mathsf{C}^0 + \mathsf{C}^{\kappa})\mathbf{u}\right\rangle_{\tau} &+ \left\langle \mu, (\mu_0\mathsf{A}^0 + \mu_c\mathsf{A}^{\kappa})\lambda\right\rangle_{\tau} &= g(\mu)\end{aligned}}$$

$$(38)$$

for all $\mathbf{v} \in \boldsymbol{H}_{\perp}^{-\frac{1}{2}}(\mathbf{curl}_{\Gamma}, \Gamma)$, $\mu \in \boldsymbol{H}_{\|}^{-\frac{1}{2}}(\mathrm{div}_{\Gamma}\, 0, \Gamma)$. The right hand side is given by

$$f(\mathbf{v}) := \frac{1}{\mu_0}\left\langle (\tfrac{1}{2}Id + \mathsf{B}^0)\gamma_N\mathbf{E}_s, \mathbf{v}\right\rangle_{\tau} + \frac{1}{\mu_0}\left\langle \mathsf{N}^0(\gamma_{\mathbf{t}}\mathbf{E}_s), \mathbf{v}\right\rangle_{\tau}\, ,$$
$$g(\mu) := \left\langle \mu, (\tfrac{1}{2}Id + \mathsf{C}^0)\gamma_{\mathbf{t}}\mathbf{E}_s\right\rangle_{\tau} + \left\langle \mu, \mathsf{A}^0(\gamma_N\mathbf{E}_s)\right\rangle_{\tau}\, .$$

$$(39)$$

Theorem 12. *The bilinear form $\mathbf{d}$ associated with the variational problem* (38) *is $\boldsymbol{H}_{\perp}^{-\frac{1}{2}}(\mathbf{curl}_{\Gamma}, \Gamma) \times \boldsymbol{H}_{\|}^{-\frac{1}{2}}(\mathrm{div}_{\Gamma}\, 0, \Gamma)$-elliptic in the sense that there is $c = c(\Gamma, \kappa, \mu_0, \mu_c) > 0$ such that*

$$\left|\mathbf{d}\left(\begin{pmatrix}\mathbf{v}\\\mu\end{pmatrix}, \begin{pmatrix}\mathbf{v}\\\mu\end{pmatrix}\right)\right| \geq c\left(\|\mathbf{v}\|^2_{\boldsymbol{H}_{\perp}^{-\frac{1}{2}}(\mathbf{curl}_{\Gamma}, \Gamma)} + \|\mu\|^2_{\boldsymbol{H}_{\|}^{-\frac{1}{2}}(\mathrm{div}_{\Gamma}, \Gamma)}\right)$$

for all $\mathbf{v} \in \boldsymbol{H}_{\perp}^{-\frac{1}{2}}(\mathbf{curl}_{\Gamma}, \Gamma)$ and $\mu \in \boldsymbol{H}_{\|}^{-\frac{1}{2}}(\mathrm{div}_{\Gamma}\, 0, \Gamma)$.

Proof. As a simple consequence of the block skew-symmetric structure of the variational problem (cf. Thm 10) we find for $\mathbf{v} \in \boldsymbol{H}_\perp^{-\frac{1}{2}}(\mathbf{curl}_\Gamma, \Gamma)$, $\mu \in \boldsymbol{H}_{\|}^{-\frac{1}{2}}(\mathrm{div}_\Gamma\, 0, \Gamma)$ that

$$\mathbf{d}((\mathbf{v}, \mu), (\mathbf{v}, \mu)) = \left\langle (\mathsf{N}^0 + \tfrac{1}{\mu_r}\mathsf{N}^\kappa)\mathbf{v}, \mathbf{v} \right\rangle_\tau + \left\langle \mu, (\mathsf{A}^0 + \mu_r\mathsf{A}^\kappa)\mu \right\rangle_\tau .$$

Subsequently, the estimates of Thm. 11 permit us to conclude ellipticity of the whole bilinear form from separate estimates for the individual terms. $\square$

Corollary 1. *The variational problem* (38) *has a unique solution* $(\mathbf{u}, \lambda) \in \boldsymbol{H}_\perp^{-\frac{1}{2}}(\mathbf{curl}_\Gamma, \Gamma) \times \boldsymbol{H}_{\|}^{-\frac{1}{2}}(\mathrm{div}_\Gamma\, 0, \Gamma).$

By the derivation of the boundary integral equations we can be certain that traces $\gamma_{\mathbf{t}}^- \mathbf{E}$ and $\gamma_\times^- \mathbf{H}$ will always give rise to solutions of (38). Their uniqueness then confirms that we get the traces of solutions of the **E**-based eddy current model (17). These traces are fixed regardless of the gauging of **E** employed in Ω_e.

7.2 Galerkin Discretization

We select a conforming Galerkin boundary element discretization of (38) and (39) that relies on finite dimensional subspaces $\mathcal{W}_h \subset \boldsymbol{H}_\perp^{-\frac{1}{2}}(\mathbf{curl}_\Gamma, \Gamma)$ and $\mathcal{V}_h \subset \boldsymbol{H}_{\|}^{-\frac{1}{2}}(\mathrm{div}_\Gamma\, 0, \Gamma)$. These should be boundary element spaces in the sense that

1. the functions in both $\mathcal{W}_h$ and $\mathcal{V}_h$ are piecewise polynomial tangential vector fields with respect to a mesh Γ_h of Γ consisting of flat triangles.
2. there are bases of $\mathcal{W}_h$ and $\mathcal{V}_h$ that only comprise locally supported functions.

For the construction of $\mathcal{W}_h$ we start from $\boldsymbol{H}(\mathbf{curl}; \Omega_c)$-conforming finite element schemes for the approximation of vector potentials. The simplest is provided by the so-called edge elements [38]. Keeping in mind that

$$\boldsymbol{H}_\perp^{-\frac{1}{2}}(\mathbf{curl}_\Gamma, \Gamma) := \gamma_{\mathbf{t}}(\boldsymbol{H}(\mathbf{curl}; \Omega_c)),$$

we simply take the tangential projections of edge element functions on a mesh Ω_h with $\Omega_{h|\Gamma} = \Gamma_h$ as space $\mathcal{W}_h$. This will give a space of piecewise linear vector fields on Γ, whose tangential components are continuous across edges of triangles. This is a well-known sufficient condition for $\mathcal{W}_h \subset \boldsymbol{H}_\perp^{-\frac{1}{2}}(\mathbf{curl}_\Gamma, \Gamma)$. The local shape functions on a triangle T are given by the formula

$$\mathbf{b}_{i,j}^T := \lambda_i \, \mathbf{grad}_\Gamma\, \lambda_j - \lambda_j \, \mathbf{grad}_\Gamma\, \lambda_i , \quad 1 \le i < j \le 3 , \qquad (40)$$

where λ_i, $i = 1, 2, 3$, are the local linear barycentric coordinate functions in T. These basis functions are sketched in Fig. 4. They are associated with the edges of Γ_h so that $\dim \mathcal{W}_h$ will agree with the total number of edges of Γ_h. Note that $\mathcal{W}_h$ can also be obtained by $90°$-rotation of the lowest order div-conforming Raviart-Thomas elements in 2D, *cf.* [11, Ch. 3]. More details can be found in [6, Sect. 2.2].

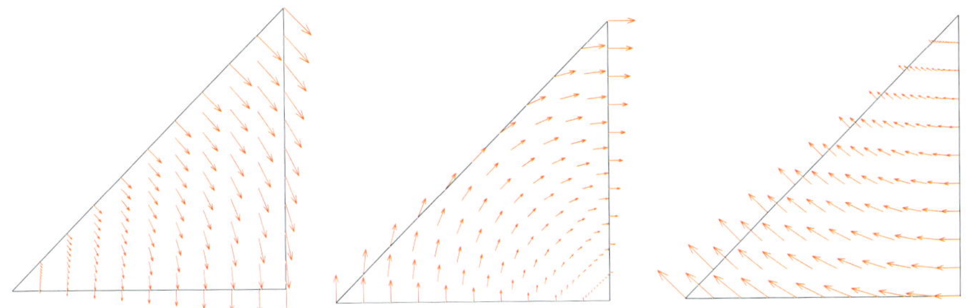

Fig. 4. Local shape functions of $\mathcal{W}_h$.

In order to find $\mathcal{V}_h$ we remember that λ is the rotated tangential trace of the magnetic field $\mathbf{H}$. As $\boldsymbol{H}(\mathbf{curl};\Omega)$ is the right function space for $\mathbf{H}$, too, we get the right boundary element space for magnetic traces by rotating functions in $\mathcal{W}_h$ by $90°$. This will give surface vector fields with continuous fluxes across edges of triangles, which is a very desirable property for discrete equivalent surface currents. However, ellipticity of (38) only holds provided that $\mathrm{div}_\Gamma \lambda = 0$. Therefore, this property has to be enforced on $\mathcal{V}_h$. Formally, we may choose

$$\mathcal{V}_h := \{\mu_h \in \mathcal{W}_h \times \mathbf{n},\ \mathrm{div}_\Gamma \mu = 0\}\ . \tag{41}$$

Using the formula (40), we readily see that $\mathcal{V}_h$ only contains piecewise constant vector fields.

By Thm. 12 and Cea's lemma [24, Thm. 2.4.1] conformity of the Galerkin method directly translates into the quasi-optimal error estimate in energy norm

$$\|\mathbf{u} - \mathbf{u}_h\|_{\boldsymbol{H}_\perp^{-\frac{1}{2}}(\mathrm{curl}_\Gamma,\Gamma)} + \|\lambda - \lambda_h\|_{\boldsymbol{H}_\|^{-\frac{1}{2}}(\mathrm{div}_\Gamma,\Gamma)} \leq$$

$$\leq C \left(\inf_{\mathbf{v}_h \in \mathcal{W}_h} \|\mathbf{u} - \mathbf{v}_h\|_{\boldsymbol{H}_\perp^{-\frac{1}{2}}(\mathrm{curl}_\Gamma,\Gamma)} + \inf_{\zeta_h \in \mathcal{V}_h} \|\lambda - \zeta_h\|_{\boldsymbol{H}_\|^{-\frac{1}{2}}(\mathrm{div}_\Gamma,\Gamma)} \right) , \tag{42}$$

where $\mathbf{u}_h$ and λ_h stand for the boundary element solutions, and $C > 0$ depends on the ellipticity and continuity constants of the continuous variational problem (38). Hence, approximation error estimates for the finite element spaces will directly provide us with rates of convergence. Let us assume quasi-uniform and shape regular families of surface meshes Γ_h, where h denotes the meshwidth. Provided that the continuous solutions $\mathbf{u}$ and λ are sufficiently smooth, we arrive at

$$\|\mathbf{u} - \mathbf{u}_h\|_{\boldsymbol{H}_\perp^{-\frac{1}{2}}(\mathrm{curl}_\Gamma, \Gamma)} + \|\lambda - \lambda_h\|_{\boldsymbol{H}_{||}^{-\frac{1}{2}}(\div_\Gamma, \Gamma)} \leq$$

$$\leq C \left(h^{\min\{\frac{3}{2}, \eta + \frac{1}{2}\}} \|\mathbf{u}\|_{\boldsymbol{H}^\eta(\mathrm{curl}_\Gamma, \Gamma)} + h^{\min\{\frac{3}{2}, \rho + \frac{1}{2}\}} \|\lambda\|_{\boldsymbol{H}^\rho(\Gamma)} \right) , \quad (43)$$

for some $\eta, \rho > 0$. The constant $C > 0$ now depends on the shape-regularity of the meshes, too. Details about approximation by functions in $\mathcal{W}_h$ can be found in [14, Sect. 4.2.2]. The possible ranges of η and ρ depend on the geometry of Γ: the presence of edges and corners will impose limits on η, ρ. At worst, these may only be slightly larger than zero.

The divergence constraint is essential in the definition (41) of the boundary element trial space for the surface currents. We cannot simply use rotated shape functions from Fig. 4 to get a locally supported basis, because the constraint has to be enforced. Two options are available:

Lagrangian Multipliers

We may take the cue from mixed finite element schemes for second order elliptic boundary value problems[11, Ch. 4] and use Lagrangian multipliers to impose the linear constraints $\div_\Gamma \lambda_h = 0$. The natural discrete Lagrangian multiplier space is

$$\mathcal{M}_h := \div_\Gamma(\mathcal{W}_h \times \mathbf{n}) \qquad (44)$$

$$= \{\nu \in L^2(\Gamma) : \mu_{h|K} \equiv \mathrm{const}\ \forall K \in \Gamma_h, \int_\Gamma \mu_h \, dS = 0\} .$$

Care must be taken when selecting the sesqui-linear form $m(\cdot, \cdot)$ that brings the Lagrangian multiplier to bear on λ_h in the sense that

$$\mathcal{V}_h = \{\mu_h \in \mathcal{W}_h \times \mathbf{n} : m(\mu_h, \nu_h) = 0 \quad \forall \nu_h \in \mathcal{M}_h\} .$$

For the sake of asymptotic stability of the discrete problem, the form m must be both h-uniformly continuous and satisfy inf-condition [11, Ch. 3]

$$\sup_{\mu_h \in \mathcal{W}_h \times \mathbf{n}} \frac{|m(\mu_h, \nu_h)|}{\|\mu_h\|_{\boldsymbol{H}_{||}^{-\frac{1}{2}}(\div_\Gamma, \Gamma)}} \geq c \|\nu_h\|_M \quad \forall \nu_h \in \mathcal{M}_h , \qquad (45)$$

where $c > 0$ should not depend on the meshwidth h. The norm $\|\cdot\|_M$ with which $\mathcal{M}$ is endowed is still at our disposal.

Next, note that the tempting choice $\|\cdot\|_M = \|\cdot\|_{L^2(\Gamma)}$ and

$$m(\mu_h, \nu_h) := (\div \mu_h, \nu_h)_{L^2(\Gamma)}$$

must be ruled out, though (45) is easily seen to hold, because this m will fail to be continuous on $\boldsymbol{H}_{||}^{-\frac{1}{2}}(\div_\Gamma, \Gamma) \times L^2(\Omega)$.

A viable option is

$$\|\cdot\|_M = \|\cdot\|_{H^{-\frac{1}{2}}(\Gamma)} \quad \text{and} \quad m(\mu_h, \nu_h) := \langle \operatorname{div}_\Gamma \mu_h, \mathsf{V}^0 \nu_h \rangle_{1/2,\Gamma} \ .$$

Here, continuity is immediate from (29) and the h-uniform inf-sup condition (45) has been shown in [22], see also [10, Sect. 5.3].

Surface Stream Functions

Another way to deal with the divergence constraint resorts to scalar *surface stream functions*. Let $\mathcal{S}_h$ stand for the space of Γ_h-piecewise linear and continuous functions on Γ. Then, if Γ is simply connected, we know from deRham's theorem [38, Cor. 3.3] that $\mathcal{V}_h = \mathbf{curl}_\Gamma \mathcal{S}_h$. Hence, we may simply use the surface rotation of the "hat basis functions" of $\mathcal{S}_h$ as a basis for $\mathcal{V}_h$, see Fig. 5 (left).

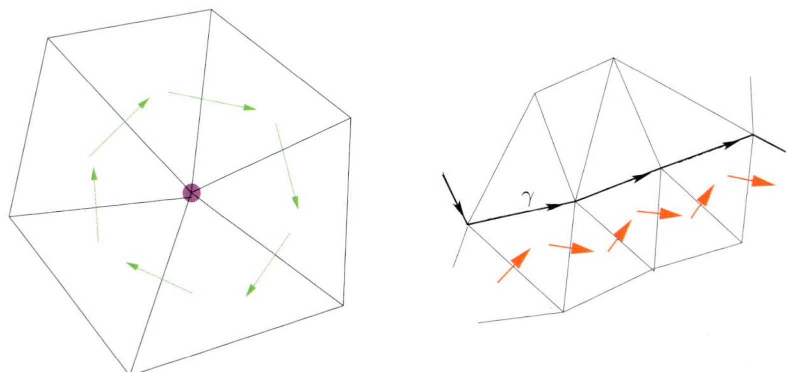

Fig. 5. Basis function of $\mathcal{V}_h$ associated with a vertex (left). Current sheet along a section of a path γ (right).

Because we have not ruled out more general topologies of Γ, surface cohomology vector fields can also contribute to the kernel of $\operatorname{div}_\Gamma$:

$$\mathcal{V}_h = \mathbf{curl}_\Gamma \mathcal{S}_h \oplus \mathcal{H}_h \quad , \quad \dim \mathcal{H}_h = \beta_1(\Gamma) \ , \tag{46}$$

where $\beta_1(\Gamma)$ is the first Betti number of Γ, which is twice the number of holes drilled through Ω_c. This means that $\dim \mathcal{V}_h$ will be equal to the number of vertices of Γ_h plus $\beta_1(\Gamma)$.

To find a basis of $\mathcal{H}_h$ we need representatives γ_k, $k = 1 \ldots, \beta_1(\Gamma)$, of a basis of the cohomology group $H_1(\Gamma_h, \mathbb{Z})$ in the form of oriented closed edge paths (cycles). In other words, we need a maximal set of closed curves on the surface that do not cut the surface into two separate parts, and cannot be deformed into each other by sweeping them over parts of Γ. Typical choices for the torus are depicted in Fig. 3. We can always find such curves that run along edges of Γ_h and this can be done with a computational effort proportional to the number of edges in Γ_h [41]. To each such path γ a "current sheet" $\boldsymbol{\eta}_\gamma$ can be associated, a circular current traveling along the path, see Fig. 5 (right).

Consider a non-bounding surface edge cycle γ that is bounding with respect to Ω_e, that is, there is an oriented surface $\Sigma \subset \Omega_e$ such that $\gamma = \partial \Sigma$. Then we get from Stokes theorem

$$\int_\gamma (\gamma_N \mathbf{E}_r \times \mathbf{n}) \cdot d\mathbf{s} = \int_\Sigma \mathbf{curl}\,\mathbf{curl}\,\mathbf{E}_r \cdot \mathbf{n}\,dS = 0 \;.$$

As $\mathbf{curl}\,\mathbf{curl}\,\mathbf{E}_r = 0$ in Ω_e, this means that, in the discrete variational problem (47), we can confine ourselves to those $\lambda_h \in \mathcal{V}_h$ that satisfy $\int_\gamma (\lambda_h \times \mathbf{n}) \cdot d\mathbf{s} = 0$ for all cycles γ bounding relative to Ω_e. This means that we only have to take into account current sheets along cycles bounding relative to the exterior. An algorithm for the construction of these cycles has been developed in [41]. The resulting basis of the relevant subspace of $\mathcal{H}_h$ will be denoted by $\iota_1, \ldots, \iota_L$, $L =: \frac{1}{2}\beta_1(\Gamma)$. Then the discrete linear variational problem arising from (38) read search for $\mathbf{u}_h \in \mathcal{W}_h$, $\varphi_h \in \mathcal{S}_h/\mathbb{R}$, $(\alpha_1, \ldots, \alpha_L)^T \in \mathbb{C}^L$ such that

$$-\left\langle \widetilde{\mathsf{N}^0}\mathbf{u}_h, \mathbf{v}_h \right\rangle_\tau \;-\; \left\langle \widetilde{\mathsf{B}^0}\,\mathbf{curl}_\Gamma\,\varphi_h, \mathbf{v}_h \right\rangle_\tau \;-\; \sum_{k=1}^L \alpha_k \left\langle \widetilde{\mathsf{B}^0}\iota^k, \mathbf{v}_h \right\rangle_\tau$$
$$= f(\mathbf{v}_h) \;,$$

$$\left\langle \widetilde{\mathsf{B}^0}\,\mathbf{curl}_\Gamma\,\psi_h, \mathbf{u}_h \right\rangle_\tau + \left\langle \mathbf{curl}_\Gamma\,\psi_h, \widetilde{\mathsf{A}^0}\,\mathbf{curl}_\Gamma\,\varphi_h \right\rangle_\tau + \sum_{k=1}^L \alpha_k \left\langle \mathbf{curl}_\Gamma\,\psi_h, \widetilde{\mathsf{A}^0}\iota_k \right\rangle_\tau$$
$$= g(\mathbf{curl}_\Gamma\,\psi_h) \;,$$

$$\left\langle \widetilde{\mathsf{B}^0}\iota^j, \mathbf{u}_h \right\rangle_\tau \;+\; \left\langle \iota^j, \widetilde{\mathsf{A}^0}\,\mathbf{curl}_\Gamma\,\varphi_h \right\rangle_\tau \;+\; \sum_{k=1}^L \alpha_k \left\langle \iota^j, \widetilde{\mathsf{A}^0}\iota_k \right\rangle_\tau$$
$$= g(\iota^j) \;,$$
$$(47)$$

for all $\mathbf{v}_h \in \mathcal{W}_h$, $\psi_h \in \mathcal{S}_h/\mathbb{R}$, $j = 1, \ldots, L$. We abbreviated $\widetilde{\mathsf{A}^0} := \mu_0 \mathsf{A}^0 + \mu_c \mathsf{A}^\kappa$, $\widetilde{\mathsf{B}^0} = \mathsf{B}^0 + \mathsf{B}^\kappa$, $\widetilde{\mathsf{N}^0} := \frac{1}{\mu_0}\mathsf{N}^0 + \frac{1}{\mu_c}\mathsf{N}^\kappa$. From (47) we can retrieve $\lambda_h = \mathbf{curl}_\Gamma\,\varphi_h + \sum_{k=1}^L \alpha_k \iota^k$.

Remark 2. If surface stream functions are used, Non-local inductive excitation can taken into account in an amazingly simple fashion: for each loop of the conductor there is basis cycle of $H_1(\Gamma, \mathbb{Z})$ that "winds around it", see Fig. 3 for an example. We realize that the circulation of the magnetic field along that fundamental cycle, which is equal to the flux of λ through it, agrees with the total current in the loop. Hence, inductive excitation amounts to fixing some of the α_k in the variational formulation (47). More details are given in [39, Sect. 8].

8 H-Based Model

For want of a transmission problem, the derivation of symmetrically coupled boundary integral equations starts from the variational problem (19).

8.1 Boundary Reduction

In order to be able to perform a reduction to the boundary through integration by parts we have to resort to scalar potentials. Therefore we use (15) to replace $\mathcal{V}$ by

$$\mathcal{V}[\mathbf{H}_s] = \{(\mathbf{V}, \Phi) \in \boldsymbol{H}(\mathbf{curl}; \Omega_c) \times H^1_\Sigma(\Omega_e), \; \gamma_\mathbf{t}^- \mathbf{V} - \gamma_\mathbf{t}^+ \widetilde{\mathbf{grad}\Phi} = \gamma_\mathbf{t}^+ \mathbf{H}_s \text{ on } \Gamma\} \, .$$

For the notations we refer to Sect. 4. Thus, (42) is converted into: Seek $(\mathbf{H}, \Psi) \in \mathcal{V}[\mathbf{H}_s]$ such that

$$\left(\sigma^{-1} \mathbf{curl}\,\mathbf{H}, \mathbf{curl}\,\mathbf{V}\right)_{\boldsymbol{L}^2(\Omega_c)} + i\omega\mu_c \left(\mathbf{H}, \mathbf{V}\right)_{\boldsymbol{L}^2(\Omega_c)} +$$
$$+ i\omega\mu_0 \left(\mathbf{H}_s + \widetilde{\mathbf{grad}\Psi}, \widetilde{\mathbf{grad}\Phi}\right)_{\boldsymbol{L}^2(\Omega_e)} = 0 \, , \quad (48)$$

for all $(\mathbf{V}, \Phi) \in \mathcal{V}[0]$. As $\operatorname{div}\mathbf{H}_s = 0$ in Ω_e, testing with functions compactly supported either in Ω_c or Ω_e shows that for $k = 1, \dots, N$

$$\mathbf{curl}\,\sigma^{-1}\mathbf{curl}\,\mathbf{H} + i\omega\mu_c\mathbf{H} = 0 \quad \text{in } \Omega_c \, , \quad (49)$$
$$-\Delta\Psi = 0 \text{ in } \Omega' \quad , \quad [\partial_\mathbf{n}\,\mathbf{grad}\,\Psi]_{\Sigma_k} = 0, \; [\gamma\Psi]_{\Sigma_k} = \text{const.} \, . \quad (50)$$

Integration by parts can be carried out on both Ω_c and Ω'. Thus, setting $\tau = (i\omega\sigma\mu_0)^{-1}$, (48) becomes

$$\tau \left\langle \sigma^{-1}\gamma_N^- \mathbf{H}, \gamma_\mathbf{t}^- \mathbf{V} \right\rangle_\tau - \left\langle \partial_\mathbf{n}'\Psi, \gamma\Phi \right\rangle_{1/2,\partial\Omega'} = \left\langle \gamma_\mathbf{n}\mathbf{H}_s, \gamma'\Phi \right\rangle_{1/2,\partial\Omega'} \, . \quad (51)$$

Here, γ' and $\partial_\mathbf{n}'$ are the standard trace and conormal derivative onto $\partial\Omega'$. The definition of $\partial_\mathbf{n}'$ relies on the interior unit normal vectorfield on $\partial\Omega'$.

Remark 3. Splitting the duality pairing $\langle \gamma_\mathbf{n}\mathbf{H}_s, \gamma'\Phi \rangle_{1/2,\partial\Omega'}$ into contributions of Γ and of the cuts cannot be done immediately, because the individual integrals are no continuous functionals on the space $H^{\frac{1}{2}}(\partial\Omega')$. This procedure must be postponed until after discretization.

8.2 Coupled Problem

For both (49) and (50) we need a realization of the Dirichlet-to-Neumann operator by boundary integral operators. For (50) we can rely on the exterior Calderon projector for the Laplacian on Ω' [58, Sect. 3.6], which gives the identities

$$\gamma'\Psi = (\tfrac{1}{2}Id + \mathsf{K}')(\gamma'\Psi) - \qquad \mathsf{V}'(\partial'_{\mathbf{n}}\Psi)\,,$$
$$\partial'_{\mathbf{n}}\Psi = \qquad -\mathsf{D}'(\gamma'\Psi) \qquad + (\tfrac{1}{2}Id - (\mathsf{K}')^*)(\partial'_{\mathbf{n}}\Psi)\,. \tag{52}$$

The integral operators match those introduced in the beginning of Sect. 6, but this time they are defined on $\partial\Omega'$ and based on a unit normal vectorfield pointing into the interior of Ω': K' is the double layer potential integral operator for Δ, $(\mathsf{K}')^*$ its $\boldsymbol{L}^2(\partial\Omega')$-adjoint, and D' stands for the hypersingular operator, see (29).

What is not reflected in the statement of the Calderon identities is the special nature of the "Dirichlet trace" $\gamma'\Psi$ and "Neumann trace" $\partial'_{\mathbf{n}}\Psi$ entailed by the transmission conditions of (50). They imply that

$$\gamma'\Psi \in H^{\frac{1}{2}}_{\Sigma}(\partial\Omega') := \{v \in H^{\frac{1}{2}}(\partial\Omega'),\, [v]_{\Sigma_j} = \mathrm{const.},\, j = 1,\ldots,N\}\,,$$

$$\partial'_{\mathbf{n}}\Psi \in H^{-\frac{1}{2}}_{\Sigma}(\partial\Omega') := \{\phi \in H^{-\frac{1}{2}}(\partial\Omega'),\, \phi^+ + \phi^- = 0 \text{ on } \Sigma_j,\, j = 1,\ldots,N\}\,.$$

For the interior problem (49) we can reuse the Calderon identities (36) with $\kappa = \frac{1}{\sqrt{2}}(1+i)\sqrt{\omega\sigma\mu_c}$ and $\mathbf{H}$ instead of $\mathbf{E}$:

$$\gamma^-_{\mathbf{t}}\mathbf{H} = \mathsf{A}^\kappa(\gamma^-_N\mathbf{H}) + (\tfrac{1}{2}Id + \mathsf{C}^\kappa)(\gamma^-_{\mathbf{t}}\mathbf{H})\,,$$
$$\gamma^-_N\mathbf{H} = (\tfrac{1}{2}Id +$$
$$mathsfB^\kappa)(\gamma^-_N\mathbf{H}) + \mathsf{N}^\kappa(\gamma^-_{\mathbf{t}}\mathbf{H})\,. \tag{53}$$

Now we can merge (51), (52), and (53), making use of $\gamma^-_{\mathbf{t}}\mathbf{V} = \mathbf{grad}_\Gamma\,\gamma^+\Phi$ and $\gamma^-_{\mathbf{t}}\mathbf{H} = \mathbf{grad}_\Gamma\,\gamma^+\Psi + \gamma_{\mathbf{t}}\mathbf{H}_s$. This results in: Seek $u \in H^{\frac{1}{2}}_{\Sigma}(\partial\Omega')/\mathbb{R}$, $\psi \in H^{-\frac{1}{2}}_{\Sigma}(\partial\Omega')$, $\boldsymbol{\eta} \in \boldsymbol{H}^{-\frac{1}{2}}_{||}(\mathrm{div}_\Gamma, \Gamma)$ such that

$$n'(u,v) + b(\boldsymbol{\eta},v) - k'(\psi,v) = f(v)\,,$$
$$-b(\mu,u) + a(\boldsymbol{\eta},\mu) \qquad\qquad = g(\mu)\,, \tag{54}$$
$$k'(\phi,u) \qquad\qquad + d'(\psi,\phi) = 0\,.$$

for all $v \in H^{\frac{1}{2}}_{\Sigma}(\partial\Omega')/\mathbb{R}$, $\phi \in H^{-\frac{1}{2}}_{\Sigma}(\partial\Omega')$, $\mu \in \boldsymbol{H}^{-\frac{1}{2}}_{||}(\mathrm{div}_\Gamma, \Gamma)$, where

$$n'(u,v) := \tau\,\langle\mathsf{N}^\kappa(\mathbf{grad}_\Gamma\,u),\mathbf{grad}_\Gamma\,v\rangle_\tau + \langle\mathsf{D}'u,v\rangle_{1/2,\partial\Omega'}\,,$$
$$b(\boldsymbol{\eta},v) := \tau\,\langle(\tfrac{1}{2}Id + \mathsf{B}^\kappa)\boldsymbol{\eta},\mathbf{grad}_\Gamma\,v\rangle_\tau\,,$$
$$k'(\psi,v) := \langle\psi,(\tfrac{1}{2}Id - \mathsf{K}')v\rangle_{1/2,\partial\Omega'}\,,$$
$$a(\boldsymbol{\eta},\mu) := \tau\,\langle\mu,\mathsf{A}^\kappa\boldsymbol{\eta}\rangle_\tau\,,$$
$$d'(\psi,\phi) := \langle\phi,\mathsf{V}'\psi\rangle_{1/2,\partial\Omega'}\,,$$
$$f(v) := \langle\gamma'_{\mathbf{n}}\mathbf{H}_s,v\rangle_{1/2,\partial\Omega'} - \langle\mathsf{N}^\kappa(\gamma_{\mathbf{t}}\mathbf{H}_s),\mathbf{grad}_\Gamma\,v\rangle_\tau\,,$$
$$g(\mu) := \tau\,\langle\mu,(\tfrac{1}{2}Id - \mathsf{C}^\kappa)\gamma_{\mathbf{t}}\mathbf{H}_s\rangle_\tau\,.$$

It is worth noting that (11) yields the identity (cf. [25, Formula (2.86)])

$$\langle\mathsf{N}^\kappa\mathbf{u},\mathbf{v}\rangle_\tau = \kappa^2\,\langle\gamma_{\mathbf{t}}\boldsymbol{\Psi}^\kappa_A(\mathsf{R}\mathbf{u}),\mathsf{R}\mathbf{v}\rangle_\tau + \langle\mathsf{V}^\kappa(\mathrm{curl}_\Gamma\,\mathbf{u}),\mathrm{curl}_\Gamma\,\mathbf{v}\rangle_{1/2,\Gamma}\,. \tag{55}$$

This leads to an alternative expression for the first contribution to $n'(u,v)$:

$$\langle \mathsf{N}^\kappa(\mathbf{grad}_\Gamma\, u), \mathbf{grad}_\Gamma\, v\rangle_\tau = \kappa^2\, \langle \gamma_\mathbf{t}\boldsymbol{\Psi}_A^\kappa(\mathbf{curl}_\Gamma\, u), \mathbf{curl}_\Gamma\, v\rangle_\tau\ .$$

The surface gradient of the u-component of the solution of (54) provides the tangential trace of $\mathbf{H}$, whereas $\psi := \partial_\mathbf{n}'\Psi$ can be viewed as the (scaled) magnetic flux through $\partial\Omega'$. The meaning of $\boldsymbol{\eta} := \gamma_N^-\mathbf{H}$ is that of a (scaled) twisted tangential trace of the electric field.

Theorem 13. *The bilinear form associated with the variational problem* (54) *is $H_\Sigma^{\frac{1}{2}}(\partial\Omega')/\mathbb{R} \times H_\Sigma^{-\frac{1}{2}}(\partial\Omega') \times \boldsymbol{H}_{||}^{-\frac{1}{2}}(\mathrm{div}_\Gamma, \Gamma)$-elliptic*

Proof. As in the proof of Thm. 12 we can exploit the block skew symmetric structure, because the bilinear forms on the diagonal are elliptic on their respective spaces, see Thm. 11 and (35). $\square$

Remark 4. Actually, the coupled variational problem for the **H**-based model fails the condition that only equations on Γ may be involved, because some integral operators rely on cutting surfaces, too. This is an enormous practical obstacle to the use of the **H**-based model, because the construction of cutting surfaces requires a triangulation of some part of the air region and can be prohibitively expensive [46, 37].

One might wonder why this drawback is inevitable with the **H**-based model but not encountered in the case of the **E**-based model. We owe this to the second nature of **E** as a vector potential. For this reason we do not have to introduce another potential to carry out boundary reduction. On top of that a vector potential always exists and is not tied to any topological constraints.

8.3 Galerkin Discretization

Assume that a combined triangulation Γ_h' of Γ and the cuts Σ_k, $k = 1, \ldots, N$, is supplied. As before, we write Γ_h for its restriction to Γ. Thanks to Thm. 13 a conforming Galerkin discretization will yield quasi-optimal approximations of solutions u, ψ, and $\boldsymbol{\eta}$ of (54).

In particular, the space $\mathcal{F}_h(\Gamma_h) \subset \boldsymbol{H}_{||}^{-\frac{1}{2}}(\mathrm{div}_\Gamma, \Gamma)$ can be reused as trial space for $\boldsymbol{\eta}$. To approximate u and ψ we can employ the usual conforming boundary element spaces for $H^{\frac{1}{2}}(\partial\Omega')$ and $H^{-\frac{1}{2}}(\partial\Omega')$. Let $\mathcal{S}_h(\Gamma_h')$ and $\mathcal{Q}_h(\Gamma_h')$ stand for these.

A common trait of the boundary element spaces is that they offer far more regularity than required by mere conformity. For instance, all boundary element functions will belong to $L^\infty(\partial\Omega')$. Then the constraints inherent in the spaces $H_\Sigma^{\frac{1}{2}}(\partial\Omega')$ and $H_\Sigma^{-\frac{1}{2}}(\partial\Omega')$ permit us to restrict the operators V', K', and, D' to Γ: Straightforward manipulations using the integral operator representations of V', D', and K' show that for $u, v \in H_\Sigma^{\frac{1}{2}}(\partial\Omega') \cap L^\infty(\partial\Omega')$ and $\phi, \psi \in H_\Sigma^{-\frac{1}{2}}(\partial\Omega') \cap L^\infty(\partial\Omega')$

$$\langle \psi, \mathsf{V}'\phi \rangle_{1/2,\partial\Omega'} = \langle \psi, \mathsf{V}^0\phi \rangle_{1/2,\Gamma} \quad , \quad \langle \mathsf{D}'u, v \rangle_{1/2,\partial\Omega'} = \langle \mathsf{D}^0 u, v \rangle_{1/2,\Gamma} \ ,$$

$$\langle \phi, \mathsf{K}'v \rangle_{1/2,\partial\Omega'} = \langle \phi, \mathsf{K}^0 v \rangle_{1/2,\Gamma} + \sum_{k=1}^{N} [v]_{\Sigma_k} \int_{\Gamma}\int_{\Sigma_k} \frac{\partial G_0(\mathbf{x},\mathbf{y})}{\partial \mathbf{n}(\mathbf{y})} \phi(\mathbf{x})\, dS(\mathbf{y}) dS(\mathbf{x}) \ ,$$

$$\langle \phi, v \rangle_{1/2,\partial\Omega'} = \langle \phi, v \rangle_{1/2,\Gamma} + \sum_{k=1}^{N} [v]_{\Sigma_k} \cdot \int_{\Sigma_k} \phi(\mathbf{x})\, dS(\mathbf{x}) \ .$$

$$(56)$$

We observe that the cuts will enter the discrete variational problem only through some global integral quantities that are not sensitive to the choice of boundary elements on the cuts. Sloppily speaking, this permits us to cover each cut by only a single surface element. More precisely, we may choose

$$\mathcal{S}_h(\Gamma_h') = \mathcal{S}_h(\Gamma_h) + \mathrm{Span}\left\{c_h^1, \dots, c_h^N\right\} \ ,$$
$$\mathcal{Q}_h(\Gamma_h') = \mathcal{Q}_h(\Gamma_h) + \mathrm{Span}\left\{\chi_1, \dots, \chi_N\right\} \ .$$

Here, c_h^k is a Γ_h'-piecewise linear function $\in C^0(\partial\Omega') \cap H_{\Sigma}^{\frac{1}{2}}(\partial\Omega')$, whose restriction to Γ has a jump of height 1 across the edge cycle σ_k and is continuous across any other σ_j, $j \neq k$. The function $\chi_k \in L^\infty(\partial\Omega') \cap H_{\Sigma}^{-\frac{1}{2}}(\partial\Omega')$ assumes the values $+1$ and -1 on Σ_k^+ and Σ_k^-, respectively, and vanishes on $\partial\Omega' \setminus \Sigma_k$.

Using the identities (56), the discrete variational problem can be rephrased as: Seek $\tilde{u}_h \in \mathcal{S}_h(\Gamma_h)$, $\tilde{\psi}_h \in \mathcal{Q}_h(\Gamma_h)$, $\boldsymbol{\eta}_h \in \mathcal{F}_h(\Gamma_h)$, $\alpha_1, \dots, \alpha_N \in \mathbb{C}$, $\beta_1, \dots, \beta_N \in \mathbb{C}$ such that

$$
\begin{aligned}
n(\tilde{u}_h, v_h) + b(\boldsymbol{\eta}_h, v_h) - k(\tilde{\psi}_h, v_h) + \sum_k \alpha_k n(c_k, v_h) &= f(v_h) \ , \\
-b(\mu_h, \tilde{u}_h) + a(\boldsymbol{\eta}_h, \mu_h) - \sum_k \alpha_k b(\mu_h, c_k) &= g(\mu_h) \ , \\
k(\phi_h, \tilde{u}_h) + d(\psi_h, \phi_h) + \sum_k \alpha_k k'(\phi_h, c_k) &= 0 \ , \\
n(\tilde{u}_h, c_j) + b(\boldsymbol{\eta}_h, c_j) - k'(\tilde{\psi}_h, c_j) + \sum_k \alpha_k n(c_k, c_j) - \sum_k \beta_k k'(\chi_k, c_j) &= f(c_j) \ , \\
\sum_k \alpha_k k'(\chi_l, c_k) &= 0 \ .
\end{aligned}
$$

for all $v_h \in \mathcal{S}_h(\Gamma_h)$, $\mu_h \in \mathcal{F}_h(\Gamma_h)$, $\phi_h \in \mathcal{Q}_h(\Gamma_h)$, $j = 1, \dots, N$, $l = 1, \dots, N$. Here we set, using [32, Thm. 7, Ch. XI],

$$n(u,v) := \frac{1}{\tau^2}\left(\mathsf{N}^\kappa(\mathbf{grad}_\Gamma u), \mathbf{grad}_\Gamma v\right)_{\boldsymbol{L}^2(\Gamma)} - \left(\mathsf{A}^0\widetilde{\mathbf{curl}_\Gamma u}, \widetilde{\mathbf{curl}_\Gamma v}\right)_{\boldsymbol{L}^2(\Gamma)} \ ,$$

$$d(\psi, \phi) := \left(\phi, \mathsf{V}^0\psi\right)_{\boldsymbol{L}^2(\Gamma)} \ , \quad k(\psi, v) := \left(\psi, (\tfrac{1}{2}Id - \mathsf{K}^0)v\right)_{\boldsymbol{L}^2(\Gamma)} \ ,$$

for bilinear forms induced by integral operators on Γ alone. The discrete solution can be obtained as $u_h = \tilde{u}_h + \sum_k \alpha_k c_k$ and $\psi_h = \tilde{\psi}_h + \sum_k \beta_k \chi_k$. A closer study of the boundary integral operators shows that the cuts only come into play through integrals of the form

$$\int\limits_{\Gamma}\int\limits_{\Sigma_k} \frac{\partial G_0(\mathbf{x},\mathbf{y})}{\partial \mathbf{n}(\mathbf{y})}\phi_h(\mathbf{x})\,dS(\mathbf{y},\mathbf{x}) \quad , \quad \int\limits_{\Sigma_k} \mathbf{H}_s \cdot \mathbf{n}\,dS \ ,$$

for $\phi_h \in \mathcal{Q}_h(\Gamma_h)$. Obviously, by Gauß' divergence theorem, Σ_k can be replaced by any other surface homologous in $H_2(\Omega_e, \mathbb{Z})$ without changing the values of the integrals. Paradoxically, information about the concrete geometry of the Σ_k seems to be indispensable for the evaluation of the integrals.

The case of lumped parameter excitation is treated in a similar fashion as in the case of the $\mathbf{E}$-based model. First, note that α_k measures the jump of the magnetic scalar potential across Σ_k. According to Ampere's law the height of this jump agrees with the total current in the loop of the conductor corresponding to Σ_k. Hence, a prescribed total current in a loop of the conductor can be taken into account by fixing the value of α_k for the related cut.

Remark 5. The intrinsic use of a (multivalued) magentic scalar potential in Ω_c paves the way for accommodating non-local inductive current excitation: by Ampere's law, we only need to fix the jump of ψ across a cut associated with a current carrying loop of the conductor. In the above variational formulation, this boils down to fixing some of the α_k.

Remark 6. The values of the β_k agree with the total magnetic flux through the cut Σ_k. By Faraday's law it is proportional to the electromotive force along σ_k. Hence, if the voltage around a loop of the conductor is to be imposed, we can do so by fixing the value of the associated β_k. The possibility to take into account lumped parameter voltage excitation is only available with the $\mathbf{H}$-based model.

9 Postprocessing

As we have remarked in the introduction, getting approximate Cauchy data $(\gamma_\mathbf{t}\mathbf{E}, \gamma_\times\mathbf{H})$ on Γ might not be the eventual goal of the computation. Thus, we have to figure out how to get (i) the total Ohmic losses in Ω_C, and (ii) the total force acting on Ω_C. Here, we focus on the $\mathbf{E}$-based formulation of Sect. 7 and assume that by solving (47) we have obtained approximate Cauchy data $(\mathbf{u}_h, \lambda_h)$.

Ohmic losses are the only mechanism for the dissipation of field energy in the eddy current model. Moreover, since all fields are harmonic in time, the total field energy inside Ω_C will not change over one period. Therefore, we get the averaged Ohmic losses by appealing to Poynting's theorem

$$\bar{P}_{\mathrm{Ohm}} = -\frac{1}{2}\Re\left\{\int\limits_{\Gamma}(\mathbf{E}\times\overline{\mathbf{H}})\cdot\mathbf{n}\,dS\right\} = \frac{1}{2}\,\Re\left\{\langle\mathbf{u},\lambda\rangle_\tau\right\} \ .$$

244   R. Hiptmair

A natural approximation is

$$\bar{P}_{\text{Ohm}} \approx \bar{P}^h_{\text{Ohm}} := \tfrac{1}{2}\Re\,\langle \mathbf{u}_h, \lambda_h\rangle_\tau \;.$$

The error can be estimated by

$$\begin{aligned}
\bar{P}_{\text{Ohm}} - \bar{P}^h_{\text{Ohm}} &= \tfrac{1}{2}\Re\left\{\langle \mathbf{u}, \lambda - \lambda_h\rangle_\tau + \langle \mathbf{u} - \mathbf{u}_h, \lambda_h\rangle_\tau\right\} \\
&\le \frac{1}{2}\left(\|\mathbf{u}\|_{H_\perp^{-\frac{1}{2}}(\mathrm{curl}_\Gamma,\Gamma)}\, \|\lambda - \lambda_h\|_{H_\|^{-\frac{1}{2}}(\mathrm{div}_\Gamma,\Gamma)}\right. \\
&\quad\left. + \|\mathbf{u} - \mathbf{u}_h\|_{H_\perp^{-\frac{1}{2}}(\mathrm{curl}_\Gamma,\Gamma)}\, \|\lambda_h\|_{H_\|^{-\frac{1}{2}}(\mathrm{div}_\Gamma,\Gamma)}\right),
\end{aligned}$$

where we have exploited the continuity of the pairing $\langle\cdot,\cdot\rangle_\tau$. This shows that $\bar{P}^h_{\text{Ohm}}$ will converge with the same rate as observed for the Cauchy data.

To compute the total force on the conductor we can resort to the magnetic Maxwell stress tensor for linear materials [43, Sect. 6.7]

$$\mathbb{T} := \mathcal{B}\cdot\mathcal{H}^T - \tfrac{1}{2}(\mathcal{B}\cdot\mathcal{H})\,\mathbb{I}\,, \tag{57}$$

where, $\mathcal{B}$ and $\mathcal{H}$ denote the real, time dependent fields. Ignoring the electric forces is consistent with the eddy current model, which rests on the assumption of negligible electric field energy. Next, we consider $\mathbb{T}$ on Γ and split both the magnetic induction $\mathcal{B}$ and $\mathcal{H}$ into tangential and normal components, *cf.* [49, Sect. 6].

$$\mathcal{B}(\mathbf{x}) = \mathcal{B}_n(\mathbf{x})\mathbf{n}(\mathbf{x}) + \mathcal{B}_{\mathbf{t}}(\mathbf{x}) \quad,\quad \mathcal{H}(\mathbf{x}) = \mathcal{H}_n(\mathbf{x})\mathbf{n}(\mathbf{x}) + \mathcal{H}_{\mathbf{t}}(\mathbf{x})\,,\quad \mathbf{x}\in\Gamma\,.$$

Using the constitutive equation $\mathcal{B} = \mu_0\mathcal{H}$, that is valid in Ω_e, we express

$$\mathcal{H}_n(\mathbf{x}) = \frac{1}{\mu_0}\mathcal{B}_n \quad,\quad \mathcal{B}_{\mathbf{t}}(\mathbf{x}) = \mu_0\mathcal{H}_{\mathbf{t}}\,.$$

and get on Γ

$$\mathbb{T}(\mathbf{x})\mathbf{n}(\mathbf{x}) = \frac{1}{2}(\frac{1}{\mu_0}\mathcal{B}_n^2(\mathbf{x}) - \mu_0|\mathcal{H}_{\mathbf{t}}|^2)\mathbf{n}(\mathbf{x}) + \mathcal{B}_n(\mathbf{x})\,\mathcal{H}_{\mathbf{t}}(\mathbf{x})\,,\quad \mathbf{x}\in\Gamma\,.$$

Hence, the total force on the conductor at a particular time is given by

$$\begin{aligned}
F_{\text{tot}} &= \int_\gamma \mathbb{T}(\mathbf{y})\mathbf{n}(\mathbf{y})\,\mathrm{d}S(\mathbf{y}) \\
&= \int_\gamma \frac{1}{2}(\frac{1}{\mu_0}\mathcal{B}_n^2(\mathbf{y}) - \mu_0|\mathcal{H}_{\mathbf{t}}(\mathbf{y})|^2)\mathbf{n}(\mathbf{y}) + \mathcal{B}_n(\mathbf{y})\,\mathcal{H}_{\mathbf{t}}(\mathbf{y})\,\mathrm{d}S(\mathbf{y})\,.
\end{aligned}$$

Let us revert to complex amplitudes $\mathbf{B}$ and $\mathbf{H}$, for which the averaged force over one period is given by

$$\bar{F}_{\text{tot}} = \int_{\gamma} \frac{1}{4}(\frac{1}{\mu_0}|B_n(\mathbf{y})|^2 - \mu_0|\mathbf{H_t}(\mathbf{y})|^2)\mathbf{n}(\mathbf{y}) + \frac{1}{2}\Re\{B_n(\mathbf{y})\,\mathbf{H_t}(\mathbf{y})\}\ \mathrm{d}S(\mathbf{y})\ .$$

From $\mathbf{B} = (i\omega)^{-1}\,\mathbf{curl}\,\mathbf{E}$ we infer $B_n = (i\omega)^{-1}\,\mathrm{curl}_\Gamma\,\mathbf{u}$, where curl_Γ stands for the scalar surface rotation (div_Γ applied to the rotated field). On the other hand, it is straightforward that $\mathbf{H_t} = -\lambda \times \mathbf{n}$. Thus, we can rewrite

$$\bar{F}_{\text{tot}} = \int_{\gamma} (\frac{1}{4\mu_0\omega^2}|\,\mathrm{curl}_\Gamma\,\mathbf{u}(\mathbf{y})|^2 - \frac{\mu_0}{4}|\lambda(\mathbf{y})|^2)\mathbf{n}(\mathbf{y})- \tag{58}$$
$$\frac{1}{2\omega}\Re\{\mathrm{curl}_\Gamma\,\mathbf{u}(\mathbf{y})\,(\lambda(\mathbf{y}) \times \mathbf{n}(\mathbf{y}))\}\ \mathrm{d}S(\mathbf{y})\ .$$

Finally, we have expressed the total force in terms of quantities that occur as unknowns in the variational problem (38). Now, it is straightforward how to compute an approximation of $\bar{F}_{\text{tot}}$ from the boundary element solution $(\mathbf{u}_h, \lambda_h)$. As far as the approximation error is concerned, the same considerations apply as for the energy flux.

It is important to be aware that the force as given by (58) is by no means a continuous functional in the natural trace norms, because the inclusions $\boldsymbol{H}_\perp^{-\frac{1}{2}}(\mathbf{curl}_\Gamma, \Gamma) \subset \boldsymbol{L}^2(\Gamma)$ and $\boldsymbol{H}_{||}^{-\frac{1}{2}}(\mathrm{div}_\Gamma, \Gamma) \subset \boldsymbol{L}^2(\Gamma)$ do **not** hold (Compare the case to the Neumann trace space $H^{-\frac{1}{2}}(\Gamma)$ for second order elliptic problems). Of course, (58) can easily be evaluated for the boundary element functions, but unlike in the case of the total energy flux, rates of convergence for $\bar{F}_{\text{tot}}$ cannot be inferred from (42).

Remark 7. We emphasize that approximations for the traces of the fields onto Γ are directly available, because we have relied on a *direct boundary element method.* If an indirect method had been used, it would have taken expensive post-processing, in order to get the same information.

References

1. A. Alonso-Rodriguez, P. Fernandes, A. Valli: Weak and strong formulations for the time-harmonic eddy-current problem in general multi-connected domains. Eur. J. Appl. Math. 14 (2003) 387–406.
2. A. Alonso-Rodriguez, R. Hiptmair, A. Valli: A hybrid formulation of eddy current problems. Num. Meth. Part. Diff. Equ. 21 (2005) 742–763.
3. H. Ammari, A. Buffa, J.-C. Nédélec: A justification of eddy currents model for the Maxwell equations. SIAM J. Appl. Math. 60 (2000) 1805–1823.
4. C. Amrouche, C. Bernardi, M. Dauge, V. Girault: Vector potentials in three–dimensional nonsmooth domains. Math. Meth. Appl. Sci. 21 (1998) 823–864.
5. E. Bänsch, W. Dörfler: Adaptive finite elements for exterior domain problems. Numer. Math. 80 (1998) 497–523.
6. S. Börm, J. Ostrowski: Fast evaluation of boundary integral operators arising from an eddy current problem. J. Comp. Phys. 193 (2003) 67–85.

7. A. Bossavit: Two dual formulations of the 3D eddy–currents problem. COMPEL, 4 (1985) 103–116.

8. A. Bossavit: Computational Electromagnetism. Variational Formulation, Complementarity, Edge Elements. vol. 2 of Electromagnetism Series, Academic Press, San Diego, CA, 1998.

9. A. Bossavit, J. Vérité: A mixed FEM–BIEM method to solve 3D eddy–current problems. IEEE Trans. MAG, 18 (1982) 431–435.

10. J. Breuer: Schnelle Randelementmethoden zur Simulation von elektrischen Wirbelstromfeldern sowie ihrer Wärmeproduktion und Kühlung. PhD thesis, Universität Stuttgart, Institut für Angewandte Analysis und numerische Simulation, Stuttgart, Germany, 2005.

11. F. Brezzi, M. Fortin: Mixed and hybrid finite element methods. Springer, 1991.

12. A. Buffa: Hodge decompositions on the boundary of a polyhedron: The multiconnected case. Math. Mod. Meth. Appl. Sci. 11 (2001) 1491–1504.

13. A. Buffa: Trace theorems on non-smmoth boundaries for functional spaces related to Maxwell equations: An overwiew. In Computational Electromagnetics (C. Carstensen, S. Funken, W. Hackbusch, R. Hoppe, P. Monk eds.), Lecture Notes in Computational Science and Engineering, Vol. 28, Springer, Berlin, pp. 23–34, 2003.

14. A. Buffa, S. Christiansen: The electric field integral equation on Lipschitz screens: Definition and numerical approximation. Numer. Math. 94 (2002) 229–267.

15. A. Buffa, S. Christiansen: A dual finite element complex on the barycentric refinement. C.R. Acad. Sci Paris, Ser I, 340 (2005) 461–464.

16. A. Buffa, P. Ciarlet: On traces for functional spaces related to Maxwell's equations. Part I: An integration by parts formula in Lipschitz polyhedra. Math. Meth. Appl. Sci. 24 (2001) 9–30.

17. A. Buffa, P. Ciarlet: On traces for functional spaces related to Maxwell's equations. Part II: Hodge decompositions on the boundary of Lipschitz polyhedra and applications. Math. Meth. Appl. Sci. 24 (2001), 31–48.

18. A. Buffa, M. Costabel, D. Sheen: On traces for $\mathbf{H}(\mathbf{curl}, \Omega)$ in Lipschitz domains. J. Math. Anal. Appl. 276 (2002) 845–867.

19. A. Buffa, R. Hiptmair: Galerkin boundary element methods for electromagnetic scattering. In: Topics in Computational Wave Propagation. Direct and inverse Problems (M. Ainsworth, P. Davis, D. Duncan, P. Martin, B. Rynne eds.), Lecture Notes in Computational Science and Engineering, Vol. 31, Springer, Berlin, pp. 83–124, 2003.

20. A. Buffa, R. Hiptmair, T. von Petersdorff, C. Schwab: Boundary element methods for Maxwell equations on Lipschitz domains. Numer. Math. 95 (2003) 459–485. Published online (DOI 10.1007/s00211-002-0407-z).

21. M. Cessenat: Mathematical Methods in Electromagnetism. Advances in Mathematics for Applied Sciences, Vol. 41, World Scientific, Singapore, 1996.

22. S. Christiansen: Mixed boundary element method for eddy current problems. Research Report 2002-16, SAM, ETH Zürich, Zürich, Switzerland, 2002.

23. S. Christiansen, J.-C. Nédélec: A preconditioner for the electric field integral equation based on Calderón formulas. SIAM J. Numer. Anal. 40 (2002) 1100–1135.

24. P. Ciarlet: The Finite Element Method for Elliptic Problems. Studies in Mathematics and its Applications, Vol. 4, North-Holland, Amsterdam, 1978.

25. D. Colton, R. Kress: Integral equation methods in scattering theory. Pure and Applied Mathematics, John Wiley & Sons, 1983.

26. D. Colton, R. Kress: Inverse Acoustic and Electromagnetic Scattering Theory. Applied Mathematical Sciences, Vol. 93, Springer, Heidelberg, 2nd ed., 1998.

27. M. Costabel: Symmetric methods for the coupling of finite elements and boundary elements. In: Boundary Elements IX (C. Brebbia, W. Wendland, G. Kuhn eds.), Springer, Berlin, pp. 411–420, 1987.

28. M. Costabel: Boundary integral operators on Lipschitz domains: Elementary results. SIAM J. Math. Anal. 19 (1988) 613–626.

29. M. Costabel, M. Dauge: Singularities of Maxwell's equations on polyhedral domains. In: Analysis, Numerics and Applications of Differential and Integral Equations (M. Bach, C. Constanda, G. C. Hsiao, A.-M. Sändig, P. Werner eds.), Longman Pitman Res. Notes Math. Ser., Vol. 379, Addison Wesley, Harlow, pp. 69–76, 1998.

30. M. Costabel, M. Dauge: Maxwell and Lamé eigenvalues on polyhedra. Math. Methods Appl. Sci. 22 (1999) 243–258.

31. M. Costabel, E. P. Stephan: A direct bounary equation method for transmission problems. J. Math. Anal. Appl. 106 (1985) 367–413.

32. R. Dautray, J.-L. Lions: Mathematical Analysis and Numerical Methods for Science and Technology, Vol. 4, Springer, Berlin, 1990.

33. H. Dirks: Quasi–stationary fields for microelectronic applications. Electrical Engineering 79 (1996) 145–155.

34. P. Fernandes, G. Gilardi: Magnetostatic and electrostatic problems in inhomogeneous anisotropic media with irregular boundary and mixed boundary conditions. Math. Models Meth. Appl. Sci. 7 (1997) 957–991.

35. V. Girault: Curl–conforming finite element methods for Navier–Stokes equations with non–standard boundary conditions in $\mathbb{R}^3$. Lecture Notes in Mathematics, Vol. 1431, Springer, Berlin, pp. 201–218, 1989.

36. V. Girault, P. Raviart: Finite element methods for Navier–Stokes equations. Springer, Berlin, 1986.

37. P. Gross: Efficient finite element-based algorithms for topological aspects of 3-dimensional magnetoquasistatic problems. PhD thesis, College of Engineering, Boston University, Boston, USA, 1998.

38. R. Hiptmair: Finite elements in computational electromagnetism. Acta Numerica 11 (2002), 237–339.

39. R. Hiptmair: Symmetric coupling for eddy current problems. SIAM J. Numer. Anal. 40 (2002) 41–65.

40. R. Hiptmair: Coupling of finite elements and boundary elements in electromagnetic scattering. SIAM J. Numer. Anal. 41 (2003) 919–944.

41. R. Hiptmair, J. Ostrowski: Generators of $H_1(\Gamma_h, \mathbb{Z})$ for triangulated surfaces: Construction and classification. SIAM J. Computing 31 (2002) 1405–1423.

42. R. Hiptmair, O. Sterz: Current and voltage excitations for the eddy current model. Int. J. Numer. Model 18 (2005) 1–21.

43. J. Jackson: Classical electrodynamics. John Wiley, 3rd ed., 1998.

44. L. Kettunen, K. Forsman, A. Bossavit: Gauging in Whitney spaces. IEEE Trans. Magnetics 35 (1999) 1466–1469.

45. P. Kotiuga: On making cuts for magnetic scalar potentials in multiply connected regions. J. Appl. Phys. 61 (1987) 3916–3918.

46. P. Kotiuga: An algorithm to make cuts for magnetic scalar potentials in tetrahedral meshes based on the finite element method. IEEE Trans. Magnetics 25 (1989) 4129–4131.

47. P. Kotiuga: Topological considerations in coupling magnetic scalar potentials to stream functions describing surface currents. IEEE Trans. Magnetics 25 (1989) 2925–2927.

48. M. Kuhn, O. Steinbach: FEM-BEM coupling for 3d exterior magnetic field problems. Math. Meth. Appl. Sci. 25 (2002) 357–371.

49. S. Kurz, J. Fetzer, G. Lehner, W. Rucker: Numerical analysis of three-dimensional eddy current problems with moving bodies by boundary element–finite element coupling. Surv. Math. Ind. 9 (2000) 131–150.

50. U. Langer, O. Steinbach: Boundary element tearing and interconnecting methods. Computing 71 (2003) 205–228.

51. R. McCamy, E. P. Stephan: Solution procedures for three-dimensional eddy-current problems. J. Math. Anal. Appl. 101 (1984) 348–379.

52. W. McLean: Strongly Elliptic Systems and Boundary Integral Equations. Cambridge University Press, Cambridge, UK, 2000.

53. J.-C. Nédélec: Acoustic and Electromagnetic Equations: Integral Representations for Harmonic Problems. Applied Mathematical Sciences, Vol. 44, Springer, Berlin, 2001.

54. J. Ostrowski: Boundary Element Methods for Inductive Hardening. PhD thesis, Fakultät für Mathematik und Physik, Tübingen, 2002.

55. M. Reissel: On a transmission boundary-value problem for the time-harmonic Maxwell equations without displacement currents. SIAM J. Math. Anal. 24 (1993) 1440–1457.

56. Z. Ren, F. Bouillault, A. Razek, J. Verité: Comparison of different boundary integral formulations when coupled with finite elements in three dimensions. IEE Proc. A 135 (1988) 501–505.

57. Z. Ren, A. Razek: New techniques for solving three–dimensional multiply connected eddy–current problems. IEE Proc. A 137 (1990) 135–140.

58. S. Sauter, C. Schwab: Randelementmethoden. Analyse, Numerik und Implementierung schneller Algorithmen. B. G. Teubner, Stuttgart, Leipzig, Wiesbaden, 2004.

59. A. Schwarz: Topology for Physicists. Grundlehren der mathematischen Wissenschaften, Vol. 308, Springer, Berlin, 1994.

60. O. Steinbach: Stability estimates for hybrid coupled domain decomposition methods. Lecture Notes in Mathematics, Vol. 1809, Springer, Heidelberg, 2003.

61. O. Steinbach, W. L. Wendland: The construction of some efficient preconditioners in the boundary element method. Adv. Comput. Math. 9 (1998) 191–216.

62. J. Stratton, L. Chu: Diffraction theory of electromagnetic waves. Phys. Rev. 56 (1939) 99–107.

63. T. von Petersdorff: Boundary integral equations for mixed Dirichlet, Neumann and transmission problems. Math. Meth. Appl. Sci. 11 (1989) 185–213.

Fast Boundary Element Methods in Computational Electromagnetism

Stefan Kurz[1], Oliver Rain[1] and Sergej Rjasanow[2]

[1] Robert Bosch GmbH, Postfach 10 60 50, 70049 Stuttgart, Germany
   `oliver.rain@de.bosch.com`
[2] Universität des Saarlandes, Fachrichtung 6.1 - Mathematik, Postfach 15 11 50,
   66041 Saarbrücken, Germany
   `rjasanow@num.uni-sb.de`

Summary. When the Boundary Element Method (BEM) is used to analyse electromagnetic problems one is able to achieve an almost linear complexity by applying matrix compression techniques. Beyond this, on symmetrical domains the computational costs can be reduced by significant factors. By using several symmetry considerations (geometry, mesh, kernel, excitation) it will be shown how the combination of the Adaptive Cross Approximation (ACA) and the symmetry exploitation allows an efficient solution of electromagnetic problems. This approach will be demonstrated on the scalar BEM formulation for electrostatics and can also be applied to the vectorial eddy current formulations. The symmetry exploting ACA algorithm not only reduces the problem size due to the symmetry but also possesses an almost linear complexity w.r.t. the number of unknowns.

1 Introduction

Electromagnetic devices can be analysed by the coupled BE-FE method, where the conducting and magnetic parts are discretised by finite elements. In contrast, the surrounding space is described with the help of the boundary element method (BEM). This discretisation scheme is well suited especially for problems including moving parts [11]. The BEM discretisation of the boundary integral operators usually leads to dense matrices without any structure. A naive strategy for the solution of the corresponding linear system would need at least $O(N^2)$ operations and memory, where N is the number of unknowns. Methods such as fast multipole [6] and panel clustering [9] provide an approximation to the matrix in almost linear complexity. These methods are based on explicitly given kernel approximations by degenerate kernels, i.e. a finite sum of separable functions, which may be seen as a blockwise low-rank approximation of the system matrix. The blockwise approximant permits a fast matrix-vector multiplication, which can be exploited in iterative solvers, and can be stored efficiently. In contrast to the methods mentioned the ACA algorithm

[2, 3] generates the low-rank approximant from the matrix itself using only few entries and without using any explicit a priori known degenerate-kernel approximation. Special emphasis is put on the handling of symmetry conditions in connection with ACA [13]. The feasibility of the proposed method is demonstrated by means of numerical examples.

2 Statement of the Problem

The electromagnetic phenomena are described by Maxwell's equations, which can be written in the form of partial differential equations as follows

$$\mathbf{curlH} = \boldsymbol{\jmath} + \partial_t \mathbf{D}\,, \tag{1}$$

$$\mathbf{curlE} = -\partial_t \mathbf{B}\,, \tag{2}$$

$$\mathrm{div}\mathbf{B} = 0\,, \tag{3}$$

$$\mathrm{div}\mathbf{D} = \rho\,. \tag{4}$$

The equations describe the correlation between the magnetic field $\mathbf{H}$, magnetic induction $\mathbf{B}$, electric field $\mathbf{E}$ and electric displacement $\mathbf{D}$. $\boldsymbol{\jmath}$ denotes the electric current density and ρ the electric charge density. The equations have to be supplemented by the material laws

$$\mathbf{B} = \mu \mathbf{H}\,, \tag{5}$$

$$\mathbf{D} = \varepsilon \mathbf{E}\,, \tag{6}$$

$$\boldsymbol{\jmath} = \kappa \mathbf{E} + \boldsymbol{\jmath}_S\,, \tag{7}$$

where μ is the magnetic permeability, ε the electric permittivity, κ denotes the electric conductivity and $\boldsymbol{\jmath}_S$ the impressed source current density.

2.1 Formulation of the problem

For the sake of simplicity we consider in the sequel the electrostatic case

$$\mathbf{curlE} = 0\,, \tag{8}$$

$$\mathrm{div}\mathbf{D} = \rho\,, \tag{9}$$

$$\mathbf{D} = \varepsilon \mathbf{E}\,. \tag{10}$$

Based on the potential ansatz

$$\mathbf{E} = -\mathbf{grad}\varphi\,, \tag{11}$$

where φ is the electric scalar potential, we obtain the potential formulation

$$\mathrm{div}\varepsilon\,\mathbf{grad}\phi = -\rho\,, \tag{12}$$

which has to be solved in the whole $\mathbb{R}^3$. In order to apply the BE-FE discretisation scheme we perform a domain decomposition (see Fig. 1) of the computational space into the bounded domain Ω_{FEM} containing dielectric components, the unbounded domain Ω_{BEM} and the coupling boundary $\Gamma = \bar{\Omega}_{\mathrm{FEM}} \cap \bar{\Omega}_{\mathrm{BEM}}$. In this paper we put the emphasis on the BE formulation.

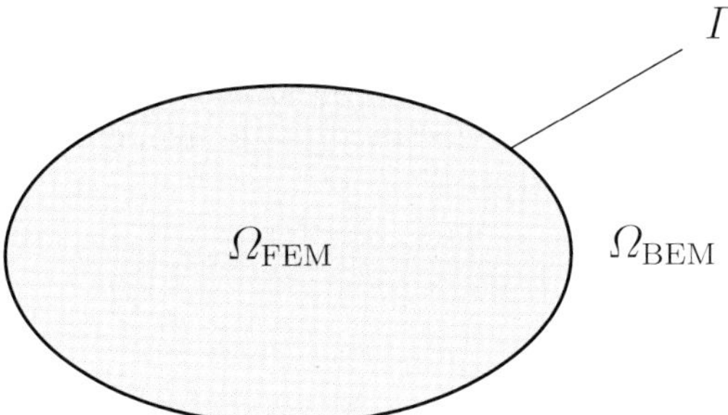

Fig. 1. Domain Decomposition.

Representation Formula

In the BE domain which usually describes the surrounding air we assume $\varepsilon \equiv \varepsilon_0$. Thus, the potential formulation (12) turns out to be the Poisson equation

$$\Delta\phi = -\frac{1}{\varepsilon_0}\,\rho\,. \tag{13}$$

Multiplying (13) by the fundamental solution of the Laplacian

$$u^*(x,y) = \frac{1}{4\pi|x-y|} \tag{14}$$

and performing integration by parts yields the representation formula for smooth boundary points $y \in \Gamma$

$$\frac{1}{2}\phi(y) = \int_\Gamma u^*(x,y)\,\partial_n\phi(x)\,dS_x - \int_\Gamma \partial_n u^*(x,y)\,\phi(x)\,dS_x$$

$$+ \varepsilon_0^{-1}\int_{\Omega_{\mathrm{BEM}}} u^*(x,y)\,\rho(x)\,dx\,. \tag{15}$$

2.2 Discretisation

First, the spatial discretisation has to be introduced. Let

$$\Gamma_h = \bigcup_{j=1}^{N_{\mathrm{El}}} \Gamma_j$$

be a union of boundary elements Γ_j approximating the coupling boundary Γ and

$$\{\phi_j\,,\ j=1,\ldots,N_\phi\} \text{ and } \{\psi_j\,,\ j=1,\ldots,N_\psi\} \tag{16}$$

systems of compact-supported ansatz functions for Dirichlet and Neumann data, respectively. The scalar electric potential and its normal derivative are discretised by the corresponding scalar ansatz functions

$$\phi(x) \approx \sum_{j=1}^{N_\phi} a_j \phi_j(x)\,, \tag{17}$$

$$\partial_n \phi(x) \approx \sum_{j=1}^{N_\psi} q_j \psi_j(x) \tag{18}$$

For the Neumann-type problem the representation formula (15) will be evaluated in N_ϕ collocation points $\{y_i,\ i = 1,\ldots,N_\phi\}$. Together with the discretisation (17-18) this yields N_ϕ discrete boundary integral equations which can be written as the matrix equation

$$\left(\frac{1}{2}I + H\right)\mathbf{a} = G\mathbf{q} + \mathbf{b} \tag{19}$$

with the matrices of the single and double layer potential

$$g_{ij} = \int_{\mathrm{supp}\,\psi_j} u^*(x, y_i)\,\psi_j(x)\,dS_x\,, \quad i = 1,\ldots,N_\phi\,,\ j = 1,\ldots,N_\psi\,, \tag{20}$$

$$h_{ij} = \int_{\mathrm{supp}\,\phi_j} \partial_n u^*(x, y_i)\,\phi_j(x)\,dS_x\,, \quad i,j = 1,\ldots,N_\phi\,. \tag{21}$$

The matrices G and H are fully populated and don't possess any structure. Thus, the computational costs when setting up the matrices and the memory consumption are both of order $O(N_\phi N_\psi)$ and $O(N_\phi^2)$, respectively. Each single matrix entry is computed by the use of a combination of analytical and numerical integration.

3 Hierarchical Matrices

The formal definition and description of hierarchical matrices as well as operations involving those matrices can be found in [7, 8]. In this section we give a more intuitive introduction to this topic.

3.1 Motivation

Let $K : [0,1] \times [0,1] \to \mathbb{R}$ be a given function of two scalar variables and $A \in \mathbb{R}^{N \times M}$ a given matrix having the entries

$$a_{k\ell} = K(x_k, y_\ell)\,, \quad k = 1,\ldots,N\,,\ \ell = 1,\ldots,M\,, \tag{22}$$

with $(x_k, y_\ell) \in [0,1] \times [0,1]$. It is obvious, that the asymptotic memory requirement for the matrix A is $\mathrm{Mem}(A) = \mathcal{O}(N\,M)$ and the asymptotic number of arithmetical operations required for the matrix-vector multiplication $\mathrm{Op}(A\,s) = \mathcal{O}(N\,M)$ if $N, M \to \infty$. This quadratic amount is too high already for moderate values of N and M. However, if we agree to store an approximation $\tilde{A}$ of the matrix A and to deal with the product $\tilde{A}\,s$ instead of the exact value $A\,s$ the situation may change. However, then it is necessary to control the error, i.e. to guarantee the inequality

$$\|A - \tilde{A}\|_F \le \varepsilon \|A\|_F , \tag{23}$$

where $\|A\|_F$ denotes the Frobenius norm of the matrix A

$$\|A\|_F = \left(\sum_{k,\ell} a_{k\ell}^2 \right)^{1/2} \tag{24}$$

for some prescribed accuracy ε. The best approximation of the matrix A is given by its partial singular value decomposition

$$A \approx \tilde{A} = \tilde{A}(r) = \sum_{i=1}^{r} \sigma_i\, u_i\, v_i^\top \tag{25}$$

where the rank $r = r(\varepsilon)$ is chosen corresponding to the condition

$$\|A - \tilde{A}\|_F^2 \le \sum_{i=r+1}^{\min(N,M)} \sigma_i^2 \le \varepsilon^2 \sum_{i=1}^{\min(N,M)} \sigma_i^2 = \varepsilon^2 \|A\|_F^2. \tag{26}$$

Unfortunately, the complete singular value decomposition of the matrix A requires $\mathcal{O}(N^3)$ arithmetical operations when assuming $N \sim M$, and therefore, is too expensive for practical computations. However, the singular value decomposition can be perfectly used for the illustration of the main ideas.

Example 1. Let us consider the following function on $[0,1] \times [0,1]$

$$K(x, y) = \frac{1}{\alpha + (x - y)^2} , \tag{27}$$

where $\alpha > 0$ is a parameter. For $\alpha \sim 1$ the function K is smooth but for small values of α the function K becomes an artificial "strong singularity" at the diagonal $\{(x, x)\}$ of the square $[0,1] \times [0,1]$.

The domain $[0,1] \times [0,1]$ is uniformly discretised using the nodes

$$(x_k, y_\ell) = \left((k - 1)h_x\, , \ (\ell - 1)h_y \right), \ h_x = \frac{1}{N-1}, \ h_y = \frac{1}{M-1} \tag{28}$$

for $k = 1, \ldots, N$ and $\ell = 1, \ldots, M$. In Fig. 2 the logarithmic plot of the singular values of the matrix (22) (i.e. the quantities $\log_{10} \sigma_i$, $i = 1, \ldots, N$)

for $N = M = 32$ (left plot) and $N = M = 1024$ (right plot) is presented for $\alpha = 1$. It is clear to see that only very few singular values are needed to represent the matrix A in its singular value decomposition (25) for moderate value of the parameter $\varepsilon = 10^{-5} - 10^{-6}$. Almost all singular values are close to the computer zero for $N = M = 1024$. Thus the behaviour of the singular values determines the quality of the low rank approximation (25).

The situation changes if the "singularity" of the function K is more serious. In Fig. 3 (left plot) the rank $r(\varepsilon)$ for $\varepsilon = 10^{-6}$ and $N = M = 256$ is shown as a function of the parameter α. The horizontal axis corresponds to the values $-\log_2(\alpha)$ while α changes from 2^0 till 2^{-8}. However, if we "separate" the variables x and y, i.e. consider only a quarter $[0, 0.5] \times [0.5, 1]$ of the square $[0, 1] \times [0, 1]$ then the situation is better. The right plot in Fig. 3 shows the same curve for separated x and y which is more or less constant now.

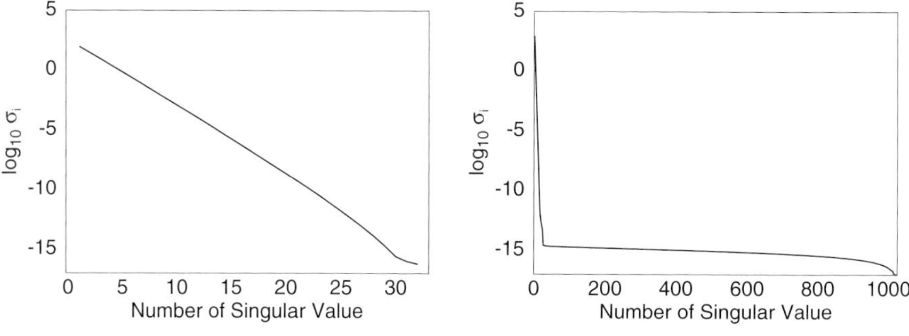

Fig. 2. Distribution of singular values for $N = 32$ (left) and $N = 1024$ (right).

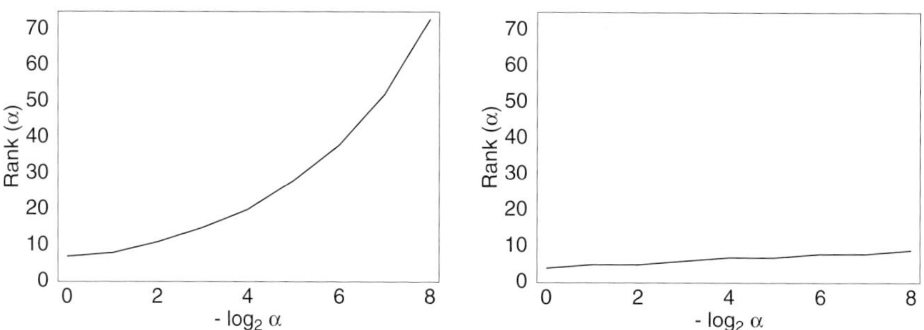

Fig. 3. Rank of the matrix $\tilde{A}$ depending on parameter α for non-separated (left) and separated (right) domains.

Now the main idea of hierarchical methods is very clear. If we decompose the whole matrix A in four blocks corresponding to the domains $[0, 0.5] \times [0, 0.5], [0, 0.5] \times [0.5, 1], [0.5, 1] \times [0, 0.5]$ and $[0.5, 1] \times [0.5, 1]$ we will be able to approximate two of these four blocks efficiently. The two remaining, main diagonal blocks have the same structure as the initial matrix but only the half of the size and their rank will be smaller. In Fig. 4, the left diagram corresponds to the whole matrix and its rank $r(\varepsilon) = 73$ is obtained for $\alpha = 2^{-9}$ and $\varepsilon = 10^{-6}$ for $N = M = 256$. The 2×2 block matrix together with ranks of the blocks is shown in the second diagram of Fig. 4. The approximation of the separated blocks is now acceptable and we continue to decompose only the blocks on the main diagonal. The results can be seen in the third and in the fourth diagram of Fig. 4. The memory requirements for these four matrices is quite different. The first matrix needs $146N$ words of memory, the second $94N$, the third $74N$ and finally we will need $72N$ words of memory for the last block matrix in Fig. 4. Thus a hierarchical decomposition in blocks and their separate approximation using a singular value decomposition leads to a drastic reduction of memory requirements even for this rather small matrix having "diagonal singularity". Note that the rank of the blocks on the main diagonal increases almost linear with the dimension: $12 - 20 - 38 - 73$ while the rank of separated blocks has at most logarithmic growth: $7 - 8 - 9$.

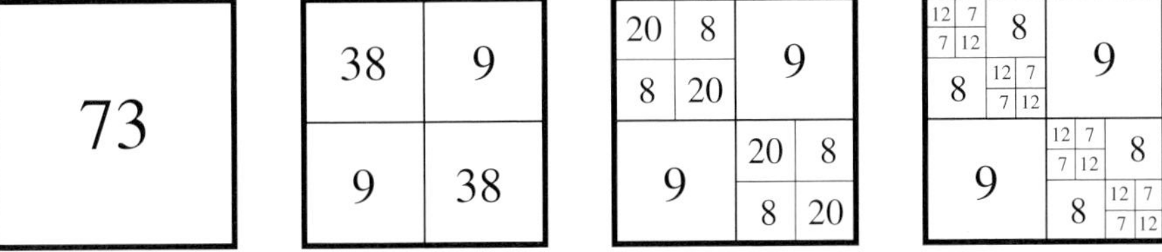

Fig. 4. Initial matrix and its hierarchical decomposition in blocks.

Thus a hierarchical approximation of large dense matrices arising from some generating function having diagonal singularity consist of three steps

- Construction of clusters for variables x and y,
- Finding of possible admissible blocks (i.e. blocks with separated x and y),
- Low rank approximation of admissible blocks.

In the above example the clusters were simply the sets of points x_k which belong to smaller and smaller intervals. The problem is more complicated for three-dimensional irregular point sets. Also the admissible blocks in the above example are very natural. They are just blocks outside of the main diagonal. In the general case we will need some permutations of rows and columns of the matrix to construct such blocks. Finally, the singular value decomposition approximation we have used is not applicable for more realistic examples. We will need more efficient algorithms to approximate admissible blocks. The approximation of the blocks for separated variables x and y in the above example is based on the smoothness of the function K for $x \neq y$. However, if the function K is degenerated, i.e. it is a finite sum of products of functions depending only on x and y

$$K(x,y) = \sum_{i=1}^{r} p_i(x)q_i(y) \tag{29}$$

then the rank of the matrix A defined in (22) is equal to r independent of its dimension. Thus for $N, M \gg r$ the matrix A is a low rank matrix. This property is independent of the smoothness of the functions p_i, q_i in (29). The low rank representation of the matrix A is now

$$A = \sum_{i=1}^{r} u_i v_i^{\top}, \tag{30}$$

with

$$(u_i)_k = p_i(x_k), \quad (v_i)_\ell = q_i(y_\ell) \tag{31}$$

for $k = 1, \ldots, N$ and $\ell = 1, \ldots, M$. Note that this representation is not the singular value decomposition (25). If the function is smooth enough then we can use its Taylor series with respect to the variable x in some point x^*

$$K(x,y) = \sum_{i=1}^{r} \frac{1}{i!} \frac{\partial^i K(x^*,y)}{\partial x^i}(x - x^*)^i + R_r(x,y) \tag{32}$$

to obtain a degenerated approximation

$$A \approx \tilde{A} = \sum_{i=1}^{r} u_i v_i^{\top}, \tag{33}$$

with

$$(u_i)_k = (x_k - x^*)^i, \quad (v_i)_\ell = \frac{1}{i!} \frac{\partial^i K(x^*,y_\ell)}{\partial x^i} \tag{34}$$

for $k = 1, \ldots, N$ and $\ell = 1, \ldots, M$. Note again that (33) is not the singular value decomposition of the matrix $\tilde{A}$. If the remainder R_r is uniformly bounded by the original function K

$$\left| R_r(x,y) \right| \leq \varepsilon \left| K(x,y) \right| \tag{35}$$

for all x and y with some $r = r(\varepsilon)$ then we can guarantee the accuracy of the low rank matrix approximation

$$\|A - \tilde{A}\| \leq \varepsilon \|A\|_F \tag{36}$$

for all dimensions N and M. The rank $r = r(\varepsilon)$ of the matrix $\tilde{A}$ is also independent of its dimension. Thus, for $N \approx M$ the matrix $\tilde{A}$ requires only $\mathrm{Mem}(\tilde{A}) = \mathcal{O}(N)$ words of computer memory. However, an efficient construction of the Taylor series for a given function in three-dimensional case is practically impossible. Thus it is rather an illustration for the fact that there exist low rank decompositions which are not based on the singular value decomposition. A further example of low rank approximation of the given function is a decomposition of the fundamental solution of the Laplace operator

$$u^*(x, y) = \frac{1}{4\,\pi} \frac{1}{|x - y|} \quad \text{for } x, y \in \mathbb{R}^3$$

in spherical harmonics which is used by multipole methods (see [6]).

3.2 Hierarchical Clustering

To find a suitable permutation, a cluster tree is constructed by recursively partitioning some weighted characteristic points

$$\left\{ (x_k, g_k),\ k = 1, \dots, N \right\} \subset \mathbb{R}^3 \times \mathbb{R}_+ \tag{37}$$

and

$$\left\{ (y_\ell, q_\ell),\ \ell = 1, \dots, M \right\} \subset \mathbb{R}^3 \times \mathbb{R}_+ \tag{38}$$

in order to separate the variables x and y. A large distance between two characteristic points results in a large difference of the respective equation numbers. While dealing with boundary element matrices the characteristic points can be the collocation points and the weights the areas of the supports of the trial functions. A given cluster

$$Cl = \left\{ (x_k, g_k),\ k = 1, \dots, n \right\}$$

with $n > 1$ can be separated in two sons using the following algorithm.

Algorithm 1

1. Mass of the cluster

$$G = \sum_{k=1}^{n} g_k \in \mathbb{R}_+ ,$$

2. Centre of the cluster

$$X = \frac{1}{G} \sum_{k=1}^{n} g_k\, x_k \in \mathbb{R}^3$$

3. Covariance matrix of the cluster

$$C = \sum_{k=1}^{n} g_k \left(x_k - X\right) \left(x_k - X\right)^{\top} \in \mathbb{R}^{3\times 3},$$

4. Eigenvalues and eigenvectors

$$C\,v_i = \lambda_i\,v_i\,,\quad i = 1,2,3\,,\quad \lambda_1 \geq \lambda_2 \geq \lambda_3 \geq 0\,,$$

5. separation
   5.1 initialisation

$$Cl_1 := \oslash\,,\quad Cl_2 := \oslash\,,$$

   5.2 for $k = 1,\ldots,n$

$$\texttt{if } \left(x_k - X, v_1\right) \geq 0 \quad \texttt{then }\; Cl_1 := Cl_1 \cup \left(x_k, g_k\right)$$
$$\texttt{else }\; Cl_2 := Cl_2 \cup \left(x_k, g_k\right).$$

The eigenvector v_1 of the matrix C corresponds to the largest eigenvalue of this matrix and shows in the direction of the longest extension of the cluster. The separation plane $\left\{x \in \mathbb{R}^3 \,:\, (x - X, v_1) = 0\right\}$ goes through the centre X of the cluster and is orthogonal to the eigenvector v_1. Thus, Algorithm 1 divides a given arbitrary cluster of weighted points in two more or less equal sons. In Fig. 5 the first two levels of separation of a simplified model of an exhaust manifold are shown. The separation of a given cluster in two sons defines a permutation of the points in the cluster. The points in the first son will be numbered first and then in the second son. Algorithm 1 will be applied recursively to the sons until they contain less than or equal to some prescribed (small and independent of N) number $n_{\min}$ of points. Next, cluster pairs which are geometrically well separated are identified. They will be

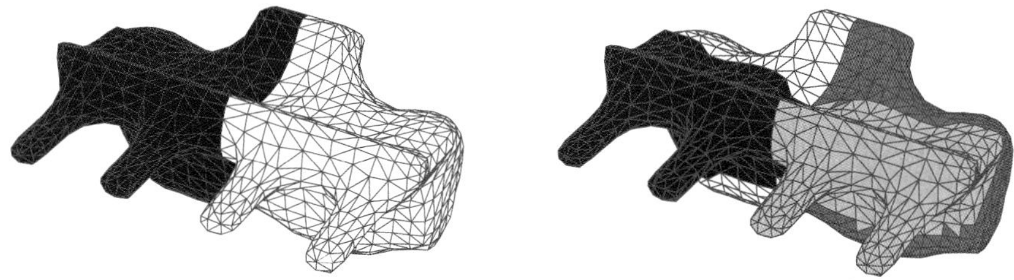

Fig. 5. Clusters of the first two levels.

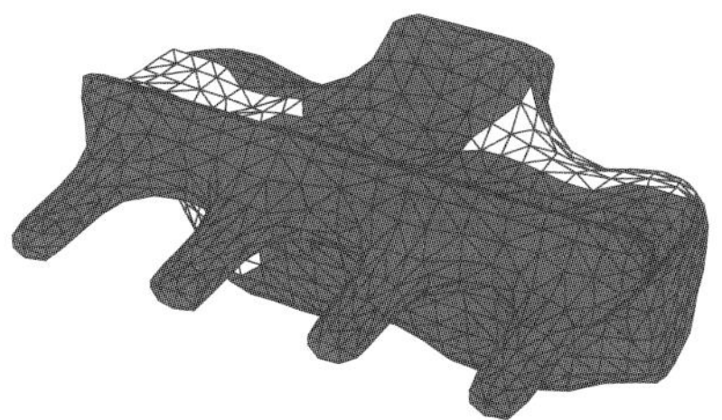

Fig. 6. An admissible cluster pair.

regarded as admissible cluster pairs, e.g. the clusters in Fig. 6. An appropriate admissibility criterion is the following simple geometrical condition. A pair of clusters (Cl_x, Cl_y) with $n_x > n_{\min}$ and $m_y > n_{\min}$ elements is admissible if

$$\min\left(\operatorname{diam}(Cl_x), \operatorname{diam}(Cl_y)\right) \leq \eta \operatorname{dist}(Cl_x, Cl_y), \tag{39}$$

where $0 < \eta < 1$ is a given parameter. Although the criterion (39) is quite simple a rather large computational effort (quadratic with respect to the number of elements in the clusters Cl_x and Cl_y) is required for calculating the exact values

$$\operatorname{diam}(Cl_x) = \max_{k_1, k_2} |x_{k_1} - x_{k_2}|,$$
$$\operatorname{diam}(Cl_y) = \max_{\ell_1, \ell_2} |y_{\ell_1} - y_{\ell_2}|,$$
$$\operatorname{dist}(Cl_x, Cl_y) = \min_{k, \ell} |x_k - y_\ell|.$$

In practice we use more rough and more restrictive but easily computable bounds

$$\operatorname{diam}(Cl_x) \leq 2 \max_k |X - x_k|,$$
$$\operatorname{diam}(Cl_y) \leq 2 \max_\ell |Y - y_\ell|,$$
$$\operatorname{dist}(Cl_x, Cl_y) \geq |X - Y| - \frac{1}{2}\left(\operatorname{diam}(Cl_x) + \operatorname{diam}(Cl_y)\right),$$

where X and Y are the already computed centres (cf. Algorithm 1) of the clusters Cl_x and Cl_y, for the admissibility condition. If a cluster pair is not admissible and $n_x > n_{\min}$ and $m_y > n_{\min}$ then there exist sons of the both clusters

$$Cl_x = Cl_{x,1} \cup Cl_{x,2}, \ \ Cl_y = Cl_{y,1} \cup Cl_{y,2}.$$

Let us assume for simplicity that the cluster Cl_x is bigger: $\operatorname{diam}(Cl_x) \geq \operatorname{diam}(Cl_y)$. In this case we check two new pairs

$$\left(Cl_{x,1}, Cl_y \right), \ \left(Cl_{x,2}, Cl_y \right)$$

for admissibility and so on. This recursive procedure stops if $n_x \leq n_{\min}$ or $m_y \leq n_{\min}$. The corresponding block of the matrix is small and will be computed exactly. The cluster trees for the variables x and y together with the set of admissible cluster pairs as well as of small cluster pairs allow to split the matrix into a collection of blocks of various sizes. The block structure of the Galerkin matrix for the single layer potential on the surface form Figs. 5–6 is shown in Fig. 7. The colour of the blocks indicates the "quality" of the approximation. The light grey colour corresponds to well approximated blocks while dark grey and especially black colour indicates less good approximation or even exact computation. Thus the main problem remains is how to approximate the big blocks without using the singular value decomposition. The corresponding procedures will be described in the forthcoming section.

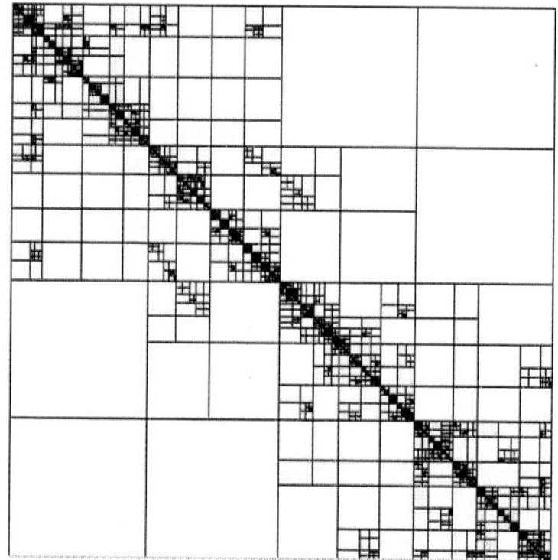

Fig. 7. Matrix decomposition.

3.3 Adaptive Cross Approximation

On the matrix level the fully pivoted ACA algorithm can be written in the following form:

Algorithm 2

1. Initialisation

$$R_0 = A, \ \ S_0 = 0.$$

2. For $i = 0, 1, 2, \ldots$ compute
   2.1. pivot element

$$(k_{i+1}, \ell_{i+1}) = \operatorname{ArgMax} |(R_i)_{k\ell}|,$$

2.2. normalising constant

$$\gamma_{i+1} = \left((R_i)_{k_{i+1}\ell_{i+1}}\right)^{-1} ,$$

2.3. new vectors

$$u_{i+1} = \gamma_{i+1} R_i e_{\ell_{i+1}} , \quad v_{i+1} = R_i^\top e_{k_{i+1}},$$

2.4. new residuum

$$R_{i+1} = R_i - u_{i+1} v_{i+1}^\top ,$$

2.5. new approximation

$$S_{i+1} = S_i + u_{i+1} v_{i+1}^\top .$$

The whole residuum matrix R_i is inspected in Step 2.1 of Algorithm 2 for its maximal entry. Thus the appropriate stopping criterion for a given $\varepsilon > 0$ at step r is

$$\|R_r\|_F \leq \varepsilon \|A\|_F .$$

Note that the crosses built from the column-row pairs with the indices k_i, ℓ_i for $i = 1, \ldots, r$ will be computed exactly, while all other elements are approximated. The number of operations required to generate the approximation $\tilde{A} = S_r$ is $\mathcal{O}(r^2 N M)$. The memory requirement for Algorithm 2 is $\mathcal{O}(N M)$ since the whole matrix A is assumed to be given at the beginning. Thus, Algorithm 2 is much faster than a singular value decomposition but still rather expensive for large matrices. If the matrix A has not yet been generated but there is a possibility of generating its entries $a_{k\ell}$ individually then the following partially pivoted ACA algorithm can be used for the approximation.

Algorithm 3

1. Initialisation

$$S_0 = 0 , \ \mathcal{I} = \varnothing , \ c = 0 \in \mathbb{R}^N ,$$

2. Recursion
2.1. Choice of the next not yet generated row

$$k_{i+1} = \min\left\{k : k \notin \mathcal{I}\right\} , \ \mathcal{I} = \mathcal{I} \cup \{k_{i+1}\} ,$$

or stop if all rows are generated, i.e. $\mathcal{I} = \{1, \ldots, N\}$,
2.2. Generation of the row

$$a = A^T e_{k_{i+1}} ,$$

2.3. Row of the residuum and the pivot column

$$R_i^\top e_{k_{i+1}} = a - \sum_{m=1}^{i} (u_m)_{k_{i+1}} v_m \,,$$

$$\ell_{i+1} = \mathrm{ArgMax} \left| (R_i)_{k_{i+1}\ell} \right| \,,$$

2.4. Test

$$\text{if } \mathrm{Max} \left| (R_i)_{k_{i+1}\ell} \right| = 0 \text{ then goto 2.1.}$$

2.5. Normalising constant

$$\gamma_{i+1} = \left((R_i)_{k_{i+1}\ell_{i+1}} \right)^{-1} \,,$$

2.6. Generation of the column, Update of the control vector

$$a = A e_{\ell_{i+1}} \,, \quad c = c + |a| \,,$$

2.7. Column of the residuum and the pivot row

$$R_i e_{\ell_{i+1}} = a - \sum_{m=1}^{i} (v_m)_{\ell_{i+1}} u_m \,,$$

$$k_{i+2} = \mathrm{ArgMax} \left| (R_i)_{k\ell_{k+1}} \right| \,,$$

2.8. New vectors

$$u_{i+1} = \gamma_{i+1} R_i e_{\ell_{i+1}} \,, \quad v_{i+1} = R_i^\top e_{k_{i+1}} \,,$$

2.9. New approximation

$$S_{i+1} = S_i + u_{i+1} v_{i+1}^\top \,.$$

2.10. Recursion

$$i := i + 1 \,, \text{ goto 2.2}$$

Since the matrix A will not be generated completely we can use the norm of its approximant S_i to define a stopping criterion. This norm can be computed recursively as follows,

$$\|S_{i+1}\|_F^2 = \|S_i\|_F^2 + 2 \sum_{m=1}^{i} u_{i+1}^\top u_m \, v_m^\top v_{i+1} + \|u_{i+1}\|_F^2 \|v_{i+1}\|_F^2 \,. \tag{40}$$

An appropriate stopping criterion in Step 2.8 is then

$$\|u_r\|_F \|v_r\|_F \leq \varepsilon \|S_r\|_F \,. \tag{41}$$

However, since the whole matrix A will not be generated while using partially pivoted ACA algorithm, it is necessary to check the control vector c updated after every column generation for zero components in not yet generated rows. If there is some index $i^* \notin \mathcal{I}$ with $c_{i_*} = 0$ then the row i^* has not yet contributed to the matrix. It can happen that this row contains relevant information and, therefore, we have to set $i := i + 1, k_{i+1} = i^*$ and to restart the algorithm in **Step 2.2**. With this trivial modification Alg. 3 can be used not only for dense matrices but also for reducible and even for sparse matrices.

Algorithm 3 requires only $\mathcal{O}(r^2(N + M))$ arithmetical operations and its memory requirement is $\mathcal{O}(r(N + M))$. Thus this algorithm is perfect for large matrices. Using the theory of polynomial multidimensional interpolation the following result was proven in [2].

Theorem 4. *Let the function $K(x, y)$ be asymptotically smooth with respect to y, i.e. $K(x, \cdot) \in C^\infty(\mathbb{R}^3 \backslash \{x\})$ for all $x \in \mathbb{R}^3$, satisfying*

$$|\partial_y^\alpha K(x, y)| \leq c_p |x - y|^{g-p}, \quad p = |\alpha| \tag{42}$$

for all multiindices $\alpha \in \mathbb{N}_0^3$ with a constant $g < 0$. Moreover, the matrix $A \in \mathbb{R}^{N \times M}$ is decomposed in blocks corresponding to the admissibility condition

$$\operatorname{diam}(Cl_y) \leq \eta \operatorname{dist}(Cl_x, Cl_y), \quad \eta < 1. \tag{43}$$

Then the matrix A with $M \sim N$ can be approximated up to an arbitrary given accuracy $\varepsilon > 0$ using a system of given points $(\tilde{x}_k, \tilde{y}_\ell)$,

$$\|A - \tilde{A}\|_F \leq \varepsilon \|A\|_F, \tag{44}$$

and

$$\operatorname{Op}(\tilde{A}) = \operatorname{Op}(\tilde{A}s) = \operatorname{Mem}(\tilde{A}) = \mathcal{O}(N^{1+\delta}\varepsilon^{-\delta}) \quad \text{for all } \delta > 0. \tag{45}$$

4 Exploitation of Symmetry

The exploitation of symmetry is another possibility to reduce computational costs and has been presented in [1, 4, 5] using linear representation theory for finite groups. The aim is a decomposition of function spaces into orthogonal subspaces of symmetric functions, such that each subproblem is defined on a so called symmetry cell. The global solution can then be reconstructed from these components. In the following we will give an overview of exploiting symmetry in the BEM. The considered procedure can easily be extended to a vector case, e.g. a magnetostatic or eddy current problem as shown in the numerical results section.

4.1 Algebraic Description

A (complete) *geometrical symmetry* of the domain Ω_{BEM} is given if there exists a finite group $\mathcal{Q}$ of isometries of $\mathbb{R}^3$, such that Ω_{BEM} is invariant w.r.t. $\mathcal{Q}$. For each element of the symmetry group $\mathcal{Q}$ there is an orthogonal matrix $Q \in \mathbb{R}^3$ (i.e. $QQ^T = Q^TQ = I$) and a symmetry point $x_0 \in \mathbb{R}^3$ such that

$$x' = x_0 + Q(x - x_0) \in \Omega_{\mathrm{BEM}} , \ \forall\, x \in \Omega_{\mathrm{BEM}} . \tag{46}$$

For the sake of simplicity we assume in the sequel that $x_0 = 0$. The geometrical symmetry of the domain Ω_{BEM} implies that geometrical symmetry also holds for its boundary Γ, i.e. the symmetry mapping Q fulfills

$$x' = Qx \in \Gamma , \ \forall\, x \in \Gamma . \tag{47}$$

For smooth boundary Γ the condition (47) implies the following connection of unit normal vectors to Γ at x and $x' = Qx$

$$Qn_x = n_{Qx} = n_{x'} , \ \forall\, x \in \Gamma . \tag{48}$$

For a symmetric problem only a part, the so called *symmetry cell* needs to be discretised and considered. The symmetry cell is the smallest subdomain which generates the entire domain under the action of the symmetry group. Let Γ_h be the discretisation of the symmetry cell of the boundary Γ. An entire boundary mesh can then be obtained by $m - 1$ consecutive applications of Q on Γ_h as shown in Fig. 8.

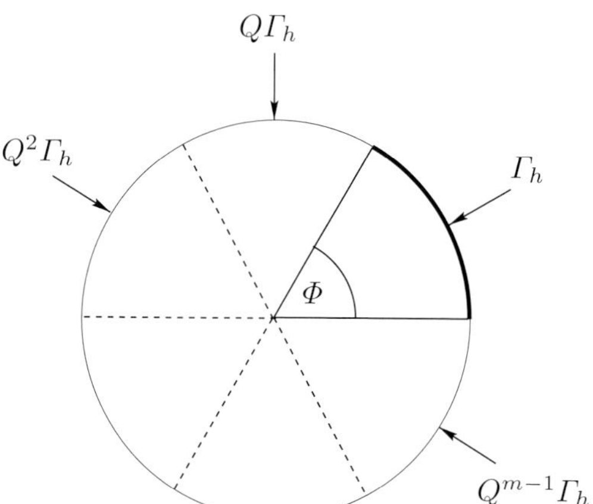

Fig. 8. Discretisation symmetry.

In the following we consider the most simple case when the system matrix A can be renumbered and partitioned into $m \times m$ blocks structure having blocks of size exactly $n = N/m$. This is the case for a piecewise constant discretisation scheme, where N is the number of unknowns, m is the size of the symmetry group and n is the number of unknowns in each symmetry cell.

All considerations can be extended to a more general discretisation scheme (17)-(18) with $N_\phi \neq N_\psi$.

The system of boundary elements, collocation points and the ansatz functions features *discretisation symmetry* if there exists a permutation σ such that the index set $\{1,\dots,N\}$ of all degrees of freedom can be written as

$$\{1,\dots,n,\sigma(1),\dots,\sigma(n),\dots,\sigma^{m-1}(1),\dots,\sigma^{m-1}(n)\} \tag{49}$$

with $\sigma^m(i) = i$, $\forall i = 1,\dots,n$, and, additionally, for collocation points and ansatz functions holds

$$Q(\operatorname{supp}\phi_j) = \operatorname{supp}\phi_{\sigma(j)}, \ j = 1,\dots,N, \tag{50}$$
$$\phi_j(x) = \phi_{\sigma(j)}(Qx), \ \forall x \in \operatorname{supp}\phi_j, \ j = 1,\dots,N, \tag{51}$$
$$Qy_i = y_{\sigma(i)}, \ i = 1,\dots,N. \tag{52}$$

The permutation σ offers the possibility for renumbering unknowns corresponding to the symmetry of the problem. For the general case $N_\phi \neq N_\psi$ two different permutations σ_ϕ and σ_ψ of the index sets $\{1,\dots,N_\phi\}$ and $\{1,\dots,N_\psi\}$, respectively, have to be introduced.

The problem features the *symmetry of the kernel* if the following condition does hold for the kernel K

$$K(Q^k x, Q^l y) = K(x, Q^{l-k} y), \ \forall x, y \in \Gamma, \ \forall k, l \in \mathbb{Z}. \tag{53}$$

Especially, for $k = l$ we obtain $K(Q^k x, Q^k y) = K(x, y)$, $\forall k \in \mathbb{Z}$. Note that the BEM matrices in (20)-(21) are both generated by symmetrical kernels.

Lemma 1. *The symmetrical BEM discretisation* (50)–(52) *of the geometrically symmetrical problem* (48) *having kernel symmetry* (53) *leads, after numbering of unknowns corresponding to* (49), *to the following property of the matrix entries:*

$$a_{ij} = a_{\sigma(i)\sigma(j)}, \ \forall i, j. \tag{54}$$

Proof. Definition of the matrix entries leads after substitution (47) to

$$
\begin{aligned}
a_{ij} &= \int_{\operatorname{supp}\phi_j} K(x, y_i)\, \phi_j(x)\, dS_x \\[2mm]
&= \int_{\operatorname{supp}\phi_j} K(Qx, Qy_i)\phi_{\sigma(j)}(Qx)\, dS_x \\[2mm]
&= \int_{Q(\operatorname{supp}\phi_j)} K(x', Qy_i)\phi_{\sigma(j)}(x')\, dS_{x'} \\[2mm]
&= \int_{\operatorname{supp}\phi_{\sigma(j)}} K(x, y_{\sigma(i)})\phi_{\sigma(j)}(x')\, dS_{x'} \ = \ a_{\sigma(i)\sigma(j)},
\end{aligned}
\tag{55}
$$

where the properties (48)–(52) have been used. $\qquad\square$

Since $\sigma^{m-k}(\sigma^k(i)) = i$ for all i the property (54) implies

$$a_{\sigma^k(i)\,j} = a_{i\,\sigma^{m-k}(j)} \, , \quad \forall \, i, j \, .$$

Thus the system of linear equations of the symmetrical BEM takes the following block-circulant form

$$\begin{pmatrix} A_1 & A_2 & \ldots & A_m \\ A_m & A_1 & \ldots & A_{m-1} \\ \ldots & \ldots & \ldots & \ldots \\ \ldots & \ldots & \ldots & \ldots \\ A_2 & A_3 & \ldots & A_1 \end{pmatrix} \begin{pmatrix} u_1 \\ u_2 \\ \ldots \\ \ldots \\ u_m \end{pmatrix} = \begin{pmatrix} b_1 \\ b_2 \\ \ldots \\ \ldots \\ b_m \end{pmatrix} . \tag{56}$$

Thus only the basis matrices $A_1, A_2, \ldots, A_m$ should be generated and stored. The amount of numerical work and of memory will therefore be reduced from N^2 to N^2/m. This factor can be very useful for practical computations. The numerical solution of the system of linear equations having a block-circulant matrix can also be implemented much more efficiently than a straightforward direct elimination method which would lead to $O(N^3)$ arithmetical operations [18]. The main property of the circulant matrices

$$A = \begin{pmatrix} a_1 & a_2 & a_3 & \ldots & a_{m-1} & a_m \\ a_m & a_1 & a_2 & a_3 & \ldots & a_{m-1} \\ a_{m-1} & \ldots & \ldots & \ldots & \ldots & \ldots \\ \ldots & \ldots & \ldots & \ldots & \ldots & \ldots \\ \ldots & \ldots & \ldots & \ldots & \ldots & a_2 \\ a_2 & a_3 & \ldots & a_{m-1} & a_m & a_1 \end{pmatrix} \in \mathbb{C}^{m \times m}$$

is that all of them are simultaneously diagonalised by the matrix of the discrete Fourier transform F_m:

$$A = \frac{1}{m} F_m \Lambda F_m^* , \tag{57}$$

$$f_{k,l} = \omega_m^{(k-1)(l-1)} = e^{i \frac{2\pi}{m}(k-1)(l-1)} \, .$$

The most simple nontrivial circulant matrix

$$J = \begin{pmatrix} 0 & 1 & 0 & \ldots & 0 \\ 0 & 0 & 1 & \ldots & 0 \\ \ldots & \ldots & \ldots & \ldots \\ \ldots & \ldots & \ldots & \ldots \\ 0 & 0 & 0 & \ldots & 1 \\ 1 & 0 & 0 & \ldots & 0 \end{pmatrix}$$

has the following eigenvalues

$$\Lambda = \operatorname{diag}\left(\omega_m^{l-1} , \, l = 1, \ldots, m \right) \, .$$

Using the Kronecker product $\otimes$ of matrices we rewrite the block-circulant matrix A of the system (56) in the form (cf. (57))

$$A = \sum_{k=1}^{m} J^{k-1} \otimes A_k = \frac{1}{m} \sum_{k=1}^{m} \left(F_m \, \Lambda^{k-1} \, F_m^* \right) \otimes A_k \,,$$

where the dimension of the matrices A_k is now $n = N/m$. Since

$$F_m \, F_m^* = F_m^* \, F_m = I_m$$

and using the known property of the Kronecker product

$$(A \otimes B)(C \otimes D) = (AC) \otimes (BD)$$

we obtain

$$\left(F_m^* \otimes I_n \right) A \left(F_m \otimes I_n \right) = \sum_{k=1}^{m} \left(F_m^* \otimes I_n \right) \left(J^{k-1} \otimes A_k \right) \left(F_m \otimes I_n \right)$$

$$= \sum_{k=1}^{m} \left(F_m^* J^{k-1} F_m \right) \otimes \left(I_n A_k I_n \right) \;=\; \frac{1}{m} \sum_{k=1}^{m} \Lambda^{k-1} \otimes A_k \,.$$

The system (56) can now be rewritten in the block-diagonal form

$$\begin{pmatrix} D_1 & 0 & \ldots & 0 \\ 0 & D_2 & \ldots & 0 \\ \ldots & \ldots & \ldots & \ldots \\ \ldots & \ldots & \ldots & \ldots \\ 0 & 0 & \ldots & D_m \end{pmatrix} \begin{pmatrix} \tilde{u}_1 \\ \tilde{u}_2 \\ \ldots \\ \ldots \\ \tilde{u}_m \end{pmatrix} = \begin{pmatrix} \tilde{b}_1 \\ \tilde{b}_2 \\ \ldots \\ \ldots \\ \tilde{b}_m \end{pmatrix} , \tag{58}$$

where

$$D_l = \frac{1}{m} \sum_{k=1}^{m} \omega_m^{(l-1)(k-1)} A_k \in \mathbb{C}^{n \times n} \tag{59}$$

and

$$\tilde{u} = (F_m^* \otimes I_n) u \,, \quad \tilde{b} = (F_m^* \otimes I_n) b \,.$$

Thus the following algorithm has been derived (similar to proposed in [1])

1. *Compute all basis matrices A_k, $k = 1, \ldots, m$*
2. *Compute*

$$\tilde{b} = (F_m^* \otimes I_n) b$$

   using n Fast Fourier Transforms (FFT).
3. *For $l = 1, \ldots, m$*

3.1 Generate the matrix

$$D_l = \frac{1}{m} \sum_{k=1}^{m} \omega_m^{(l-1)(k-1)} A_k$$

3.2 Solve the system

$$D_l \tilde{u}_l = \tilde{b}_l$$

4. Compute

$$u = (F_m \otimes I_n)\tilde{u}$$

using n FFT's.

The straightforward implementation of this algorithm leads to $O(mn^2)$ operations and memory units in Step 1., $O(nm\log(m))$ operations in Step 2., $O(mn^2)$ operations and memory units in Step 3.1, $O(mn^3)$ operations for solving all systems in Step 3.2 and finally $O(nm\log(m))$ operations in the last Step 4.. Thus Step 3.2 is the most expensive and defines the final amount of numerical work for the whole algorithm $O(mn^3) = O(N^3/m^2)$. This amount remains of the same capital order of $O(N^3)$, but it is reduced by a remarkable factor m^2.

4.2 Symmetry of Excitation

As described in Section 2.1 in the BE domain the equation

$$\Delta\phi = -\frac{1}{\varepsilon_0}\rho$$

is to be solved, where ϕ is the electric scalar potential and ρ is the electric charge density. Discretisation by nodal ansatz functions and point collocation leads in case of symmetry to the equation system of the form (56). Electromagnetic devices often possess the *symmetry of excitation*, which means for the electrostatic case, that the symmetry mappings Q fulfill

$$\rho(Q^k x) = \alpha_{k+1}\rho(x)\,, \ \forall x \in \Gamma\,, \ k = 0,\ldots,m-1\,, \tag{60}$$

for some $\alpha_k \in \mathbb{R}$. In case of an excitation symmetry we don't perform the Fourier transform as described in the previous section but simplify the equation system (56) in a different way. As a consequence from (60) we obtain a linear dependency of the components of the r.h.s in (56)

$$b_k = \alpha_k b_1\,, \ k = 1,\ldots,m\,. \tag{61}$$

Additionally, we require the following condition to be fulfilled

$$\frac{\alpha_1}{\alpha_2} = \frac{\alpha_2}{\alpha_3} = \ldots = \frac{\alpha_{m-1}}{\alpha_m} = \frac{\alpha_m}{\alpha_1}. \tag{62}$$

Thus the equation system can be reduced to one subsystem

$$(\alpha_1 A_1 + \alpha_2 A_2 + \cdots + \alpha_m A_m)\, u_1 \;=\; b_1 \tag{63}$$

of dimension N/m, where N is the total number of unknowns. The remaining solution components can be computed by

$$u_k \;=\; \alpha_k u_1\,,\ \ k = 1,\ldots,m\,.$$

Thus the exploitation of the excitation symmetry leads to reduction of computational costs from N^2 to N^2/m^2.

5 Numerical Experiments

5.1 Asymptotic Behaviour

We start the numerical studies considering the most simple smooth surface $\Gamma = \partial\Omega$ for $\Omega \subset \mathbb{R}^3$, namely the surface of the unit sphere,

$$\Gamma = \left\{ x \in \mathbb{R}^3 \,:\, |x| = 1 \right\}. \tag{64}$$

As an appropriate discretisation of Γ we consider the icosahedron that is uniformly triangulated before being projected onto the circumscribed unit sphere. On this way we obtain a sequence $\{\Gamma_N\}$ of almost uniform meshes on the unit sphere which are shown in Fig. 9 for different numbers of boundary elements N. This sequence allows to study the convergence of boundary element methods for different examples. In Fig. 10 the clusters of the levels 1 and 2 obtained

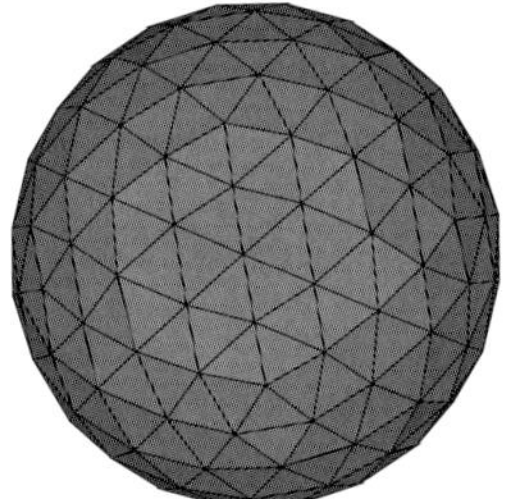
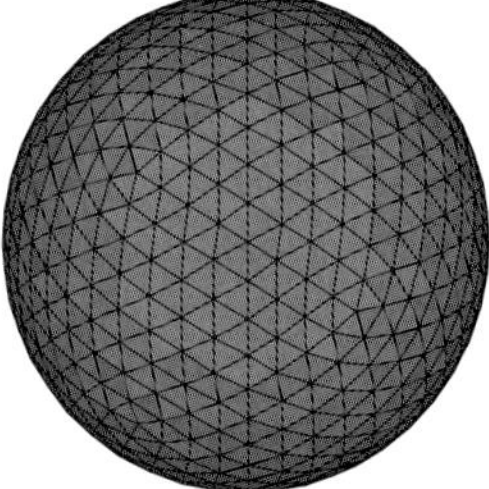

Fig. 9. Discretisation of the unit sphere with $N = 320$ and $N = 1280$.

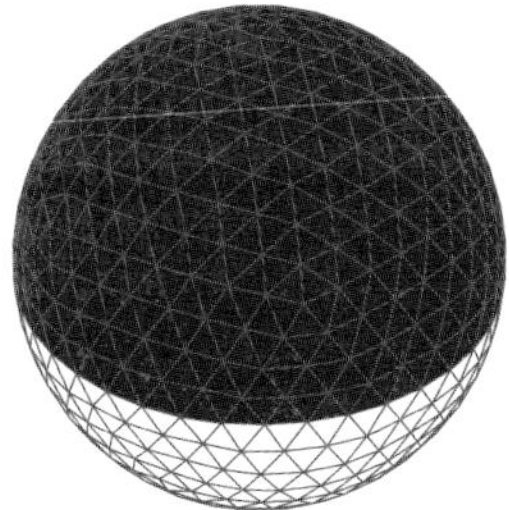 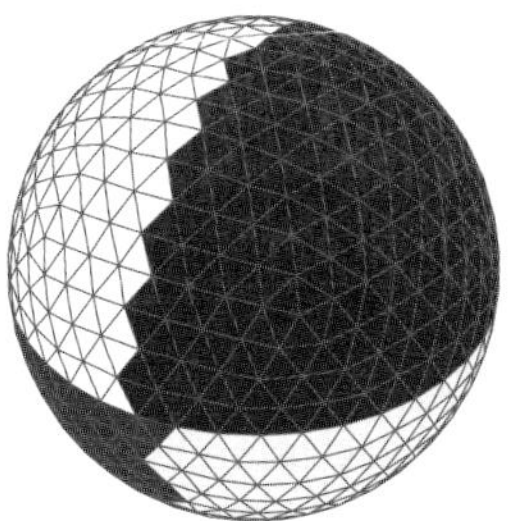

Fig. 10. Clusters of the level 1 and 2 for $N = 1280$.

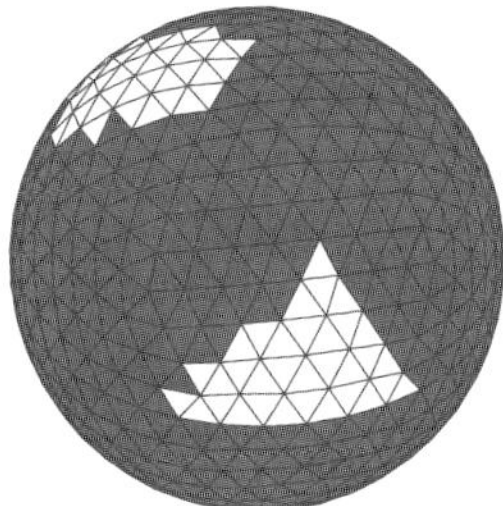

Fig. 11. An admissible cluster pair for $N = 1280$.

with Alg. 1 for $N = 1280$ are presented. In Fig. 11 a typical admissible cluster pair is shown. We solve the interior Dirichlet boundary value problems for the Laplace equation using a Galerkin boundary element method. The piecewise linear basis functions will be used for approximation of the Dirichlet datum and piecewise constant basis functions for approximation of the Neumann datum. We will use the L_2 projection for the approximation of the given part of the Cauchy data. The boundary element matrices G and H are generated in approximative form using the partially pivoted ACA algorithm with a variable relative accuracy ε_1 depending on the expected discretisation error. The resulting systems of linear equations are solved using some variants of the Conjugate Gradient Method (CGM) with or without preconditioning up to a relative accuracy $\varepsilon_2 = 10^{-8}$. The analytical solution is a harmonic function

$$\phi(x) = (1 + x_1) \exp(2\pi\, x_2) \cos(2\pi\, x_3). \tag{65}$$

The results of the computations are shown in Tables 1 and 2. The number of boundary elements is listed in the first column of these tables. The second column contains the number of nodes while in the third column of Table 1 the prescribed accuracy for the ACA algorithm for approximation of both matrices $H \in \mathbb{R}^{N \times M}$ and $G \in \mathbb{R}^{N \times N}$ is given. The fourth column of this table shows the memory requirements in MByte for the approximate double

Table 1. ACA approximation of the matrices H and G, Dirichlet problem.

N	M	ε_1	MByte(H)	%	MByte(G)	%
80	42	$1.0 \cdot 10^{-2}$	0.03	97.8	0.02	48.7
320	162	$1.0 \cdot 10^{-3}$	0.26	65.6	0.21	27.2
1280	642	$1.0 \cdot 10^{-4}$	2.45	39.1	1.94	15.5
5120	2562	$1.0 \cdot 10^{-5}$	20.05	20.0	15.72	7.9
20480	10242	$1.0 \cdot 10^{-6}$	149.19	9.3	115.83	3.6
81920	40962	$1.0 \cdot 10^{-7}$	1085.0	4.2	837.50	1.6

layer potential matrix H. The quality of this approximation in percentage of the original matrix is listed in the next column. The corresponding values for the single layer potential matrix G can be seen in the columns six and seven. The partitioning of the matrix for $N = 5120$ as well as the quality of the approximation of single blocks is shown in Fig. 12. The left diagram in Fig. 12 shows the symmetric single layer potential matrix G while the rectangular double layer potential matrix H is depicted in the right diagram. The legend indicates the percentage of memory needed for the ACA approximation of the blocks compared to the full memory. Further numerical results are shown in Table 2. The third column shows the number of Conjugate Gradient iterations needed to reach the prescribed accuracy ε_2. The relative L_2-error for the Neumann datum ψ

$$Error_1 = \frac{\|\psi - \tilde{\psi}\|_{L_2(\Gamma)}}{\|\psi\|_{L_2(\Gamma)}}, \tag{66}$$

where $\tilde{\psi}$ denotes the numerical solution, is given in the fourth column. The next column represents the rate of convergence for the Neumann datum, i.e.

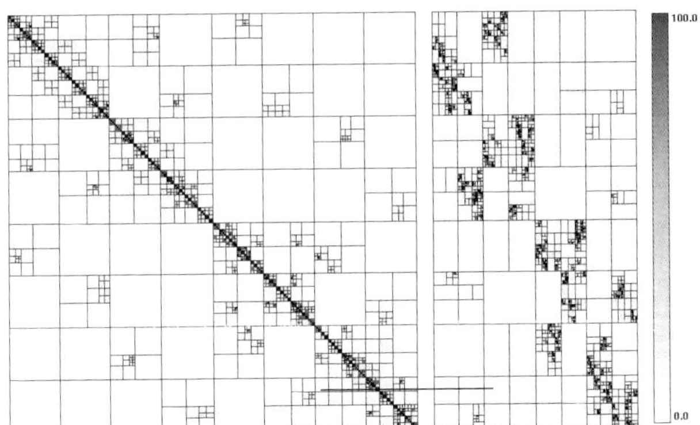

Fig. 12. Partitioning of the BEM matrices for $N = 5120$ and $M = 2562$.

Table 2. Accuracy of the Galerkin method, Dirichlet problem.

N	M	$Iter$	$Error_1$	CF_1	$Error_2$	CF_2
80	42	22	$9.34 \cdot 10^{-1}$	–	$7.29 \cdot 10^{-0}$	–
320	162	32	$5.06 \cdot 10^{-1}$	1.85	$3.29 \cdot 10^{-1}$	22.16
1280	642	45	$2.23 \cdot 10^{-1}$	2.27	$3.53 \cdot 10^{-2}$	9.32
5120	2562	56	$1.04 \cdot 10^{-1}$	2.14	$3.54 \cdot 10^{-3}$	9.97
20480	10242	72	$5.11 \cdot 10^{-2}$	2.03	$4.11 \cdot 10^{-4}$	8.61
81920	40962	94	$2.53 \cdot 10^{-2}$	2.02	$4.30 \cdot 10^{-5}$	9.56

the quotient between the errors in two consecutive lines of column four. Finally, the last two columns show the absolute error in a prescribed inner point $x^* \in \Omega$,

$$Error_2 = |\phi(x^*) - \tilde{\phi}(x^*)|, \quad x^* = (0.250685, 0.417808, 0.584932)^\top \quad (67)$$

for the value $\tilde{\phi}(x^*)$ obtained using an approximate representation formula. Table 2 obviously shows a linear convergence $\mathcal{O}(N^{-1/2}) = \mathcal{O}(h)$ of the Galerkin boundary element method for the Neumann datum in the L_2 norm. It should be noted that this theoretically guaranteed convergence order can already be observed when approximating the matrices H and G with much less accuracy as it was used to obtain the results in Table 1. However, this high accuracy is necessary in order to be able to observe the third order (or even better) pointwise convergence rate within the domain Ω presented in the last two columns of Table 2. Especially for $N = 81920$ a very high accuracy of $\varepsilon_1 = 1.0 \cdot 10^{-7}$ of the ACA approximation is necessary.

5.2 Examples with Symmetries

For numerical tests we consider TEAM workshop problem 10 [14] (TEAM= Testing Electromagnetic Analysis Methods). An exciting coil is set between two steel channels, and a steel plate is inserted between the channels. The geometry is symmetrical with respect to all three coordinate planes. In order to examine the behaviour of the ACA algorithm and the full BEM method when exploiting symmetries, we consider along with the full model three further meshes exploiting one, two and all three symmetries respectively (Fig. 13). Additionally, for each mesh of this mesh sequence we gradually perform two refinements to show the linear behaviour of the ACA algorithm with respect to the problem size. Thus we obtain three mesh sequences with altogether 12 meshes. Hexahedral second order FEM elements (20 nodes) are used in connection with rectangular second order BEM elements (8 nodes) for both Dirichlet and Neumann data.

In the case when there are some fixed collocation points (e.g. points on a symmetry face in case of a mirror symmetry), the size of each subblock in (56) is close to, but not exactly equal to, $n = N/m$ and the matrix blocks

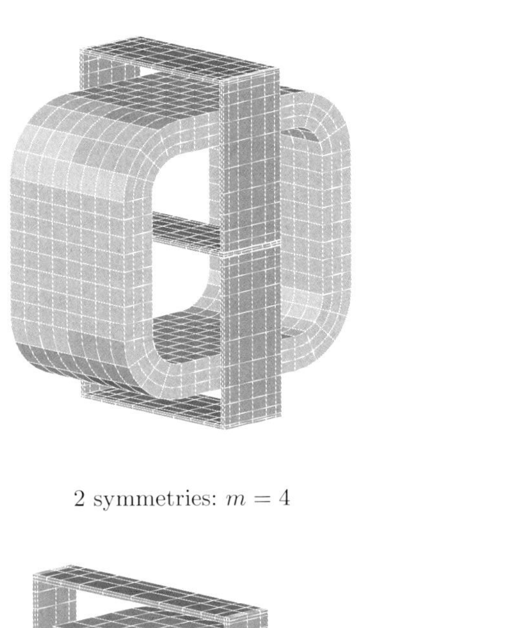

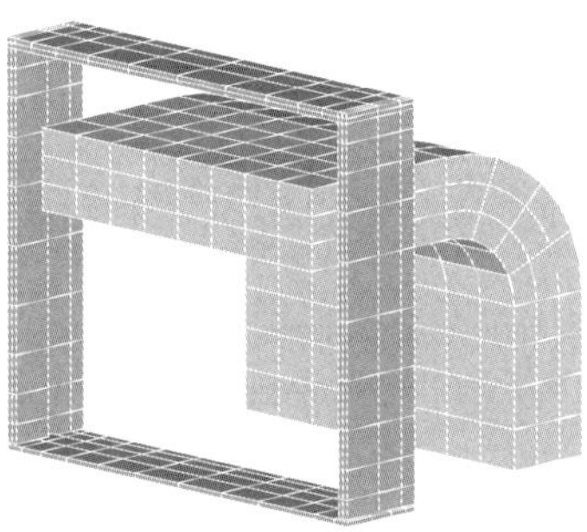

Fig. 13. TEAM problem 10. An exciting coil is set between two steel channels, and a steel plate is inserted between the channels. This geometry is symmetrical with respect to all coordinate planes.

of the single layer potential become singular. However, the global system has a unique solution [1]. There are several methods to handle the subsystems via regularisation or via projections proposed in [1]. Since in our solver no inversion of approximated matrices takes place, also singular matrices can be handled and the unique global solution can still be reconstructed without further difficulties.

TEAM problem 10 is treated as a magnetostatic problem (for details see [11, 16]). For the numerical solution the potential approach is used, so that in the BE domain the equation

$$\Delta \mathbf{A} = -\mu_0 \boldsymbol{\jmath}_S$$

has to be solved, where $\mathbf{A}$ is the Coulomb gauged magnetic vector potential and $\boldsymbol{\jmath}_S$ is the impressed source current density. This equation decouples into three scalar Laplace equations, so that a componentwise discretisation with nodal elements leads to the equation system (56) for each Cartesian component of the vector potential. Additionally, the problem features the excitation symmetry described in Section 4.2, i.e. each Cartesian component of the excitation given by the impressed source current density $\boldsymbol{\jmath}_S$ satisfies the symmetry conditions (60)-(62). Thus, depending on whether the coefficient sets $\{\alpha_k\}$ are different for some Cartesian components, we obtain up to three different system matrices for reduced systems of equations (63). Let us denote them by D_x, D_y and D_z.

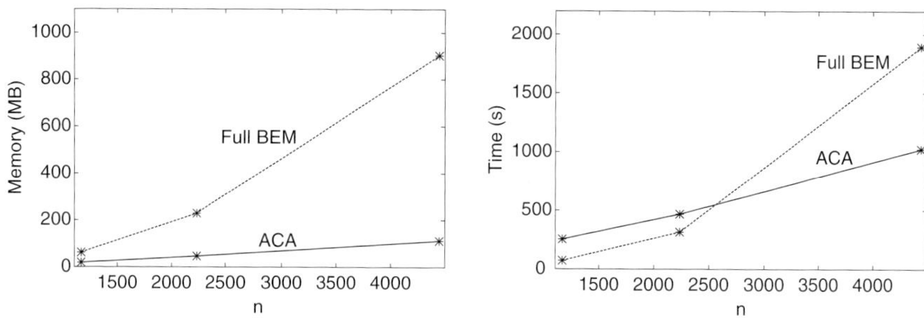

Fig. 14. Memory requirements (left) and CPU times (right) versus problem size for fixed $m = 4$ and variable mesh.

In all computations we set the ACA accuracy $\varepsilon = 10^{-4}$. The problem is solved using both the ACA algorithm and the full BEM method. Figure 14 shows for both algorithms the memory requirement of BEM matrices as well as the CPU time needed for the solution. All values refer to a 450-MHz Sun Ultra workstation. We compared the average magnetic induction in the centre of the inner steel plate ($\bar{B}_z = 1.663\ T$) with measurements ($\bar{B}_z = 1.654\ T$) [16] and found good agreement. The difference of the computed flux densities with and without ACA is neglectable ($\Delta\bar{B}_z \approx 3 \cdot 10^{-4}\ T$).

One can observe for any kind of symmetry that the increasing problem size due to the mesh refinements results in a linear behaviour of the memory consumption and the CPU time for the ACA algorithm. Fig. 14 shows the comparison between the ACA and the full BEM for one kind of symmetry. Although the ACA algorithm is slower for coarse meshes, its linear complexity makes it superior for large n.

Now we examine the effect of the symmetry exploitation. It is clear that the profit using full BEM should be of order $O(n^2)$ whereas the memory requirement and CPU time reduction using ACA is expected to be linear.

Fig. 15 shows the behaviour of the memory usage and CPU time for the medium mesh sequence.

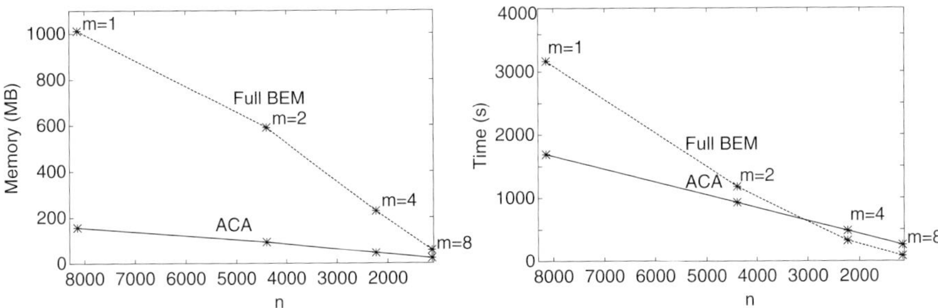

Fig. 15. Memory requirements (left) and CPU times (right) with respect to the symmetry for variable m and fixed mesh (medium discretisation).

As mentioned above in the case of ACA the individual approximation and storage of all m basis matrices will be performed. The relative size of the basis matrices resulting from the single layer potential is shown in Table 3. The assembly of the system matrices in (63) by means of linear combination is carried out in the matrix-vector multiplication.

Table 3. Relative size of BEM matrices coming from the single layer potential for the medium mesh sequence. The percentage gives the relative size after compression obtained by the ACA algorithm for each individual submatrix compared to a fully populated block. Submatrices which involve transformed nodes show a very good compression.

Block matrix	$m = 1$ $n = 8142$	$m = 2$ $n = 4399$	$m = 4$ $n = 2234$	$m = 8$ $n = 1131$
A_1	12.5%	15.4%	20.5%	32.8 %
A_2	-	10.1%	12.9%	18.4 %
A_3	-	-	8.7%	12.3 %
A_4	-	-	6.3%	8.2 %
A_5	-	-	-	6.1 %
A_6	-	-	-	5.1 %
A_7	-	-	-	5.1 %
A_8	-	-	-	3.1 %
Total memory	63.2 MB	37.6 MB	18.4 MB	8.9 MB

The full BEM method performs the assembly of system matrices D_x, D_y, D_z during the matrix computation. The number of different matrices depends on the kind of geometrical and excitational symmetry. For the TEAM

10 example it holds that $D_x = D_y = D_z$ in the case without symmetry, $D_x \neq D_y = D_z$ in the case of one symmetry, and three different matrices arise in case of two or three symmetries. For this reason the curve corresponding to the total memory requirements of the full BEM method in Fig. 15 does not actually decrease like $O(n^2)$ but the memory requirements for each single matrix do.

The numerical example considered here exhibits the property of excitation symmetry. Note that in the general case of non-symmetric excitation the memory requirements would decrease linearly w.r.t. the size m of the symmetry group, as can be seen from the equation (58), and therefore like $O(n)$.

5.3 Industrial Application

In this Section a claw-pole alternator, nowadays a mass-produced article used for the generation of electrical power in vehicles, is considered as an example. The complex magnetic flux guidance requires a three-dimensional modelling of this electrical machine. For an alternator with $p = 6$ pole pairs Fig. 16, left shows a 60°-sector of the solid rotor that coincides with one pole pair. For the same sector Fig. 16, right depicts the supplementary stator part with the inlying stator coils. The entire geometry of the alternator and the magnetic fields are obtained by consecutive rotation of the discretised part by an angle $\Phi = 2\pi/p$ around the machine axis. This periodic symmetry concerning the transformation (47) is obvious.

It is well known that modelling of one pole-pitch ($\Phi = \pi/p$), that is a 30°-sector in our case, is sufficient for the computation of the magnetic field

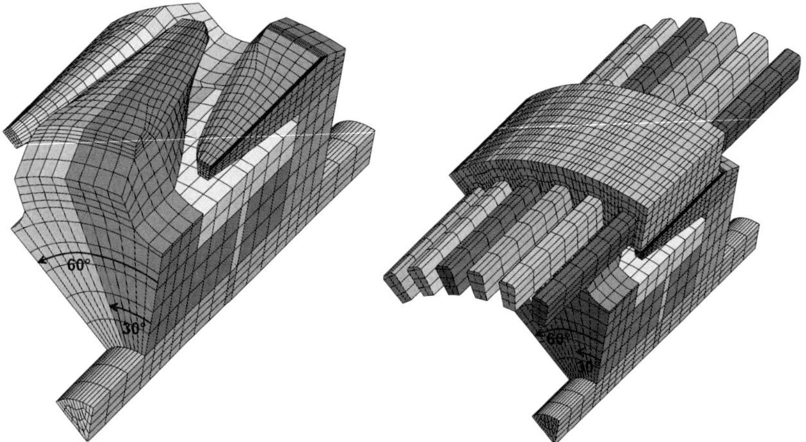

Fig. 16. Discretised rotor part (left) and stator with inlying coils (right) of a claw-pole alternator.

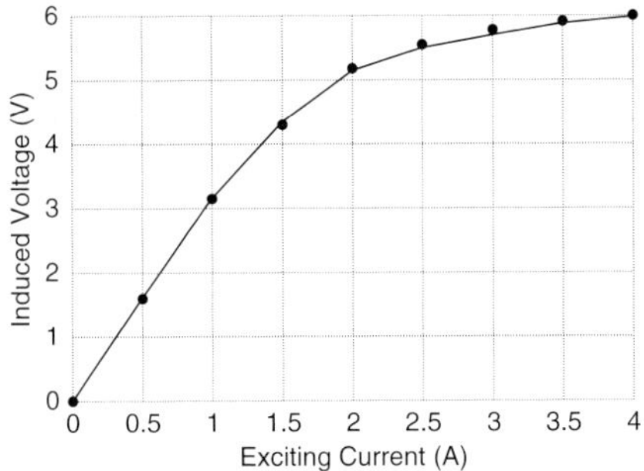

Fig. 17. No-load characteristics at 1000 1/min. The agreement of computed values (solid line) and measurements (dots) is very good.

[10]. For the solution of the eddy current problem hexahedral second order nodal FEM elements have been used, coupled to rectangular second order BEM elements. Fig. 17 shows the induced voltage in the stator coils versus the exciting current at a rotational speed of 1000 rounds per minute.

Table 4. Relative size of matrix blocks of the single layer potential. Submatrices that describe remote interactions show excellent compression.

Block matrix	$m = 3$ (**120°**) $n = 27751$	$m = 6$ (**60°**) $n = 13999$	$m = 12$ (**30°**) $n = 7123$
A_1	13.8%	16.6%	18.0%
A_2	2.8%	5.8%	8.9%
A_3	2.9%	1.8%	4.3%
A_4	-	1.1%	2.7%
A_5	-	1.9%	1.7%
A_6	-	5.9%	1.3%
A_7	-	-	1.2%
A_8	-	-	1.3%
A_9	-	-	1.8%
A_{10}	-	-	2.7%
A_{11}	-	-	4.3%
A_{12}	-	-	8.9%
Total memory	1145.7 MB	494.8 MB	221.1 MB

In order to examine the effect of symmetry exploitation a sequence of three meshes with different symmetry angles has been analysed. The number of boundary nodes n approximately bisects from a 120°- to a 60°- and to a 30°-mesh, respectively, while m reduplicates. For an implementation according to Section 4, Table 4 shows that it is reasonable to approximate and store the submatrices A_k individually. Although interactions between more sectors have to be represented while increasing symmetry exploitation emerging farpoint-interaction gives rise to good approximation of the respective submatrices leading to high-grade compression rates. This is especially important when solving time-dependent problems with motion.

6 Conclusions

The memory consumption of the standard BEM turns out to be the limiting factor in many practical applications. The above results show that the ACA technique is a feasible means to overcome these limitations. ACA can be applied to several BEM formulations [12, 15, 17] discretised by nodal or edge elements where matrices are generated by asymptotically smooth kernels. The combination of the ACA algorithm and the exploitation of symmetry yields an asymptotically optimal and practically feasible procedure for efficient solution of electromagnetic problems.

References

1. E. L. Allgower, K. Georg, R. Miranda, J. Tausch: Numerical exploitation of equivariance. Z. Angew. Math. Mech. 78 (1998) 795–806.
2. M. Bebendorf: Approximation of boundary element matrices. Numer. Math. 86 (2000) 565–589.
3. M. Bebendorf, S. Rjasanow: Adaptive Low-Rank Approximation of Collocation Matrices. Computing 70 (2003) 1–24.
4. M. Bonnet. Exploiting partial or complete geometrical symmetry in 3D symmetric Galerkin indirect BEM formulations. Int. J. Num. Meth. Engrg. 57 (2003) 1053–1083.
5. A. Bossavit: Symmetry, groups and boundary value problems: a progressive introduction to noncommutative harmonic analysis of partial differential equations in domains with geometrical symmetry. Comp. Meth. in Appl. Mech. Engrg. 56 (1986) 167–215.
6. H. Cheng, L. Greengard, V. Rokhlin: A fast adaptive multipole algorithm in three dimensions. J. Comput. Phys. 155 (1999) 468–498.
7. W. Hackbusch: A sparse matrix arithmetic based on $\mathcal{H}$-matrices. Part I. Computing 62 (1999) 89–108.
8. W. Hackbusch, B. N. Khoromskij: A sparse $\mathcal{H}$–matrix arithmetic. Part II. Application to multi–dimensional problems. Computing (2000) 64 (2000) 21–47.
9. W. Hackbusch, Z. P. Nowak: On the fast matrix multiplication in the boundary element method by panel clustering. Numer. Math. 54 (1989) 463–491.
10. S. Küppers, G. Henneberger, I. Ramesohl: The influence of the number of poles on the output performance of a claw-pole alternator. Proc. of the ICEM, pp. 268–272, 1996.
11. S. Kurz, J. Fetzer, G. Lehner, W. M. Rucker: Numerical analysis of 3D eddy current problems with moving bodies using BEM-FEM coupling. Surv. Math. Industry 9 (1999) 131–150.

12. S. Kurz, O. Rain, V. Rischmüller, S. Rjasanow: Discretization of boundary integral equations by differential forms on dual grids. IEEE Trans. Magnetics 40 (2004) 826–829.
13. S. Kurz, O. Rain, S. Rjasanow: Application of the adaptive cross approximation technique for the coupled BE-FE-solution of symmetric electromagnetic problems. Comp. Mech. 32 (2003) 423–429.
14. T. Nakata, N. Takahashi, K. Fujiwara: Summary of results for benchmark problem 10 (steel plates around a coil). COMPEL 11 (1992) 335–344.
15. J. Ostrowski, Z. Andjelić, M. Bebendorf, B. Crânganu-Creţu, J. Smajić: Fast BEM-solution of Laplace problems with $\mathcal{H}$-matrices and ACA. Proceedings of the IEEE, 2005.
16. K. Preis, I. Bardi, O. Biro, C. Magele, W. Renhart, K. R. Richter, G. Vrisk: Numerical analysis of 3D magnetostatic fields. IEEE Trans. Magnetics 27 (1991) 3798–3803.
17. O. Rain: Kantenelementbasierte BEM mit DeRham-Kollokation für Elektromagnetismus. PhD thesis, Universität des Saarlandes, 2004.
18. S. Rjasanow: Effective algorithms with block circulant matrices. Linear Alg. Appl. 202 (1994) 55–69.

BEM-Based Simulations in Engineering Design

Zoran Andjelić, Jasmin Smajić, and Michael Conry

ABB Switzerland Ltd., Corporate Research, 5405 Baden-Dättwil, Switzerland
{zoran.andjelic,jasmin.smajic,michael.conry}@ch.abb.com

Summary. The simulation of the real-world industrial problems is nowadays faced with a number of the challenging requirements, mainly arising in the daily design praxis of power engineering devices. Complex structures, complex physics, huge dimensions and huge aspect ratio in model dimensions are just some of the critical modelling issues that need to be encountered by the simulation tools. Thanks to the advances achieved in the last several years, BEM become a powerful numerical technique for the simulations of such industrial products. Until recent time this technique has been recognized as a technique offering from one side some excellent features (2D instead of 3D discretization, open-boundary problems, etc.), but from the other side having some serious practical limitations, mostly related to the full-populated, often ill-conditioned matrices. The new, emerging numerical techniques like MBIT (Multipole-Base Integral Technique), ACA (Adaptive Cross-Approximations), DDT (Domain-Decomposition Technique) seems to bridge some of these known bottlenecks, promoting those the BEM in a high-level tool for even daily-design process of the 3D real-world problems.

The aim of this Chapter is to illustrate how this numerical technique can be used for the simulation of both single-physics problems appearing in the Dielectric Design (Electrostatics), and multi-physics problems in Thermal Design (coupling of Electromagnetic-Heat Transfer) and Electro-Mechanical Design (coupling of Electromagnetic-Structural Mechanics) of power engineering devices like power transformers or switchgears.

1 Introduction

The simulation of real-world engineering problems is nowadays faced with a number of challenging requirements, mainly arising in daily design praxis through:

- **huge dimensions** of the problem to be simulated, especially stressed when going towards *Simulation-Based Design*, including *assembly simulation* as opposed to *component simulation*.

- **huge aspect ratio** in model dimensions. For example, in the *dielectric analysis* of power transformers, in order to correctly perform a simulation analysis one is forced to evaluate simultaneously handle massive parts like windings including shielding rings having a radius of $1 - 2[m]$, alongside the very thin paper insulation around them of thickness $1 - 10e^{-3}\ [m]$. Another example is the *thermal* analysis of transformers, whereby the huge aspect ratio in the dimensions of the tanks (enclosures) $1 - 10[m]$ versus their wall-thickness of 10-20 millimeters can lead to difficulties on both the meshing and numerics sides.
- **complex physics** requiring well founded mathematical formulations and proper numerical evaluation. As an example, consider the diffusion problem in low-frequency electromagnetics, where in typical devices ranging up to several meters in dimensions, the electromagnetic field penetrates into the magnetic material for just a couple of millimeters. A proper representation of this diffusion problem is a real challenge, and for the analysis of 3D problems this requires highly sophisticated numerics.
- **evaluation time**, which for the realization of a complete *simulation chain* for practical problems can be very long, but from the other side needs to be as short as possible for a *daily design* process.

Thanks to advances in numerics made in the last several years, the Boundary Element Method (BEM) has become a powerful numerical technology for 3D simulation of complex practical problems. In spite of some limitations of BEM, for certain classes of problems this method possesses several important advantages in comparison with the classical differential methods like FEM (Finite Element Method) or FDM (Finite Difference Method).

- Probably the most important feature of BEM is that for *linear* classes of problems the discretization needs to be performed only over the interfaces between different media. This excellent characteristic of BEM makes the discretisation/meshing of complex 3D problems morestraightforward and usable for simulations in a *daily design* process.
- Also, this feature is of utmost importance when dealing with the simulation of *moving boundary problems*. Thanks to the fact that the space between the moving objects does not need to be meshed, BEM offers an excellent platform for the simulation of *dynamics*, especially in 3D geometry.
- Furthermore, the *open boundary problem* is treated easily with BEM, without needing to take into account any additionally boundary condition. When using tools based on the differential approach (FEM,FDM), the *open boundary problem* requires an additional *bounding box* around the object of interest, which has a negative impact on both mesh size and computation error.
- Another important feature of BEM is its *accuracy*. Contrary to differential methods, where *adaptive mesh refinement* is almost imperative to achieve the required accuracy, with BEM it is frequently possible to obtain good results even with a relatively rough mesh. But, at this point we also don't

want to say that "adaptivity" could not make life easier even when using BEM.

If we limit ourselves to BEM-based simulations of industrial problems, the classification of the application areas can be done in a variety of different ways. Here we will follow the classification frequently used in **engineering** design, where we often speak about:

- **Dielectric Design** (Electrostatic problems),
- **Thermal Design** (Coupled Electromagnetic/Thermal Problems),
- **Electro-Mechanical Design**
  (Coupled Electro-Magnetic/Structural-Mechanics problems).

Under **Dielectric Design** we usually understand the *Simulation-Based Design* (SBD) of configurations consisting of one or more electrodes loaded with either *fixed* or *floating* potential and being in contact with one or more dielectric media. From the physics point of view, here we deal with a *single-physics* problem, which can be described either by a Laplace or Poisson equation.

In *multi-physics* problems we deal with the coupling of two or more different physical phenomena. The most representative problems appearing in almost all current-carrying devices are coupled Electro-Magnetic/Structural-Mechanic problems (**Electro-Mechanical Design**) or coupled Electro-Magnetic/Thermal problems (**Thermal Design**).

Although in the scope of this material we intend to focus on some practical aspects of BEM usage in industrial design, we shall try to accompany it with brief descriptions of the corresponding formulations, pointing the reader to the relevant references for more background on the theoretical fundamentals or the numerics involved.

The material in this contribution is structured in the following way. Section 2 is devoted to BEM applications for **Dielectric Design**, where alongside the basic formulation we point to some important aspects of simulations for dielectric design in real-world practice.

In Section 3 we present the simulation workflow for coupled *Electro-Magnetic / Structural-Mechanics* problems, where we limit ourselves to *linear elasticity* problems on the mechanics side. We show the possible approaches for the calculation of the physical quantities on both electromagnetic and mechanics side using BEM. As a knowledge of the electromagnetic force density distribution is an essential part of the simulation of coupled *Electro-Magnetic / Structural-Mechanics* problems, we shall give here a brief introduction into the *force analysis*. Additionally, special attention will be paid to the BEM simulation of *eddy-current* problems including *skin-effects*, probably one of the most complicated physical problems in low-frequency electromagnetics. We also stress the importance of the usage of acceleration methods such as MBIT and ACA for practical simulations tasks.

Finally, in Section 4 we give a brief overview of coupled *Electro-Magnetic/Thermal problems*. In this case it is shown how when using an integrated environment based on BEM one can efficiently simulate complex practical problems like power transformers and switchgear.

2 Dielectric Design Using BEM

Dielectric design generally covers the design of devices influenced by the action of *electrostatic*, or in certain applications, *quasi-electrostatic*, fields. In the scope of this section we shall limit ourselves to *electrostatic* problems. These appear in all devices being subjected to different potential loads.

The *simulation workflow* of such problems consists of several steps, shown for illustration in Fig. 1. The first step is *Geometry design*. In the real design process we usually speak about CAD-based[1] design, assuming that the apparatus geometry is modeled by one of the popularly available CAD tools. This step is usually the most time consuming step in the whole simulation workflow. The next step is *discretization* of the model, i.e. the *Mesh generation*. As mentioned before, when working on linear problems with BEM we need to discretize only the interfaces between different media. For non-linear problems the volume of the parts whose materials posses non-linear features, Krstajic [37], must also be discretized. In the *Analysis* step we perform the solution of the problem, usually resulting in the calculation of the distribution of significant physical quantities (fields, potentials, stresses,...). The knowledge of those distributions may be called *primary* information, i.e. it is a **necessary** but not a **sufficient** condition for the designers. The quality of the real design is judged in the next step: by the evaluation of *Design criteria*, mentioned briefly in Example 2, page 292. Finally, the ultimate goal of each simulation process, or more correctly of each design process, is to achieve a

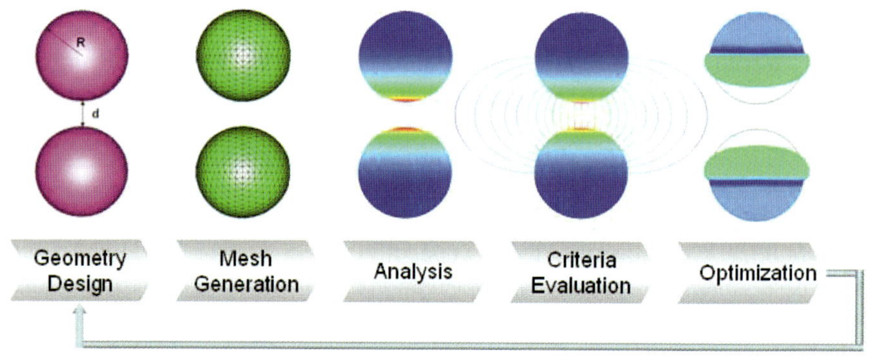

Fig. 1. Simulation workflow.

---

[1]Computer-Aided Design

degree of *Design optimization*. Thanks to the inherent features of BEM outlined at the beginning of this chapter, this method offers an excellent platform for 3D *automatic* procedures for design optimization, Andjelic [1].

In the scope of the following section we shall give a brief illustration of some of these steps, starting with the **Formulation** of the problems used for the *Analysis* step in **Dielectric Design**.

2.1 Formulation

Staying with *electrostatic* problems, the equations that must be satisfied by the field caused by a stationary charge distribution can be derived directly from the Maxwell equations, assuming that all time derivatives are equal to zero. Then, for all regular points in space:

$$\nabla \times \mathbf{E} = 0 \tag{1}$$

and

$$\nabla \cdot \mathbf{D} = \rho \tag{2}$$

i.e. divergence of the electrostatic flux density $\mathbf{D}$ is equal to the charge ρ. This is a *differential form* of Gauss's law of electrostatics. The conservative nature of the field is necessary and sufficient condition for existence of scalar potential whose gradient can be expressed as:

$$\mathbf{E} = -\nabla\varphi \tag{3}$$

If the medium is homogeneous and isotropic, i.e. described by a single dielectric constant ε, we can write:

$$\mathbf{D} = -\varepsilon\mathbf{E} = -\varepsilon\nabla\varphi \tag{4}$$

For homogeneous media the Poisson equation should then be satisfied:

$$\nabla^2\varphi = -\frac{\rho}{\varepsilon} \tag{5}$$

If there are no space charges ($\rho = 0$), the equation can be reduced to the Laplace equation:

$$\nabla^2\varphi = 0 \tag{6}$$

There are basically two fundamental approaches for the translation of the partial differential equation to the integral equation formulation used in BEM.

The first one is the so called direct method (ansatz) whereby an integral formulation is achieved by the application of Green's second theorem to the Laplace equation. The second approach is the indirect method, based on the assumption that the solution can be expressed via equivalent source density functions prescribed over the boundaries, Banerjee [10]. Although the direct formulation is usually preferred, we use here some advantages of the indirect

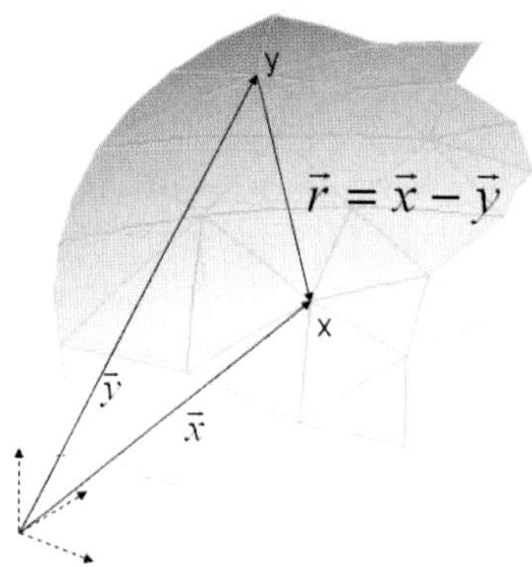

Fig. 2. Position vectors in $\Re^3$.

approach for the formulation of electrostatic problem explained later in this section.

Suppose for now that the system consists only of electrodes in free space. In this case the relation between the potential φ in some point x and the electric surface charge densities $\sigma^e(y)$ is described by *I Fredholm integral equation,* Tozoni [61], Koleciskij [35]:

$$\varphi(x) = \varphi^{ext}(x) + \frac{1}{4\pi\varepsilon_0} \sum_{m=1}^{M} \int_{\Gamma_m} \sigma^e(y) K_1 d\Gamma_m(y) \tag{7}$$

where $K_1 = \frac{1}{r} = \frac{1}{|\mathbf{x}-\mathbf{y}|}$ is a *weakly singular* kernel, r is a distance between the calculation point x and integration point y as shown in Fig. 2, M is the number of closed surfaces $\Gamma := \partial\Omega$ around the electrodes[2] and $\varphi^{ext}(x)$ is the potential of an external electrostatic field in the point x. If we apply (7) to all calculation nodes lying on the electrodes, we obtain a system of equations that can be written as:

$$[A] \cdot [\sigma^e] = [V] \tag{8}$$

where σ^e is the vector of unknown charge densities, and V are the known values of the applied potential. After solving (14), we obtain a vector with the calculated equivalent surface charge densities σ^e. At this point it must be noted that, if for a node lying on the electrode and assuming that the electrode is not on a *floating* potential[3], the electrical field on the electrode is

---

[2] As we shall see later, in cases where we also have dielectric interfaces in the model, M is the number of <u>all</u> closed surfaces in the model, including those between different dielectrics.

[3] For the electrodes on a *floating* potential, a modified approach for the field calculation has to be used, Bachmann [9], Andjelic [3].

directly related to those charge densities through[4] :

$$\varepsilon \mathbf{E} = 4\pi\sigma^e \mathbf{n} \tag{9}$$

This fact leads to two important conclusions:

- The *indirect ansatz* delivers as the matrix solution not only the equivalent charges, but also (indirectly) the values of the electric fields on the electrode surfaces,
- The direct consequence of this is that the field is calculated *without any differentiation* of the potential, which enables more accurate field computation compared to other methods.

Additional steps should be taken in cases where the space of interest is filled with multiple different dielectric materials. On the boundary surfaces between different materials it leads to the induction of *polarized charge*, which causes changes in field distribution. Maxwell [40] has in his early work shown that the analysis of the field in the space containing conductive bodies and different dielectrics can be reduced to the analysis of the field in space containing the bodies positioned in the vacuum. In that case it is necessary to introduce so called *equivalent charges* distributed over the boundary surfaces between different dielectrics. The distribution of equivalent charges must be such that the field produced by those charges and the charges of previously charged bodies situated in vacuum is the same as the field produced by the charged bodies in presence of different dielectrics. Thus, the distribution of *equivalent charges* may not be arbitrary, i.e. it has to satisfy some integral equations.

Let surface Γ be the interface surface between two different dielectrics having dielectric permittivities ε_e and ε_i, respectively, (Fig. 3a). The choice of the equivalent charges (*singe layer* charges or *double layer* charges) is determined by the choice of the observed field quantities that should *not* be altered when

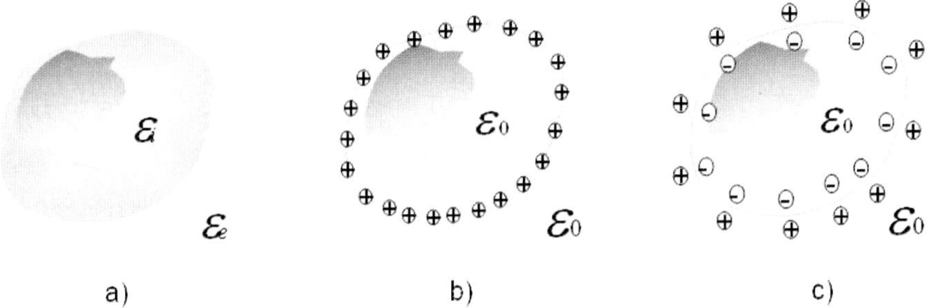

Fig. 3. Equivalent Charges.

---

[4]See Tamm [60] - *Dielectrics* for more detailed discussion.

replacing the real charges with the equivalent charges: electrical field $\mathbf{E}$ or flux density $\mathbf{D}$. If we prefer to keep the field $\mathbf{E}$ unchanged, then we use the *single layer* charge distribution, Fig. 3b, thus forcing the following condition (10) to be fulfilled:

$$\varepsilon_e \cdot \mathbf{E}_n^e = \varepsilon_i \cdot \mathbf{E}_n^i \tag{10}$$

alternatively, in the second case where the flux $\mathbf{D}$ is to be unchanged, we use the *double layer* charge distribution, Fig. 3c, fulfilling the condition

$$\frac{\mathbf{D}_t^e}{\varepsilon_e} = \frac{\mathbf{D}_t^i}{\varepsilon_i}. \tag{11}$$

Here ε_e and ε_i represent dielectric permittivities outside and inside of the boundary Γ, $\mathbf{E_n}^e$ and $\mathbf{E_n}^i$ are the external and internal normal components of the electrostatic field $\mathbf{E}$, and $\mathbf{D}_t^e$ and $\mathbf{D}_t^i$ are the external and internal tangential components of the flux densities $\mathbf{D}$.

It can be easily shown, Tozoni [61], that fulfilling the condition (10) leads to the *II Fredholm integral equation* of the form[5]

$$\sigma^e(x) = \frac{\lambda}{2\pi} \sum_{m=1}^{M} \int_{\Gamma_m} \sigma^e(y) \frac{\mathbf{r} \cdot \mathbf{n}}{r^3} d\Gamma_m(y) \tag{12}$$

where $\mathbf{n}$ is a unit normal vector in point x directed into the surrounding medium, and $\lambda = \frac{\varepsilon_i - \varepsilon_e}{\varepsilon_i + \varepsilon_e}$.

This expression provides the relationship between the *unknown* values of surface charge densities $\sigma^e(x)$ and surface charge densities $\sigma^e(y)$, whereby the point x is a point lying on the boundary between two different dielectrics, and point y can be positioned anywhere in space, including the electrode surfaces. We already saw that equation (7) gives us the relation between the known potential in the points lying on electrode surfaces and surface charge densities over both surfaces with known potential - electrodes and surfaces with polarized charges - surfaces between different dielectrics. Thus, using equations (7) and (12) it is possible to obtain an equation system which enables us to calculate the equivalent charge densities over all surfaces in the space of interest

$$\begin{bmatrix} A_1 & B_1 \\ A_2 & B_2 \end{bmatrix} \begin{bmatrix} \sigma_1^e \\ \sigma_2^e \end{bmatrix} = \begin{bmatrix} V \\ 0 \end{bmatrix} \tag{13}$$

where σ_1^e and σ_2^e are the electric surface charge densities on the electrodes and dielectric interfaces, respectively. $A1, B1, A2$ and $B2$ are the matrix blocks representing the electrode, electrode-dielectric, dielectric-electrode and dielectric coefficients, respectively. Knowing the equivalent charge distributions $\sigma^e = (\sigma_1^e, \sigma_2^e)$, the electrostatic field strength $\mathbf{E}$ in any point of space of interest[6] can be determined as

---

[5]In the presence of the *bound* surface charges, the equation (12) should be modified, see Andjelic [3].

[6]We already saw that in the points on the electrodes we do not need to perform this derivation of potential, but rather use the expression (9) for the field $\mathbf{E}$.

$$\mathbf{E}(x) = -\nabla\varphi(x) = -\frac{1}{4\pi\varepsilon_0} \sum_{m=1}^{M} \int_{\Gamma_m} \sigma^e(y) \cdot \nabla K_1 d\Gamma_m(y) \qquad (14)$$

whereby the position vector $\mathbf{r} = \mathbf{x} - \mathbf{y}$ in K_1 is pointed towards the collocation point x.

Example 1: **Spherical capacitor**

To illustrate briefly the above procedure, we have solved an academic example, with known analytical solution for the field and potential distribution. The example presents a spherical capacitor with two types of dielectrics, $\varepsilon_1 = 1$ and $\varepsilon_2 = 5$, Fig. 4. The analytical solution for the maximal field strength appearing on the $\varepsilon = 1$ side of the interface between two dielectrics is E = 190.47 $[V/m]$, and for the maximal field appearing on the inner electrode E = 85.74 $[V/m]$. To illustrate the dependency between the calculation error and the mesh size, we performed a calculation of this example for different mesh sizes[7], starting with the minimal mesh size possible (1 element per sphere octant, i.e. per 90 degrees curvature).

The graph in Fig. 6 shows the behavior of the relative error as a function of the mesh size. It can be seen that for the nodes on electrode the error is smaller, thanks to the fact that the field can be calculated without performing the derivation of the potential. Already with the relatively rough mesh (5 elements per 90 degrees curvature) the calculation error on the electrode is less than 1% , whereby for the dielectric interface we need 8 elements to reach the same accuracy.

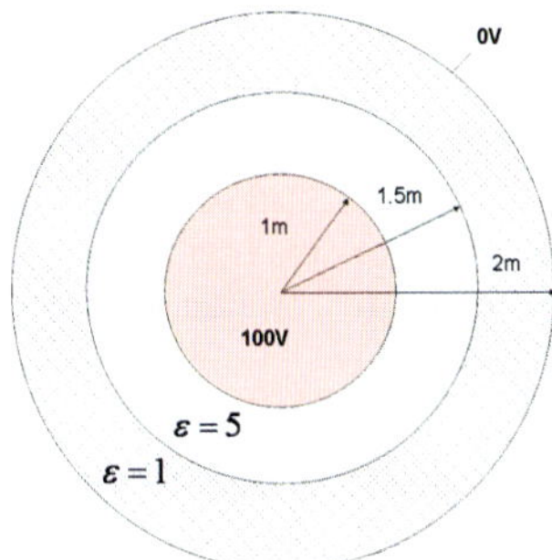

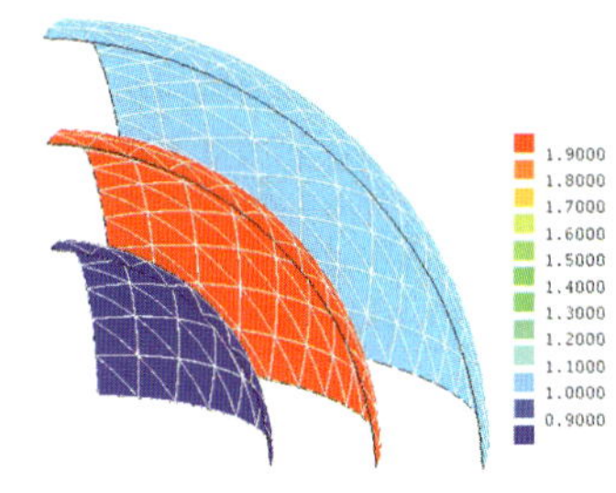

Fig. 4. Spherical capacitor.

Fig. 5. Electrical field strength E[V/m] on the spherical capacitor.

---

[7]Thanks to the symmetry, only 1/8 of the whole model needs to be taken for the calculation.

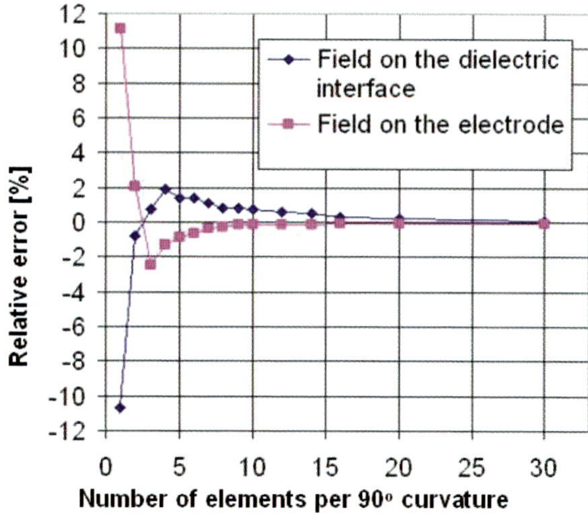

Fig. 6. Relative error vs. mesh size.

Simulation-Based Design

A general trend in almost all industrial branches is to replace costly *Experimentally-Based Design* (*EBD*) with *Simulation-Based Design* (*SBD*). In other words, the aim is to replace the process of experimentally-based prototyping with *digital prototyping* using numerical simulations. Then, at the final stage, experimental validation should be performed. Again BEM offers an excellent platform for *SBD*, thanks to the features already mentioned at the beginning of this chapter.

Regarding *SBD* for **Dielectric Design**, the main question that needs to be answered by the simulation is: *whether the analyzed design fulfils the required safety margin*? Experience shows that often the judgement about the *safety margin* of an entire device cannot be based only on the separate analysis of the device's components. An important implication of this is that in the scope of *SBD* for **Dielectric Design**, the entire assembly needs to be considered in order to take into account the mutual field interferences between the device's components. Analysis of real-world apparatus in the frame of such stringent demands imposes very rigorous requirements on the numerical tools employed:

- **accuracy** - *the tools need to delivery accurate, reliable, and reproducible results.* Qualification of the tool for *SBD* assumes the successful completion of a series of *validation* pre-tests, usually involving comparison with experimental data or known analytical solutions in the simpler cases.

- **robustness** - *the tool needs to deliver correct results independent of the mesh quality.* Designers are usually not trained to "play" with the mesh in order to produce good results. It is then necessarily the task of the tool to provide numerically correct treatment of all "tough" cases arising from irregular meshes. In the BEM case this means that the underlying numerics should be well tuned to handle on one side the mesh irregularities arising from high aspect-ratio elements, skewed elements, *nearly-singular* elements, etc. and from other side the peculiarities encountered with some physical problems, for example the *field penetration* appearing in diffusion problems (more in section "BEM-based Eddy-Currents Analysis, page 312).

- **speed** - *the tool needs to deliver results in a time-frame that is acceptable for the design process.* Simulation of complex practical structures involves large meshes. Even when working with BEM, i.e. when meshing only the interfaces between different media, it can lead to meshes that range from several tens of thousands to several hundreds or even million of elements. Solving this with classical BEM, which results in a fully-populated matrix, would be practically impossible on a standard, single-node machine. One solution to solve the problem is to use a parallel version of the code, see the Example 2, page 292. Another solution is to perform accelerated matrix computation and matrix compression using the emerging numerical techniques that have appeared in the last several years such as Fast Multipole Methods, Greengard [25] or ACA[8], Bebendorf [13]. The application of Fast Multipole is discussed briefly in the following section, and the application of the ACA method is discussed in more details in "ACA for eddy-current computation", page 322.

Fast BEM in Electrostatic Problems

The discretization of equations (7) and (12) yields a densely populated matrix, which is well known as the major bottleneck in BEM computations. The amount of storage is of order $O(N^2)$, with N being the number of unknowns. Furthermore, the essential step at the heart of the iterative solution of this system is a matrix-vector multiplication and the cost of such a multiplication is also of order $O(N^2)$. Thus a reduction of the complexity to $O(N \log N)$ or $O(N)$ would naturally be very desirable. Developments started with a seminal paper by Greengard [25] that proposed a *Fast Multipole Method*, which became highly popular in several numerical communities. Another fundamental development was brought about by Hackbusch [27, 28]. In the following we present a brief description of the MBIT[9] algorithm that is used in our computations. The central idea is to split the discretized boundary integral operator into a *far-field* and a *near-field* zone. The singularity of the kernel of the integral operator is then located in the *near-field*, whereas the kernel is continuous and smooth in the *far-field*. Compression can then be achieved by a separation of variables in the *far-field*. In order to reach this goal, the boundary in the first stage is subdivided into clusters of adjacent panels that

---

[8]ACA stands for Adaptive Cross-Approximation.
[9]Multipole-Based Integral Technique

are stored in a hierarchical structure called the panel-cluster tree. Then, in the second stage we collect all admissible pairs of clusters, i.e. pairs that fulfill the admissibility condition $|\mathbf{x} - \mathbf{x}_0| + |\mathbf{x}^c - \mathbf{x}_0^c| \leq \eta\,|\mathbf{x}_0 - \mathbf{x}_0^c|$ where $0 \leq \eta < 1$ into the far-field block. The centers of gravity of the clusters are here denoted by $\mathbf{x}_0$, $\mathbf{x}_0^c$. All other pairs of clusters (the non-admissible ones) belong to the *near-field*. Then the matrix entries corresponding to the *near-field* zone are computed as usual, whereas the matrix blocks of the *far-field* are only approximated. This is achieved by an expansion of the kernel function $k(\mathbf{x}, \mathbf{x}^c)$ that occurs in the matrix entries

$$a_{ij} = \int\limits_{\Gamma} \int\limits_{\Gamma} \hat{\varphi}_i(\mathbf{x}) k(\mathbf{x}, \mathbf{x}^c) \varphi_j(\mathbf{x}^c) d\Gamma(x) d\Gamma(x^c). \tag{15}$$

The expansion

$$k(\mathbf{x}, \mathbf{x}^c) \approx k_m(\mathbf{x}, \mathbf{x}^c; \mathbf{x}_0, \mathbf{x}_0^c) = \sum_{(\mu,\nu) \in I_m} k_{(\mu,\nu)}(\mathbf{x}, \mathbf{x}_0^c) X_\mu(\mathbf{x}, \mathbf{x}_0) Y_\nu(\mathbf{x}^c, \mathbf{x}_0^c) \tag{16}$$

decouples the variables $\mathbf{x}$ and $\mathbf{x}^c$ and must be done only in the *far-field*. Then, the matrix-vector products can be evaluated as

$$\nu = \tilde{A} \cdot u = N \cdot u + \sum_{(\sigma,\tau) \in F} X_\sigma^T (F_{\sigma,\tau}(Y_\tau \cdot u)). \tag{17}$$

Several expansions can be used for this purpose: Multipole-, Taylor- and Cebyshev-expansion. The procedures lead to a low rank approximation of the *far-field* part and it is shown in Schmidlin [50] that one obtains exponential convergence for a proper choice of parameters. A more detailed elaboration and comparison of all three type of expansions can also be found in the same reference.

Example 2: **SBD** *for a generator circuit-breaker design*

In this example it is briefly shown how Simulation-Based Design of the Generator Circuit-Breaker (GCB) is performed using a BEM[10] module for electrostatic field computation. Generator circuit-breakers, Fig. 7, are important components of electricity transmission systems. Some of the main tasks they have to fulfil are for example, Zehnder [63]:

- Synchronize the generator with the main system
- Separate the generator from the main system
- Interrupt currents
- Interrupt system-fed and generator-fed short-circuit currents
- Interrupt currents under out-of-phase conditions

---

[10]This BEM module is a sub-module for electrostatic analysis in POLOPT, a 3D BEM-based simulation package for single and multi-physics computation.

Fig. 7. ABB generator circuit-breaker.

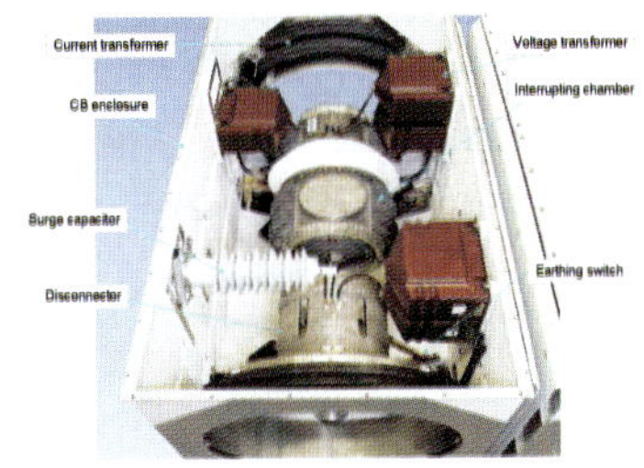

Fig. 8. Main assembly of GCB.

Fig. 8 shows the complete assembly of a GCB containing, beside the interrupting chamber as a key component, all other parts such as current and voltage transformers, earthing switches, surge capacitors, etc. The detail of the corresponding CAD model is shown in Fig. 9, while the triangle mesh is shown in Fig. 10. From Fig. 9 and 10 it can be also be seen that such small details as screws and thin connecting wires have been included in the input to the simulation model.

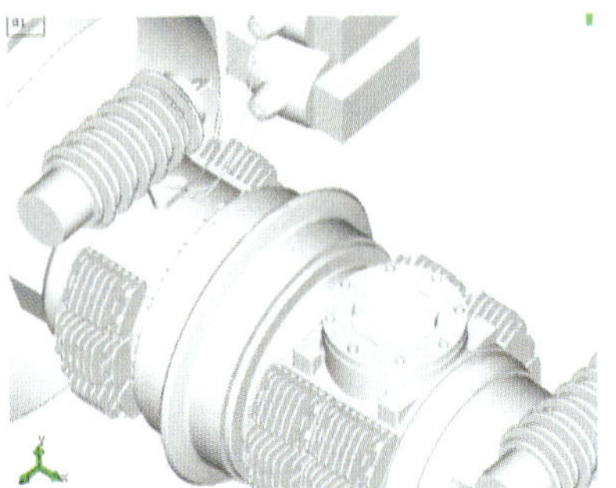

Fig. 9. CAD-model of the GCB.

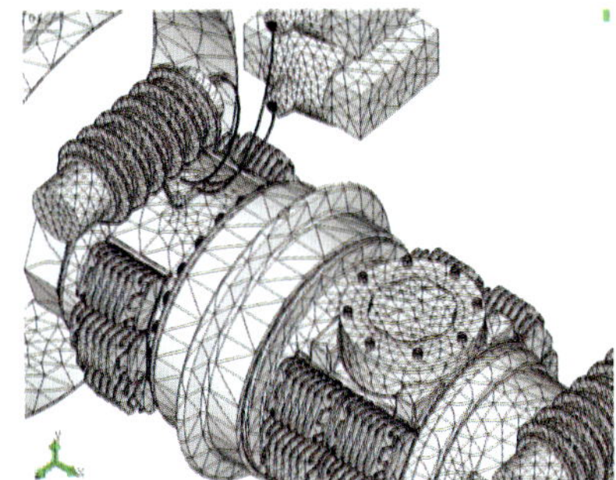

Fig. 10. Meshed model of the GCB.

Step I: Primary Analysis

The simulation details for the above shown generator circuit-breakers case were:

- The discretization of the model has been performed using *second order* triangle elements.
- The stiffness matrix has been assembled using an *Indirect* Ansatz with *collocation* in the main triangle vertices, formulas (7) and (12), page 285.

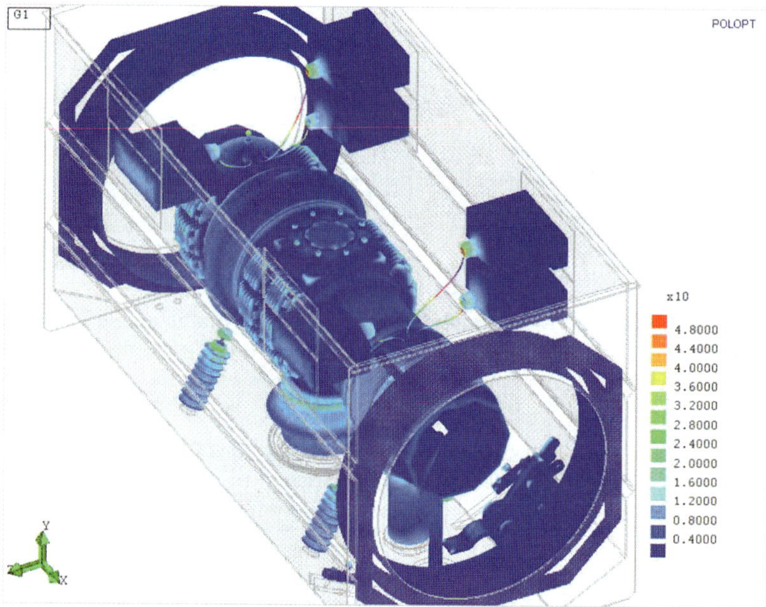

Fig. 11. ABB Generator Circuit-Breaker: Electrostatic field distribution, E[V/m].

It has to be mentioned here that in both the real design and consequently then in the simulation model, geometrical singularities like edges and corners have been removed through *rounding*. In the real design this is a common practice in all high-voltage devices in order to prevent the occurrence of *dielectric breakdown*. On the numerics side, this fact enables usage of the *nodal collocation* method - which is also the fastest one - without violating the mathematical correctness of the problem. The *nodal collocation* method is often questionable when being applied to models containing geometrical singularities. A useful discussion on geometrical singularities can be found in Bladel [16].

- The coefficients of the stiffness matrix have been calculated using the **multipole approach**, Greengard [25], with *monopole, dipole* and *quadropole* approximations for the far-field treatment, Andjelic [5]. *Diagonal matrix preconditioning* has been used, which enables fast and reliable matrix solution using GMRES. This run has been accomplished without any matrix compression, but using a parallelized version of the code, Blaszczyk [17]. For a parallel run we used a PC cluster with 22 nodes. The data about memory and CPU time are given in Table 1.
- The calculated electrostatic field distribution is shown in Fig. 11. It can be seen that the highest field strength appears on the small feature details, such as screws.

Table 1. The analysis data for GCB example.

Elements	Nodes	Main vertices	Memory	CPU
145782	291584	80230	42GByte	2h20'

Step II: Criteria Evaluation

As mentioned before, the field and potential distribution are only the first level of information needed by the designers , i.e. they are a *necessary* but not a *sufficient* condition by which to judge the *safety margin* of a design. The complete information satisfying both conditions usually calls for additional **criteria evaluation**, based on the potential and field distribution. There are a number of different criteria required for proper judgement about the *safety margin* of a design, Andjelic [2]. One possibility is to use so called FLC (Field Line Criteria), whereby the specific design criteria are evaluated based on the field/potential distribution along the *field lines*. This requirement requires the additional computation of the field lines, starting from the locations with critical field strengths. As is visible from Fig. 12, the critical points appear on the screws, emphasizing the point that such small details should not be omitted from the simulation model.

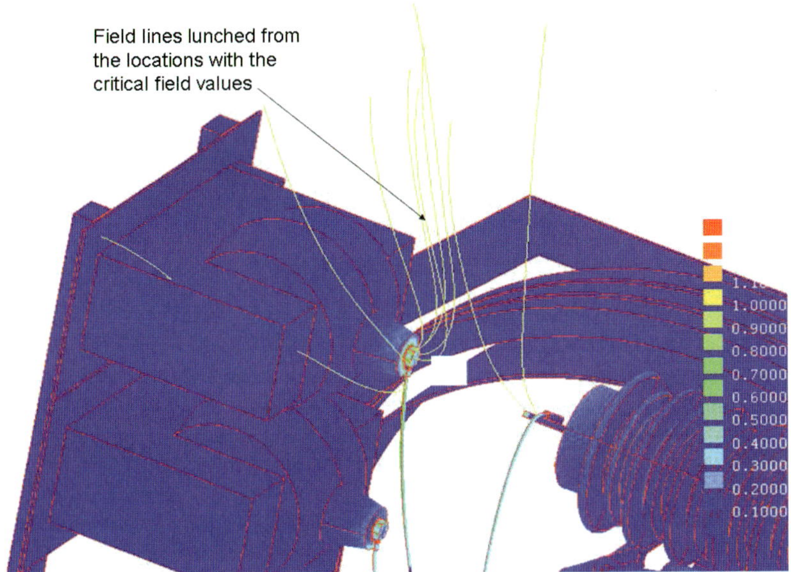

Fig. 12. Field lines starting from the locations having critical field strength.

Again, BEM in combination with fast multipole method offers an excellent platform for the fast and accurate evaluation of post-processing steps such as field line calculation. The field line computation is performed using formulas (7) and (14), together with multipole acceleration for the sub-integral coefficients calculation.

3 Electro-Mechanical Design

Under **electro-mechanical Design** we usually understand the design of electrical apparatus and devices, whereby of primary interest is the *mechanical response* of structures under the influnece of acting electromagnetic forces. This problem appears in any device conducting DC or AC current, or subjected to the action of an external electrostatic or magnetic field[11]. For example, in transformer design we are interested in the mechanical stability of the windings' structures, when subjected to short-circuit forces. Similarly, in circuit-breaker design we are interested in the mechanical response of the bus-bar structures, caused again by short-circuit stresses. By contrast with the above examples, where we are usually looking for a design with *minimal* mechanical response to electromagnetic forces, in some devices such as MEMS[12] sensors/transducers we are looking for *maximal* mechanical response, producing measurable *output* information about the *input* physical quantities.

From the physics point of view, here we deal with the coupling of the electromagnetic (or electrostatic) problems and structural-mechanics problems. The coupling can be either *weak* or *strong*. Under *weak* coupling we understand the sequential analysis of each phenomena separately, coupled together via an iterative scheme. In *strong* coupling we usually deal with the simultaneous solution of both problems, whereby the coupling is preserved on the equations level. With the exception of quantum mechanics and particle physics, there are quite few approaches where *strong* coupling is used in engineering practice. In most industrial applications, which are the main subject of the present material, we deal with *weak* coupling.

Weak coupling usually assumes two main steps:

- **calculation of electromagnetic (EM)/electrostatic (ES) forces,**
- **calculation of mechanical response** (displacement, stresses)

---

[11]In certain applications (force sensors, pressure sensors, accelerometers) we are not looking for *mechanical response* caused by the electromagnetic forces, but rather for *electrical response* caused by the mechanical forces (*piezoelectric problem*). This case will not be covered in the scope of this material. More information about BEM treatment of these classes of problems can be found in Gaul [23], Hill [30]. Here we shall also not cover the topic of coupled Electro-Magnetic/Mechanics problems related to magnetostriction phenomena (change of the shape of magnetostrictive material under the influence of a magnetic field). More information for example in Whiteman [64].

[12]Micro Electro-Mechanical Systems

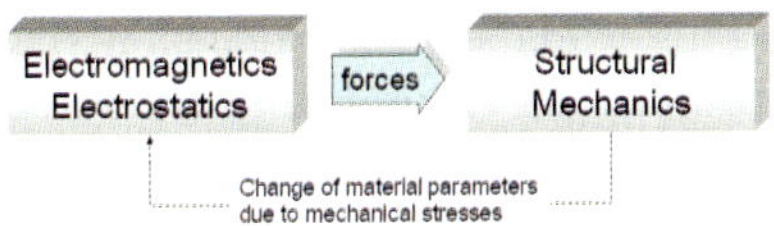

Fig. 13. Weak coupling scheme for EM-SM problems.

In such coupled schemes the role of the electromagnetic (electrostatic) solver is to deliver correctly calculated forces, which are then further passed as an *external load* to the mechanical solver, Fig. 13. In the following sections we shall briefly elaborate the main steps for the BEM-based simulation of these types of problems. As the "forces" are the main link between the electromagnetics and mechanics, let us start with a short overview of force analysis.

3.1 Force Analysis

Force analysis covers an extremely broad field, and in certain areas is still a subject of contradictory discussions in the research community, Reyne [48]. In the scope of this section we shall focus on the analysis cases most commonly appearing in industrial praxis.

Generally speaking, if we limit ourselves to the macroscopic view of material properties, then the force acting on some macroscopic system is a sum of the partial forces acting on each volume element that constitute the system. This means that the total force can be expressed as

$$\mathbf{F} = \int_{\Omega} \mathbf{f}(\mathbf{r})d\Omega \tag{18}$$

where $\mathbf{f}$ is a *volume force density* in $[N/m^3]$. The local forces $\mathbf{f}$ between the completely internally positioned volume elements cancel, thus the only contribution to the total force comes from the volume elements interfacing with the

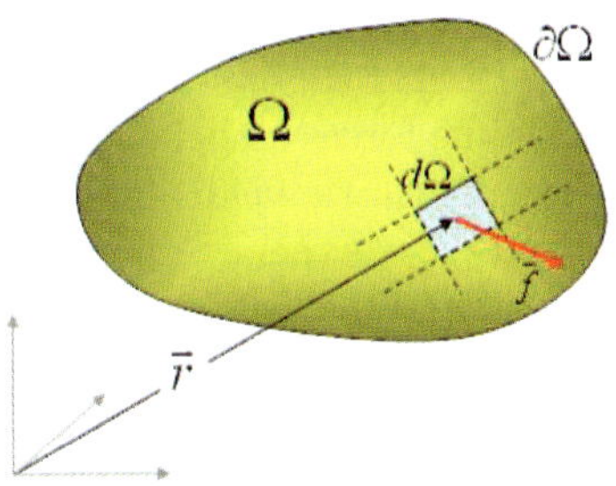

Fig. 14. Force on the volume element dV.

surrounding medium. This leads to the conclusion that it could be possible to replace the integration over an volume with the integration over the closed surface $\Gamma := \partial\Omega$ around the body Ω. In other words, the aim is to replace the ponderomotive (body) forces with tensile forces acting on the surface (stress). From the physics point of view, it is described by the well know principle of *conservation of momentum* saying that the momentum does not change (is conserved), i.e. can be neither created nor destroyed, except through the action of external forces, Jackson [33].

Using vector analysis, we can express the volume integral of the force density vector $\mathbf{f}$ as a surface integral in the case that $\mathbf{f}$ is expressible as the divergence of the stress tensor $\bar{\bar{S}}$ of rank two, Melcher [45],

$$\mathbf{f} = \nabla \cdot \bar{\bar{S}}. \tag{19}$$

So for the total force $\mathbf{F}$ in [N]

$$\mathbf{F} = \int_\Omega \mathbf{f}(\mathbf{r})d\Omega = \oint_\Gamma \bar{\bar{S}} \cdot \mathbf{n}d\Gamma = \oint_\Gamma \mathbf{t}d\Gamma \tag{20}$$

where $\mathbf{t}$ is a *traction* vector or the tensile forces per unit area acting on the closed surface Γ[13].

Forces in the Stationary Fields

Assuming an isotropic, homogeneous solid body, having permittivity ε and permeability μ that are not field dependent, it can be shown, Stratton [59], Melcher [45], that the volume force densities in the body exposed to the electro- and magneto-static fields can be expressed for the electrostatic case as

$$\mathbf{f}_e = \rho\mathbf{E} - \frac{1}{2}E^2\nabla\varepsilon + \mathbf{f}_e^s, \tag{21}$$

and

$$\mathbf{f}_m = \mathbf{J} \times \mathbf{B} - \frac{1}{2}H^2\nabla\mu + \mathbf{f}_m^s \tag{22}$$

for the magnetostatic case. In (21) the ρ stands for a *net* (unpaired) charge density.[14] The $\mathbf{f}_e^s$ and $\mathbf{f}_m^s$ are the forces associated with the pure strain caused by the electrostatic and magnetostatic fields, Stratton [59].

Representing the (21) and (22) as the divergence of a tensor, we can obtain the expressions for the same forces using the surface integration

$$\mathbf{t}_e = (\varepsilon + \frac{a_2 - a_1}{2})\mathbf{E}(\mathbf{E} \cdot \mathbf{n}) - \frac{\varepsilon + a_2}{2}E^2\mathbf{n} \tag{23}$$

---

[13]should not be understood as a *surface force density*

[14]$\rho = N_+q_+ - N_-q_-$, where N_- and N_+ stand for the number of negative and positive charges q_- and q_+, respectively.

for the forces exerted by the electrostatic field $\mathbf{E}$, and as

$$\mathbf{t}_m = (\mu + \frac{b_2 - b_1}{2})\mathbf{H}(\mathbf{H} \cdot \mathbf{n}) - \frac{\mu + b_2}{2}H^2\mathbf{n} \tag{24}$$

for the forces exerted by the magnetostatic field $\mathbf{H}$, respectively. The constants a_1, a_2 in (23) and b_1, b_2 in (24) have to be taken into account if the elastic deformation of a body needs to be calculated[15]. If we are interested in the total force acting on the body, than due to their local compensation by induced elastic stresses, they can be omitted.

Thus, the total force acting on the body in an electrostatic field can be obtained as

$$\mathbf{F}_e = \oint_{\Gamma} \left(\varepsilon\mathbf{E}(\mathbf{E} \cdot \mathbf{n}) - \frac{\varepsilon}{2}E^2\mathbf{n} \right) d\Gamma \tag{25}$$

and in a magnetostatic field as

$$\mathbf{F}_m = \oint_{\Gamma} \left(\mu\mathbf{H}(\mathbf{H} \cdot \mathbf{n}) - \frac{\mu}{2}H^2\mathbf{n} \right) d\Gamma. \tag{26}$$

Note again, that when calculating the total force, the term containing the constants a_1, a_2 and b_1, b_2 are not present due to the compensation by elastic stresses mentioned above.

Some Special Cases

Volume force densities in a non-permeable conductor (μ=1)

If the body being exposed to the magnetostatic field is made of a material with $\mu = 1$, then equation(22) for volume force density reduces to

$$\mathbf{f}_m = \mathbf{J} \times \mathbf{B}. \tag{27}$$

The total force on the conductor is then simply

$$\mathbf{F}_m = \int_{\Omega} (\mathbf{J} \times \mathbf{B})d\Omega. \tag{28}$$

Equation (28) is a well-know form of the Lorentz equation for the forces on the current-carrying conductor.

---

[15]These constants usually have to be determined by measurement. Physically, for example, the parameter a_1 expresses the increment of e corresponding to an elongation parallel to the lines of field intensity, while a_2 determines this increment for strains at right angles to these lines.

Force on body immersed in a medium with $\mu = \mu_0$

If the body is positioned in the medium with the permeability $\mu = \mu_0$, than the tensile forces acting on the body surface can be obtained after integrating the equation (24) on either side of the body surface

$$\mathbf{t}_m = \frac{\mu_2 - \mu_0}{2}\left[H_2^2 + \left(\frac{\mu_2 - \mu_0}{\mu_0}\right)\cdot H_{n2}^2\right]\mathbf{n}_1 + \frac{b_2}{2}H_2^2\mathbf{n}_1 + \frac{b_1 - b_2}{2}H_{n2}\mathbf{H}_2 \quad (29)$$

whereby the $\mathbf{n}_1$ is a normal vector pointed to the surrounding medium with $\mu = \mu_0$, $\mathbf{H}_2$ is the magnetic field inside the body surface, H_{n2} is the normal component of the magnetic field within the body, and b_1 and b_2 are the constants applied to the magnetic body. For a linear, homogeneous, isotropic and incompressible medium, it can be further simplified, Reyen [48]:

$$\mathbf{t}_m = \mathbf{H}_1(\mathbf{B}_1.\mathbf{n}_1) - ((\mathbf{B}_1\mathbf{H}_1)/2)\mathbf{n}_1 - \mathbf{H}_2(\mathbf{B}_2.\mathbf{n}_1) + ((\mathbf{B}_2\mathbf{H}_2)/2)\mathbf{n}_1 \quad (30)$$

Up to now we have spoken about the forces appearing in *stationary* cases. In practice it is also often asked to determine the forces, and consequently then also the mechanical response, in the *time-varying* field.

Forces in Time-Varying Fields

In the previous section, we saw that for stationary fields the total force transmitted by the electromagnetic field across any closed surface can be obtained by combining equations (25) and (26)

$$\mathbf{F} = \mathbf{F}_e + \mathbf{F}_m = \oint_\Gamma \left(\varepsilon\mathbf{E}(\mathbf{E}\cdot\mathbf{n}) - \frac{\varepsilon}{2}E^2\mathbf{n} + \mu\mathbf{H}(\mathbf{H}\cdot\mathbf{n}) - \frac{\mu}{2}H^2\mathbf{n}\right)d\Gamma \quad (31)$$

whereby all quantities were only functions of the position in space. While in that case the force has been interpreted as the force exerted by a field on the matter, in the case of time-varying fields we deal with the inward flow of momentum per unit time through the surface Γ, whereby all the physical quantities are functions of both space and time. In Stratton [59] it has been shown that the total force on a body in time-varying field can be calculated as[16]

$$\mathbf{F}(t) = \int_\Omega \left(\rho\mathbf{E} + \mathbf{J}\times\mathbf{B} - \frac{1}{2}E^2\nabla\varepsilon - \frac{1}{2}H^2\nabla\mu + \frac{\mu_r\varepsilon_r - 1}{c^2}\frac{\partial\mathbf{S}}{\partial t}\right)d\Omega \quad (32)$$

or, when integrating over the surface

---

[16]The bold-typed vectorial quantities are in time-varying case the complex-valued vectors.

$$\mathbf{F}(t) = \oint_{\Gamma} \left(\varepsilon(\mathbf{E}\cdot\mathbf{n})\mathbf{E} - \frac{\varepsilon}{2}E^2\mathbf{n} + \mu(\mathbf{H}\cdot\mathbf{n})\mathbf{H} - \frac{\mu}{2}H^2\mathbf{n} \right)d\Gamma - \qquad (33)$$

$$-\frac{1}{c^2}\frac{d}{dt}\int_{\Omega}(\mathbf{E}\times\mathbf{H})d\Omega$$

where the constant $c^2 = (\varepsilon_0\mu_0)^{-1}$, and $\mathbf{S} = \mathbf{E}\times\mathbf{H}$ is a Poynting vector. The last term in (33) is very small due to factor $c^2 = (\varepsilon_0\mu_0)^{-1}$, and for many practical calculations can be omitted. If we are considering the force on a conductor with $\mu=1$, than (33) is reduced to

$$\mathbf{F}(t) = \int_{\Omega}(\mathbf{J}\times\mathbf{B})d\Omega. \qquad (34)$$

Usually in praxis we are interested in the time-averaged force density $\bar{f}\,[N/m^3]$

$$\bar{f} = \frac{1}{2}\mathrm{Re}\left\{\rho^e\mathbf{E}^* + \mathbf{J}\times\mathbf{B}^* + \rho^m\mathbf{H}^* + \mathbf{M}\times\mathbf{D}^*\right\} \qquad (35)$$

where $\mathbf{M} = i\omega\mathbf{P}^m = i\omega\mu_0(\mu_r - 1)\mathbf{H}$ are the bounded magnetic currents, and ρ^m are the bounded magnetic charges.

The time-average Lorentz force density in a non-permeable current-carrying conductor ($\mu=1$) is

$$\bar{f} = \frac{1}{2}\mathrm{Re}\left\{\mathbf{J}\times\mathbf{B}^*\right\} \qquad (36)$$

where (*) denotes a complex conjugate.

Force analysis - summary

- To sum up the results elaborated above, if we need to find the **total force** on a body in a stationary field, we shall either $i)$ sum up the volume forces inside the observed body by integrating over its volume Ω_+ using (21) for electrostatic case or (22) for magnetostatic case, or $ii)$ integrate the divergence of the tensor $\bar{\bar{T}}$ over the surface Γ, completely containing the observed body Ω_+, using (25) or (26) for electrostatic and magnetostatic case respectively[17].

  If we have a body in a time-varying field, than the total force should be calculated either using volume integration, formula (32), or using surface integration, formula (33).

---

[17] Here two more facts need to be noted: $i)$ for equivalence of the volume forces and traction, it is necessary that not only the *resultant* of the forces applied to an arbitrary volume remain constant, but also that the *moment* of these forces remains constant when body forces are replaced with equivalent stresses; $ii)$ an additional condition for above equivalence is that the stress tensor is symmetrical, see Tamm [60] for more information.

- When performing a coupled electromagnetic-structural mechanics analysis, we are not interested in the *total force*, but rather in the *local force* density distribution, i.e. in:
  i) **forces per unit volume**, $[N/m^3]$, formulas (21) or (22) for stationary case or sub-integral function in (32) for time-varying fields,
  or in the:
  ii) **forces per unit area**, $[N/m^2]$, formulas (25) or (26) for stationary case or sub-integral function in (33) for time-varying fields.

These local forces are then further passed as an *external load* for the analysis of the mechanical quantities, see section "BEM for Structural Mechanics", page 329.

In the following we shall limit ourselves to the coupled problems appearing in electrical apparatus, whereby the forces are caused either by the magnetostatic or time-varying electromagnetic field[18]. Fig. 15 shows a workflow for the BEM-based coupled analysis of Electro-Magnetic/Structural-Mechanics problems. Following the steps shown in this workflow, let us first introduce some numerical procedures and examples related to the calculation of forces in both stationary and time-varying fields.

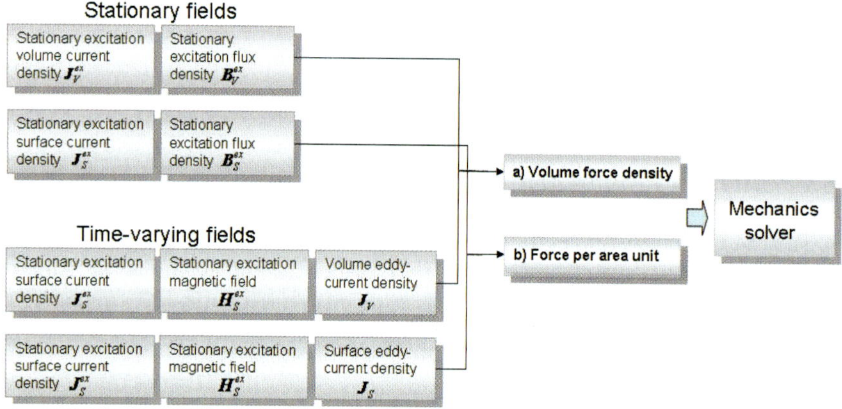

Fig. 15. Coupled EM-EM workflow in the current-carrying structures.

3.2 BEM for Force Analysis in the Stationary Fields

Forces on the Non-permeable Structures

As mentioned before, for the analysis of coupled electro-magnetic / structural mechanics problems it is necessary to calculate the *local* forces. In this section

---

[18] In the scope of this material we shall not cover forces due to electrostatic fields.

let us limit ourselves to the analysis of forces on a conductor made of material with permeability $\mu = 1$. This is typical of copper or aluminium bus-bars or windings in switchgear and transformers. In that case we can use the formula $\mathbf{f}_m = \mathbf{J} \times \mathbf{B}$ for volume force density, where $\mathbf{J}$ is a volume stationary current density in $[A/m^3]$, and $\mathbf{B}$ is a corresponding magnetic flux density in $[T]$. As can be seen from Fig. 15, the knowledge of the stationary current distribution is an initial step, required for both stationary or time-varying case. In the following section we shall briefly elaborate some procedures to compute it.

Computation of the Stationary Current Distribution

Current distribution in homogeneous structures (Single-material case)

The calculation the stationary current distribution in the conductors assumes the solution of the Laplace problem, analogous to the previously described electrostatic case. If the conductor occupying the volume Ω is filled with a uniform and isotropic medium, we solve the Laplace equation (37)

$$\Delta\varphi(x) = 0; \qquad \forall x \in \Omega \tag{37}$$

with boundary conditions (38) in the case that we work with **known potential** φ at the inlet and outlet surfaces $\Gamma_\pm$ (Dirichlet type of boundary conditions),

$$\frac{\partial\varphi(x)}{\partial n_x} = 0 \quad \forall x \in \Gamma \backslash \Gamma_\pm, \qquad \varphi(x) = U \quad \forall x \in \Gamma_\pm, \tag{38}$$

or (39) in the case that we work with **known current** I boundary conditions on the inlet and outlet surfaces $\Gamma_\pm$ (Neumann type of boundary conditions),

$$\frac{\partial\varphi(x)}{\partial n_x} = 0 \quad \forall x \in \Gamma \backslash \Gamma_\pm, \qquad \frac{\partial\varphi(x)}{\partial n_x} = \pm\frac{I}{\sigma \cdot \Gamma_\pm} \quad \forall x \in \Gamma_\pm \tag{39}$$

where σ is the electrical conductivity, Γ_+ i.e. Γ_- stand for *inlet* and *outlet* conductor's surfaces respectively and I is the value of the current in $[A]$ flowing through the conductor.

Using the *direct* collocation approach described in Lean [44], we can obtain the following equation for the unknown potential φ in the conductor,

$$\oint_\Gamma G(x,y)\frac{\partial\varphi}{\partial n_x}(x)d\Gamma(x) - \oint_\Gamma \varphi(x)\frac{\partial G}{\partial n_x}(x,y)d\Gamma(x) = \frac{\theta(y)}{4\pi}\varphi(y), \quad y \in \Gamma, \tag{40}$$

where G is the fundamental solution associated with the Laplace equation in 3D: $G(x,y) = 1/4\pi|x-y|$, x and y are respectively the source and field points and Θ is the solid angle which is defined as follows,

$$\Theta(y) = \int_\Gamma \frac{x-y}{|x-y|^3}d\Gamma(x). \tag{41}$$

For a Neumann-type of boundary condition, the matrix needs to be regularized, usually by means of the Lagrange multipliers technique, Bochev [18].

Using the potential φ obtained by (40), one can finally obtain the stationary current density as

$$\mathbf{J} = -\sigma\nabla\varphi. \tag{42}$$

Example 3: **Three-phase industrial transformer**

Here we have used the above formulation for the force analysis in a three-phase industrial transformer. One of the typical duties of industrial transformers is to serve as a steel arc furnace transformer, operating under very severe conditions with regard to frequent over-currents and over-voltages generated by short-circuits in the furnaces and the operation of connected HV circuit breakers. Fig. 16 shows the bus-bar structure, together with the three windings (U,V,W) of a three-phase industrial transformer.

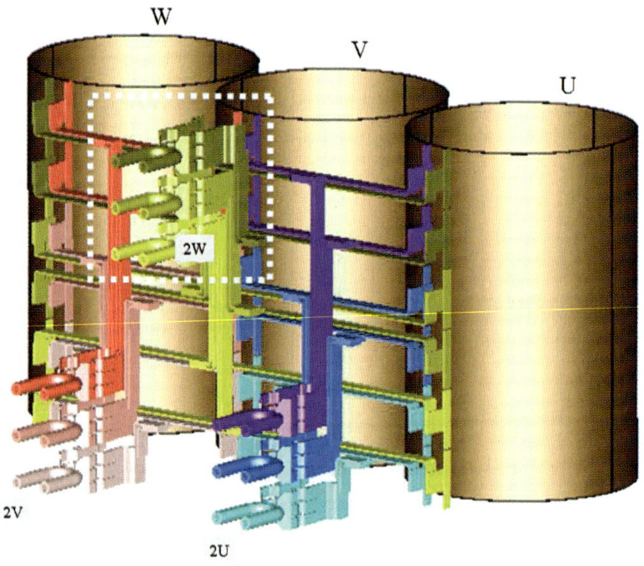

Fig. 16. CAD model of the three-phase industrial transformer's busbar structures.

In order to illustrate the major advantage of BEM as a boundary method over other domain methods (FEM for example) this complicated geometry has been calculated without any advanced technique for matrix compression or preconditioning. The surface of the conductors has been simply meshed, the boundary conditions for inlet and outlet surfaces have been defined and the Galerkin BEM has been used for the discretization of the integral equation. The resulting large dense linear system has been solved using a direct solver [32]. Since the bus-bars from Fig. 16 contain several bodies insulated from each other, the solution process can be accelerated by solving only one conductor (body or bus-bar) at the time. The mesh size, number of degrees of freedom (DOFs) and matrix memory allocation for each body (conductor) is given in Table 2. As one can see the amount of memory needed for matrix storage is relatively small and complete bus-bar can be solved on standard PC. The number of DOFs (the fourth column of Table 2 is always bigger by one than the number of nodes. This comes from the fact that our integral formulation was applied for a pure Neumann problem. Hence, regularization has been used by simply augmenting the original system of equations.

Table 2. Numerical data for the current distribution analysis of the industrial transformer bus-bar presented in Fig. 16 are given.

Bus-bar ID	NE	NN	NDOF	MDENSE (MB)
1 phase U	4292	2134	2135	35
2 phase U	4050	2013	2014	31
3 phase U	4022	1999	2000	31
4 phase V	5320	2648	2649	54
5 phase V	5534	2755	2756	58
6 phase V	7626	3801	3802	110
7 phase W	4226	2101	2102	34
8 phase W	4486	2231	2232	38
9 phase W	4052	2014	2015	31

NE   number of elements; NN   number of nodes; NDOF   number of DOFs (matrix dimension); MDENSE (MB)   memory of dense matrix given in MB

The calculation of the stationary current distribution has been performed in order to estimate the forces acting in the presence of the short-circuit currents. Fig. 17 gives a detail of the calculated current distribution in one of the phases shown in the dotted polygon of Fig. 16. The force densities are calculated using (27), whereby the magnetic flux density $\mathbf{B}$ is calculated as $\mathbf{B} = \mathbf{B}_{windings} + \mathbf{B}_{bus-bars}$, i.e. taking into account the influence of the magnetic field of windings. A force density distribution on the one of the phases is shown on Fig. 18.

Current distribution in non-homogeneous structures (Multi-material case)

The formulation (40) is restricted to structures composed of one single homogeneous material. In praxis it is often asked to analyze problems in multi-material structures. In the following we give a formulation suited for the stationary current density calculation in the multi-material structures, Smajic [55]. It must be noted that this formulation results in a well-conditioned

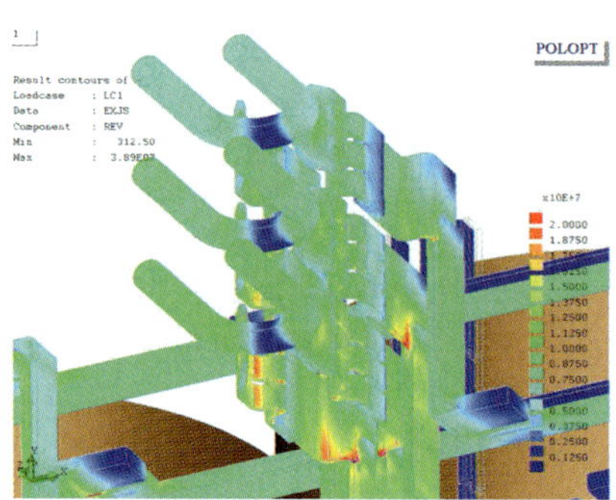

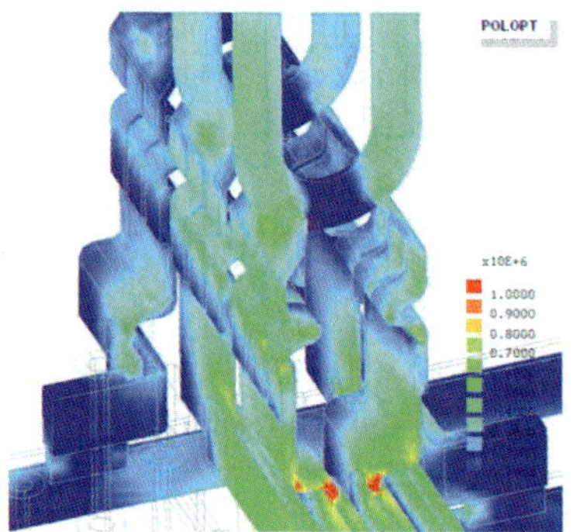

Fig. 17. Transformer busbars: current density distribution $[A/m^2]$.

Fig. 18. The complex magnitude of the volume force density $[N/m^3]$ for the electric current phase $\boldsymbol{W}$ is presented; One can clearly recognizes the parts with large mechanical tension.

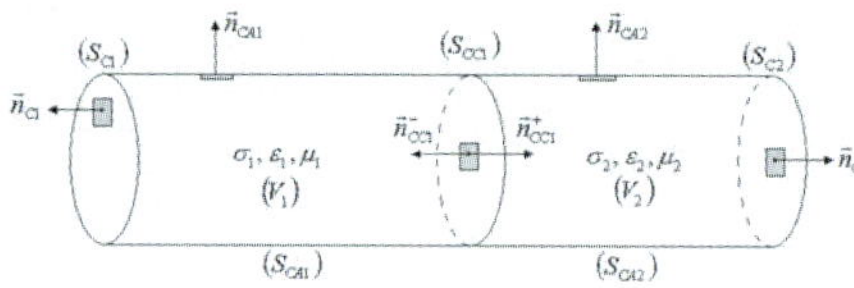

Fig. 19. Conductor made of two different materials.

matrix, i.e. no any regularization is needed. Further, it also enables the "symmetrization" of the matrix, explained later in this section. Fig. 19 shows a single conductor made of two different conductive materials and with two contacts (Neumann null-boundary condition on the lateral surface Γ_{CA} and Dirichlet on the contact surface Γ_C),

$$\frac{\partial \varphi_{1,2}}{\partial n_x}(x) = 0, \quad x \in \Gamma_{CA1}, \Gamma_{CA2}, \quad \varphi_{1,2}(x) = F_{1,2}, \quad I \in \Gamma_{C1}, \Gamma_{C2}. \quad (43)$$

On the interface between the two materials (Γ_{CC1}) the following potential continuity conditions hold,

$$\varphi_1(x) = \varphi_2(x), \quad \sigma_1 \frac{\partial \varphi_1}{\partial n_x}(x) = \sigma_2 \frac{\partial \varphi_2}{\partial n_x}(x), \quad x \in \Gamma_{CC1} \quad (44)$$

One starts by writing an integral equation (40) for the boundary of each different material. As a result, on Γ_{CC1} both the potential and its derivative are unknown. To eliminate this, a linear combination of the integral equations is performed (σ_1 times integral equation on first domain plus σ_2 times integral equation on second domain). Then continuity (44) is embedded, yielding the following integral formulation

$$\sigma_1 \frac{\theta_1(y)}{4\pi} \varphi_1(y) + \sigma_2 \frac{\theta_2(y)}{4\pi} \varphi_2(y) - \tag{45}$$

$$-\sigma_1 \int\limits_{(\Gamma_{C1})} G(x,y) \cdot \frac{\partial \varphi}{\partial n_x}(x) \cdot d\Gamma(x) - \sigma_2 \int\limits_{(\Gamma_{C2})} G(x,y) \cdot \frac{\partial \varphi}{\partial n_x}(x) \cdot d\Gamma(x)$$

$$+\sigma_1 \int\limits_{(\Gamma_{CA1})} \varphi(x) \cdot \frac{\partial G}{\partial n_x}(x,y) \cdot d\Gamma(x) + \sigma_2 \int\limits_{(\Gamma_{CA2})} \varphi(x) \cdot \frac{\partial G}{\partial n_x}(x,y) \cdot d\Gamma(x) +$$

$$+(\sigma_1 - \sigma_2) \int\limits_{(\Gamma_{CC1})} \varphi(x) \cdot \frac{\partial G}{\partial n_x}(x,y) \cdot d\Gamma(x) =$$

$$= -F_1\sigma_1 \int\limits_{(\Gamma_{C1})} \frac{\partial G}{\partial n_x}(x,y) \cdot d\Gamma(x) - F_2\sigma_2 \int\limits_{(\Gamma_{C2})} \frac{\partial G}{\partial n_x}(x,y) \cdot d\Gamma(x), \quad y \in \mathbb{R}^3.$$

Equation (45) is written for the problem consisting of two different materials. The generalization of equation (45) for an arbitrary number of materials is given in (46):

$$(1 - K(y)) \sum_{i=1}^{N_B} \frac{\theta_i(y)}{4\pi} \sigma_i \cdot \varphi_i(y) - \sum_{i=1}^{N_{CON}} \sigma_i \int\limits_{(\Gamma_{Ci}^+)} G(x,y) \frac{\partial \varphi_i}{\partial n_x}(x) d\Gamma(x) + \tag{46}$$

$$+ \sum_{i=1}^{N_{LAT}} \sigma_i \int\limits_{(\Gamma_{CAi}^+)} \varphi_i(x) \frac{\partial G}{\partial n_x}(x,y) d\Gamma(x) +$$

$$\sum_{i=1}^{N_{INT}} (\sigma_{cni} - \sigma_{ni}) \int\limits_{(\Gamma_{CCi}^+)} \varphi_i(x) \frac{\partial G}{\partial n_x}(x,y) d\Gamma(x) =$$

$$-K(y) \sum_{i=1}^{N_B} \frac{\theta_i(y)}{4\pi} \sigma_i \varphi_i(y) - \sum_{i=1}^{N_{CON}} F_i \sigma_i \int\limits_{(\Gamma_{Ci}^+)} \frac{\partial G}{\partial n_x}(x,y) d\Gamma(x); \quad y \in \mathbb{R}^3$$

where: N_B - number of conductive bodies; N_{INT} - number of interfaces between different conductive materials; N_{LAT} - number of lateral surfaces (interface air/conductor); N_{CON} - number of contacts; σ_{CNI} - conductivity of the material in the direction counter to the normal vector; σ_{NI} - conductivity of the domain in the direction of the normal vector,

$$K(y) = \begin{cases} 1, y \in \bigcup\limits_{i=1}^{N_{CON}} \Gamma_{Ci}, \\ 0. \end{cases} \tag{47}$$

If there are no interfaces, i.e. no different materials involved, the formulation (45) or its simpler counterpart (44) reduces to its special case, i.e. the classical formulation (40). To sum up there are two strong point of this approach:

firstly, one can conveniently compute problems with conductors excited with prescribed voltages (Dirichlet BC) or with prescribed currents (Neumann BC) or with a combination of those boundary conditions. Secondly, the formulation allows one to compute complex configurations involving several different conductive materials, with multiple interfaces and contacts.

Nearly-symmetric formulation

The integral formulation (40) of the classical single-material problem can also be used as starting point for the derivation of a *symmetric* formulation, Sirtori [54], Costabel [20]. Namely, the integral form (40) can be written in the operator notation as

$$\frac{\Theta(y)}{4\pi}\varphi(y) = (V\partial_n\varphi)(y) - (K\varphi)(y) \quad y \in \Gamma \tag{48}$$

where

$$V(\partial_n\varphi)(y) = \oint_{(S)} G(x,y)\frac{\partial\varphi}{\partial n_x}(x)dS(x)$$

is a single-layer boundary integral operator acting on the Neumann data of type $\partial_n\varphi$, and

$$K(\varphi)(y) = \oint_{(S)} \varphi(x)\frac{\partial G}{\partial n_x}(x,y)dS(x)$$

acting on Dirichlet type of data φ. The BEM discretization of the integral equation (45) will produce an asymmetric matrix. Following [20, 54], the Calderon projector (i.e. the Dirichlet trace over the Neumann boundary and Neumann trace over the Dirichlet boundary) applied to the integral form (47) leads us to the symmetric system

$$\begin{bmatrix} -V & K \\ K^T & D \end{bmatrix} \begin{bmatrix} \partial_n\varphi \\ \varphi \end{bmatrix} = \begin{bmatrix} f_D \\ f_N \end{bmatrix} \tag{49}$$

where D is the hypersingular operator arising from the application of the Neumann trace to the double-layer integral operator, and the right hand side f_D, f_N is determined by known values of potential at contacts (Dirichlet BC) and zero normal current on remainder of the boundary (Neumann BC). Using the symmetric formulation (48) instead of the asymmetric one (47) allows one not only to use the advantages of symmetric matrices but also to produce better conditioned systems of equations.

At this point it is worth mentioning that there is the possibility to make a large part of the matrix arising from multimaterial formulation (40) symmetric (reducing memory requirements), while at the same time keeping the algorithm simple enough for practical implementation. The symmetrization improves the algorithm not only from a memory viewpoint, but also with regard to the stability and convergence behaviour of the iterative solver. For the simple multimaterial configuration depicted in Fig. 19 it is possible to write the following integral equation based on (40)

$$\sigma(y)\varphi(y) = \int_{(\partial\Omega-\partial\text{int})} \sigma(x)G(x,y)\partial_{n_x}\varphi(x)dS_x - \tag{50}$$

$$\int_{(\partial\Omega-\partial\text{int})} \sigma(x)\varphi(x)\partial_{n_x}G(x,y)dS_x + \int_{\partial\text{int}} [\sigma_2(x)-\sigma_1(x)]\varphi(x)\partial_{n_x}G(x,y)dS_x$$

where y represents the field point somewhere in the volume of the first or second material, ∂_{int} is the surface interface between the domains and $\partial\Omega$ is the union of boundaries of both material domains. Equation (50) is obtained by adding the equations written for the same point y for two different domains from Fig. 19. Before adding, each equation has been multiplied by the corresponding electric conductivity $\sigma(y)$. Regarding equation (50), it is also important to notice that the Neumann data over the interface are eliminated using the boundary conditions (43). The procedure of symmetrization goes as is usual by applying the Calderon projector on equation (50), i.e. by applying Dirichlet and Neumann trace on (50). In addition to this standard procedure here we have an "Interface trace" that has been performed in Dirichlet's sense. According to the notation used earlier it is useful to introduce the following integral operators:

$$(V_{(\partial\Omega-\partial Int)}\partial_n\varphi)(y) = \int_{(\partial\Omega-\partial Int)} \sigma(x)G(x,y)\partial_{n_x}\varphi(x)dS(x) \tag{51}$$

$$(K_{(\partial\Omega-\partial Int)}\varphi)(y) = \int_{(\partial\Omega-\partial Int)} \sigma(x)\varphi(x)\partial_{n_x}G(x,y)dS(x) \tag{52}$$

$$(K_{(\partial Int)}\varphi)(y) = \int_{(\partial Int)} (\sigma_2(x)-\sigma_1(x))\,\varphi(x)\partial_{n_x}G(x,y)dS(x) \tag{53}$$

$$(D_{(\partial\Omega-\partial Int)}\varphi)(y) = \int_{(\partial\Omega-\partial Int)} \sigma(x)\varphi(x)\partial_{n_x}\partial_{n_y}G(x,y)dS(x) \tag{54}$$

$$(D_{(\partial Int)}\varphi)(y) = \int_{(\partial Int)} (\sigma_2(x)-\sigma_1(x))\,\varphi(x)\partial_{n_x}\partial_{n_y}G(x,y)dS(x) \tag{55}$$

After Dirichlet-, Neumann- and "Interface-" traces of equation (50) by taking into account an assumption that the boundary was everywhere smooth $(\Theta(y) = 2\pi)$ the following system of equations has been obtained for $y \in \partial\Omega_D$

$$\frac{\sigma(y)}{2}\varphi(y) = (V_{(\partial\Omega-\partial Int)}\partial_n\varphi)(y) - (K_{(\partial\Omega-\partial Int)}\varphi)(y) + (K_{(\partial Int)}\varphi)(y), \tag{56}$$

for $y \in \partial\Omega_N$

$$\frac{\sigma(y)}{2}\partial_n\varphi(y) = (K'_{(\partial\Omega-\partial Int)}\partial_n\varphi)(y) + (D_{(\partial\Omega-\partial Int)}\varphi)(y) + (D_{(\partial Int)}\varphi)(y), \quad (57)$$

and for $y \in \partial\Omega_{Int}$

$$\frac{\sigma_1(y) + \sigma_2(y)}{2}\varphi(y) = (V_{(\partial\Omega-\partial Int)}\partial_n\varphi)(y) - (K_{(\partial\Omega-\partial Int)}\varphi)(y) + (K_{(\partial Int)}\varphi)(y). \quad (58)$$

After employing a set of auxiliary functions (a standard procedure for obtaining a symmetric formulation in the case of a single material), the system (56)-(58) can be written in the following way

$$\begin{bmatrix} V_{(\partial\Omega-\partial Int)} & -K_{(\partial\Omega-\partial Int)} & K_{(\partial Int)} \\ K'_{(\partial\Omega-\partial Int)} & D_{(\partial\Omega-\partial Int)} & D_{(\partial Int)} \\ V_{(\partial\Omega-\partial Int)} & -K_{(\partial\Omega-\partial Int)} & K_{(\partial Int)} \end{bmatrix} \cdot \left\{ \begin{array}{c} \partial_n\tilde{\varphi}_{\partial\Omega_D} \\ \tilde{\varphi}_{\partial\Omega_N} \\ \tilde{\varphi}_{\partial\Omega_{Int}} \end{array} \right\} = \left\{ \begin{array}{c} f_{\partial\Omega_D} \\ f_{\partial\Omega_N} \\ f_{\partial\Omega_{Int}} \end{array} \right\} \quad (59)$$

where

- $\partial_n\tilde{\varphi}_{\partial\Omega_D}$- unknown Neumann-data over the Dirichlet-boundary,
- $\tilde{\varphi}_{\partial\Omega_N}$- unknown Dirichlet-data over the Neumann-boundary,
- $\tilde{\varphi}_{\partial\Omega_{Int}}$- unknown Dirichlet-data over the Interface-boundary.

In is easy to recognize a large symmetric part of the system matrix arising from classical Calderon projector. The discretisation of the system (59) has been performed using Galerkin algorithm in continuous sense. Constant shape functions are used for Neumann data and linear ones are used for Dirichlet data. After the system (59) has been solved, the Neumann data over the interface are still not known. Based on known Dirichlet data everywhere, and the integral equation (40), one can easily obtain those data over interfaces by solving a system of equations that is usually much smaller than the size of the system (59). However in the case of stationary current distribution, the Neumann data over interfaces (normal component of the current density) are not needed at all. For postprocessing (magnetic field computation) the potential distribution is enough. Although the system of equations (59) has been derived for two-material configuration, the structure will be the same for multimaterial conductors involving several interfaces. The only difference will be that the part belonging to "interface trace" will grow as interface area increases and the jump of conductivity in the integral operators (4, 6) will be different for different interfaces.

It is useful also to note that there is also a way to produce a fully symmetric formulation by keeping unknown Neumann data over the interface and by doing the interface trace in the Neumann sense in addition to the existing interface trace in the Dirichlet sense.

The integral operators in (50) have the same meaning as in (49), as do the right hand side terms. The formulation (50) is not fully symmetric as one can observe comparing (50) with (49). The symmetry is broken by the presence of the interface equation. However, according to our experience, in industrial applications the "interface" type of boundary is rather small compared to the size of the Dirichlet and Neumann boundaries. Hence, a large block of the matrix arising from the equation (50) is symmetric and hence memory requirements can be significantly reduced (by almost as much as half of the matrix in case of the asymmetric formulation). In addition to this, working with the symmetric formulation we usually obtain a system that is much better conditioned when compared to an asymmetric one. The aim of this symmetric approach is to show that one has also possibilities to improve the asymmetric method and to finally find a fully symmetric formulation for the multimaterial case.

Example 4: **Multi-material conductor with Voltage/Current excitation**

To illustrate the method described above we have calculated a simple problem and compared it with the solution obtained by FEM. The geometry of the massive conductor involving inhomogeneous material properties, boundary conditions and surface mesh (9232 triangular elements) can be seen in the Fig. 20. The potential calculation is done using equation (50), and then the current density using equation (42). Results for the current density distribution are shown in Fig. 21. One can observe that the current distributes according to the conductivity of the material (almost no current flows through the conductor with $\sigma_2 = 0.5e^6[Sm^{-1}]$). The maximum appears at inner corners, also as expected. In Table 3 we present a comparison with a FEM simulation. Agreement between the two methods is good.

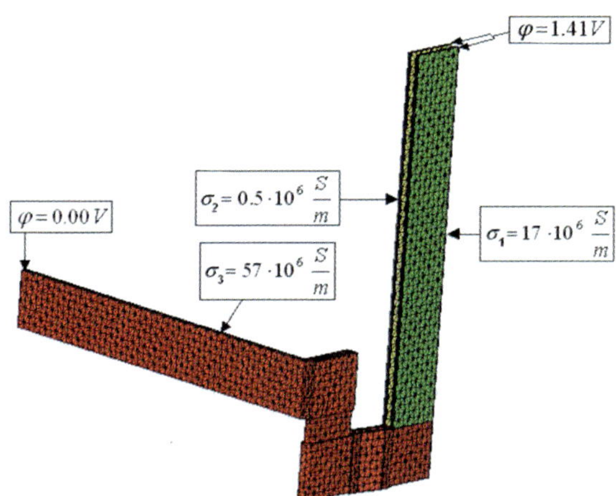

Fig. 20. Conductor composed of three different materials.

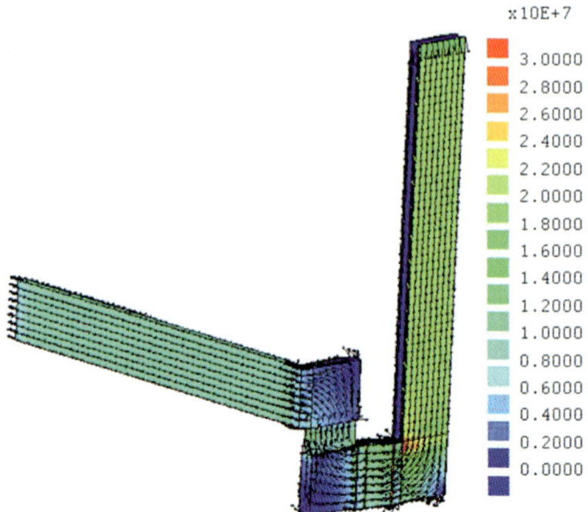

Fig. 21. Conductor composed of three different materials.

Table 3. FEM-BEM comparison.

	Ne	$I\sigma_1[A]$	$I\sigma_2[A]$	$I\sigma_3[A]$	$\sum[A]$	ε
BEM	9232	27867	891	-28862	-104	0.3
FEM	25073	27845	865	-28779	-69	0.1
	(179619)					

Forces in the Time-Varying Fields

BEM-based eddy-currents analysis

As written at the beginning, the final goal of this Section is to illustrate BEM-based procedures for the analysis of coupled Electro-Magnetic and Structural-Mechanics problems. In the Sect. 3.2 it has been shown that the calculation of forces in the current-carrying regions requires knowledge of the current and magnetic field distribution. In the previous section we have shown how the *stationary* current distribution can be calculated in current-carrying structures using BEM. In the following section we shall briefly elaborate the BEM approach for the analysis of *eddy-currents*, appearing in metallic structures under the influence of time-varying fields.

Eddy-current formulation

In most industrial applications the low frequency condition is valid ($\sigma \gg \omega \cdot \varepsilon$), i.e. the displacement current term $\partial\mathbf{D}/dt$ in the Maxwell equations can be neglected. Then, when the exciting current is sinusoidal time periodic, the eddy-current problem is described by the *reduced* Maxwell's equations

$$\nabla \times \mathbf{E} = -j\omega\mathbf{B}, \tag{60}$$

$$\nabla \times \mathbf{H} = \mathbf{J} + \mathbf{J}_0, \tag{61}$$

$$\nabla \cdot \mathbf{B} = 0, \tag{62}$$

$$\nabla \cdot \mathbf{D} = 0 \tag{63}$$

where

$$\mathbf{B} = \mu\mathbf{H}, \tag{64}$$

$$\mathbf{D} = \varepsilon\mathbf{E}, \tag{65}$$

$$\mathbf{J} = \sigma\mathbf{E}. \tag{66}$$

There are a number of the formulations in computational electromagnetics for the treatment of eddy-current problems. Each of these approaches has its own strengths and weaknesses, many of which are problem dependent. A useful overview of the available eddy-current formulations can be found in Kost [36]. In the following we shall focus on the $H - \varphi$ method, as one of the formulations having some comparative advantages for the simulation of large practical problems.

The $H - \varphi$ formulation is based on the *indirect* Ansatz, leading thus to the minimal number of 4 degrees of freedom (DoF) per node[19]. This nice feature makes this formulation suitable for the eddy-current analysis of complex, real-world problems. The $H - \varphi$ formulation need to be used with a care in cases where the problem is *multi-valued,* i.e. when the model belongs to the class *multi-connected problems,* Tozoni [61], discussed later in this session.

Furthermore, we shall give a brief description of the $H - \varphi$ approach for the treatment of eddy-current problems in current-carrying conductors, whereby particular attention needs to be paid to the modeling of *skin-effect* phenomena.

Skin-Effect Treatment

In Mayergoyz [42] it was proposed to treat the skin-effect problem by introducing a *virtual filament,* carrying the same total current as the conductor itself. Although Mayergoyz's approach offers an excellent way to treat skin-effect problems, it is linked with some inherent modeling problems for complex 3D geometries, i.e. the filament needs to be modelled and added to the original model geometry. The formulation presented here, Andjelic [3], differs the from the formulation given in [42] in so far that it is not necessary to model such additional filaments.

If besides the external source, (the coil excited by the current I_{ex} in Fig. 22), there is also an exciting current $\mathbf{J}_0^{in}$ flowing in the interior of the conductor Ω^+ and producing a magnetic field $\mathbf{H}_0^{in}$, then the exterior magnetic field

---

[19]With $H - \varphi$ formulation it is possible to work even with only 3 DoF/node, whereby the eddy-currents on the surfaces are described in a *surface* coordinate system instead of Cartesian, Yuan [62].

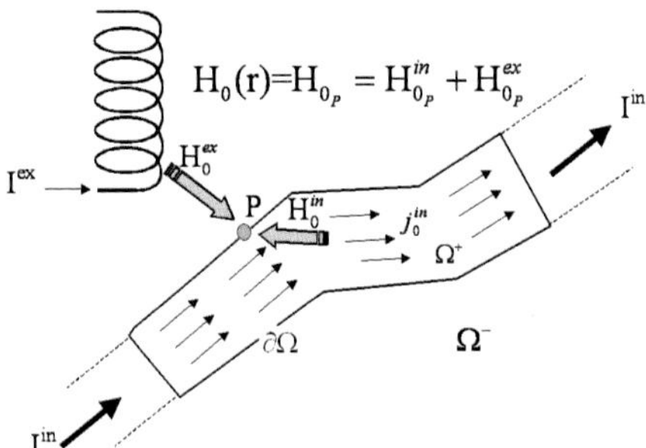

Fig. 22. Field $\boldsymbol{H}_0$ in the conductor's surface point(P) is composed of the field produced by the external sources $\mathbf{H}_0^{ex}$, and the field produced by the conductor current itself, $\mathbf{H}_0^{in}$.

$\mathbf{H}^-$ in any point P lying on the conductor surface Γ, can be written as a superposition of the scalar magnetic potential φ and the exciting field (67),

$$\mathbf{H}^- = -\nabla\varphi + \mathbf{H}_0 \tag{67}$$

whereby the exciting field $\mathbf{H}_0$ is composed of the exciting field $\mathbf{H}_0^{ex}$ produced by the external sources, and the field $\mathbf{H}_0^{in}$ produced by the internal exciting current. Both components of $\mathbf{H}_0$, $\mathbf{H}_0^{ex}$ and $\mathbf{H}_0^{in}$, can be calculated by Biot-Savart's law. For the field $\mathbf{H}_0^{in}$ it yields

$$\mathbf{H}_0^{in}(x) = \int\limits_\Omega \frac{\mathbf{J}_0^{in}(y) \times \mathbf{r}_0}{4\pi r^3}\, d\Omega, \quad \forall x \in \mathbb{R}^3 \tag{68}$$

where $\mathbf{J}_0^{in}$ is a stationary current distribution, pre-calculated by the procedures described in the previous section, $r = |\mathbf{x} - \mathbf{y}|$, and vector $\mathbf{r}_0$ is a unit vector directed from the current element towards the point of observation. The interior field $\mathbf{H}^+$ remains unchanged, and we can write

$$\nabla \times \nabla \times \mathbf{H}^+ = -i\omega\sigma\mu\mathbf{H}^+ \quad \forall x \in \Omega \tag{69}$$

$$\Delta\varphi = 0; \quad \forall x \in \Omega \tag{70}$$

with the boundary conditions

$$\mathbf{n} \times (\mathbf{H}^+ + \nabla\varphi) = \mathbf{n} \times \mathbf{H}_0 \quad \forall x \in \Gamma \tag{71}$$

$$\mathbf{n} \cdot (\mu\mathbf{H}^+ + \mu_0\nabla\varphi) = \mu_0 \cdot \mathbf{n} \cdot \mathbf{H}_0 \quad \forall x \in \Gamma. \tag{72}$$

The advantage of the $H - \varphi$ formulation is that the vector-valued equations are only used inside the conductor Ω^+. Furthermore, only a scalar potential equation has to be solved on the exterior of the conductor. The main idea is to separate the (virtual) sources on the surface for the field inside and outside of the conductor Ω^+. Accordingly, the virtual surface current density $\mathbf{j}$ distributed over Γ will be employed for the calculation of the magnetic field within the conductor, while the magnetic surface charges σ^m distributed over the same *Gamma* will be used for the calculation of the scalar potential outside of the conductor. The following integral representation is used

$$\mathbf{H}^+(x) = \nabla \times \left(\frac{1}{4\pi} \oint_\Gamma \mathbf{j}(y) \frac{e^{-(1+i)k \cdot r}}{r} d\Gamma \right), \quad x \in \Omega^+, \tag{73}$$

$$\Phi^-(x) = \frac{1}{4\pi} \oint_\Gamma \frac{\sigma^m(y)}{r} d\Gamma, \quad x \in \Omega^- \tag{74}$$

where $k = \sqrt{\omega \mu \sigma / 2}$ denotes the skin-depth and $\mu = \mu_0 \mu_r$, σ and ω are the magnetic permeability, conductivity and frequency of the time harmonic fields, respectively. For any $\mathbf{j}(y)$ and $\sigma^m(y)$ the magnetic field $\mathbf{H}^+$ given by equation (73) and the scalar potential given by equation (74) satisfy the equations (69) and (70). Using the jump relation of the boundary integral operators (73) and (74), the boundary conditions equations (71) and (72) are fulfilled if the virtual current $\mathbf{j}$ and the virtual charge σ^m are the solutions of the following system of boundary integral equations:

$$\frac{1}{2}\mathbf{j}(x) + \frac{1}{4\pi} \oint_\Gamma \mathbf{n}(x) \times \left(\mathbf{j}(y) \times \nabla \frac{e^{-(1+i)kr}}{r} \right) d\Gamma(y) \tag{75}$$

$$- \frac{1}{4\pi} \oint_\Gamma \sigma^m(y)(n(x) \times \nabla \frac{1}{r} d\Gamma(y) = -\mathbf{n}(x) \times \mathbf{H}_0(x)$$

$$\frac{1}{2}\sigma^m(x) + \frac{1}{4\pi} \oint_\Gamma \sigma^m(y) \mathbf{n}(x) \cdot \nabla(\frac{1}{r}) d\Gamma(y) \tag{76}$$

$$+ \frac{\mu}{4\pi\mu_0} \oint_\Gamma \mathbf{n}(x) \left(\mathbf{j}(y) \times \nabla \frac{e^{-(1+i)kr}}{r} \right) d\Gamma(y) = -\mathbf{n}(x) \cdot \mathbf{H}_0(x)$$

These equations are valid if the points x and y do not lie on an edge or corner. If the nodes are on edges or corners, the coefficient $1/2$ in the above equations has to be replaced by the appropriately calculated space angle.

This boundary integral equation system can be written in operator form

$$\begin{bmatrix} A_1 & B_1 \\ B_2 & A_2 \end{bmatrix} \begin{pmatrix} \mathbf{j} \\ \sigma^m \end{pmatrix} = \begin{pmatrix} -2\mathbf{n} \times \mathbf{H}_0 \\ -2\mathbf{n} \cdot \mathbf{H}_0 \end{pmatrix}. \tag{77}$$

For more details on a numerical side of this approach the reader is referred to Schmidlin [51]. Solution of the equation system (77) gives the virtual magnetic charges σ^m and virtual current density $\mathbf{j}$. Then, the magnetic field in conductive materials can be expressed as

$$\mathbf{H}^+(x) = \frac{1}{4\pi} \oint_\Gamma \nabla \times [\mathbf{j}(y)K(x,y)]\,d\Gamma(y); \quad x \in \Omega^+; y \in \Omega^+ \tag{78}$$

and

$$\mathbf{H}^-(x) = \mathbf{H}o(x) - \frac{1}{4\pi} \oint_\Gamma \sigma_m(y)\nabla_x G(x,y)d\Gamma(y); \quad x \in \Omega^-; y \in \Omega^- \tag{79}$$

in the non-conductive materials. $\mathbf{H}_0$ is the primary magnetic field produced by the exciting current $\mathbf{J}_0$ and $K = e^{-(1+i)k \cdot \mathbf{r}}/r$, $G = 1/r$.

To sum up, following the above procedure we are able to obtain expressions for the computation of skin-effect problems in current-carrying regions. The price we had to pay in order to avoid the usage of the filament, Mayergoyz [42], was a pre-calculation of the field $\mathbf{H}_0^{in}$, that is later incorporated in the right-hand side of (77). Further, we have to note here that the same procedure is applicable also for the calculation of eddy-current problems in *passive* structures, i.e. in structures that do not carry the excitation current. The only difference is that the term $\mathbf{H}_0^{in}$ will then be zero.

*Example 5: **Skin-effect in a cylindrical conductor***

To test the above described approach we used an analytical solution, Simonyi [53], for a cylindrical copper conductor, of radius 10 $[mm]$ excited with a current value of $I_{RMS}=$ 1000 $[A]$, frequency $f=$ 50 $[Hz]$, Fig. 23. It must be noted that the analytical solution assumes an infinite-length cylinder, and in this test example we took a finite length cylinder. The analytical values are compared with the calculated values in the middle of the cylinder

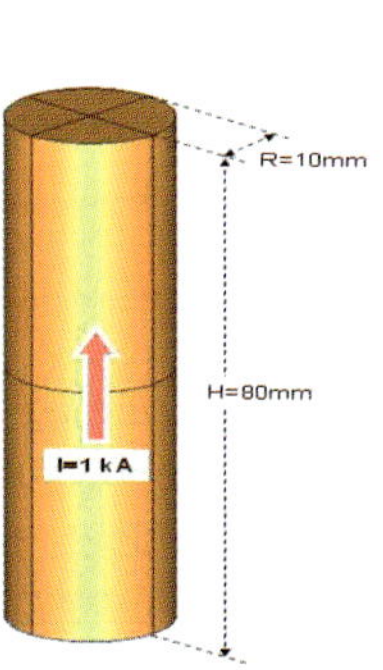

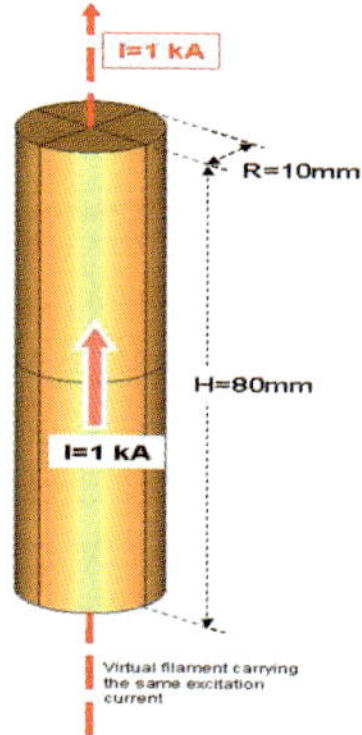

Fig. 23. Copper-made cylindrical conductor of finite length.

Fig. 24. Copper-made cylindrical conductor of finite length. Additional virtual filaments added to compensate the influence of the infinite long conductor.

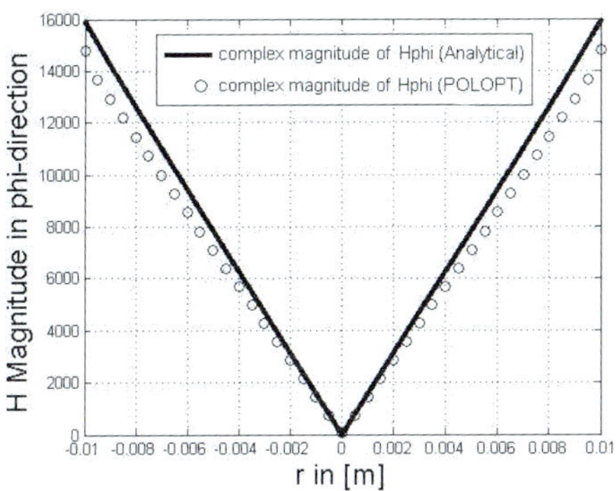

Fig. 25. Magnetic field distribution along the cylinder radius of a Cu conductor; r=0 corresponds to the center of the cylinder.

where end-effects are at a minimum. To calculate the skin-effect problem, we followed the procedure described in the previous section, whereby the skin-effect is calculated in a manner such that the H_0 in equations (75) and (76) is pre-calculated as a function of the excited stationary currents. The graphs of complex magnitude of magnetic field H for the example given in Fig. 23 are shown in Fig. 25. It can be noted that there exist a deviation of the calculated field comparing to the analytical solution. This deviation is caused because in this example we neglected the rest of the infinite conductor used for analytical calculation. Even when comparing the values in the middle of the cylinder, these missing parts cause some error of cca. 10%. The contribution of the field of the missing part of the conductor can be taken into account by virtual filaments carrying the same current, Fig. 24. For such a case the graphs of

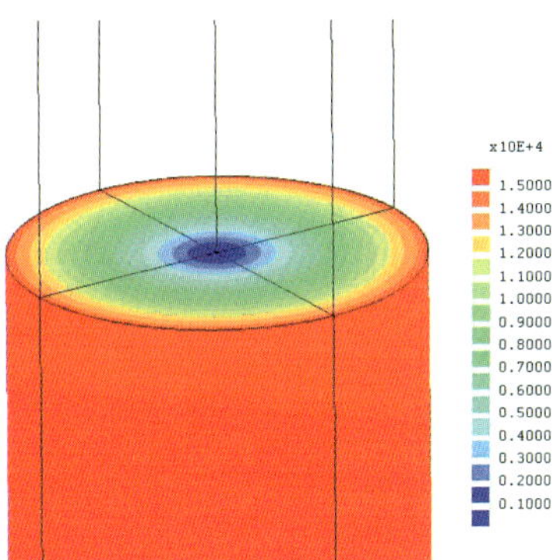

Fig. 26. Magnetic field distribution (real component) over the cross-section plane of a Cu-made cylinder.

complex magnitude of magnetic field $\boldsymbol{H}$ and eddy-current $\boldsymbol{j}$ along the radial line of the cross-section plane in the middle of the conductor are shown in Fig. 27 and Fig. 28, indicating a good agreement with the analytical solution.

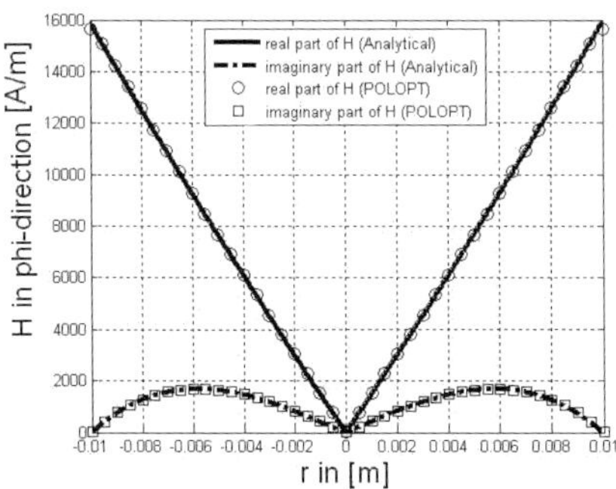

Fig. 27. Magnetic field distribution along the cylinder radius of a Cu conductor; r=0 corresponds to the center of the cylinder.

Fig. 28. Eddy current distribution along the cylinder radius of a Cu conductor

For the case that the conductor is made of a magnetic steel, having for example permeability $\mu = 200$ and conductivity $\sigma = 6.66e^6 [Sm/m^2]$, the graphs for the same physical quantities are shown in Fig. 29 and Fig. 30, respectively.

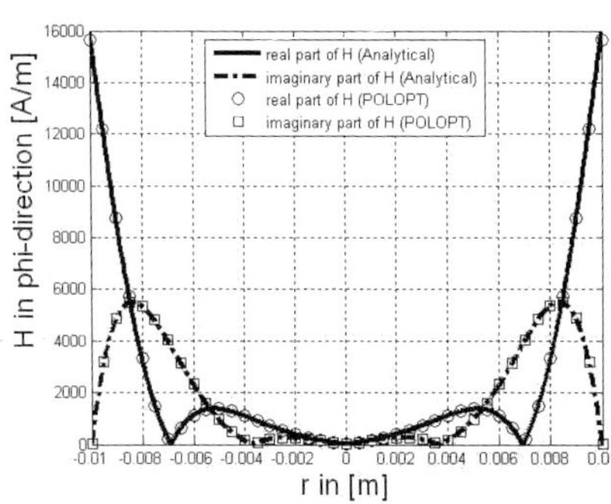

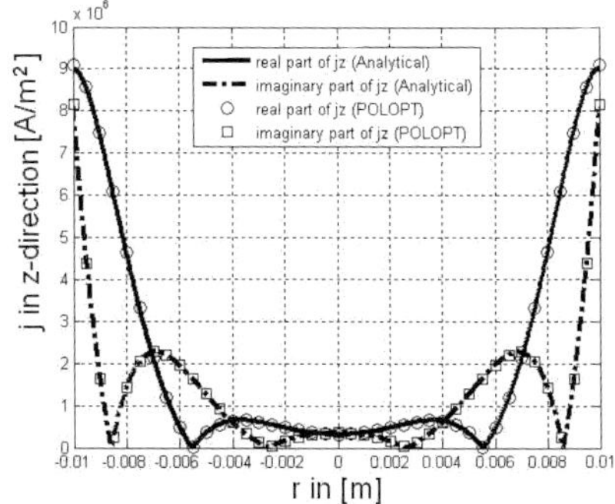

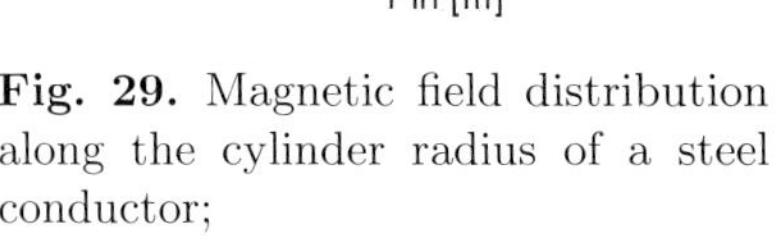

Fig. 29. Magnetic field distribution along the cylinder radius of a steel conductor;
$\mu = 200$, $\sigma = 6.66e^6 [Sm/m^2]$; r = 0 corresponds to the center of the cylinder.

Fig. 30. Eddy current distribution along the cylinder radius of a steel conductor;
$\mu = 200$, $\sigma = 6.66 [Sm/mm^2]$

It can clearly be seen that due to the smaller penetration depth for steel (≈ 1.95 [mm]), the current tends to concentrate more towards the conductor's surface then in the copper conductor case.

Example 6: **Eddy-current calculation in the "passive" structures**

In the previous example we have illustrated how eddy-currents, including skin-effect, can be calculated in current-carrying structures, i.e. structures carrying some impressed (injected) current. Very often it is required to calculate the eddy-currents that are induced in "passive" metallic structures, i.e. in the structures that are not subject to an injected current. In the previous section we have already mentioned that for both type of the problems we can use the same $H - \varphi$ formulation described before, with a small difference explained. The following example is a well known TEAM-Bechmark[20] problem No. 7, Fujiwara [22]. The problem consists of an aluminium plate with an asymmetrically positioned hole. The plate is illuminated by an electro-magnetic field produced by a time-varying current flowing through a coil, Fig. 31. From the mathematical point of view, this is a typical *multi-valued* problem, well elaborated in the literature Tozoni [61], and used to test the capabilities of codes treating these classes of problems. Fig. 32 shows the distribution of the calculated magnitude of the y-components of complex eddy-current j, used as a reference in Fujiwara [22]. Fig. 33 shows the vector flow of the real part of the calculated eddy-currents. The graph shown in Fig. 34 gives a comparison between the measured and calculated values of the magnitude of j. It has to be

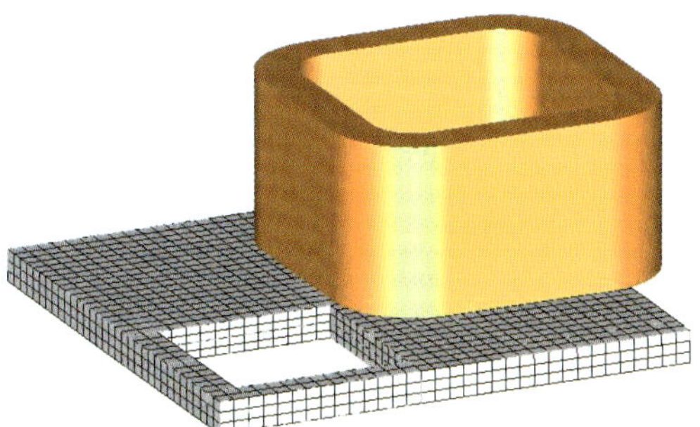

Fig. 31. Asymmetrical conductor with a hole.

---

[20]TEAM-Benchmark problems are the series of the electromagnetic problems defined in the electromagnetic community around COMPUMAG (*Int. Conf. on Computational Electromagnetics*), with the goal to compare different codes and procedures on well-defined problems validated by experimental results.

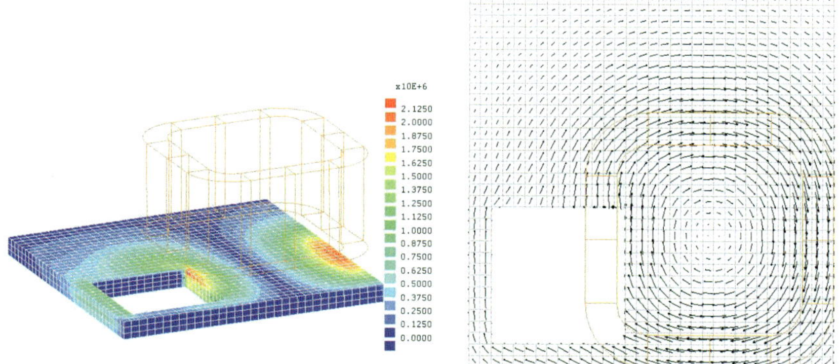

Fig. 32. Calculated magnitude of the y-components of j.

Fig. 33. Vector flow of the real part of current j.

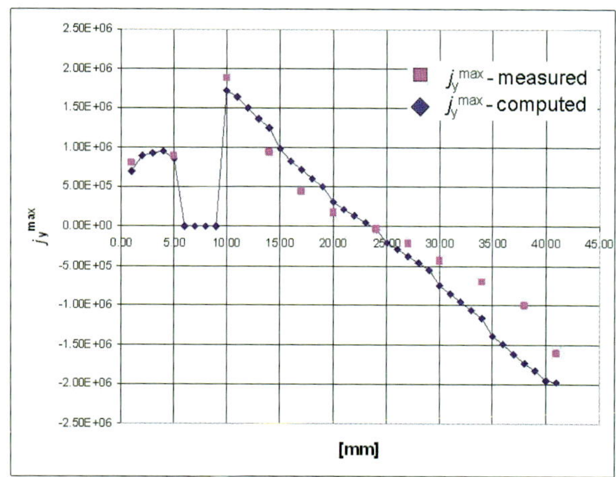

Fig. 34. Calculated vs. measured values of the magnitude of $\mathbf{j}_y$.

noted that following [22], if the real part of j is negative, than the magnitude is also shown negative in the graph.

Fast BEM for Eddy-Current Computation

Fast multipole for eddy-current computation

As mentioned at the beginning of this Chapter, the Fast Multipole algorithm is a well known approach for BEM matrix compression, Greengard [25], and can be applied also for eddy-current analysis. This algorithm reduces the memory required for matrix storage, as well as the complexity of matrix vector multiplication which is of crucial importance for iterative solvers. It can be

also successfully applied for acceleration of the BEM integration in the case of
the right-hand side (RHS) computation and postprocessing. This is depicted
on the left-hand side of Fig. 35. There are basically two main drawbacks of
fast multipole technique:

- necessity to have a suitable kernel expansion. That means that different
  types of kernels require different types of expansions, as mentioned already
  in the section "Fast BEM in Electrostatic problems, page 291,
- not easy to perform the matrix pre-conditioning.

The ACA technique described in the following overcomes both of these difficulties.

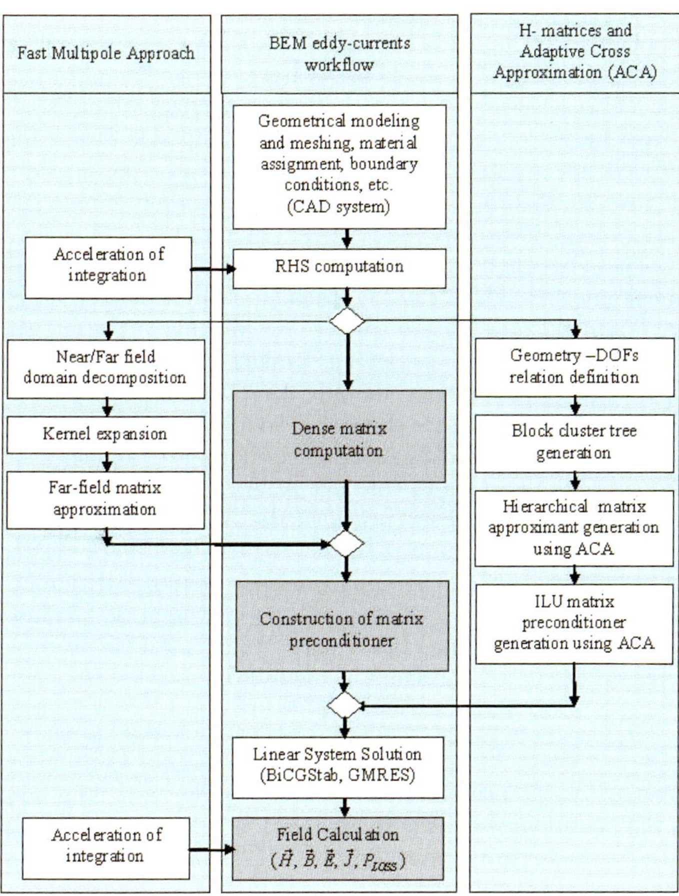

Fig. 35. Fast BEM for eddy-current computation.

ACA for eddy-current computation

ACA - Adaptive Cross-Approximation is another more recent alternative technique for matrix compression. The workflow of ACA for eddy-current analysis is depicted on the right-hand side of Fig. 35. This method is based on the Hierarchical Matrices (H-matrices) Arithmetic Grasedyck [24] and Adaptive Cross Approximation (ACA) Technique Bebendorf [11], Bebendorf [12]. The name Adaptive Cross Approximation comes from the basic algorithm for a low-rank approximation of matrix block. Namely, it uses elements of the original matrix in a sequence of pivot-columns and pivot-rows making every time a sort of cross over the block. Similar to the previously discussed technique, the ACA technique is also based on the low-rank approximation of far-field blocks of system matrix. By contrast with the Fast Multipoles method, the ACA technique computes low-rank approximants from the matrix entries themselves without explicitly dealing with the kernel, Bebendorf [11], i.e. the kernel expansion is not needed. Therefore, the matrix entries computation routines can be used from existing codes without any change, which is of course a major advantage. As one can see on the right hand side of the Fig. 35, the ACA approach consists of several steps that have to be performed. It is very significant that the required inputs for ACA are relatively simple and are usually already present in every code for field computations. Namely, the ACA routines need the geometry description of the degrees of freedom DOFs, i.e. the system unknowns, which can be obtained easily from the mesh with respect to the integral formulation that we are dealing with (one single node can be related to more than one DOF). Besides the geometry description of DOFs, ACA requests from our code only computation of certain matrix entries, and these functions are the same as for dense matrix computation. From a practical viewpoint this makes ACA much more programmer friendly than fast multipoles and allows us to consider ACA almost as a black box. If the geometry description of DOFs is provided the ACA module starts with matrix partitioning. At the beginning the cluster tree and block cluster tree are generated [1-3] which is of crucial importance for matrix partitioning and generation of the hierarchical low-rank matrix approximant. It has been already proved and published that the matrix partitioning algorithm complexity and memory requests of partitioning are of the following order, Bebendorf [11]:

$$Complexity_{Matrix-partitioning} = O(\eta^{-(d-1)} N \log N) \tag{80}$$

The number of blocks generated is of order N. The next step according to Fig. 35 is the hierarchical matrix approximation generation. According to the available theoretical results, Bebendorf [11], [12], this can be done using ACA with the following numerical effort:

$$Complexity_{Matrix-approximation} = O(\eta^{-(d-1)} k^2 N \log N) \tag{81}$$

$$Memory_{h-Matrix-approximant} = O(\eta^{-(d-1)} k N \log N) \tag{82}$$

$$Complexity_{Matrix-vector-muptiplication} = O(\eta^{-(d-1)} kN \log N) \qquad (83)$$

where k is the maximum rank of approximant, N is the matrix dimension, d is the number of dimensions of problem (for example 3 for 3D problem) and η is a parameter of the following admissibility condition:

$$\min\{diam X_{t_1}, diam X_{t_2}\} \leq \eta \cdot dist(X_{t_1}, X_{t_2}) \qquad (84)$$

where t_1 and t_2 are the index subsets (clusters) that determine certain matrix block and X_{t_1} , X_{t_2} are the supports of the clusters t_1 and t_2 , i.e. as follows:

$$X_{t_1} = \bigcup_{i \in t_1} X_i, X_{t_2} = \bigcup_{i \in t_2} X_i \qquad (85)$$

where X_i=**supp**(φ) and φ_i - an ansatz function. After all these theoretical results it is possible to see that almost linear complexity is achieved. The final step of the ACA approach is the construction of a cheap (in the sense of time and memory) and efficient matrix preconditioner. Due to the fact that the spectral equivalence of the matrix approximant and original matrix is preserved even for relatively rough approximations Bebendorf [14], ACA can generate an ILU preconditioner with L and U factors much smaller (and less precise, of course) than the memory needed for the matrix approximant Bebendorf [14].

Table 4. Comparison of the theoretical abilities of ACA and Fast Multipols.

Property	ACA	Fast Multipoles
Generality with respect to the kernel	+	-
Compression efficiency	+	+
Acceleration of the BEM integration	-	+
Efficient and cheap preconditioning	+	-

Furthermore our LU factor can be used as a direct solver by setting the same accuracy for the LU factors as for the matrix approximation Bebendorf [14]. This appears to be a significant advantage of ACA versus Fast Multipoles and is very important for real-life problems that typically produce ill-conditioned matrices due to bad mesh quality or high contrast between different materials. On the other hand, fast multipole has one significant advantage versus ACA. Namely, that fast multipoles can significantly increase the speed of the integration routines for both computing the RHS and post-processing which is, typically for BEM, very time-consuming. Up to now, in the existing ACA solutions Bebendorf [15] this possibility does not exist.

As a summary in the following table the abilities of the fast multipole algorithm and the ACA are compared: In order to check these theoretical statements in the case of BEM based eddy-current analysis, the example of conductive sphere in homogenous magnetic field has been chosen. An analytical solution exists for this example and this is used as a reference in order to verify the same level of solution accuracy for fast multipoles and ACA. The test has been performed for several different meshes, i.e. for different number of DOFs. The mesh quality is kept the same in order to prevent different conditioning of the linear system due to different mesh quality. For each case the memory requirements for matrix storage are compared. The condition number is not estimated but computed directly using SVD because the matrices are relatively small. In the following Fig. 36 numerical results of this eddy-current example are presented. Fig. 36 confirms the theoretical statements that ACA typically has worse compression than Fast Multipoles, but still much better than the dense matrix. At this point it is worth mentioning that successful ACA application on the BEM-based eddy-current analysis can not be done working directly with the system matrix arising from the integral formulation (74, 75). Usually, it is not good idea to construct one single approximant for more than one integral operator. Hence a separate approximant is generated for each matrix block of the block matrix structure (76). In order to illustrate the efficiency of the ACA approximation typical H-matrix block approximants of the eddy-current matrix are depicted in Fig. 2 along with its rank distribution. The matrix blocks from equation (76) are separated with black solid lines in order to distinguish the blocks produced by different integral operators. The block (0,0) and the block (1,0) are related to the Helmholtz like kernels with complex exponent $(e^{-\beta r}/r^2)$. Consequently, as one can see from Fig. 36 the compression is excellent for both blocks. The block (0,1) and the block (1,1) are related to the double-layer like kernels and the matrix compression is much worse for this particular example of a sphere.

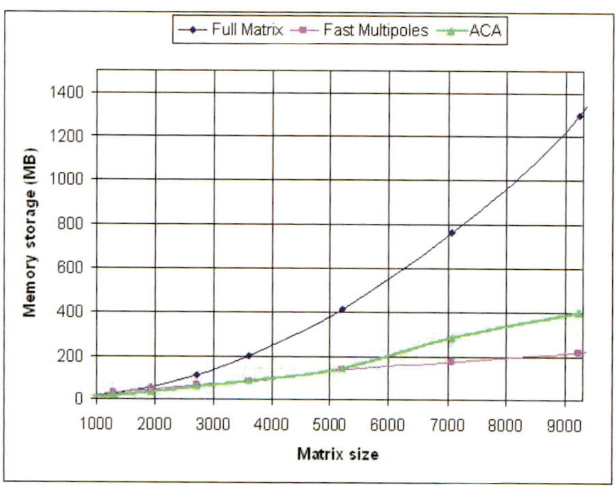

Fig. 36. Fast Multipole versus ACA.

In order to illustrate the ability of the ACA concerning matrix preconditioning one has to look at an example with an ill-conditioned matrix, i.e. with problematic convergence. For an eddy-current analysis this occurs when we encounter bad mesh quality or with a material with a high value of magnetic permeability, Schmidlin [51]. Fig. 37 and Fig. 38 shows the spatial distribution of the calculated matrix coefficients for copper with $\mu = 1$ and for iron with $\mu = 200$, respectively. The maximal coefficients are presented in red color. It can be seen that in the first case we have a very well conditioned matrix, by contrast with the second case where the matrix is poorly conditioned due to the presence of highly permeable material, or in the formulation sense, due to the term μ/μ_0 in the equation (76).

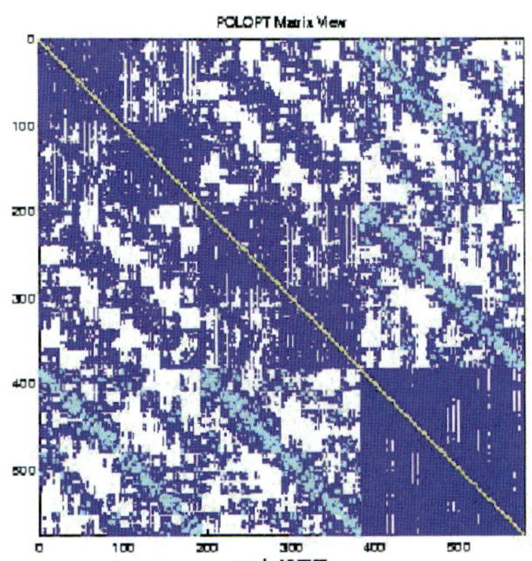

Fig. 37. Size of the matrix coefficients for copper sphere (red color represents the maximal size coefficients).

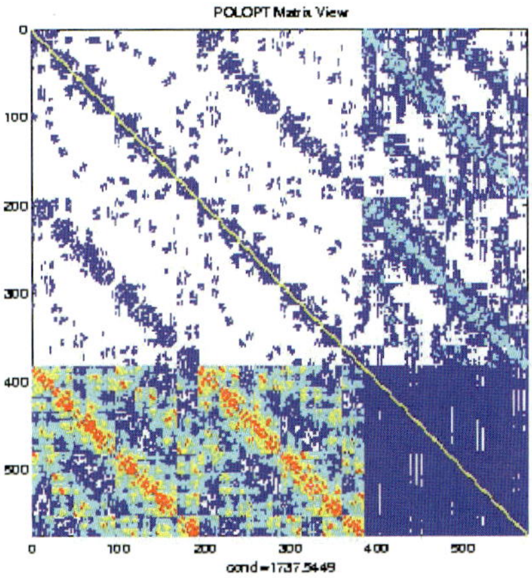

Fig. 38. Size of the matrix coefficients for iron sphere (red color represents the maximal size coefficients).

If the fast multipole algorithm is applied in that case a special block preconditioner based on the Schur complement has to be constructed, Schmidlin [51], which appears to be very cumbersome. The eddy-current example with conductive sphere is appropriate for this analysis as well. The magnetic permeability of the material has been increased and the condition number of the matrix is calculated. The efficiency of the fast multipole algorithm along with the Schur complement based preconditioner has been compared against the efficiency of the ACA algorithm with ILU hierarchical preconditioner.

3.3 ACA Approach for Matrix Preconditioning in BEM Based Eddy-Currents Analysis

In the previous section concerning the ACA compression of BEM matrices it has been mentioned that the ACA scheme can be very successfully used for

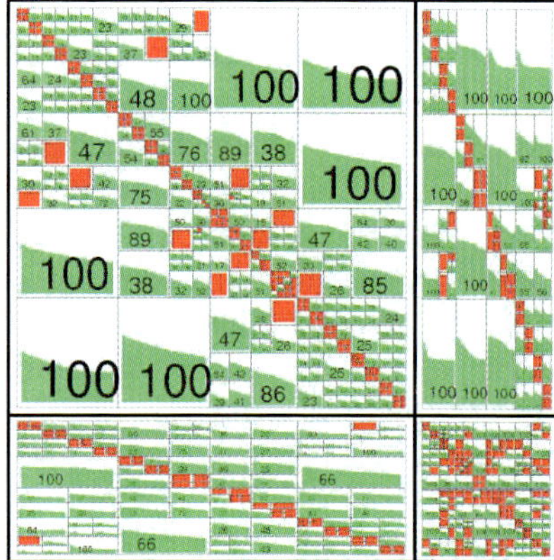

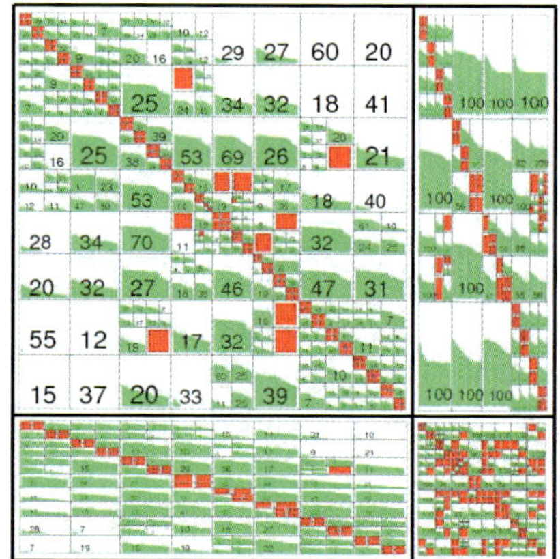

Fig. 39. The rank distribution for the copper sphere.

Fig. 40. The rank distribution over the blocks of a typical BEM matrix approximated by H-matrices and ACA (an eddy-current example from the Table 1 with 9224 DOFs).

the construction of cheap and efficient pre-conditioners. In order to demonstrate this theoretical statement, Bebendorf [14], let us apply this method to the ill-conditioned complex matrices arising from BEM-based eddy current analysis. As has been already pointed out, the eddy-current matrix is a block structure that can be very well compressed block-wise using either fast multipoles or ACA (the compression quality of ACA is depicted in the Fig. 39 and Fig. 40). According to the reference Bebendorf [14] our block structure of eddy-current matrix is appropriate for ACA preconditioning based on the Incomplete LU decomposition (ILU). Namely, for construction of an approximative LU decomposition the ACA algorithm is almost ideal due to the fact that even low precision LU factors will preserve spectral equivalence with the original dense matrix, [14]. Having this spectral equivalence of the rough ILU approximant, the quality of preconditioner is guaranteed. At the same time the computational effort for the construction of the preconditioner can be neglected due to the rough approximation, [14]. That is practically all that is needed for practical BEM-based eddy-current analysis.

For the same reasons as in the previous section, the conductive sphere in homogenous time-harmonic magnetic field will be used. Ill-conditioning of the matrix will be produced by the large value of relative magnetic permeability. The preconditioning quality of the ACA algorithm will be compared against the Schur complement based block matrix preconditioner used together with fast multipoles.

According to the existing literature Bebendorf [14], Golub [26], the block structure of eddy-current matrix can be efficiently used for ILU preconditioner construction in the following way

$$\begin{bmatrix} A & B \\ C & D \end{bmatrix} \approx P = LU = \begin{bmatrix} L_{11} & 0 \\ L_{12} & L_{22} \end{bmatrix} \begin{bmatrix} U_{11} & U_{12} \\ 0 & U_{22} \end{bmatrix} \qquad (86)$$

where A, B, C, D are original blocks of the matrix, P is a rough approximation of the original matrix that will be computed in the factorized LU form.

From equation (86), the following can be verified:

$$L_{11}U_{11} = \tilde{A} \qquad (87)$$

$$L_{11}U_{12} = \tilde{B} \qquad (88)$$

$$L_{21}U_{11} = \tilde{C} \qquad (89)$$

$$L_{22}U_{22} = \tilde{D} - L_{21}U_{12} \qquad (90)$$

Obviously, from the first equation one can compute factors L11 and U11 as a LU decomposition of $\tilde{A}$ (the rough approximation of block A). The next step would be to compute U_{12} from (87) using the already computed L11, which is inexpensive due to the fact that this is a triangular system with multiple right-hand sides (88). After that, using (89), it is possible to compute L_{21} by solving the problem of similar structure as in (88). Having those factors already computed, the last and final step is relatively simple - using (90) the factors L_{22} and U_{22} are computed. This method is used for all the calculations presented here and it appears to be very efficient.

The ILU factorisation (86) of the matrix from the Table 4, (the example with 2712 DOFs) of the previous chapter, is depicted in the Fig. 41 as a H-matrix with its rank distribution.

Approximation of the blocks used for ILU factorisation can be very rough, and so accuracy is chosen to be 0.1, as opposed to the accuracy of the matrix

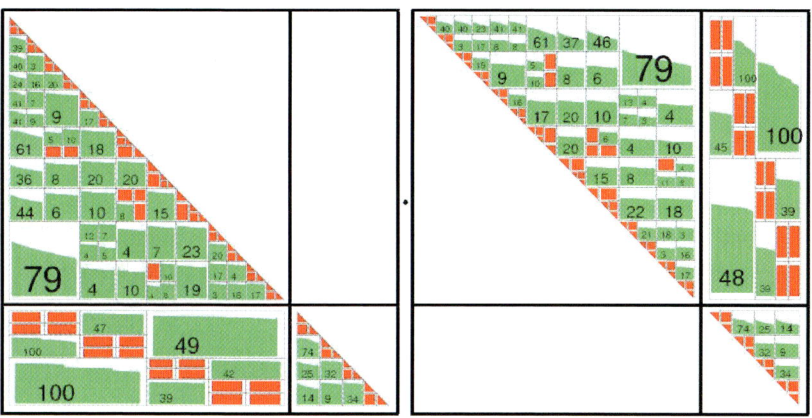

Fig. 41. ILU factorisation of the Eddy-current matrix (an eddy-current example with 2712 DOFs, Fig. 36).

approximation which is 0.0001. In the following Table 3, we see results of the eddy-current analysis (conductive sphere) with various different matrix conditioning. The condition number has been increased using a high value for the magnetic permeability. The condition number has been calculated (not estimated) using the SVD algorithm Golub [26]. Since the SVD is very expensive in terms of memory and CPU time, the matrices under consideration are kept relatively small. Some of the examples shown in the graph in Fig. 36 in the previous section have been taken, the relative magnetic permeability and the electric conductivity have been changed from $\mu_r{=}1$ and $\sigma = 5.7e^7[S/m]$ to $\mu_r{=}200$ and $\sigma = 6.66e^6[S/m]$, respectively (the ferromagnetic iron). At the beginning, all examples are solved with GMRES and diagonal preconditioning Golub [26]. Then the fast multipoles and the Schur complement-based block-preconditioning are applied and finally the problems have been solved by using the H-matrices/ACA compression and ACA/ILU preconditioning. The results are given in Table 5.

Table 5. ACA vs. MBIT – Data comparison.

Problem Size	Condition $_\kappa$ Number $\kappa_2 = \dfrac{\sigma_1}{\sigma_N}$	Dense Matrix		Fast Multipoles			Adaptive Cross Approx. (ACA)		
DOFs		Matrix (MB)	GMRES + DP (iter)	Matrix (MB)	Schur Comple. (MB)	GMRES + SC (iter)	Matrix (MB)	ILU (MB)	GMRES + ILU (iter.)
2712	977	112	128	56	49	4	46	48	8
3608	1227	199	134	71	59	6	67	53	10
5192	1603	411	148	104	130	6	113	82	10
7064		761		129	176	7			

It can be seen that for equal mesh quality and geometry the spectral condition number increases with matrix size. This is a clear sign that this method is producing ill-conditioned matrices, Steinbach [57]. Hence the matrix preconditioning is of paramount importance. In the 4-th column of the Table 3 one can see that the number of iterations of GMRES with diagonal preconditioning is above 100, even though our geometry is a simple sphere, the mesh is almost ideal, and only one material is being used. This confirms the previous statement and shows that the method will have very problematic convergence in the case of bad mesh quality or high contrast between different materials. In the case of fast multipoles for matrix compression along with a Schur-complement based preconditioner Schmidlin [51], one can see that the compression of the matrix is very good. On the other hand, the Schur-complement preconditioner size is not so promising and it is typically larger than near-field matrix itself. This appears to be a bottleneck in the case of real-life problems. The second option and more recent one, is H-matrices/ACA compression with ILU preconditioning. The results of this approach are presented on the right-hand side of the Table 5. It is obvious that rough approximation of the matrix for ILU preconditioner construction preserves spectral equivalence with the dense matrix and the number of iterations of the iterative solver GMRES is dramatically reduced. It shows and confirms the preconditioner efficiency. Furthermore, the preconditioner size is typically well under the size of the matrix approximant (rough approximation) which means that the preconditioner itself is cheap in the sense of memory requirements, as well as in the sense of computational effort and time. The matrix approximation accuracy used here was and the accuracy of the ILU preconditioner generation was, which explains the smaller size of preconditioner. In order to make this clearer the data from Table 3 are graphically presented in Fig. 42. As one can see the size of the ILU preconditioner based on ACA is significantly smaller than the size of the Schur complement based block preconditioner for the fast multipole method. It seems that the size difference increases as matrix size becomes larger. This is of great importance for large real-life problems. The compression quality is about the same for both methods.

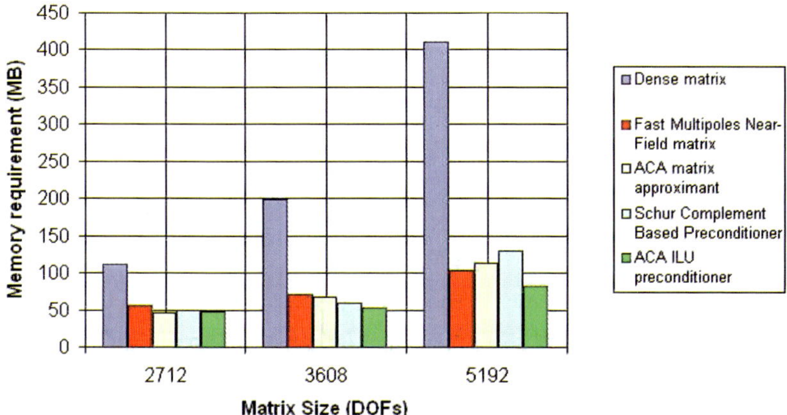

Fig. 42. Memory requirements for various matrix compression and preconditioning methods.

3.4 BEM for Structural Mechanics

As mentioned before, our goal is to perform coupled Electro-Magnetic / Structural Mechanics run using BEM. So far we have elaborated how to treat the electromagnetic part with BEM in order to compute the electromagnetic forces that should then be passed further to the mechanics module. Let us now give a brief introduction about the BEM formulation for the linear elasticity problems.

Basic Equations

The equilibrium behavior of a linear elastic, generally anisotropic, solid can be conveniently expressed using tensorial notation as shown below in equation (91), where indices i and j vary over 1,2,3. The usual tensor summation convention is assumed

$$\frac{\partial s_{ij}}{\partial x_j} = f_i. \tag{91}$$

Here, s_{ij} is the stress tensor (a symmetric quantity), while f_i is a body force per unit volume acting on the elastic medium. For an elastic medium, the stress tensor can be related to the strain tensor for a general anisotropic material using the Hooke stiffness tensor, c_{ijkl}. This gives the constitutive relation shown in (92),

$$s_{ij} = c_{ijkl} e_{kl}. \tag{92}$$

The definition of the strain tensor links it directly to the displacment via the partial derivatives presented in (93),

$$e_{kl} = \frac{1}{2} \left(\frac{\partial u_l}{\partial x_k} + \frac{\partial u_k}{\partial x_l} \right). \tag{93}$$

Clearly this quantity is again symmetric and $e_{ij} = e_{ji}$.

The stiffness tensor c_{ijkl} has $3^4 = 81$ entries, but even for the most general material many of these terms are not independent. Due to the symmetry of s_{ij} and e_{kl}, we have $c_{ijkl} = c_{jikl} = c_{ijlk} = c_{jilk}$ By energy considerations Nayeh [46], Auld [7], it can be shown that there is further symmetry in the stiffness tensor, leading to the result that $c_{ijkl} = c_{klij}$ These simplifications mean that rather than having $3 \times 3 \times 3 \times 3 = 81$ independent values, c_{ijkl} has at most 21 independent coefficients, even for the most generally anisotropic materials.

In many engineering design problems, we can at least approximate the material behaviour as being isotropic. That is to say, we can assume that material properties, and in particular the stiffness, are the same in all directions. Equivalently, we could say that the stiffness tensor entries c_{ijkl} are the same irrespective of the coordinate system used. The consequence of this condition is that only two independent stiffness constants remain. There are various possibilities for the expression of these two constants. Young's modulus (E) and Poisson's Ratio (ν) are a common choice, used frequently in engineering calculations. Another common choice, convenient in formulating expressions describing material behavior, are the Lamé constants: λ and μ. For a given material, all such pairs of constants are in any case equivalent. In this way, the pair of Lamé constants can be expressed in terms of Young's modulus and Poisson's ratio as shown in equation (94),

$$l = \frac{\nu E}{(1 + \nu)(1 - 2\nu)}, \qquad m = \frac{E}{2(1 + \nu)}. \tag{94}$$

If we start from the equilibrium and constitutive relations above, specialised for an isotropic material and utilizing the Lamé constants, we get the following equilibrium relation:

$$-m\Delta\mathbf{u}(x) - (l + m)\,\nabla\nabla \cdot \mathbf{u}(x) = \mathbf{f}(x) \tag{95}$$

Representation formulae

In order to construct a Boundary Element Method approximation of our problem, we must formulate a representation formula Steinbach [56]. This formula, for displacement component $u_k(\tilde{x})$ where $\tilde{x}$ is a point within the domain Ω, bounded by the boundary Γ, gives

$$u_k(\tilde{x}) = \int_\Gamma U^*_{kl}(\tilde{x}, y)t_l(y)d\Gamma(y) - \int_\Gamma T^*_{kl}(\tilde{x}, y)u_l(y)d\Gamma(y) + \int_\Omega U^*_{kl}(\tilde{x}, y)f_l(y)d\Omega(y). \tag{96}$$

In this expression, $U^*(x, y)$ is the displacement associated with the fundamental solution for an infinite body (the Kelvin solution, shown in equation (97)), while $T^*(x, y)$ is the associated stress,

$$U^*_{kl} = \frac{1}{8\pi}\frac{l + m}{m\,(l + 2m)}\left[\frac{l + 3m}{l + m}\frac{\delta_{kl}}{|x - y|} + \frac{(y_k - x_k)\,(y_l - x_l)}{|x - y|^3}\right]. \tag{97}$$

Following a limiting process, where $\tilde{x}$ is brought to the boundary Γ, $\tilde{x} \in \Omega \to x \in \Gamma$, introduces a jump-term (which is $1/2$ due to the fact that we are working with a Galerkin formulation) and so the representation formula becomes

$$u_k(x) = \int_\Gamma U^*_{kl}(x, y)t_l(y)d\Gamma(y) + \frac{1}{2}u_k(x) - \int_\Gamma T^*_{kl}(x, y)u_l(y)d\Gamma(y) \tag{98}$$

$$+ \int_\Omega U^*_{kl}(x, y)f_l(y)d\Omega(y).$$

This representation formula is used on the regions of the boundary where displacement is prescribed, that is to say, where we have Dirichlet boundary conditions,

$$x \in \Gamma_D \Rightarrow u(x) = g_D. \tag{99}$$

Shortening the notation somewhat, we rewrite the representation on Γ_D as

$$(Vt)\,(x) = \frac{1}{2}g_D(x) + (Ku)\,(x) - (N_0 f)\,(x), \quad x \in \Gamma_D. \tag{100}$$

In this expression, V is the single-layer potential, K is the double-layer potential, N_0 is the Newton Potential (applied to body forces f), while g_D denotes the given Dirichlet data.

The complementary region of the boundary is where we have prescribed Neumann boundary conditions (i.e. prescribed traction),

$$x \in \Gamma_N \Rightarrow t(x) = g_N. \tag{101}$$

On this region, we use a different representation formula, formed by taking the normal derivative of the already presented representation. The results of this operation are shown in (102),

$$(Du)(x) = \frac{1}{2}g_N(x) + (K't)(x) - (N_1 f)(x), \quad x \in \Gamma_N. \tag{102}$$

In this expression, two new terms have been introduced: D denotes the hypersingular operator, while N_1 is the first Newton potential. These are given by

$$(Du)_k(x) = -T_x \int_{\Gamma} T_{kl}^*(x,y)u_l(y)d\Gamma(y), \tag{103}$$

$$(N_1 f)_k(x) = T_x \int_{\Omega} U_{kl}^*(x,y)f_l(y)d\Omega(y). \tag{104}$$

Here the T_x operator represents the normal derivative operation just mentioned.

Now, with appropriate representation formulae established for $\Gamma = \Gamma_D \cup \Gamma_N$, the system under consideration that will be solved is comprised of (100) and (102) (see Steinbach [57], page 130).

Discretisation

The discretisation applied is based on a surface triangulation using 3-node planar elements. Tractions are approximated with constant basis functions, while displacements are approximated using linear basis-functions.

Galerkin integration is employed throughout. This results in double integration over the solution domain's surface, but with the benefit of improved smoothness in the results by comparison with more straight-forward collocation methods.

The evaluation of the system matrix terms is done using the OSTBEM [58] integration routines. This means that the elasticity single layer potential, V, is obtained from the Laplace single layer potential $(n \times n)$, and 7 further $n \times n$ sub-matrices. The elasticity double layer potential, K, is obtained from V, the Laplacian Double Layer, and the Surface Curls. The Elasticity Hypersingular is evaluated using V and the Surface Curls.

The Newton Potential, N_0, which appears when body forces are encountered (e.g. in coupled electromagnetic-mechanical systems) is evaluated by an integration over the volume of prescribed Body Forces multiplied by the fundamental solution displacments.

The N_1 Newton Potential is not calculated directly from its definition as presented earlier. Before computing N_1, some manipulations are helpful. From the boundary integral equations we can rearrange the two representations presented earlier to get expressions for $t(x)$

$$t(x) = V^{-1}\left(\frac{1}{2}I + K\right)u(x) - V^{-1}\left(N_0 f\right)(x),\tag{105}$$

$$t(x) = (Du)(x) + \left(\frac{1}{2}I + K'\right)t(x) + (N_1 f)(x).\tag{106}$$

Substituting from the first expression (105) into the second (106) for $t(x)$ gives the result

$$t(x) = \left(D + \left(\frac{1}{2}I + K'\right)V^{-1}\left(\frac{1}{2}I + K\right)\right)u(x)$$
$$- \left(\frac{1}{2}I + K'\right)V^{-1}\left(N_0 f\right)(x) + (N_1 f)(x).\tag{107}$$

Comparing (107) with (105), and noting that these relations are true in general, it follows from considering $u(x) = 0$ that:

$$-\left(\tfrac{1}{2}I + K'\right)V^{-1}\left(N_0 f\right)(x) + (N_1 f)(x) = -V^{-1}\left(N_0 f\right)(x)\tag{108}$$
$$\Longleftrightarrow (N_1 f)(x) = \left(K' - \tfrac{1}{2}I\right)V^{-1}\left(N_0 f\right)(x)\tag{109}$$

The quantity of present interest, coming from the equation shown in (102), is $N_1 f$. From (109), it is seen that this can be calculated from the elasticity double-layer potential, an identity operator (mass matrix) and the N_0 Newton potential. The remaining quantity is the inverse of the elasticity single layer potential. Rather than directly calculating this quantity, it is better to solve the linear system

$$Vw = (N_0 f).\tag{110}$$

Then $w \equiv V^{-1}(N_0 f)$ can be substituted immediately into (109) to evaluate the N_1 Newton potential.

The resulting system of equations can be solved in various ways. For the mechanical examples presented in this chapter, a preconditioned Schur-complement, conjugate gradient scheme is used. The preconditioners are based on the underlying integral operators [56].

Adaptive Cross-Approximation

Adaptive Cross Approximateion (ACA) [11] has been used to accelerate the operations required to solve a given mechanical system, and to reduce the storage requirements. This has been done at the level of the component matrices described in Section 3.4. Based on these approximations, the solution scheme and all operations involving the components of the system matrix can be easily re-implemented using ACA-based matrix-vector multiplication. In fact, by careful programming, it is possible to develop code that allows the solution of the system using either ACA or Dense storage by a simple redefinition of the fundamental vector-matrix multiplication routines. This compression/acceleration process is vital if one hopes to implement a practical solver as otherwise computational cost and storage requirements grow according to the problem size squared. A comparison of matrix storage requirements for uncompressed (Dense) and compressed (ACA) systems is shown in the next section.

Example 7: **Prism**

It is useful first to consider a test-case to evaluate the performance and accuracy of the described solver when dealing with a simple mechanical problem. For this purpose, consider a prism, with square cross-section of 1.0 [m] × 1.0 [m], *and* 2.0 [*m*] long.

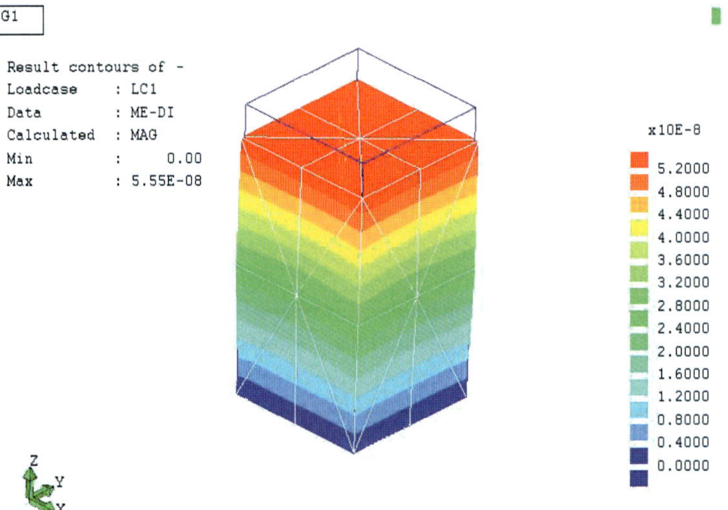

Fig. 43. Displacement Profile for Aluminium Prism in Compression (2 [kPa]).

This prism is loaded at its end by a compressive pressure of 2 [kPa], while the other end is fully constrained. The prism has the material properties of aluminium (Young's Modulus E = 70[GPa], Poisson's Ratio $\nu = 0.333$). From a quick analytical calculation, we would expect to have a strain of a little less than

$$e_{33} = \frac{s_{33}}{E} = \frac{2e^3}{70.7e^9} = 2.857e^{-8}. \tag{111}$$

This would give a distortion of $l \times e_{33} = (2.0)(2.857e^{-8}) = 5.7e^{-8}$[m]. Even with a very coarse mesh (48 elements), as shown in Fig. 43, we see that this result is very nearly exactly obtained. Notable also is that the displacement is very smooth, as would be hoped, and that the "bellying" of the prism (due to the constraint at its base) is captured.

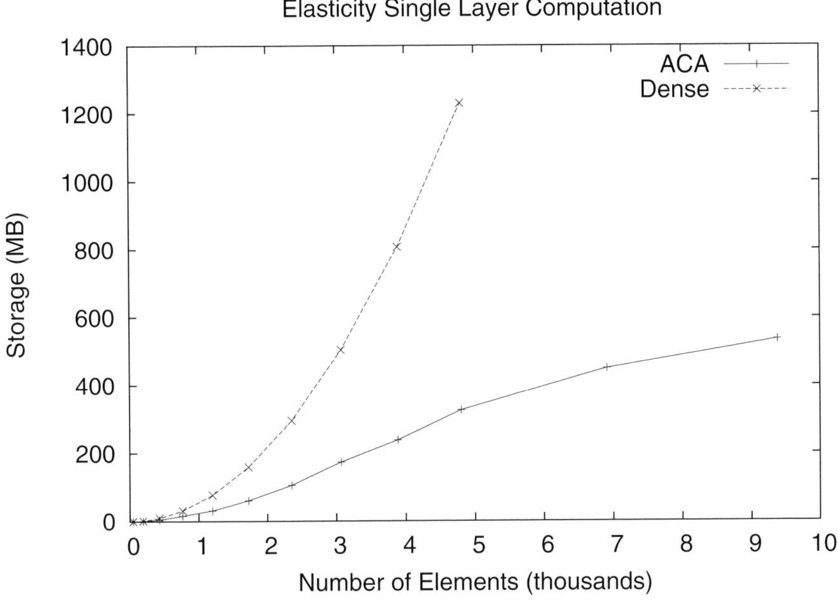

Fig. 44. Comparison of Dense and ACA Single Layer Computation.

By progressively remeshing the prism, and running the model both with and without ACA compression, it is possible to assess the effectiveness of ACA in accelerating the analysis of the problem (and simultaneously reducing storage requirements). Fig. 44 shows the variation of the memory required to store the single layer potential for elasticity, this being the largest and slowest part of the problem analysis. The time performance follows essentially the same trends as memory use, both for dense storage and for the case of using ACA approximations.

Note that when using dense matrices, it is not possible to compute problems larger than about 5000 elements on the computer used. Even looking just at the results up to this point, we see that there is a clear quadratic increase in computation time as the number of elements increases. By contrast, when computing an ACA comparison there is a steadily increasing computational efficiency as the number of elements increases.

Example 8: **Conducting cylinder**

In order to examine the basic principles of coupled electromagnetic and mechanical interactions, it is useful to look at a simple example. One such case is a current-carrying cylindrical conductor. The cylinder that will be presented here has an inner radius of 5 [cm], a wall-thickness of 1 [cm] and a height of 9 [cm]. The current passes circumferentially through the conductor, and has a current density of 5 [A/mm^2] (equivalent to a total current of 4.5 [A]). After performing a current-distribution analysis on this component, distributed induced body-forces are produced. Based on this input, the mechanical response can be evaluated. The displacement profile resulting is shown in Fig. 45.

Barely visible, at the rear of the model, is the "cut" in the cylinder introduced in order to allow current to be injected into and to exit the conductor. Naturally, in this region the conductor is far less stiff than it would be if it had not been cut. However, at the front of the model, we can see the displacement profile that would be found in an intact cylinder. This allows easy comparison with the results reported by Euxibie et al [21]. In the results shown here, the maximum displacement reported on the cylinder directly opposite the cut is 1.97×10^{-9}[m], which compares very well with the value of 1.95×10^{-9}[m] calculated by Euxibie et al.

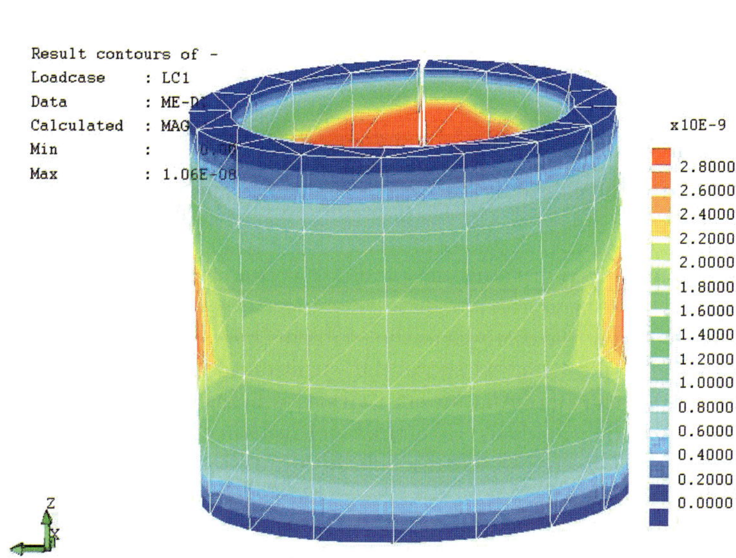

Fig. 45. Coupled Electromagnetic and Mechanical Loading on a Cylindrical Conductor.

Example 9: **Two parallel bus-bars**

In the following example we want to apply the previously described scheme for coupled Electro-Magnetic / Structural Mechanics analysis to a simple configuration consisting of two parallel, finite length, copper bus-bar like conductors. The injected currents is I = 10 [kA], flowing in opposite directions through the two conductors. Our aim is to compare how the mechanical responses (displacement, traction) differ if we treat the above example from the electromagnetic side first as a static case (stationary current distribution), and than as a quasi steady-state case (eddy-current distribution, including the skin-effect).

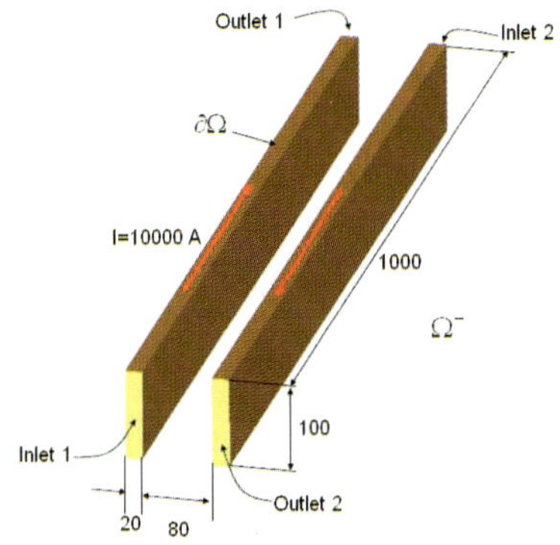

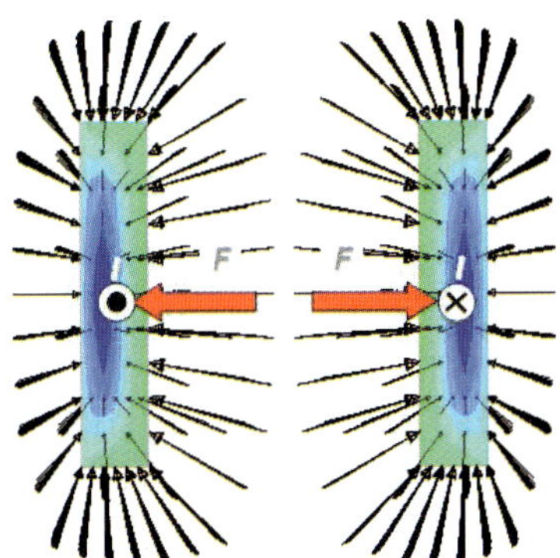

Fig. 46. Two conductors with Inlet and Outlet surfaces.

Fig. 47. Forces acting on two bus-bars with opposite current directions (top view).

Case 1: Static case

Assuming a constant distribution of the excitation (injected) current over the cross-section of the conductor, the analytic value of the current density for given dimensions is $5e^6$ [A/m^2]. For a conductor of given length, and with rectangular cross-section, it is possible to obtain the value of the magnetostatic field in closed-form, Andjelic [6]. Fig. 48 a shows the analytically calculated distribution of the magnetic flux density, with the maximal value of $7.48e^{-2}[T]$.

The calculated values for magnetic induction $\boldsymbol{B}$, current density $\boldsymbol{j}$, volume volume *nodal* forces $\boldsymbol{f}$, displacement $\boldsymbol{u}$ and traction $\boldsymbol{t}$ are shown in the Fig. 48.

Case 2: Quasi steady-state case

The same example is next solved as a quasi steady-state case, i.e. taking the time-varying excitation current with $I_{RMS} = 10$ [kA]. Eddy-current problem taking into account the skin-effect is solved exactly in the same way as in

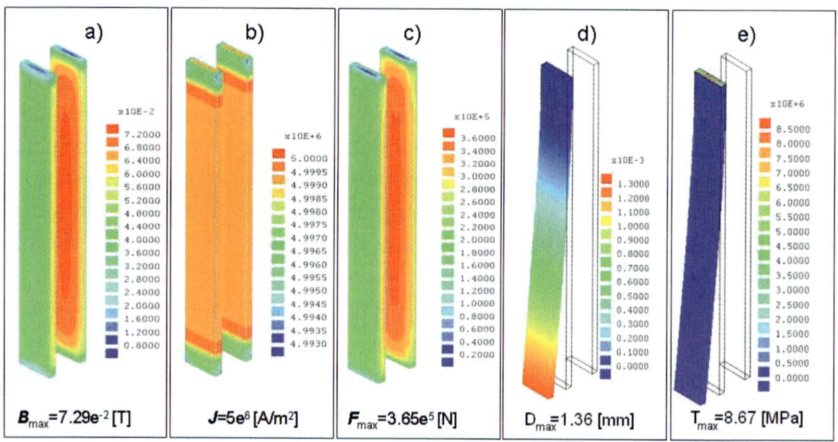

Fig. 48. Calculated B, j, f, u, t for a steady-state case.

Example 5, page 316, following the formulation given in Section 3.2. Fig. 49 shows the calculated values for magnetic induction B, eddy-current density j, volume *nodal* forces[21] f, displacement u and traction t. For B, j and f the complex magnitudes are shown.

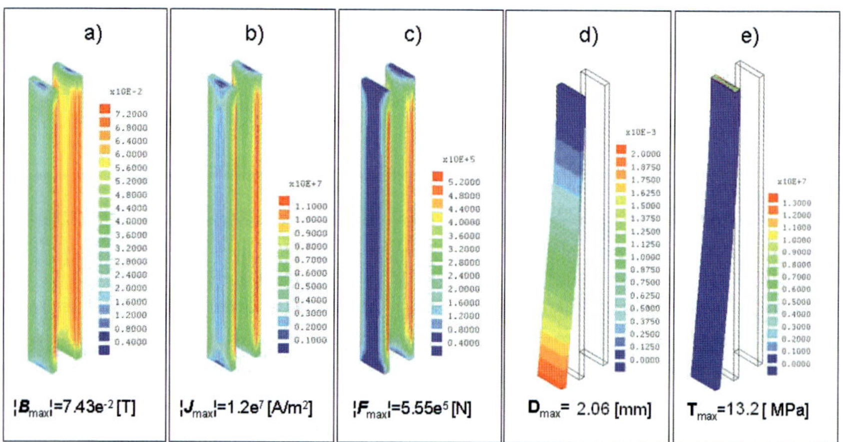

Fig. 49. Calculated B, j, f, u, t for a quasi-steady state case.

---

[21] Nodal forces are obtained after dividing the force densities in the volume nodes with the corresponding volume around each node.

When comparing the two cases, the following conclusions can be drawn:

- The spatial distribution of the physical quantities is as expected different. In the steady-state the flux, and therefore also the forces, are concentrated in the middle of the conductors.
- In the quasi-steady state the maximal flux, current and force appear along the vertical edges of the conductors.

Example 10: **Earthing switch in GCB**

Turning our attention to a full engineering problem, we can now consider the coupled electromechanical loading of a switch found in the generator circuit breaker (GCB) seen already in Example 3 (*SBD* for a Generator Circuit-Breaker design). Following a current-distribution and eddy-current analysis, it is possible by Biot-Savart calculation to find the body-forces arising out of Lorentz interactions, page 302. In fact, these forces are often of interest only in a limited region of the entire engineering system, typically in moving parts. In the GCB case presented here, a point of particular interest is the "knife" switch, where there is a tendency for the generated Lorentz forces to act so as to open the switch. The position of the switch is highlighted in magenta in Fig. 50.

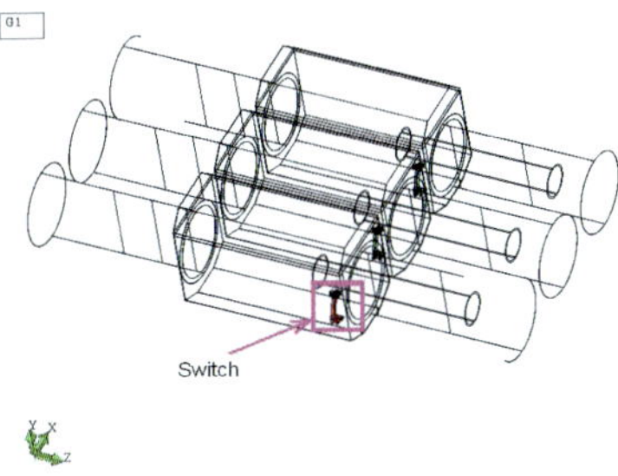

Fig. 50. Location of One-Knife Switch in overall breaker assembly.

Taking the example from earlier, for the mechanical part of the analysis only a limited portion of the mesh needs to be evaluated. Results were calculated using a mesh comprising 4130 triangular planar surface elements and 2063 nodes.

The volume discretization (necessary for the body-force coupling) comprises 14000 tetrahedra. This model has been analyzed taking advantage of the ACA approximation for the single and double layer potentials described earlier in the outline of the formulation. Results from this analysis are shown in Fig. 51.

Clearly visible is the effect of the coupling forces on the switch, which has a tendency to move out of its closed position under the action of the

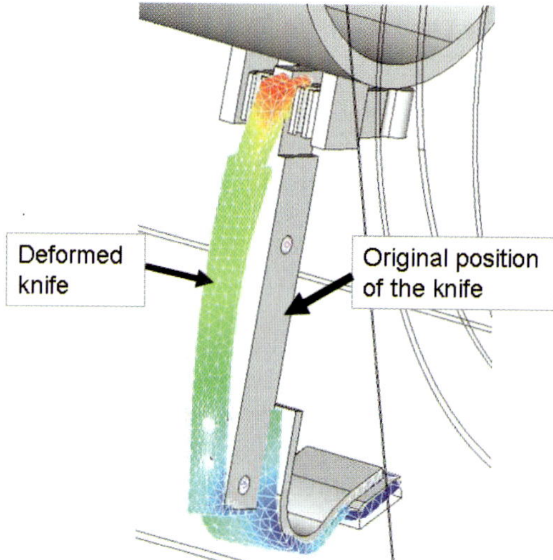

Fig. 51. Deformation of the earthing knife (overscaled), caused by the action of the short-circuit forces.

electromagnetic loading. This quantitative and qualitative information is a valuable input into the design process leading to the development of complex electromechanical systems.

4 BEM for Thermal Design

As mentioned at the beginning, under **Thermal Design** we assume only *weak* coupling of Electro-Magnetic process with the Heat-Transfer process providing the thermal response of the structure. In reality, the physics describing this problem is rather complex. There are three major physical phenomena that should be taken into account simultaneously: the electromagnetic part responsible for the losses generation, a fluidic part responsible for the cooling effects and thermal part responsible for the heat transfer. The simulation of the such problems, taking into account both complex physics and the complex 3D structures found in the real-world apparatus is still a challenge, especially with respect to the requirements mentioned in the Section "Simulation-Based Design", page 290: *accuracy-robustness-speed*.

A somehow simplified approach, that for many practical problems can still produce acceptably good results, is shown in Fig. 53. Instead of performing a complex analysis of the cooling effects by a fluid-dynamics simulation, a common practice is to introduce the *Heat-Transfer Coefficients* (HTC) obtained either by simple analytical formulae, (see for example Boehme [19]) or based on experimental observations. For this type of analysis the link be-

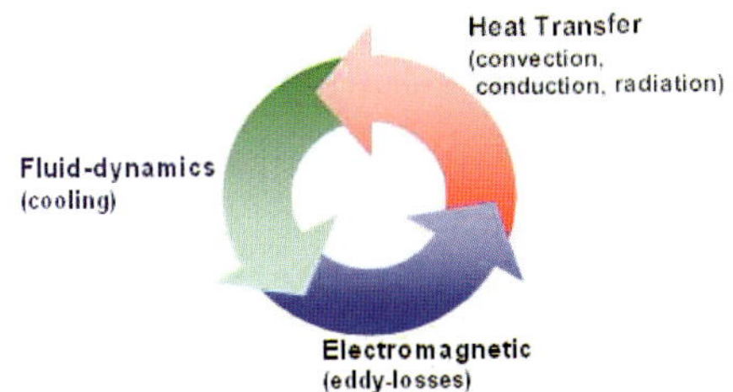

Fig. 52. Real physical cycles existing in the current-carrying structures.

tween the electromagnetic solver and heat-transfer solver is true the *losses* calculated on the electromagnetic side that are then passed as *external loads* to the heat-transfer module.

As it is well known, heating of the material leads to changes of the material parameters. The proper simulation of such phenomena would then require the iterative update of the material parameters, leading to a nonlinear process. In most of the industrial apparatus the increase of temperature has to be kept under certain prescribed limits. Mostly the temperature changes within these limits do not lead to significant changes in material properties. In that case, for certain applications we can work only with uni-directional coupling, ignoring changes of the material parameters due to temperature change. In our simulation workflow this means dropping the feed-back loop shown by the dotted line in Fig. 53.

The Electromagnetic part of *thermal* simulation is exactly the same as already described in the previous section for force analysis.

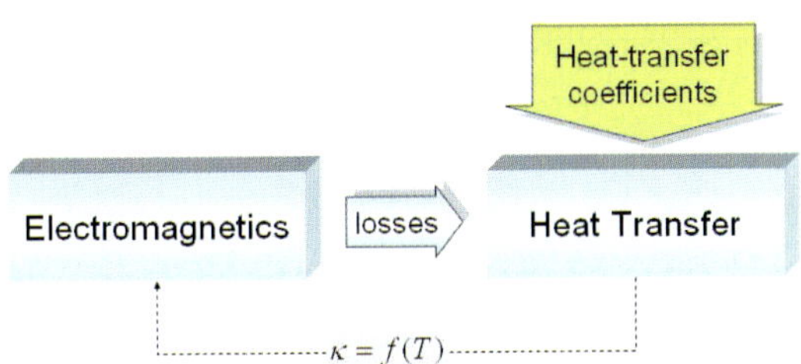

Fig. 53. Simplified Electro-Magnetic/Heat-Transfer workflow.

Losses calculation

For both stationary and time-varying fields, the losses are calculated using the Poynting vector. For stationary case the losses are then calculated as the integral of the Poynting vector over the surface of conductor

$$P = \oint_S (\mathbf{E} \times \mathbf{H}) dS \tag{112}$$

or over the volume of the conductor using integral conservation law, Haus [29],

$$P = \int_V \sigma \mathbf{E} \cdot \mathbf{E} dV \tag{113}$$

where $\mathbf{E}$ and $\mathbf{H}$ are the electric and magnetic fields for the stationary case.

In the case of the time-varying fields, the losses are calculated as the time-average of the total energy $\bar{P}$ dissipated through the surface of surrounding the conductor

$$\bar{P} = \frac{1}{2} \mathrm{Re} \oint_S (\mathbf{E} \times \mathbf{H}^*) \cdot \mathbf{n} dS \tag{114}$$

where $\mathbf{E}$ and $\mathbf{H}$ are now the electric and magnetic time-varying fields (complex vectors), and $*$ stands for conjugate complex.

4.1 Temperature Calculation

The weak coupling between electro-magnetic and thermal calculations is accomplished by using the eddy current losses as source terms for the thermal computations with the aid of the heat conducting equation. The heat conduction equation can be reduced to the simple conservation law

$$\nabla \cdot (k \nabla T(\mathbf{x})) = 0, \quad \forall \mathbf{x} \in \Omega \tag{115}$$

for steady states with the thermal conductivity coefficient λ. For the solution of the above Laplace-equation a direct approach was used, Andjelic [4]. A solution can be found by using Green's representation formula for $x \in \partial\Omega$

$$\frac{1}{2}\Theta(x)T(x) = \int_\Gamma \partial_n T(y) \frac{1}{4\pi |x-y|} d\Gamma(y) - \int_\Gamma T(y) \frac{(x-y)}{|x-y|^3} n(y) d\Gamma(y) \tag{116}$$

with the solid angle Θ.

In the AC case, the power that flows into the item is supposed to be generated in a region very close to the surface of the conductor due to the influence of the skin-effect. Therefore it can be taken into account via the boundary condition

$$\lambda \cdot \partial_n T(\mathbf{x}) = \dot{q}(\mathbf{x}) \tag{117}$$

for all points $\mathbf{x} \in \Gamma$ of the boundary. The energy flux $\dot{q}(\mathbf{x})$ is computed by means of the Poynting vector, equations (112) or (114). Convection and radiation can be approximated by a term proportional to the difference of the temperature T_0 of the item and the temperature of the environment. The resulting boundary condition is

$$\lambda \partial_n T(\mathbf{x}) = \dot{q}(\mathbf{x}) - \alpha(T(\mathbf{x}) - T_0); \quad \forall \mathbf{x} \in \Gamma \qquad (118)$$

with the heat transfer coefficient α. Plugging the boundary condition (118) into the representation (117) yields

$$\lambda \left[\frac{1}{2}\Theta(x)T(x) + \int_\Gamma T(y)\frac{(x-y)}{4\pi |x-y|^3}\mathbf{n}(y)d\Gamma(y) \right] + \qquad (119)$$

$$\alpha \int_\Gamma T(y)\frac{1}{4\pi |x-y|}d\Gamma(y) = \int_\Gamma (\dot{q}(x) + T)\frac{1}{4\pi |x-y|}d\Gamma(y); \qquad \forall x \in \Gamma.$$

Example 11: **Thermal analysis of earthing switch in GCB**

Thermal Design in circuit-breaker production, together with Dielectric and Electro-Mechanical Design, is one of the most important tasks accompanying the design of any new member of the breaker family. Very strict international norms like IEC and IEEE, require careful layout of the breakers components in order to stay within the prescribed limits.

In the case of thermal analysis, special care must be paid to hot-spots appearing on and around the *contact areas* between the different parts. For example, the maximum temperature allowed of any hot spots on the silver-plated contacts is 105 $^o C$. The simulation of such phenomena is connected with different difficulties and usually accompanied with certain simplifications. For example, consider the fact that the size of the *contact surfaces area* is usually a function of different influencing quantities (mechanical pressure, material characteristics, friction, etc.) and is not easy to define in real-world configurations. From the other side, the value of the *contact resistance* is one of the key parameters influencing the overheating of contact areas, and therefore of the entire apparatus.

In the present example we have performed a coupled Electro-Magnetic / Fluid-Dynamic / Thermal analysis of a circuit-breaker component according to the workflow shown in Fig. 54. The geometry has been generated using Pro/Engineer[22]. The electromagnetic part of the analysis was performed by POLOPT/em[23]. The fluid dynamic analysis, including the thermal simulation has been done by STAR-CD[24]. Fig. 55 shows the component of the circuit-breaker model used for the thermal simulation. The calculated results are compared with experimentally obtained temperature on the selected positions, Fig. 56. The simulation and the measurement have been conducted for four different current levels (2, 4, 13, and 18 $[kA]$). The graph in Fig. 57 shows the time development of the temperature in the selected point indicated by the arrow. The markers are the positions in time where the experimental

---

[22]Pro/Engineer is a trademark for the PTC product line (http://www.ptc.com).

[23]POLOPT/em is a module for eddy-current analysis in POLOPT multi-physics environment.

[24]STAR-CD is a trademark for the CD-adapco product line (http://www.cd-adapco.com).

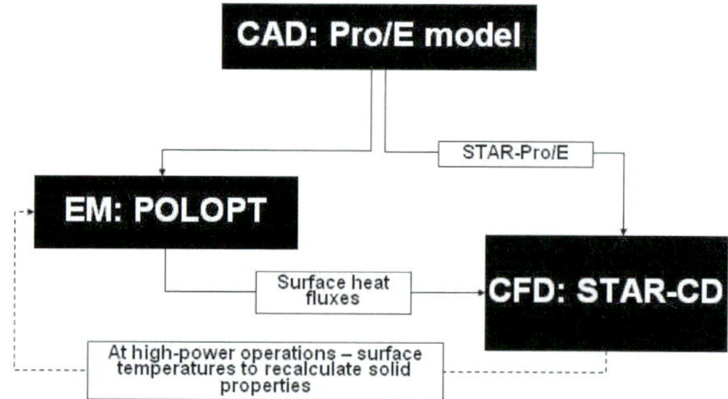

Fig. 54. Coupled EM-CFD simulation for thermal analysis of GCB.

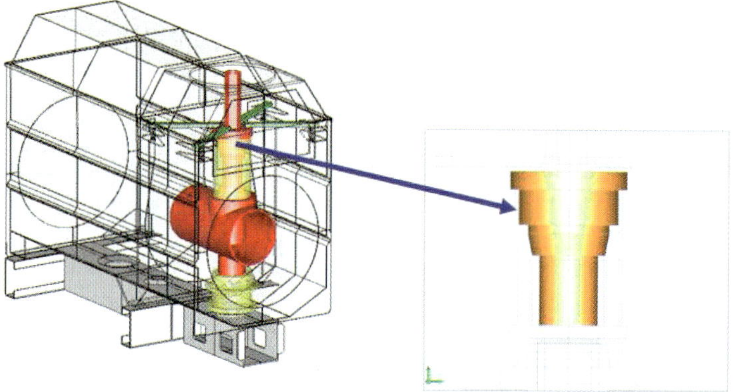

Fig. 55. Model of the breaker's component used for the simulation.

Fig. 56. Experimental set-up used for the temperature measurement.

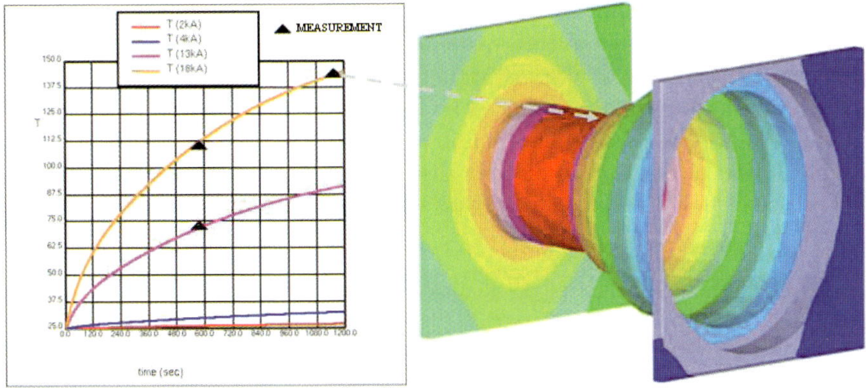

Fig. 57. Measured vs. calculated temperature distribution on the breaker component.

results were available. A good agreement between simulation and experiment confirms the validity of the coupled analysis performed.

Example 12: **Thermal design of power transformers**

In the previous example we performed the coupled Electro-Magnetic / Fluid-Dynamic / Thermal analysis of a circuit-breaker component. In the current example we shall illustrate how using the previously described BEM formalism for electro-magnetic and thermal calculation one can in an integrated environment analyze the thermal problems in power transformers, Fig. 58. In this simulation run we perform the coupled Electro-Magnetic / Thermal analysis, using estimated values for heat transfer coefficients instead of a full

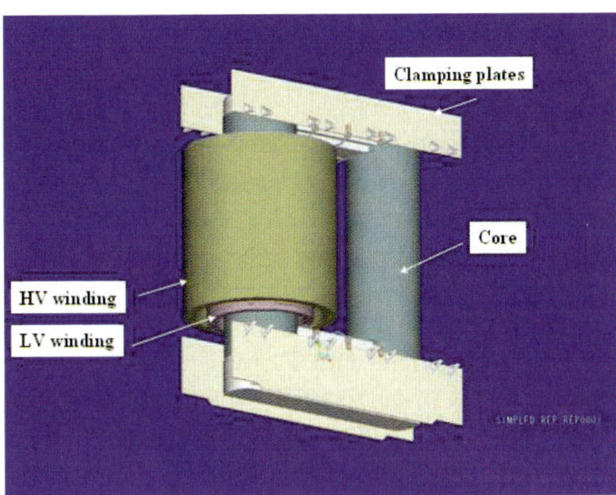

Fig. 58. ABB Power transformer: Test Unit, Vaasa, Finland (ProE model).

fluid-dynamic run. The simulation workflow is shown on Fig. 59. The goal
was to obtain the temperature distribution over the transformer clamp. In
this case the clamp is a "passive" structure, i.e. the eddy-current are induced
by the external stray field produced by the windings. The first step after completing the pre-processing steps (meshing, boundary conditions, materials,...)
is calculation of the excitation field distribution, Section 3.2, page 303. It is
followed by the eddy-currents analysis based on the formulation described in
section Section 3.2, page 312.

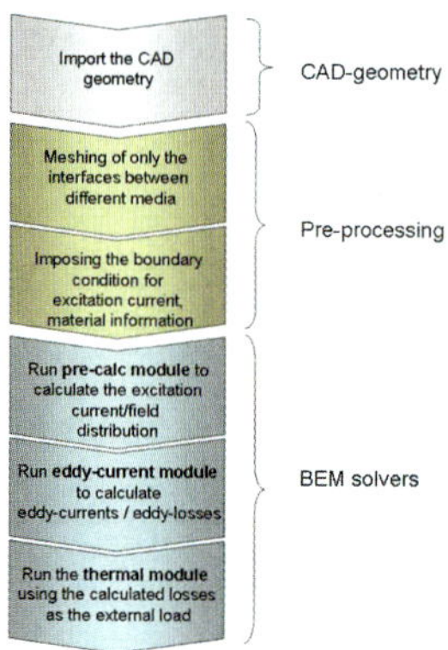

Fig. 59. Integrated environment for thermal analysis of power transformers.

The material used for the clamp is a magnetic steel with permeability
$\mu = 200$[25] and electrical conductivity $\sigma = 6.66e6 \ [\Omega/m]^{-1}$. With a frequency
of 50 $[Hz]$, the penetration depth of the magnetic field in magnetic material
is

$$\delta = \sqrt{\frac{2}{\omega\mu\sigma}} = 1.95e^{-3}[m]. \tag{120}$$

This small penetration depth of less than 2 $[mm]$ would require appropriate
mesh quality in order to capture the diffusion phenomena, especially taking
into account that the overall clamp length is more than 4 [m]. Here we want
to stress once again the excellent capabilities of BEM, enabling the solution
of this complex diffusion problem using only a boundary mesh[26].

---

[25] we assume a linear material with a constant value of the material permeability.

[26] When solving the same problem with FEM, this would require a volume mesh
both in magnetic material and in surrounding media. An work-around when using
FEM would be the usage of IBC (Impendence Boundary Condition) to avoid the
meshing of the inner volume of magnetic materia. But, this approach leads to additional errors caused by the fact that IBC assumes i) geometry without geometrical
singularites, ii) the electromagnetic field is assumed to be a plane wave, iii) the penetration depth need to be very small i.e. the current flows only on the surface of the
conductors.

Fig. 60 shows the calculated eddy-currents in the transformer clamping plate. Using the eddy-losses as inputs in the thermal module, one can obtain the temperature distribution over the clamp's surfaces, Fig. 61. The calculation has been performed using the MBIT module for eddy-current analysis, Section 3.2, page 320, i.e. using both accelerated matrix generation and fast-multipole based matrix compression.

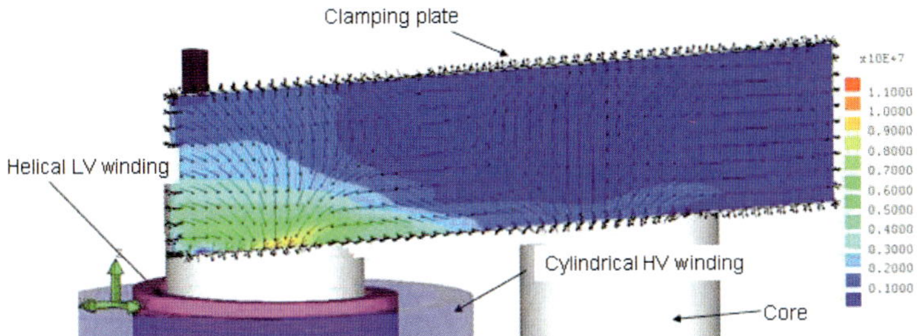

Fig. 60. Eddy-current vectors flow along the transformer clamping plate.

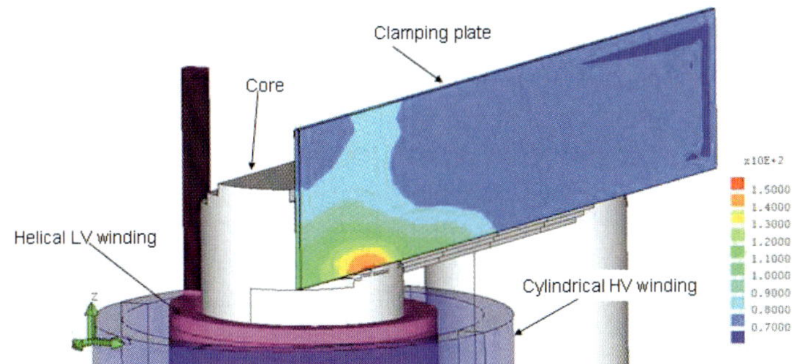

Fig. 61. Temperature distribution on the clamping plate.

5 Some Concluding Remarks

In this chapter we have tried to illustrate some BEM-based approaches for the simulation of different problems appearing in engineering design praxis. The excellent features of BEM for both single and multi-physics tasks are highlighted, together with some emerging numerical techniques like MBIT and ACA, recognized as the major drivers leading to the real breakthrough in BEM usage for practical design tasks.

Beside these and many other good features of BEM, and staying at the level of *static* or *quasi-static* simulation tasks, there are still a number of potential improvements that could be made to achieve the "best in the class" tool desired for the advanced simulation tasks in the industrial design. Let us mention just some of them.

- At the **formulations** level there is still a piece of work for the mathematicians to offer sound, BEM-oriented, formulations for **strongly coupled** physical problems. With the exception of some examples mentioned at the beginning of this Chapter, to our knowledge there is still a lack of properly founded, BEM-based mathematical formulations for coupled Electro-Magnetic / Thermal or Electro-Magnetic/Structural-Mechanics problems.
- The recent achievements in the field of the "Matrix symmetrization" highlight the new possibilities for the acceleration of matrix generation or/and solution. Extending of such (or similar) approaches to the most important mathematical formulations (part of them mentioned within the scope of this material) could contribute a lot to the advanced treatment of complex real-world problems, especially with regard to the required memory usage and matrix "conditioning".
- *Non-linearity* treatment is one of the topics often recognized as the "weakest" point of BEM. The authors experience with the BEM[27] treatment of non-linear magnetostatic problems indicates that for certain application BEM can successfully (and efficiently!) be used also for non-linear classes of problems. Appropriate formulations of non-linear problems via BEM in some other areas are still an open issue. For example, the more advanced formulations of non-linear problems in eddy-current analysis, or even more in coupled eddy-current / thermal analysis could be of great benefit for certain classes of practical problems, touched also within this material in the examples given in Section 4.1, page 345 or in Section 4.1, page 343.
- Speaking about the BEM-based analysis in the Structural Mechanics, there exist now the well elaborated formulations for the standard tasks appearing in the *linear elasticity* analysis. The treatment of the non-linear problems in mechanics is also rather well elaborated, but probably not enough highlighted, with some exceptions like in automotive industry.
  But, there are some important and, in the practical design often unavoidable, class of the problems where to authors knowledge the BEM-based formulations are still not enough promoted. As an example, there are just a few available publications where BEM-based treatment of *contact* problems is analyzed! This important area, together with the efficient treatment of *eigenvalue* problems via BEM would significantly expand the applications area of this method, especially when doing the coupled Electro-Magnetic/Structural-Mechanics type of analysis.

Finally, the authors general opinion is that the BEM already offers an excellent platform for the simulation of 3D real-world industrial problems.

---

[27]Here we still use the "BEM" notation, although in the reality, for non-linar problems we have to treat a "volume" as well, Krstajic [37].

Especially when speaking about some of the major requirements appearing in the Simulation-Based Design nowadays, like:

- *assembly* instead of *component simulation,*
- simulation for the *daily* design process,
- *user-friendly* simulation, but still preserving the *full geometrical and physical complexity,*

BEM-based numerical technology seems to fulfil the major requirements demanded of simulation tools.

References

1. Z. Andjelić, B. Krstajić, S. Milojković, Z. Haznadar: A Procedure for Automatic Optimal Shape Investigation of Interfaces Between Different Media. IEEE Trans. Mag. 24 (1988) 415–418.
2. Z. Andjelić, E. Henrikson, O. Johendal, H. Nordman: 3D Simulation in Transformer Design. 10th Int. Sym. on High Voltage Engineering, Montreal, 1997.
3. Z. Andjelić, B. Krstajić, S. Milojković, A. Blaszczyk, H. Steinbigler, M. Wohlmuth: Integral Methods for the Calculation of Electric Fields. Scientific Series of the International Bureau, Vol. 10, Forschungszentrum Jülich GmbH, Germany, 1992.
4. Z. Andjelić, J. Ostrowski, B. Cranganu-Cretu: MBIT for Skin-effect Problems in Power Transformers. 4th European Congress on Computational Methods in Applied Sciences and Engineering - ECCOMAS, Jyväskylä, Finland, 2004.
5. Z. Andjelić, P. Marchukov: Acceleration of the Electrostatic Computation Using Multipole Technique. ABB internal report, Heidelberg, 1992.
6. Z. Andjelić, B. Krstajić, S. Milojković, S. Babić: Boundary Integral Method for Electromagnetic Field Computation. NSF project, Code JF824, USA, 1991.
7. B. A. Auld: Acoustic Fields and Waves in Solids, Vol. 1. John Wiley and Sons, New York, 1973.
8. S. Babić, Z. Andjelić, B. Krstajić, S. Salon: Analytical Magnetostatic Field Calculation for a Conductor with Uniform Current in the Longitudinal Direction. J. Appl. Phys. 67 (1990) 5827–5829.
9. B. Bachmann: Freies Potential beim Ladungsverfahren. ETZ-A 94 (1973) 12.
10. P. K. Banerjee: The Boundary Element Methods in Engineering. McGraw-Hill Book Company, 1981.
11. M. Bebendorf, S. Rjasanow: Adaptive low rank approximation of collocation matrices. Computing 70 (2003) 1–24.
12. M. Bebendorf: Approximation of Boundary Element Matrices. Numer. Math 86 (2000) 565–589.
13. M. Bebendorf, S. Rjasanow, E. E. Tytyshnikov: Approximations using Diagonal-Plus Skeleton Matrices. In: Mathematical Aspects of Boundary Element Methods (M. Bonnet, A.–M. Sändig, W. L. Wendland eds.), Chapman & Hall/CRC Research Notes in Mathematics, Vol. 414, pp. 45–53, 1999.
14. M. Bebendorf: Hierarchical LU decomposition-based preconditioners for BEM. Computing 74 (2005) 225–247.

15. M. Bebendorf: Another software library on Hierarchical Matrices for elliptic differential equations (AHMED). Universität Leipzig, Fakultät für Mathematik und Informatik, 2005.
16. J. van Bladel: Singular Electromagnetic Fields and Sources. IEEE Press Series on Electromagnetic Wave Theory, 1995.
17. A. Blaszczyk, Z. Andjelić, P. Levin, A. Ustundag: Parallel Computation of Electric Fields in a Heterogeneous Workstation Cluster. In: HPCN Europe 95, Lecture Notes on Computer Science, Springer, Berlin, Heidelberg, pp. 606–611, 1995.
18. P. Bochev: On the Finite Element Solution of the Pure Neumann Problem. SIAM Review 47 (1992) 50–66.
19. H. Boehme: Mittelspannungstechnik. Verlag Technik GmbH, Berlin, München, 1992.
20. M. Costabel: Symmetric methods for the coupling of finite elements and boundary elements. In: Boundary Elements IX (C. A. Brebbia, G. Kuhn, W. L. Wendland eds.), Springer, Berlin, pp. 411–420, 1987.
21. E. Euxibie, J.–L. Coulomb, G. Meunier, J.–C. Sabonnadière: Mechanical deformation of a conductor under electromagnetic stresses. IEEE Trans. Mag. 22 (1986) 828–830.
22. K. Fujiwara, T. Nakata: Results for Benchmark Problem 7, COMPEL 9 (1990) 137–154.
23. L. Gaul, M. Kögl, M. Wagner: Boundary Element Methods for Engineers and Scientists. Springer, Berlin, 2003.
24. L. Grasedyck, W. Hackbusch: Construction and Arithmetics of $\mathcal{H}$–matrices. Computing 70 (2003) 295–334.
25. L. Greengard, V. Rokhlin: A Fast Algorithm for Particle Simulations. J. Comp. Phys. 73 (1987) 325–348.
26. G. H. Golub, C. F. van Loan: Matrix Computations. The Johns Hopkins University Press, Third edition, Baltimore, London, 1996.
27. W. Hackbusch, Z. P. Nowak: On the fast matrix multiplication in the boundary element method by panel clustering. Numer. Math. 54 (1989) 463–491.
28. W. Hackbusch: The Panel Clustering Technique for the Boundary Element Method. 9th Int. Conf. on BEM, pp. 463–473, 1987.
29. H. A. Haus, J. R. Melcher: Electromagnetic Fields. MIT Publishing.
30. L. R. Hill, T. N. Farris: Three-Dimensional Piezoelectric Boundary Element Method. AIAA Journal 36, 1998.
31. R. Hiptmair: Coupled Boundary Element Scheme for Eddy Current Computation. 2nd Kolloquium Elektromagnetische Umformung, ETH Zürich, 2003.
32. Intel Math Kernel Library 8.0, Intel Software Development Product. http://www.intel.com/cd/software/products/asmo-na/eng/perflib/mkl
33. J. D. Jackson: Classical Electrodynamics. John Wiley, 1999.
34. A. B. Kemp, M. T. Grzegirczyk M., K. A. Kong: Ab inito study of the radiation pressure on dielectric and magnetic media. Optics Express 13 (2005).
35. E. C. Koleciskij: Rascet eletriceskih poljei ustroistv visokog naprezenija. Energoatomizdat, Moskva, 1983.
36. A. Kost: Numerische Methoden in der Berechnung elektromagnetischer Felder. Springer, 1994.
37. B. Krstajić, Z. Andjelić, S. Milojković, S. Babić: Nonlinear 3D Magnetostatic Field Calculation by the Integral Equation Method with Surface and Volume Magnetic Charges. IEEE Trans. Mag. 28 (1992).

38. C. Lage: Software development for Boundary Element Method: Analysis and design of efficient techniques. Ph.D. thesis, University of Kiel, Germany, 1995.

39. L. D. Landau, E. M. Lifshitz: Theory of Eleasticity. Pergamon Press, New York, 1959.

40. J. C. Maxwell: The Treatise on Electricity and Magnetism. Dover Publication, New York, 1954.

41. I. D. Mayergoyz: Boundary Integral Equations of Minimal Order for the Calculation of Three-Dimensional Eddy Current Problems. IEEE Trans. Mag. 18 (1982).

42. I. D. Mayergoyz: A New Approach to the Calculation of Three-Dimensional Skin Effect Problems. IEEE Trans. Mag. 19 (1983).

43. I. D. Mayergoyz: Nonlinear Magnetostatic Calculation Based on Fast Multipole Method. IEEE Trans. Mag. 39 (2003).

44. W. McLean: Strongly Eliptic Systems and Boundary Integral Equations. Cambridge University Press, Cambridge, UK, 2000.

45. J. R. Melcher: Continuum Electromechanics. MIT Press, Massachusetts, 1981

46. A. H. Nayfeh: Wave Propagation in Layered Anisotropic Media with Applications to Composites. Elsevier, Amsterdam, 1995.

47. S. Peaiyoung, S. J. Salon: Some Technial Aspects of Implementing Boundary Element Technique. IEEE Trans. Mag. 25 (1989).

48. G. Reyne, J. C. Sabonnadiere, J. L. Coulomb, P. Brissonneau: A Survay on the main aspects of magnetic forces and mechanical behaviour of ferromagnetic materials under magnetisation. IEEE Trans. Mag. 23 (1987).

49. A. H. Schatz, V. Thomée, W. L. Wendland: Mathematical Theory of Finite and Boundary Element Methods. Birkhäuser, Basel, 1990.

50. G. Schmidlin: Fast Solution Algorithms for Integral Equations in $\mathbb{R}^3$. Ph.D. thesis, ETH Zürich, 2003.

51. G. Schmidlin, U. Fischer, Z. Andjelić, C. Schwab: Preconditioning of the second-kind boundary integral equations for 3D eddy current problems. Int. J. Numer. Meth. Engrg. 51 (2001) 1009–1031.

52. K. R. Shao, K. D. Zhou, J. D. Lavwers: Boundary Element Analysis Method for 3-D Multiply Connected Eddy Current Problems Based on the Second Order Potential Formulation. IEEE Trans. Mag 28 (1992).

53. K. Simonyi: Theoretische Elektrotechnik. VEB Deutscher Verlag der Wissenschaften, Berlin, 1973.

54. S. Sirtori: General stress analysis method by means of integral equations and boundary elements. Meccanica 14 (1979) 210–218.

55. J. Smajic, B. Cranganu-Cretu, J. Ostrowski, Z. Andjelić: Stationary Voltage and Current Excited Complex System of Multimaterial Conductors with BEM. IEEE Trans. Mag. (2006).

56. O. Steinbach: Numerische Näherungsverfahren für elliptische Randwertprobleme. B. G. Teubner, Stuttgart, 2003.

57. O. Steinbach: Lösungsverfahren für lineare Gleichungsysteme. Algoritmen und Anwendungen, B. G. Teubner, Wiesbaden, 2005.

58. O. Steinbach: OSTBEM. A Boundary Element Software Package, Universität Stuttgart, TU Graz, 1992–2006.

59. J. A. Stratton: Electromagnetic Theory. McGraw–Hill, 1941.

60. I. E. Tamm: Fundamentals of the Theory of Electricity. Mir Publisher, Moscow, 1965.

61. O. B. Tozoni, I. D. Maergoiz: Rascet trehmernih elektromagnetnih polei. Tehnika, Kiev, 1974.
62. J. Yuan, A. Kost: A Three-Component Boundary Element Algorithm for Three-Dimensional Eddy Current Calculation. IEEE Trans. Mag. 30 (1994).
63. L. Zehnder, J. Kiefer, D. Braun, T. Schoenemann: SF6 generator circuit-breaker system for short-circuit currents up to 200 kA, ABB Review 3, 2002.
64. J. R. Whiteman, L. Demkowicz: Mathematics of Finite Elements and Applications XI. Computer Methods in Applied Mechanics and Engineering, Vol. 194, Issues 2-5, 2005.

List of Contributors

Prof. Dr. Z. Andjelić
ABB Switzerland Ltd.
Corporate Research
CH 5405 Baden–Dättwil
zoran.andjelic@ch.abb.com

Prof. Dr. M. Bonnet
Ecole Polytechnique
Solid Mechanics Laboratory
F 91128 Palaiseau
bonnet@lms.polytechnique.fr

Prof. Dr. C. Carstensen
Humbold–Universität zu Berlin
Institut für Mathematik
Unter den Linden 6
D 10099 Berlin
cc@math.hu-berlin.de

Dr. M. Conry
ABB Switzerland Ltd.
Corporate Research
CH 5405 Baden–Dättwil
michael.conry@ch.abb.com

Prof. Dr. M. Costabel
Université de Rennes 1, IRMAR
Campus de Beaulieu
F 35042 Rennes
costabel@univ-rennes1.fr

Prof. Dr. W. Hackbusch
Max–Planck–Institut für
Mathematik in den
Naturwissenschaften
Inselstrasse 20–26
D 04103 Leipzig
wh@mis.mpg.de

Prof. Dr. R. Hiptmair
ETH Zürich, Seminar für
Angewandte Mathematik
Rämistrasse 101
CH 8092 Zürich
hiptmair@sam.math.ethz.ch

Dr. W. Kress
Max–Planck–Institut für
Mathematik in den
Naturwissenschaften
Inselstrasse 20–26
D 04103 Leipzig
kress@mis.mpg.de

Prof. Dr. S. Kurz
Robert Bosch GmbH
Postfach 106050
D 70049 Stuttgart

Prof. Dr. U. Langer
Johannes Kepler Universität Linz
Institut für Numerische Mathematik
Altenberger Strasse 69
A 4040 Linz
ulanger@numa.uni-linz.ac.at

PD Dr. M. Maischak
Universität Hannover
Institut für Angewandte Mathematik
Welfengarten 1
D 30167 Hannover
maischak@ifam.uni-hannover.de

N. Nemitz
Ecole Polytechnique
Solid Mechanics Laboratory
F 91128 Palaiseau
nemitz@lms.polytechnique.fr

Prof. Dr. N. Nishimura
Kyoto University
Graduate School of Informatics
Kyoto 606–8501, Japan
nchml@i.kyoto-u.ac.jp

Dr. G. Of
Technische Universität Graz
Institut für Numerische Mathematik
Steyrergasse 30
A 8010 Graz
of@tugraz.at

Y. Otani
Kyoto University
Graduate School of Informatics
Kyoto 606–8501, Japan
otani@mbox.kudpc.kyoto-u.ac.jp

Prof. Dr. D. Praetorius
Technische Universität Wien
Institut für Analysis und
Wissenschaftliches Rechnen
Wiedner Hauptstrasse 8–10
A 1040 Wien
dirk.praetorius@tuwien.ac.at

Dr. O. Rain
Robert Bosch GmbH
Postfach 106050
D 70049 Stuttgart
oliver.rain@de.bosch.com

Prof. Dr. S. Rjasanow
Universität des Saarlandes
Fachrichtung Mathematik
Postfach 151150
D 66041 Saarbrücken
rjasanow@num.uni-sb.de

Prof. Dr. S. A. Sauter
Universität Zürich
Institut für Mathematik
Winterthurerstrasse 190
CH 8057 Zürich
stas@math.unizh.ch

Dr. J. Smajić
ABB Switzerland Ltd.
Corporate Research
CH 5405 Baden–Dättwil
jasmin.smajic@ch.abb.com

Prof. Dr. O. Steinbach
Technische Universität Graz
Institut für Numerische Mathematik
Steyrergasse 30
A 8010 Graz
o.steinbach@tugraz.at

Prof. Dr. E. P. Stephan
Universität Hannover
Institut für Angewandte Mathematik
Welfengarten 1
D 30167 Hannover
stephan@ifam.uni-hannover.de

Dr. T. Takahashi
University of Tokyo
Earthquake Research Institute
Tokyo 113–0032, Japan
ttaka@postman.riken.go.jp

Printing: Krips bv, Meppel
Binding: Stürtz, Würzburg